Water Resources Development and Management

Indexed by Scopus

Each book of this multidisciplinary series covers a critical or emerging water issue. Authors and contributors are leading experts of international repute. The readers of the series will be professionals from different disciplines and development sectors from different parts of the world. They will include civil engineers, economists, geographers, geoscientists, sociologists, lawyers, environmental scientists and biologists. The books will be of direct interest to universities, research institutions, private and public sector institutions, international organisations and NGOs. In addition, all the books will be standard reference books for the water and the associated resource sectors.

S. Suriyanarayanan · H. P. Shivaraju · David Jenkins
Editors

Water Management in Developing Countries and Sustainable Development

Springer

Editors
S. Suriyanarayanan
JSS Science and Technology University
Mysuru, Karnataka, India

H. P. Shivaraju
JSS Academy of Higher Education and Research
Mysuru, Karnataka, India

David Jenkins
School of Engineering
University of Plymouth
Plymouth, UK

ISSN 1614-810X ISSN 2198-316X (electronic)
Water Resources Development and Management
ISBN 978-981-99-8638-5 ISBN 978-981-99-8639-2 (eBook)
https://doi.org/10.1007/978-981-99-8639-2

This Springer imprint is published by the registered company Springer Nature Singapore Pte Ltd.
The registered company address is: 152 Beach Road, #21-01/04 Gateway East, Singapore 189721, Singapore

Paper in this product is recyclable.

Foreword

According to the World Water Report 2023, approximately 26% of the world's total population does not have access to safe drinking water, and 46% lack access to safely managed sanitation. While water constitutes 70% of earth's matter; only 2.5% of it is freshwater and barely 1% of that is accessible to the masses. Population explosion and rapid urbanization have led to an increasing demand of this vital resource, while its recycle rate remains fairly constant. Many major cities across the world are on the verge of losing their freshwater reserves and might be running completely dry in the future decades.

Water crises have enormous global financial implications, including decreased agricultural productivity, increased health costs, and economic losses due to water-related disasters. The "UN Agenda—2030" recognizes the necessity of water resources in sustainable development and the vital role that safe drinking water; sanitation and hygiene play in progress to health, education, and poverty eradication. Hence, it is evident that sustainable management of water resources and access to safe water and sanitation are essential for unlocking economic growth and productivity especially in the Global South, and provide significant leverage for existing investments in health and education.

To avoid water crisis and to ensure the availability of freshwater for future generations; there is a need to reform the existing policies regarding water conservation and also to implement them effectively. Such policy reformulation should consider all the sectors consuming water such as agriculture, industries, and other domestic uses. We must acknowledge the fact that the water we have today can only be recycled and therefore, there is an urgent need to take appropriate measures, to conserve water and avoid its overuse or wastage. It is estimated that at current rates, progress toward achieving all the targets of the Sixth Sustainable Development Goal (SDG 6): "*access to clean water and sanitation for all*," is off-track and the rate of implementation needs to be quadrupled, or even more. Therefore, cooperation of governments, corporations, civil society, and other stakeholders is required to promote water resource management and subsequently achieve the water-related Sustainable Development Goals (SDGs).

We must further realize that in this "Anthropocene Epoch," we are the only species possessing the capacity to reflect on our behavior and change that is required to secure our health, economic, and social well-being. It is clearly understood that providing healthy and meaningful livelihoods is a major challenge and therefore, all of us must act together and identify the technological, economic, political, and social measures that will set a significant step toward the achievement of a desirable and more sustainable future.

Under this backdrop, I believe that a book of this kind brought out by the NAM S&T Centre will serve as a valuable source of knowledge, insight, and innovative solutions for water management. Chapters written by a diverse group of specialists from different parts of the world, take us on a journey through the intricate subject of water management, demonstrating various issues and optimal solutions that are within our capabilities. The authors have not only analyzed the problems but have also offered pragmatic solutions through looking into the fundamentals of water sustainability, highlighting the contemporary issues of water pollution, guiding us through novel approaches and infrastructure and lastly providing the best practical applications and prospective future directions toward combating water scarcity and establishing a sustainable water secure future.

I strongly believe that this work toward water conservation would show a guiding path for the water management sector and will lead to the adoption of several innovative and salient strategies and techniques in eliminating water crises, and that the information provided in the book will help the academicians, decision and policy-makers, practitioners, and concerned citizens to collaborate and strive for a more equitable and sustainable future for all in regard to water security.

I wholeheartedly congratulate the efforts put in by the NAM S&T Centre, the editors, and the authors for their commitment in bringing out such a book of utmost global importance which will enable the stakeholders and others engaged in water management sector to share their ideas and practices with a view to attaining water security across the globe.

Dr. Satish R. Wate
Former Director, CSIR-National
Environmental Engineering
Research Institute (NEERI)
Nagpur, India

Former Chairman, Recruitment
and Assessment Board (RAB)
Council of Scientific and Industrial
Research (CSIR)
New Delhi, India

Preface

Proper access to freshwater is essential for sustaining human health, prosperity and security, poverty eradication, and preservation of ecosystems. Yet, billions of people worldwide are confronted with serious freshwater challenges, arising from water scarcity, poor water quality, lack of sanitation facilities—leading to severe water-borne diseases and water-related disasters. Population growth, increasing living standards, changing trade policies, and adverse climate disruption have resulted in an increasing demand for freshwater. Estimates show that with current practices, by 2030, the world will face a 40% shortfall between forecast demand and available supply of water and as much as two-thirds of the world's population could be living in countries subject to water stress, a majority of them being in the developing countries.

While the countries are working very hard together toward the achievement of water-related goals of the "United Nations' 2030 Agenda for Sustainable Development," especially Goal 6: Access to Water and Sanitation for All; the trends and present status of access to water and sanitation remains a cause for global concern. According to the SDG Report 2022, approximately two billion individuals globally lack adequate access to safe drinking water, while around half of the world's population is currently facing significant water scarcity. Further, in the developing world, more than 90% of sewage water is discharged directly into rivers, lakes and coastal waters without any treatment. Degraded water quality is also decreasing the amount of safe water; hence proving a clear linkage between water quantity and water quality. Therefore, there is an urgent need to minimize the water consumption and develop approaches to effectively manage and sustainably use the existing water resources to satisfy social, economic, and environmental needs.

To strengthen water security against this backdrop of increasing water demand, water scarcity, growing uncertainties, greater extremes, and fragmentation challenges; we need to invest in strengthening of institutional infrastructure, clear assessment of the freshwater resources, application of integrated water resources management, common and shared interest among the stakeholders and their effective participation, comprehensiveness and global coverage of all forms of freshwater, information management as well as infrastructure development. Ensuring the rapid dissemination and appropriate adaptation or application of these advances will be a key toward strengthening global water security.

The book, through its 24 chapters contributed from 14 countries namely, Cameroon, China, Cyprus, India, Indonesia, Italy, Namibia, the Netherlands, Norway, Portugal, South Africa, UK, USA, and Zimbabwe, intends to provide sustainable water resource management strategies that present innovative ways to conserve both quality and quantity of water in developing countries. The issues addressed through these approaches are sustainable water management which means using water resources in a way that meets current, ecological, social, and economic needs without compromising the ability to meet those needs in the future, reducing wastewater contamination, enhancing and restoring water quality and quantity, and management of environmental flows.

The concept of developing such a book was put forward considering the alarming cases of water crises that are being witnessed by several developing countries and therefore, providing sustainable water management solutions that could aid in promoting water security across the globe. The contents of this book will be of great interest to scientists and researchers, practitioners, water resource managers, environmentalists, policy and decision-makers, international institutions, governmental and non-governmental organizations, educators, as well as students.

We would like to appreciate the sincere efforts of all the authors who have kindly accepted our invitation to write the chapters, revise the contents based on reviewers' comments and submit the final version well in time in spite of their busy schedule.

The Editors are thankful to Dr. Amitava Bandopadhyay, Director General; Mr. Madhusudan Bandyopadhyay, Senior Adviser; and Ms. Abhirami Ramdas, Research Associate, NAM S&T Centre and other staff members at the Centre for all the technical and administrative support rendered toward the publication of this book.

We are grateful to Dr. Satish R. Wate, Former Director, CSIR-National Environmental Engineering Research Institute (NEERI), Nagpur and Former Chairman, Recruitment and Assessment Board (RAB), Council of Scientific and Industrial Research (CSIR), New Delhi, India for writing the "Foreword" of this book in spite of his extremely busy schedule. We also express our gratitude to Dr. Loyola D'Silva, Executive Editor, Springer Nature, Singapore and his team for making this endeavor a success.

Mysuru, Karnataka, India — S. Suriyanarayanan, Ph.D.
Mysuru, Karnataka, India — H. P. Shivaraju, Ph.D.
Plymouth, UK — David Jenkins, Ph.D.

Introduction

According to the UNESCO report presented during the United Nations 2023 Water Conference in New York, 26% of the global population lacks safe drinking water, while 46% do not have access to well-managed sanitation facilities. Freshwater is a limited resource, making water management a very important factor for societal survival and functioning. Water scarcity is becoming a major global challenge due to rapid population growth, urbanization and increasing pressure from agriculture, industry, and energy sectors. The changes in landscapes, adverse effects of climate change, growth in food and energy production, mass migration of people to urban centers and severe competition for water by the municipal, agricultural, industrial, and energy sectors are altering the quantity and quality of the freshwater resources on which we solely depend to survive, both physically and economically. To face the looming water crises, we need to tackle the strategic policy and management issues as an integrated part of the solution toward accomplishing water sustainability.

Ever since COVID-19 pandemic started, it is clear that safely managed drinking water, proper sanitation, and hygiene issues are vital to sustain human health. Decades of misuse, poor management and, over-extraction and contamination of freshwater and groundwater supplies have exacerbated water stress and deteriorated aquatic ecosystems. Therefore, there is an urgent need to establish strong national and international mechanisms in order to prevent severe consequences of the existing global water crises and to ensure sustainable and equitable distribution of water to all.

Gaps in access to water supply and sanitation, growing population, water-intensive patterns of growth, increasing rainfall variability, and pollution are making water as one of the greatest risks to economic progress, poverty eradication, and sustainable development. Achieving the UN Sustainable Development Goals (SDGs) for water, especially Goal 6—"Access to Water and Sanitation for All," requires an integrated approach by the stakeholders and policymakers to manage and allocate water resources.

In order to address the above issues, the Centre for Science and Technology of the Non-Aligned and Other Developing Countries (NAM S&T Centre), New Delhi is bringing out this publication for dissemination of knowledge and information to the scientists, researchers, managers, policymakers and other stakeholders engaged in the "Sustainable Water Management" sector.

The book through its twenty-four chapters underscores the concept and principles of sustainable water resources management, recent advances in water treatment and management, case studies and directions for future research and development, and policymaking for sustainable water resources management and highlights the best management practices for water resource allocation, groundwater protection and water quality assurance for the Global South. The book intends to provide information on sustainable water resource planning and governance, water infrastructure planning and implementation, optimum regulations to conserve both quality and quantity of water in developing countries.

The book brings together the scientific communities from 14 countries namely, Cameroon, China, Cyprus, India, Indonesia, Italy, Namibia, the Netherlands, Norway, Portugal, South Africa, UK, USA, and Zimbabwe to share their knowledge and expertise on various water management practices and technologies needed to meet water requirements while maintaining and restoring the ecosystems that provide us with life sustaining services, so that the utilization of water resources are planned and protected for the present and future.

I am immensely grateful to Dr. Satish R. Wate, Former Director, CSIR-National Environmental Engineering Research Institute (NEERI), Nagpur and Former Chairman, Recruitment and Assessment Board (RAB), Council of Scientific and Industrial Research (CSIR), New Delhi, India for kindly agreeing to our request to write the Foreword of this book in spite of his very busy schedule.

I am thankful to Dr. Loyola D'Silva, Executive Editor, Springer Nature, Singapore for his kind support and guidance toward bringing out this book and Ms. Shalini Monica C. Selvam, Project Coordinator, Springer Nature, India for managing all the technical and administrative tasks for the publication process.

I would like to express my sincere gratitude to the Editors of this book, Dr. S. Suriyanarayanan, Associate Dean (Research), JSS Science and Technology University, Mysuru, Karnataka, India; Dr. Shivaraju H. P., Associate Professor, Department of Environmental Sciences, Faculty of Life Sciences, JSS Academy of Higher Education and Research (JSS AHER), Mysuru, Karnataka, India; and Dr. David Jenkins, Associate Professor, Nanomaterials and Devices, School of Engineering, Computing and Mathematics, Faculty of Science and Engineering, University of Plymouth, UK, for their initiatives and efforts and sparing their valuable time in reviewing the papers for this book and taking charge of this publication project.

We express our gratitude to Prof. B. Suresh, Pro-Chancellor, JSS Academy of Higher Education and Research (JSS AHER), Mysuru and Dr. Surinder Singh, Vice-Chancellor, JSS AHER, Mysuru for their encouragement and kind support in bringing out this publication. We are also thankful to Dr. Vishal Kumar Gupta, Dean (Academics), JSS AHER, Mysuru for his cooperation and support in this publication project.

I also acknowledge the valuable support of the entire team at the NAM S&T Centre and am especially thankful to Mr. Madhusudan Bandyopadhyay, Senior Adviser for his support and guidance, and Ms. Abhirami Ramdas, Research Associate for her contributions in taking this publication project forward.

I also record my appreciation for the assistance and support rendered by my colleagues Mr. Rahul Kumra, Assistant Administrative Officer and Mr. Pankaj Buttan, Data Processing Manager, NAM S&T Centre toward bringing out this book.

I believe that this book would serve as a valuable resource and reference material for water resource managers, scientists and researchers from various R&D institutions, hydrologists, ecologists, environmentalists, administrators, professionals from the water industry, government officials, policymakers, students, and people in general concerned with water resources management to make decisions to allocate water resources and to develop innovative and cost-effective measures and approaches to achieve sustainable water management.

Amitava Bandopadhyay, Ph.D.
Director General
NAM S&T Centre
New Delhi, India

Contents

Chapter 1
Sustainable Water Management in Developing Countries

S. Suriyanarayanan, H. P. Shivaraju, and D. Jenkins

1.1 Introduction

Water is a vital commodity that is necessary for everyone who lives in this world. Water is essential for socioeconomic progress and environmental sustainability. There are 2 billion people (26%) who lack access to safe drinking water 3.2 billion (42%) lack basic sanitation and 23 billion people live in water-stressed nations. According to the United Nations, 68% of the world population is projected to live in urban areas by 2050 with close to 90% of this increase taking place in Asia and Africa. An increase in population growth and accommodating them will raise the demands of water consumption, a decline in freshwater resources and an increase in energy and water pollution (UNEP 2021). Lack of clean water for drinking and sanitation will cause significant challenges in the human quality of life, economic development, productivity and sustainability of ecosystems (Cosgrove and Loucks 2015). Developing countries often face many water-related issues that affect their growth and sustainable development. Water Scarcity and Water pollution are among the major problems.

S. Suriyanarayanan (✉)
JSS Science and Technology University, JSS TI Campus, Mysuru 570006, Karnataka, India
e-mail: sunsjk@gmail.com

SCOPE (Scientific Committee On Problems of the Environment), JSS Academy of Higher Education and Research, Mysuru, India

H. P. Shivaraju
JSS Academy of Higher Education and Research, India (Deemed to Be University), Mysuru 570015, India

D. Jenkins
Wolfson Nanomaterials & Devices Laboratory, School of Computing, Electronics and Mathematics, Faculty of Science & Engineering, University of Plymouth, Devon 4 8AA, UK

S. Suriyanarayanan et al. (eds.), *Water Management in Developing Countries and Sustainable Development*, Water Resources Development and Management,
https://doi.org/10.1007/978-981-99-8639-2_1

The United Nations Sustainable Development Goals were developed to ensure the end of hunger & poverty, to protect the planet and enjoy peace and prosperity by 2030. In SDGs, water management plays a significant role through SDG 6 for ensuring access to clean Water and Sanitation, SDG 9 for fostering sustainable industrialisation, SDGs 1 & 2 for eliminating hunger and poverty and SDG 15 Protect the ecosystems. To achieve sustainable development water management is to be addressed at all levels from local, to national and regional. To overcome the water-related challenges and improve water resources Integrated water resources management (IWRM) was designed and is widely acknowledged by the international Community. IWRM is a cross-sectoral approach and the framework was established based on the understanding that water resources are an integral component of socio-economic and environmental aspects.

1.2 Water challenges in Developing countries

In low-income and developing countries, the problems and difficulties related to the accessibility and availability of safe drinking water are extremely complex. Merely 18% of low-income residents in developing countries possess a household connected to a water supply. Those who lack access to a safe water supply must purchase water from vendors at significantly higher prices than the piped city supply. Additionally, it is approximated that merely 8% of low-income dwellers have a house sewer connection. The majority of these low-income dwellers reside in suburban areas (Björklund 2001). A comprehensive and problem-based approach is needed for sustainable water management. Mainly the development of adequate infrastructure, effective and efficient water governance system, Community-based programs, and effective global-level cooperation are to be established. In developing countries, people should be empowered through participatory decision-making, sensitization programs, and literacy so that sustainable water management is to be ensured.

Water scarcity: About 1.42 billion people live in areas of high or extreme water vulnerability areas (UNICEF 2021). Two-thirds of the world's population is under severe water vulnerability for at least one month during the year. Globally 2.1% of land is covered by freshwater. In developing countries, 1.4% of land is covered by freshwater which increases the vulnerability to climate change and water scarcity (UNSTAT 2020). To overcome this issue UNICEF works at various levels through introducing context-specific technologies for increasing access to safe water and addressing the impacts of water scarcity. Furthermore, it is expected that the situation would deteriorate quickly in the years to come, with half of the world's population perhaps living in areas that may struggle with water scarcity by the year 2025. The effects of this catastrophe are extensive and estimated that by 2030, 700 million people could be forced to relocate due to severe water scarcity (UNICEF 2023).

Water Pollution: A lack of clean water is a major challenge facing developing countries in the 21st Century that is critical to the environment, human health, economic and social developments. Water quality is among the critical challenges for human health care especially the vulnerable and young people who are susceptible to illness. Up to 80% of illnesses are linked to inadequate water and hygiene in developing nations (WHO 2019, 2020). Increasing population, urbanisation, industrial growth, climate change and related anthropogenic activities directly affect the water quality. Water pollution causes severe risks to human health. About 8,29,000 people die every year due to diarrhoea mainly due to unsanitary conditions, and contaminated water. Among them, Nearly, 2,00,000 children that hold 5.3 % are under the age of Five (UNESCO 2021). Improper healthcare facilities in developing countries cause significant risk to patients, and health care staff and increased disease transmission. Research conducted by the World Health Organization in partnership with UNICEF points to the lack of clean water and proper hygiene protocols as major contributing factors to the prevalence of these infections (WHO 2020; Saanvi Vemareddy 2021). To realise the sustainable development goals Raising scientific knowledge and the importance of clean water and awareness among the people, the impact of water pollution and related consequences are most important. Developing and least-developed countries need multifaceted approaches including individual and collective responsibility, effective governance, regulations, awareness among the Community, education and training, multilevel cooperation, and adopting appropriate cleaner technologies and sustainable practices.

1.3 Integrated Water Resources Management (IWRM)

Since 1990, to maximize economic and social well-being coordinated development and management of water, land and related resources have been developed and that has been widely acclaimed by the international organisations (UNEP 2022). IWRM is an integrated comprehensive holistic approach considering all sectors and resources. Implementation of IWRM still has its hurdles in implementing in developing nations due to their differences in political and social systems, knowledge level, financial constraints, qualified management etc. Agyenim and Gupta (2012).

1.4 Water Governance and Policy in Developing countries

Given the increasing demands on water resources and the need to ensure equitable access to clean and safe water for all, water governance and policy in developing nations is a significant topic of concern. Due to causes such as fast population increase, urbanisation, industrialization, and climate change, developing countries confront particular challenges in managing their water supplies. Effective water governance and policy frameworks are required to address these concerns. The social and political dimensions of water governance, such as government effectiveness, influence the occurrence of inequality of access to sanitation services in developing countries, whereas the economic aspect of water governance, such as aid absorption, strongly influences the occurrence of inequality of access to basic water services (Bayu et al. 2020).

Water governance refers to the set of political, social, economic, and administrative systems that control the allocation, implementation, and regulation of water resources. It includes a wide range of players, including governments, local communities, civil society organizations, the commercial sector, and international organizations. Water governance and policy in developing nations necessitate comprehensive and adaptable approaches to the unique difficulties they face. Collaboration among many stakeholders, as well as the adoption of sustainable practices, are critical to ensuring equitable access to water resources and the conservation of this crucial and precious resource for future generations.

1.5 Water and Agriculture in Developing countries

Agriculture stands as the world's largest consumer and polluting sector in developing countries. Worldwide, 70% of water use accounts for agricultural irrigation, of which 80% in developing countries. Despite the scarcity and decrease in water availability, securing water for food production remains of utmost importance (Björklund 2001). In developing countries, the economy and livelihoods are significantly related to water and agriculture. The availability and effective use of water resources pose substantial obstacles to these countries' agricultural practices. Nearly 70% of the populations in least developed countries are engaged in agriculture. The majority of poor and food insecurity are based in rural areas. Therefore eliminating poverty and ensuring food security should begin with these areas (UNCTAD 2011). Investments in drought-resistant crops, irrigation infrastructure, training for farmer communities, policies for sustainable use of water, implementation of best practices and sustainable water management practices are essential for sustainable agriculture in Developing nations. Also, developing countries must come forward to improve their food security, and improve the farmer's livelihood and economic growth by ensuring water availability.

1.6 Clean Water, Sanitation, and Health:

Management of Water, Sanitation and Health is crucial in developing countries. Implementing the proper water management techniques is important for ensuring clean and safe water and in turn, promoting public health. Providing a safe drinking water supply, necessary sanitation facilities, wastewater treatment and promoting hygiene are the important requirements in water management in developing countries. In SDG 6, safe Water and Sanitation are a key component i.e. Ensure access to Water and Sanitation for all by 2030. To achieve this goal it is estimated that approximately 29 billion people require improved water supply and about 42 million people require enhanced sanitation facilities (Mara 2003). To maintain a clean environment, healthy ecosystems and economic development, developing countries need to implement sustainable water management plans especially on clean and improved sanitation facilities.

1.7 Climate Change and Water Management

A growing population and its related consequences lead to climate change resulting in floods, rising sea levels, shrinking ice glaciers, droughts and wildfires. Climate change leads to many consequences in developing countries, mainly the availability of water resources, and water quality, and changes in rainfall patterns lead to water scarcity, drought, and effects on agriculture. All these factors are causing significant obstacles in water management in developing countries. To overcome these obstacles developing countries need to cope with the change in climate patterns and use proactive water management practices such as increased investment in water management practices, capacity building and knowledge sharing, Community participation, risk assessment and monitoring and resilience and effective governance.

1.8 Financing and Capacity Building

Financing and capacity building are pressing needs in developing country's water management. Water pricing and pollution charges are essential economic instruments in water management. Qualified manpower and strategic financial planning at local, basin level and national levels can support effective water management. An estimate shows around one billion additional urban inhabitants will require water services in the next 30 years and 30 million people are prone to flood risks between 2010 and 2030 the area under irrigation will be expanded significantly which will lead to increased demand for water and decrease in water availability (Leckie et al. 2021). To promote sustainable water management practices and improve water quality and access to water the developing countries can increase their ability to address the issues related to. Capacity building is essential for establishing public-private partnerships, institutional strengthening and creating favourable policies and water resource assessment. These activities certainly help to achieve SDG 6, which ensures the availability and sustainable management of Water and Sanitation for all.

1.9 Way Forward

SDG 6 emphasises that to ensure the availability and sustainable management of Water and Sanitation for all it is essential to focus on the aspect of water resources and its ecosystem mainly on water availability and quality and its sustainable management. Given the above, it is understood that to achieve the SDGs goals, innovative solutions are urgently needed. Such as risk assessment, policies and their action plans, investments in water management, affordable and sustainable water solutions; including rainwater water harvesting, water and wastewater treatments, the emerging trends such as smart water management, Artificial Intelligence (AI), Internet of Things (IoT) are vital components in Sustainable Water Management. Adopting these rising water management trends and technology can greatly improve water efficiency, increase access to clean water, and encourage sustainable water practices in developing nations. To address the particular problems and opportunities in each location, however, successful implementation necessitates collaboration among governments, local communities, corporate sectors, and international organisations.

References

Agyenim JB, Gupta J (2012) IWRM and developing countries: implementation challenges in Ghana. Phys Chem Earth Parts a/B/C 47–48:46–57. https://doi.org/10.1016/j.pce.2011.06.007

Bayu T, Kim H, Oki T (2020) Water governance contribution to water and sanitation access equality in developing countries. In: Water resources research (vol 56, Issue 4). American Geophysical Union (AGU). https://doi.org/10.1029/2019wr025330

Björklund G (2001) Water management in developing countries–policy and priorities for EU development cooperation. Stockholm International Water Institute (SIWI). ISBN 91–974183–2–3. https://www.ircwash.org/sites/default/files/SIWI-2001-Water.pdf

Cosgrove WJ, Loucks DP (2015) Water management: current and future challenges and research directions. In: Water resources research (vol 51, Issue 6, pp 4823–4839). American Geophysical Union (AGU). https://doi.org/10.1002/2014wr016869

IWRM (2007) IWRM challenges in developing countries: lessons from India and elsewhere. Policy brief. http://www.iwmi.cgiar.org/Publications/Water_Policy_Briefs/PDF/WPB24.pdf

Leckie H, Smythe H, Leflaive X (2021) Financing water security for sustainable growth in Asia and the Pacific. OECD Environment Working Papers, No. 171. OECD Publishing, Paris. https://doi.org/10.1787/3bc15c5b-en

Mara DD (2003) Water, sanitation and hygiene for the health of developing nations. Public Health 117(6):452–456. https://doi.org/10.1016/s0033-3506(03)00143-4

UNCTAD (2011) https://unctad.org/system/files/official-document/presspb20116_en.pdf

UNEP (2021) Progress on integrated water resources management. Tracking SDG 6 series: global indicator 6.5.1 updates and acceleration needs. ISBN No: 978–92–807–3878–0

Vemareddy S (2021) The water quality crisis in developing countries. Pre-Collegiate Global Health Review. https://www.pghr.org/post/the-water-quality-crisis-in-developing-countries

Water scarcity: Addressing the growing lack of available water to meet children's needs. https://www.unicef.org/wash/water-scarcity

World Health Organization (2019) Drinking-water. World Health Organization. https://www.who.int/news-room/fact-sheets/detail/drinking-water

World Health Organization (2020) Almost 2 billion people depend on health care facilities without basic water services. World Health Organization. https://www.who.int/news/item/14-12-2020-almost-2-billion-people-depend-on-health-care-facilities-without-basic-water-services-who-unicef

Chapter 2
Progress, Challenges, and Prospects for Sustainable Water Resources Management and Development in Zimbabwe

Kraposy Kujinga and Innocent Nhapi

2.1 Introduction

Developments in the Zimbabwean water sector since independence were guided by the Water Act of 1976, a revision of the 1927 act, which legally disenfranchised the majority of the black population with respect to accessing agricultural water (Manzungu and Machiridza 2009). As of 1999, some 85% of agricultural water in the country was used by 4 500 white large-scale commercial farmers (representing less than 1% of the country's estimated 13 million people (Manzungu 2002a).

Under the 1927 and 1976 Water Acts, water was allocated using the priority date system, that is, "first in time, first in right" principle (Kujinga and Manzungu 2004). This system disadvantaged the indigenous population in terms of access to productive resources such as land and water. When the settlers promulgated the water right system in 1927, the indigenous population was not part of this process (Manzungu and Derman 2016). Water rights which were tied to land rights, were issued in perpetuity (Chikozho and Latham 2005). This entrenched the interests of the settlers, as there was little scope to redress the situation to the benefit of the disadvantaged communities. The fact that water rights were attached to land ownership, practically excluded the indigenous people who resided in what were

K. Kujinga (✉)
WaterNet, North-West University, Mahikeng Campus, Potchefstroom, South Africa
e-mail: kkujinga@waternetonline.org

Unit for Environmental Sciences and Management, North-West University, Potchefstroom, South Africa

I. Nhapi
Innozimtec Pvt Ltd, Plot 16311 Chirorodziva Heights, Chinhoyi, Zimbabwe

S. Suriyanarayanan et al. (eds.), *Water Management in Developing Countries and Sustainable Development*, Water Resources Development and Management,
https://doi.org/10.1007/978-981-99-8639-2_2

called "tribal trust lands" (communal areas) where they held no title to the land they lived on and cultivated (Matinenga 2000). After independence in 1980, people in communal and resettlement areas still did not hold any title to the land they farmed (Dancaescu 2003; Moyo 2013).

The administration of the 1976 Water Act was the responsibility of the Water Court which was empowered to grant water rights, investigate the use of water granted in a right, and revise or cancel a right for failing to beneficially use the right (Kujinga and Jonker 2006). In addition, the management of water was vested in River Boards (RBs), which were based on the sub-hydrological zones and Intensive Conservation Areas (Latham 2002). The main functions of the River Boards were to supervise the day-to-day management of water and provide technical advice to mainly white commercial farmers on water issues and application of water rights. In terms of representation, River Boards were composed of representatives of the white commercial farming sector, urban and rural local authorities, and manufacturing industry (Mtisi and Nicol 2003). Only water right holders could participate in water management under the framework of River Boards. Thus, participation in River Boards was restricted to mainly large-scale white commercial farmers. The River Boards' charter allowed them to raise levies on water right holders in their areas of jurisdiction and to monitor the use of water by right holders. River Boards were also involved in local informal arbitration and mediation in situations where conflicts would have arisen. In the event that local informal arbitration failed, the parties had to resort to the courts because the River Boards had no legal authority to enforce arbitration (Latham 2002).

Against the background of inequitable access to water for productive purposes and limited stakeholder participation in water resources management, the country embarked on a reform process for the water sector in the early 1990s (Manzungu 2002b). The impetus toward water reform was accelerated by the drought of 1991/92 that was the worst in the country's history. The drought posed a very serious challenge to the agricultural sector, which is the mainstay of Zimbabwe's economy, for the country was reduced from being self-sufficient in food production and a net exporter, to a net importer of maize (Nyabeze 2004).

The water reform process which involved some consultations around the country had three major objectives (Latham 2002), i.e.,

- To repeal the Water Act of 1976 and replace it with one more appropriate to contemporary Zimbabwe;
- To reform institutions involved in the water sector, increase stakeholder participation and devolve water resources management to the lowest appropriate level; and
- To improve access to water for all Zimbabweans through the application of equitable access and a recognition that the environment was also a user of water.

The results of the reform process manifested themselves in two major forms (Latham 2002):

- A new Water Act of 1998 was drafted and passed by parliament.

- New institutions were established for the better management of water resources in the country, namely Zimbabwe National Water Authority (ZINWA), Catchment Councils (CC) and Sub-catchment Councils (SCC).

2.2 Legislation for Water Resources Management

Water resources governance in Zimbabwe is now guided by two pieces of legislations, namely the Water Act [Chapter 20:24] and the Zimbabwe National Water Authority (ZINWA) Act [Chapter 20:25]. The Water Act provides for the development and utilization of Zimbabwe's water resources (Manzungu 2001). The Act removed the concept of water rights which were tied to land rights and all water now belongs to the state and everyone has the right to apply for a water permit in the case of private water (Government of Zimbabwe 1998a, b). This is a milestone provided by the Act as permits to use water for productive water are issued at catchment level and these are valid for a maximum period of twenty years.

Under the Water Act, Zimbabwe adopted a catchment management approach which saw the country being divided into seven catchment areas that are based on the seven major river basins in the country. The seven Catchment Councils are Gwayi, Manyame, Mazowe, Mzingwane, Runde, Sanyati, and Save (Fig. 2.1). Each catchment is administered by an elected catchment council, with technical support from ZINWA (Kujinga and Manzungu 2004). Each Catchment area is further divided into sub-catchment areas based on the major tributaries of the main river system. Each sub-catchment area is administered by a sub-catchment council composed of the representatives of the stakeholder groups in that area (Kujinga and Manzungu 2004). Two representatives from each SCC are seconded to the Catchment Council to represent its interests. Both the CCs and SCCs are established by statutory instruments under the Water Act and the ZINWA Act. The Catchment Manager's office provides technical and secretarial services to the respective catchment councils. The main responsibilities of each CC are:

- To prepare a River System Outline Plan (RSOP) for its river system;
- To determine and grant water use permits under criteria set by the Department of Water Resources Planning and Management;
- To regulate and supervise the exercise of rights to, and use of water in respect of its river system; and
- To ensure proper compliance with the Act and to supervise sub-catchment councils.

The sub-catchment councils are established by the Minister through a statutory instrument under the Water Act for any part of a declared river system that falls under the catchment council. The SCC is the operational arm of the CC. Its main function is to regulate and supervise the exercise of rights to water within the area for which it was established. It also performs any other functions that may be conferred upon it in terms of the Water Act. The stakeholders comprise the water users, members

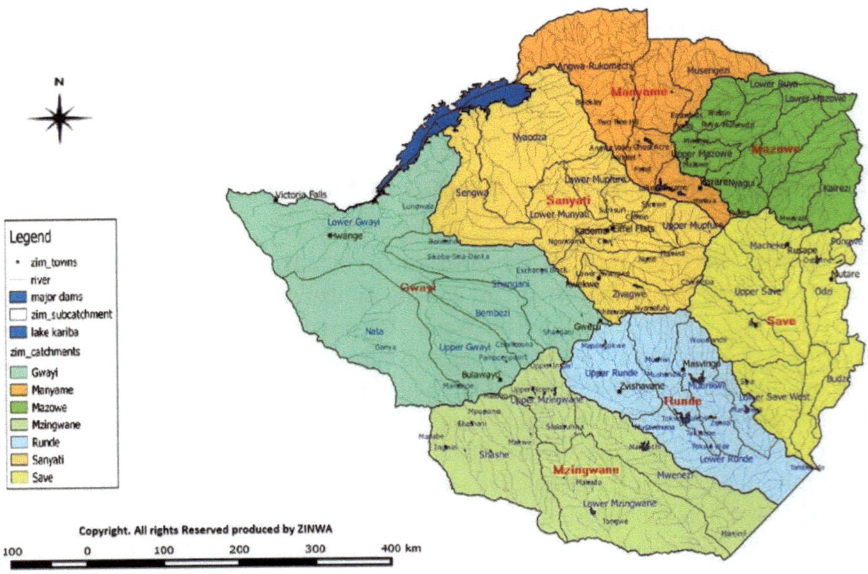

Fig. 2.1 Catchment councils and sub-catchment councils of Zimbabwe. (*Source* Zimbabwe National Water Authority)

of government departments with legal responsibilities in the management of natural resources, and private organizations that represent interests in the basin or otherwise have a direct stake in water management in the catchment. The membership of the SCC is elected from representatives of water users.

The ZINWA Act established a national water authority and provides for its functions as set out in Clause 5 of the Act and includes:

- To advise the Minister on the formulation of national water policies and standards;
- To exploit, manage, and conserve water resources in order to ensure security of supply and to facilitate equitable access to water by all sectors, and its efficient utilization, while minimizing the impacts of drought, floods, and other hazards;
- To provide specialist advice and technical assistance to local authorities and catchment councils in matters concerning the development, management, and environmental protection of water resources;
- To provide design and construction services for new works and to operate and maintain water supply facilities owned or managed by ZINWA;
- To carry out and publish hydrological and hydrogeological surveys, including water-related research, for the purposes of planning, development, and exploitation of water resources; and;
- To affect the joint management of international water resources, as directed by the Minister.

One of the major challenges confronting the water management institutions in Zimbabwe relates to water tariffs which do not reflect the cost of producing potable or supplying productive water, and hence services have not been sustainable because the revenue from users is less than the cost of running the services (Government of Zimbabwe 2013). In addition, use of nationwide blended prices leads to cross-subsidies that may constrain growth in some parts of the country while promoting uneconomic growth in others, and encouraging inefficient water use. The National Water Policy therefore advocates for demand management of water resources where tariffs will be set to reflect the true cost of production and scarcity of water to enhance sustainability.

Urban water authorities purchase bulk water from ZINWA and these have a range of other social responsibilities, in addition to water supply and sanitation (WSS), that demand high levels of funding. Revenues from water supply and sanitation have been occasionally diverted to fund other council activities, resulting in underfunding and failure of water supply and sanitation services. The Zimbabwe National Water Policy of 2012 proposes that Water Services Authorities and Water Service Providers will be required to be transparent and legally accountable to consumers. Revenue derived from water sales will be applied to the costs of providing water services in order to maintain consumer confidence and not compromise cost recovery and the use of the user pays principle. Local Authorities will undertake cost accounting and will ring-fence revenue from water sales (Government of Zimbabwe 2012).

2.3 National Water Policy Framework

The Zimbabwe National Water Policy (NWP) was launched in March 2013 (Government of Zimbabwe 2012). The private sector and civil society have an important role to play in the water sector. Government ministries or departments or local authorities should restrict themselves to and focus on policymaking, establishing regulatory regimes that provide incentives for achieving goals and sanction for serious failures (Tom and Munemo 2015). They should leave operations to parastatals and private sector based on sufficient autonomy to control outcomes. All investment plans must consider economic feasibility and infrastructure services must be affordable. Other important provisions of the Water Policy include:

- User pays and polluter pays principles.
- Subsidies are the exception only for users not able to pay for services.
- Water pricing is used as demand management instrument.
- Billing shall be done on volumetric readings.
- Payment of water services from public consumers, including government institutions, is vital.
- Local authorities are responsible for efficient and affordable water supply services and are free to delegate services to a dedicated service provider.
- Local authorities shall apply ring-fencing to revenues collected from water and sanitation services and ensure that such revenues are used for service provision.
- Tariffs should be set to achieve full cost recovery (Tom and Munemo 2015).

The NWP proposes a distinction between Water Service Authorities (WSA) and Water Service Providers (WSP) in order to address a range of issues that include: accountability for service provision, capacity, technical expertise, and the ring-fencing of water-related revenues. Water services include both water supply and sanitation (Government of Zimbabwe 2012). The policy designates Urban Local Authorities (LA) and Rural District Councils as WSA who have a duty to ensure that efficient, affordable, and sustainable access to water services are provided for all their current and potential consumers. The responsibility at operational level of providing water supply and sanitation services may be delegated by a WSA to a designated WSP. Service Authorities will have the power and authority to enter into contractual agreements with Service Providers if they do not supply the services themselves. Service Providers will be legal (public, private or mixed) entities that have the capacity to provide water supply and sanitation services to Service Authorities. A Service Provider could be ZINWA, private sector or any other legal entity. Individual LAs will have the flexibility to decide on the model they want (Government of Zimbabwe 2012).

In line with the Water Policy, ZINWA is supposed to be split into two entities:

- The National Water Supply Services Utility is dedicated to potable water supply operated from appropriate administrative locations, and
- An entity for Water Resources Management to concentrate on water resources planning, development and raw water supplies as envisaged in the Water Act of 1998 (Government of Zimbabwe 2012).

In line with the National Water Policy, the government has been actively working on setting up a regulator for Water and Wastewater Services which will be under the Ministry of Lands, Agriculture, Fisheries, Water and Rural Resettlement (Government of Zimbabwe 2012). The regulator will monitor all water resources, water supply and sanitation services and receive and assess tariff applications in collaboration with the ministries responsible for local government and agriculture. It will also oversee the licensing of Water Service Providers by Water Service Authorities (Government of Zimbabwe 2012).

Some aspects of sanitation are under Ministry of Health, which has developed a Sanitation and Hygiene Strategy 2011–2015 and 2016–2020. At present, most municipalities choose to provide their own WSS services, while smaller urban centres are served by ZINWA. In future, there may also be private WSPs, or various forms of public–private partnerships to provide WSS services. Once the regulator comes into being, licences will be required for WSS service provision, and these will be issued only to WSPs registered under the Companies Act. Hence, in order to be licensed, municipalities will be required to spin off WSS departments into stand-alone companies.

Service Agreements will be signed between WSAs and WSPs: these will be instruments to promote continuous improvements in efficiency of service delivery and to provide a framework for a high level of accountability by WSPs to WSAs.

2.3.1 Institutional Framework

Government addressed coordinating challenges in the water sector by introducing National Action Committee (NAC) on Water to have an overall oversight on the management of water resources in June 2010 (Dhoba 2022). Water and sanitation leadership is the function of the minister responsible for water. The committee also includes ministers for:-

- Health and Child Welfare;
- Transport, Communications and Infrastructural Development;
- Local Government and Public Works; and
- Energy and Power Development.

Overall coordination and management is the responsibility of the NAC made up of permanent secretaries and chaired by the Permanent Secretary for the Ministry of Lands, Agriculture, Fisheries, Water, and Rural Development (Dhoba 2022) (Figs. 2.2 and 2.3).

The Ministry of Local Government and Public Works (MoLG&PW) is responsible for supervising local authorities and regional development including both the rural and urban areas. These functions are governed by three pieces of legislation:

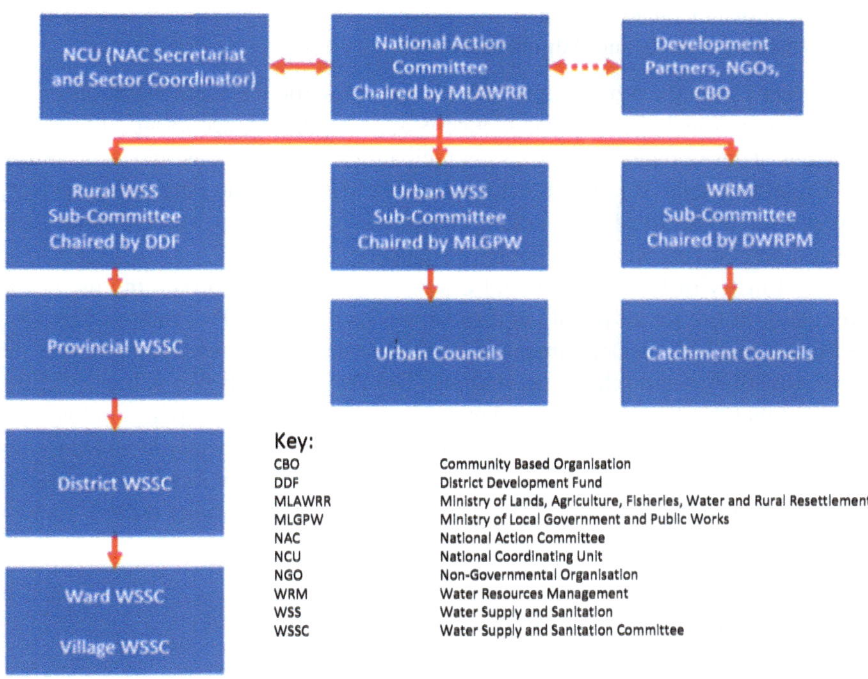

Fig. 2.2 Organogram of water resources management, water supply and sanitation in Zimbabwe

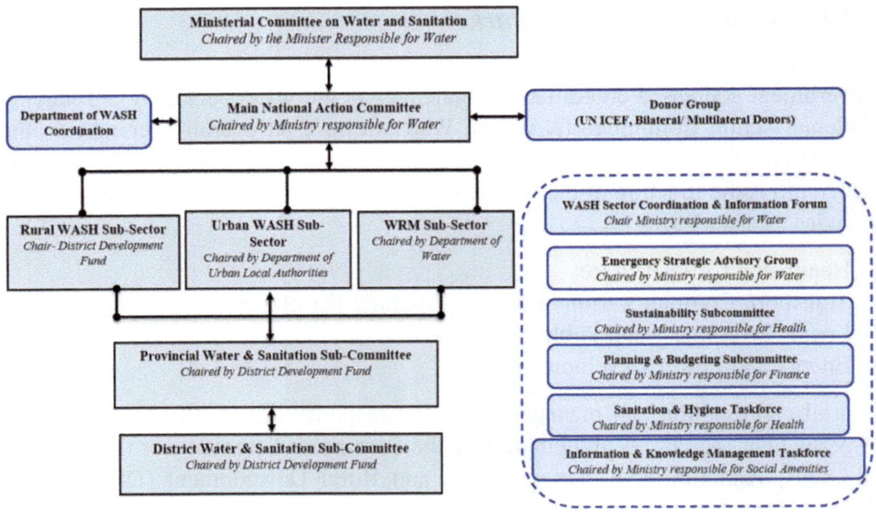

Fig. 2.3 Institutional Framework for water sector in Zimbabwe—chairing of organs. (*Source* National Action Committee on WASH)

- Urban Councils Act,
- Rural District Councils Act and
- Provincial Councils and Administration Act.

The Urban Councils Act (Chapter 29:15) provides for the administration of the country's towns and cities. Urban Councils are categorized, according to their size and levels of development, into cities, municipalities, town councils, and local boards. The Urban Councils Act affords the responsible Minister wide ranging supervisory powers over urban local authorities. The provision for input by civil society, in contrast, is relatively weak and poorly developed.

The Ministry of Health and Child Care is responsible for promoting the health and quality of life for all people in Zimbabwe. As part of its Preventative Services, the Ministry includes a Department of Environmental Health which administers the Public Health Act and related regulations. The Public Health Act (Chapter 15:09) stipulates that it is the duty of a local authority to furnish water supplies in line with health requirements. It also requires other agencies providing water in an area, such as ZINWA, to comply with this requirement. The Public Health Act also requires the local authority to maintain existing water supplies in good working order.

2.4 Capacity Issues at Different Levels

The major capacity issues for water governance at different levels mainly relate to the following:

- Financial resources which are required to fund activities as well as to invest in water security infrastructure and water governance;
- Human resources for efficiently and effectively managing water governance activities;
- Material resources are needed to execute water governance activities at the different levels.

2.4.1 Capacity Issues at Catchment Level

The Zimbabwe National Water Authority (ZINWA) which runs catchments in different parts of the country is relatively capacitated from a financial, human and material resources point of view as compared to other institutions which are involved in water governance at various levels. ZINWA has some capacity mainly because it has a legal backing to manage water on a commercial basis and, thus, collects revenue which enables it to employ human resources with different expertise for governance purposes and also to purchase the requisite material resources.

The Save Catchment Council, for example, had a total of 122 agreements (in 2017) of various quantities of water supply from which it derived its revenue (Kujinga and Nhapi 2017). About 60 of these agreements were for supplying raw water for irrigation purposes to the Lowveld for irrigation farming. According to the agreements, the client paid for the gross amount of water released from the storage facilities. This was despite the fact that there were losses along the way. Water released from Osborne Dam flows down Odzi and Save Rivers; which is a distance of more than 200 km. It is estimated that losses in these cases are about 35% of what is released and this is borne by the clients (Kujinga and Nhapi 2017).

On the other hand, the Runde Catchment supplies approximately 1 200 000 ML/annum of water to the Lowveld for sugarcane production (Kujinga and Nhapi 2017). The bulk of this water is supplied to Hippo Valley and Triangle Limited Estates. For large-scale estates, ZINWA charges USD9/ML. Assuming that all this water is supplied to the large-scale sugarcane estates, ZINWA collects something in the region of US10,800,000.00 per annum (Kujinga and Nhapi 2017). However, part of this money goes to the Zimbabwe Revenue Authority for tax, to central government for the Water Fund and Sub Catchment Councils from where the water is abstracted, for their levies. ZINWA remains with a substantial amount of money to enable it to fund its operations.

The ZINWA Catchments have clear organizational structures meant to enhance water security and water governance. The catchment is headed by a Catchment Manager who should be a qualified engineer. Below the Catchment Manager are the Operations Manager, Technical Manager, Hydrologist, Human Resources Administrator. The structure goes down to the lowest level where there are positions such as water supply operatives and hydro observers.

In terms of resources, Save Catchment has resources which include more than 10 vehicles for off road use, 15 motorbikes and 10 bicycles. These resources are used for accessing the different parts of the catchment to ensure that different clients access water and also for the management of water resources.

2.4.2 Capacity Issues at Sub-Catchment Level

Sub-catchment Councils (SCCs) are relatively capacitated financially and materially as they have a legal mandate to levy all those who have water permits within their areas of jurisdiction. The SCCs also receive part of the money collected by ZINWA in the form of water agreements. Permits are usually given to those who want to abstract water from river points and boreholes. The Lower Save West Sub-Catchment Council (LSWSCC) has more than 350 permits and has an annual budget of USD50,000. However, the SCC collects around USD30,000 per annum. On the other hand, the Chiredzi SCC administers a total of 44 permits (Kujinga and Nhapi 2017). What the two SCCs are collecting on an annual basis is inadequate in light of the activities that have to be undertaken. Moreover, some of the payments come in local currency at bank rate which make the financing of their operations difficult. The following are the budget line items for both Chiredzi SCC and LSWSCC:

- Travel and subsistence for the councils when they come for meetings;
- Sitting allowances for councillors;
- Staff salaries;
- Office Rentals;
- Contributions for CCs which is 10% of income from permits
- National Social Security Authority (NSSA) and Zimbabwe Revenue Authority (ZIMRA) contributions;
- Fuel and transport for outreach and other activities;
- Community projects

Travel and subsistence allowances and staff salaries take up about 60% of the budget. The other line items take up the remaining 40%.

The LSWSCC and Chiredzi SCC have two full-time staff members who include an administrator and one training officer each who interacts with the communities. The training officers use motorbikes to travel within the sub-catchments when carrying out their activities. The challenge is that for the two sub-catchments, it is only one training officer each who cannot attend to issues if they arise simultaneously. The LSWSCC also has limited access to a vehicle which is owned by the CC. The vehicle is also used by the other SCCs within the catchment (Kujinga and Nhapi 2017).

Ordinary stakeholders have also not been capacitated to be able to play an active role in selecting their representatives who sit on SCCs. The process is not clear and transparent as the ordinary stakeholders do not simply know the existence of SCCs and their representatives as well. Moreover, the participation of women both at SCC level in both Chiredzi SCC and LSWCC is very low and there has not been any attempts to ensure that this is improved.

2.4.3 Capacity Issues at Provincial Level

At the Provincial level, the Provincial Water Supply and Sanitation Sub-Committee (PWSSC) which coordinates all WASH issues in the province does not have a dedicated budget for its activities which include meetings and monitoring and evaluation. In most cases, for meeting the different departments of the PWSSC have to carter for themselves from their own votes. For monitoring and evaluation, they have to rely on their specific departments and development partners operating in the specific Districts they will be visiting. The greatest challenge for the PWSSC is transport which they do not have.

At the Provincial level, all the relevant ministries involved in WASH are also represented as they have the mandate to supervise those at the district level. For example, the District Development Fund (DDF) has a hydrologist who is responsible for siting boreholes in the province as well as a water engineer who is responsible for ensuring quality in the equipping of boreholes.

The DDF has been using obsolete borehole drilling equipment until recently when the Government acquired new equipment in its ambitious programme to drill 35,000 boreholes (one borehole per village). Before that, DDF was heavily reliant on development partners for hiring drilling machinery for their projects and only providing technical advice where necessary. Both the DDF provincial and district offices entirely depend on central government for funding. However, funding for development projects has not been coming to the districts.

2.4.4 Capacity Issues at District Level

At the District level, is the District Water Supply and Sanitation Committee (DWSSC), composed of officials from different ministries and chaired by the district administrator. The DWSSC does not have a budget to operate with. With regards to meetings of the DWSSC, these do not need much financial resources mainly because the members are all in a particular location like Chiredzi Town. They do not have to spend much on transport as some of them can actually walk to the meeting venue.

Resources for undertaking projects on the ground, including monitoring and evaluation and undertaking surveys related to water supply, sanitation and hygiene(WASH) are the ones which are a challenge. Most DWSSCs across the country mainly rely on development partners for project implementation and for transport to enable their staff to go out and monitor projects and activities in the wards. Vehicles which a particular DWSSC might have would have been donated by NGOs whose projects would have ended and such vehicles often lack maintenance. For example, the Chiredzi DWSSC has an old vehicle which is a left over from one project.

In terms of the personnel for monitoring, this is available from the different departments involved in WASH. For example, personnel for the Ministry of Health monitors issues related to water quality as well as sanitation and hygiene; those from District

Development Fund (DDF) monitor and evaluate issues related to functionality of water points; while those from the Rural District Councils (RDCs) focus on issues related to community participation.

Government departments involved in WASH at the district level have different capacities for water security and water governance. At the district level, DDF is headed by a technician and leads a team of water supply operators who are hands-on as they do the actual work. Since there are financial challenges with the central government related to finances, DDF now gets a bit of resources mainly for maintenance. Usually, within a particular district, there would be a number of boreholes that are down mainly because there are no financial resources to rehabilitate them.

The Ministry of Health which is also represented at the district level has a District Environmental Health Technician in each district who is responsible for supervising ward environmental health technicians within the district. There are 25 of these in Chiredzi, for example. Each technician services a catchment area for a clinic. The catchment area might include several wards. Each technician has a motorbike which is fueled and serviced by the government for easy mobility. However, at times mobility is hampered by lack of fuel. In most cases, if there are development partners operating in particular areas, these usually assist with fuel and servicing of the motorbikes.

Most rural district councils, such as Chiredzi RDC, face severe financial constraints as they cannot fund development projects related to WASH. The RDCs are supposed to have their own budgets for borehole drilling as well as equipment but this cannot take place due to financial challenges. The small budget which the RDCs can afford is mainly related to training communities in community-based management (CBM) of water points as well as monitoring. Development partners are the ones funding most projects related to WASH.

2.4.5 Capacity Issues at the Community Level

There are several capacity issues at the local level which affect water security and water governance. In terms of coordination at the ward level, there are ward councillors as well as the Ward Water Supply and Sanitation Sub-Committee (WWSSC) which all coordinate activities related to WASH. Though ward councillors and the WWSSCs are in place across the country, these do not have budgets which they control regarding the implementation of WASH activities. These institutions cannot drill or rehabilitate boreholes. They would have to simply watch as their people struggle with issues related to water supply, sanitation, and hygiene. A ward councillor simply forwards all challenges to the RDC full council where they are simply told that there are no financial resources for borehole drilling or rehabilitation.

Structures for water governance are put in place for water points and for irrigation schemes but these face various challenges which lead to water insecurity. These structures usually have constitutions in place which are supposed to ensure their continuity. This is a positive step as it is meant to ensure sustainability of infrastructure which community members benefit from. Most water points across the country have water point committees. It is commendable that the majority of water users are women who constitute the bulk of the water users and those on the water point committees.

However, for water points, some boreholes are not functional and one of the factors behind this is ineffective committees. At times water points break down permanently mainly because the committees would have engaged untrained pump minders. In some cases, the boreholes develop major problems which need DDF maintenance teams, but such problems are not reported on time.

In some instances, beneficiaries refuse to contribute either labor or financial resources in order to enhance the functionality of the water points. Some of the boreholes develop minor problems and beneficiaries fail to contribute money to buy some spare parts. With time, the boreholes breakdown permanently. This results in those beneficiaries with broken down boreholes either going to the nearest functional water point, thereby, putting pressure on such points, or resorting to fetching water from protected sources.

In order to ensure functionality of water points, DDF and development partners have been involved in the training of village pump minders (VPMs). The VPMs are trained to do simple maintenance routines such as changing leather cups, greasing, and fishing out rods which would have fallen down the boreholes. The VPMs are provided with tools which are kept at a central point like a school or a clinic where each VPM who needs them has to sign for them.

Though there are pump minders who have been trained in the communities, most of these are men who can be highly mobile. Due to the difficult prevailing economic environment, some of the pump minders once they receive technical training, they cross the border either to South Africa or Mozambique where they can be marketable. This leaves the communities vulnerable as broken-down boreholes are not fixed on time. However, the pump minders who move out of the communities are not able to go away with the tools as they do not stay in their custody. However, there are breakdowns which require major repairs which the communities cannot handle and would require spare parts and expertise from DDF. In such cases, DDF might not have the spares to fix the boreholes and this results in the water point going for longer times without being repaired.

2.5 Awareness Creation in Water Resources Management

As many water issues need broad public support and understanding, raising awareness on issues surrounding water resources is increasingly seen as important. Public awareness means the general level of understanding of a certain topic. Thus, raising awareness for water issues is a way to build a common understanding of water issues and to create shared values on how water should be used and managed.

Awareness for water resources management in Zimbabwe is raised by different actors using different platforms and channels. For example, SCCs and CCs are involved in awareness programmes related to catchment protection. Some of the awareness programmes have an element of training. The training programmes undertaken by the SCCs with the aim of raising awareness mainly focus on:

- Catchment protection, i.e., undertaking activities which involve conservation of rivers and growing vetiver grass;
- General awareness meetings;
- Conservation of water;
- Canal protection;
- General water governance.

However, these activities have generally been targeted at a few stakeholders due to limited financial resources on the part of SCCs. The SCCs have not been able to build capacity for effective conflict resolution related to access to water. Most SCCs also have limited vehicles for mobility to enable them to reach different parts of the sub-catchment areas.

The NGOs working on raising awareness on water issues in Zimbabwe include the Community Water Alliance, Monavale Wetland Development Trust, and Dabane Trust. Development partners which include NGOs working at the grassroots to promote WASH activities across the country work with relevant government departments which include health, RDCs and DDF to raise awareness on critical water issues in communities. Most NGOs would have specific programmes related to WASH and these include borehole drilling and rehabilitation, sanitation and community participation. Within these programmes and projects, there are activities related to making the communities aware of the importance of water conservation, and management of water quality at the household level (i.e., storage and use). There is also awareness and training.

2.6 Successes and Challenges in Gender Mainstreaming

There is a need to have a gender sensitive approach to water resources management since men and women have different needs, opportunities, and power. Thus, men and women should actively participate in water resources management and development and not undermine each other. Apart from constituting more than 51% of the population with 80% of them living in rural areas and 60% of them being household heads in Zimbabwe, women provide water for the fulfillment of human needs, food preparation, washing clothes, and family cleanliness (Chigusiwa et al. 2023). They are also custodians of community water supply facilities, identifying sources for drinking, bathing and animal watering. Women are involved in water quality monitoring and coming up with water conservation strategies. They are pivotal in using water for food production in agriculture.

One of the goals for the water reform process for the water sector which Zimbabwe embarked on in the 1990s was to broaden access to water resources for both productive and domestic uses in general and for women in particular, and to enhance their participation in water management. This Water Act of 1998 is premised on principles of integrated water resources management which include equity, stakeholder participation, and decentralization of water resources management to the lowest appropriate level (Government of Zimbabwe 1998a, b).

Despite the water reform process that was undertaken, the country's legal instruments do not explicitly address gender (Adeoti 2021). The 1998 Water Act is silent on a whole host of gender issues which include the participation of either women or men in the management of water resources in the country, neither does it highlight any gender considerations in the allocation, access, and utilization of water resources especially given the fact that women have been a marginalized group (Government of Zimbabwe 1998a, b). Though Water Act provides for the establishment of stakeholder institutions in the form of Catchment and Sub-Catchment Councils, it is silent on the representation of women and their active participation on these platforms (Kujinga and Manzungu 2004). The fact that the Zimbabwe water legislation is gender neutral, it makes it challenging to enforce gender issues in water resources management even if this could be provided for in policy. Women who have been disadvantaged in taking an active role in decision-making related to water resources management do not have any legal backing for their participation. This shows that women's strategic needs in water resources management such as rights of access to water and decision-making are not legislated.

Gender inclusivity through legislation can enhance equality and provide opportunities to the different gender groupings to participate in decision-making processes and management of water resources. Legislation, thus, provides for the right to equality before the law and equal protection of the law for all gender groupings in terms of access, use and management of water resources. Gender sensitive legislation, thus, ensures that the strategic gender needs of all gender groupings in terms of participation in decision-making, management of water resources and access to productive water are upheld as both men and women will be treated equally before the law and not be discriminated against on the basis of their gender.

Gender issues in water resources management and development are, however, addressed in the Zimbabwe Water Policy of 2012 (Government of Zimbabwe 2012). However, though there are gender provisions in water resources management and development in the water policy, these are not enforceable since they are not backed up by any legislation. The policy recognizes and promotes gender equity in allocation, access, and utilization of water (Government of Zimbabwe 2012). It acknowledges the disproportionate burden placed on women and the girl child when fetching water and taking care of the sick. The policy highlights that targeted programming and implementation of water resources activities should be gender sensitive. It further highlights the importance of gender-based budgeting (Government of Zimbabwe 2012). The Zimbabwe water policy also highlights that at least 30% of the Catchment and Sub-Catchment Councils positions should be reserved for women and the youth (Government of Zimbabwe 2012).

Regarding the ZINWA board, the water policy states that, at least 3 board members are supposed to be women, youth, and worker representative. Though the current ZINWA Board is chaired by a women and there are a total of four women out of the 12 board members that are appointed by the Minister responsible for water resources, the ZINWA Act of 1998 [Chapter 20:25] is silent on the gender composition of the board. However, the gender provisions of policy are a step in the right direction in terms of addressing the strategic gender needs of women in decision-making on water resources management (Government of Zimbabwe 2012). Since women are a key stakeholder group in the water sector in Zimbabwe, their representation on the ZINWA Board needs also to be re-looked at so that their composition can be increased to 50% (Government of Zimbabwe 2012). The most important issue will be the appointment of both men and women on to the board who can participate equally in taking key decision on the sustainable management of the nation's water resources.

Though the legislation on water resources management is silent on gender issues, the government of Zimbabwe together with a number of international and non-governmental organizations have played a pivotal role in promoting the participation of both men and women in water resources management especially at the grassroots level. The process has mainly been aimed at addressing the practical gender needs as emphasis has been on access to water for domestic and to some extent productive purposes (Chifamba et al. 2013). Organizations such as Plan International, Practical Action, and United Nations Children's Fund (UNICEF) promote access to water in both rural and urban areas. Programmes by NGOs and international development agencies have involved the rehabilitation and drilling of boreholes and rehabilitation and establishment of irrigation schemes. For example, in 2019, Plan International rehabilitated 20 boreholes in Chiredzi District in order to promote access to water by households. Each water point established or rehabilitated has its own water point committee which is composed of both men and women. The committee manages and maintains the water source for the benefit of the community. Some members of the community are trained as pump minders and they repair the water point when it breaks down. Households are expected to contribute to the purchasing of any spare parts.

As a way of promoting gender mainstreaming CCs in Zimbabwe try and ensure that all units and all personnel within the water resources management system have both men and women involved in all aspects of programmes and policies. Runde and Manyame Catchments have plans on quarter allocation of gender to posts. All planning activities in Manyame catchment are done with a balance on gender representation as a way of promoting implementation of Integrated Water Resources Management (IWRM), gender sensitization at all levels especially from decision-making levels. However, when it comes to heading catchments, out of the seven catchments in the country, two have been headed by a woman since the inception of these institutions. There is, therefore, a need for ZINWA to balance the positions of catchment managers in terms of gender by identifying capable women who can fill in the posts.

In Matebeleland South, Practical Action and Dabane Trust have been working with communities to build sand dams for productive purposes. The dams are made from locally available, low-cost materials and are constructed across dry riverbeds. Sand builds up behind the dam wall. When heavy downpours finally come, water is filtered through the porous sand and stored underneath for the drier months. With the water captured, solar-powered pumps and irrigation systems are developed to channel it to where it is needed to irrigate crops.

2.7 Research and Development for Water Resources Management

2.7.1 Role of Higher Education Institutions

Research within the water sector is key as it enhances the development of water resources. There is a need for research which focuses on different aspects of water resources related, for example, to hydrology, hydro-geology water pollution, water supply, and sanitation (i.e., aquifer type, storage, abstraction, and quality), socio-economic activities as they are related to water resources. Universities in Zimbabwe have played a key role in undertaking research that can inform policy and ultimately development.

Higher learning institutions such as the University of Zimbabwe and Midlands State University have programmes which focus on integrated water resources management. These programmes are key in the generation of knowledge that can be used at various levels as well as policy and also in producing human resources that are key for the water sector. The Master of Science (MSc) Programme in Water Resources Engineering and Management (WREM) and the MSc in IWRM which are run at the University of Zimbabwe since 1998 and 2002 respectively, have produced over 500 graduates and around 200 of these are from Zimbabwe. Some of these graduates have been employed in government, private sector and NGOs.

The students from the above-mentioned programmes have undertaken research in the following key areas:

- Non-revenue water assessment;
- Water pollution and water quality assessment;
- Water resources assessment;
- Groundwater management;
- Hydrological modelling in ungauged basins;
- Flood and disaster monitoring and early warning systems;
- Climate-proof water infrastructure;
- Crop-water requirements;
- Development and commissioning of river system outline plans.

There have been over 200 journal publications produced from the research projects undertaken by the students. This scientific work is mostly available online and can inform development programmes and projects in the water sector.

2.7.2 ZINWA and Water Research

ZINWA has a Data and Research department which has the responsibility of undertaking research and gathering data that is crucial for water resources management and development. The research and data gathered by the department include river flows, water quality, and rates of sedimentation and siltation. However, the department faces a number of viability challenges which include limited and old infrastructure which include flow gauging stations. Other hydrometric data is collected by the Metrological Services Department, which is also struggling to get sufficient financial and technical resources to maintain and expand its monitoring networks. The Environmental Management Agency (EMA) also collects and maintains a database on water quality in the country. All these three organizations do not readily release their data to the public or are charged for at very exorbitant rates. The country used to hold annual symposia for the water sector up until the 1980s, which allowed for the internal exchange to research data.

The country, however has standards related to the construction of reservoirs which should be informed by pre-feasibility and feasibility studies. All the dams that have been constructed in recent years such as Tugwi-Mukosi and Marovanyati and those which are under construction such as Gwayi-Shangani and Kunzvi underwent these processes (The Herald 2018; 2020). The feasibility study for Gwayi-Shangani dam showed that on completion the structure will hold 634 million cubic meters of water which will be supplied to the second largest city of Bulawayo which faces perennial water problems (The Herald 2020). The dam was expected to have been completed in December 2022 but was delayed by the COVID-19 pandemic.

2.7.3 Groundwater Research and Development

One area in which the country is lagging behind is on research and data on groundwater resources. There is currently inadequate data on groundwater aquifers across the country, recharge rates, storage, and abstraction. There is no strict enforcement of rules on borehole drilling especially in urban areas where households in affluent areas are turning to groundwater as an alternative source since most local authorities are failing to provide adequate potable water. Monitoring of groundwater is mainly done in three strategic aquifers of the country, i.e., Nyamandlovu, Save Alluvial, and Lomagundi Dolomite (Masocha et al. 2020).

The Nyamandhlovu area lies about 40 km north-west of Bulawayo on the edge of the Kalahari sedimentary basin. Thin patches of Kalahari aeolian sands cover Karoo basalts in some places. The basalts overlay the Forest Sandstone which rests directly on the granitic gneiss basement complex (Shumba et al. 2009). The basalts have been heavily weathered and fractured in places with thickness varying between 15 and 30 m (Gwaze et al. 2000). Only the Upper Karoo formation is identified in Nyamandhlovu with a thickness of up to 200 m (Shumba et al. 2009). The Nyamandlovu Aquifer has been supplying Bulawayo, a city with about 650,000 population with water since the 1990s.

The Save Alluvial aquifer is a multi-2-layered system. The top of the aquifer is generally unconfined and the bottom aquifer is generally confined by a sandy silt or sandy clay aquitard. Data is not available on the average depth of the rest water level (Altchenko and Villholth 2013). The average aquifer thickness is in the order of 35 m. The predominant source of recharge is through runoff and through direct infiltration of rain water through precipitation over the aquifer area. Within Zimbabwe the deeper confined aquifer is primarily recharged by rainfall high in the eastern mountains, while the unconfined aquifer is recharged largely through a series of shallow pans located in the western part of the plain (Altchenko and Villholth 2013). Discharge is through river base flow. The major discharge mechanism is probably through river base flow. Data was not available on the amount of groundwater abstractions, nor on the amount of fresh water that was abstracted over the aquifer area (Altchenko and Villholth 2013).

2.8 Conclusions

In terms of water legislation, Zimbabwe made phenomenal progress by reforming the water sector and adopted an act which recognizes IWRM, a current approach recognized globally. It is commendable that the country decentralized the management of water resources to the lowest appropriate level. There are, thus, prospects for the country to refine the process and ensure that all the water management structures are appropriately capacitated to carry out their various mandates. Lack of finances at different levels hamper the implementation of projects and programmes.

Though water policy recognizes the importance of gender, i.e., the role of men and women in the management of water resources, it is important that water legislation be explicit about gender issues; especially the participation of women. It is important to have women adequately represented at all levels including on the ZINWA Board, CCs, and SCCs. Having such representations backed by legislation contributes greatly to enhancing the strategic needs of the different gender groupings.

Regarding research and development, it is important to note that higher learning institutions such as the University of Zimbabwe have played a pivotal role in enhancing research and development. More resources need to be availed to such institutions so that they can continue playing a pivotal role in the generation of knowledge. However, there is a need for more resources for organizations like ZINWA so that they can also play a pivotal role in research and development. It is also important for financial resources to be channeled in groundwater studies.

References

Adeoti O (2021) Barriers to mainstreaming gender in water resources management in Nigeria. Water Sci 35(1):127–134

Altchenko Y, Villholth KG (2013) Transboundary aquifer mapping and management in Africa: a harmonised approach. Hydrogeol J 21(7):1497–1517

Chigusiwa L, Kembo G, Kairiza T (2023) Drought and social conflict in rural Zimbabwe: Does the burden fall on women and girls? Rev Dev Econ

Chikozho C, Latham J (2005) Shona customary practices in the context of water sector reforms in Zimbabwe. In: International workshop on African water laws: Plural legislative frameworks for rural water management in Africa

Dancaescu N (2003) Land reform in Zimbabwe. Florida J Int Law 15(4):5

Dhoba L (2022) Strengthening water, sanitation and hygiene governance: a critical review of Zimbabwe's WASH sector institutional arrangements. H2Open J 5(2):248–263

Ephraim Chifamba E, Shoko M, Nhodo L (2013) Women participation in community water management projects inBuhera, Ward 13, Zimbabwe. Basic Res J Agric Sci Rev 2(3). ISSN 2315–6880

Government of Zimbabwe (1998a) Water Act [Chapter 20: 24] 1998. Government Printers, Harare

Government of Zimbabwe (1998b) Zimbabwe National Water Authority Act [Chapter 20:25] 1998. Government Printers, Harare

Government of Zimbabwe (2012) National water policy. Ministry of Water Resources Development and Management, Harare, Zimbabwe

Government of Zimbabwe (2013) Constitution of Zimbabwe Amendment (No. 20). Government Printers, Harare

Gwaze P, Dahlin T, Owen R, Gwavava O, Danielsen J (2000) Geophysical investigations of the Karoo Aquifer at Nyamandhlovu, Zimbabwe. In: Proceedings for the EEGS-ES** 2000 annual meeting

Kujinga K, Nhapi (2017) Gap analyisis study on water security and governance issues: Chiredzi water security and wash project. Consultancy report submitted to Plan International

Kujinga K, Jonker L (2006) An analysis of stakeholder knowledge about water governance transformation in Zimbabwe. Phys Chem Earth Parts a/b/c 31(15–16):690–698

Kujinga K, Manzungu E (2004) (2004) Enduring contestations: Stakeholder strategic action in water resource management in the save catchment area, Eastern Zimbabwe. East Afr Soc Sci Res Rev 20(1):67–91

Latham J (2002) A review of the water reform process in Zimbabwe: the case of the Manyame Catchment Council. In: Manzungu E (ed) The process and dynamics of catchment management in Zimbabwe. Save Africa Publications, Harare, pp 21–43

Manzungu E (2001) A lost opportunity of the water reform debate in the fourth parliament of Zimbabwe. Zambezia 28(1):97–120

Manzungu E (2002a) More than a headcount: towards strategic stakeholder representation in river basin management in South Africa and Zimbabwe. Phys Chem Earth 27:927–933

Manzungu E (2002b) Global rhetoric and local realities: the case of Zimbabwe's water reform. In: Chikowore G, Manzungu E, Mushayavanhu D, Shoko D (eds) Managing common property in an age of globalisation: experiences from Zimbabwe. Weaver Press, Harare, pp 31–44

Manzungu E, Derman B (2016) Surges and ebbs: national politics and international influence in the formulation and implementation of IWRM in Zimbabwe. Water Alternatives 9(3):493

Manzungu E, Machiridza R (2009) Economic-legal ideology and water management in Zimbabwe. Econ Manag Financ Mark 4(1):66–102

Masocha M, Dube T, Owen R (2020) Using an expert-based model to develop a groundwater pollution vulnerability assessment framework for Zimbabwe. Phys Chem Earth Parts a/b/c 115:102826

Matinenga ET (2000) A new water act for Zimbabwe? University of Zimbabwe (UZ) Publications

Moyo S (2013) Land reform and redistribution in Zimbabwe since 1980. In: Land and agrarian reform in Zimbabwe: Beyond white-settler capitalism. pp 29–78
Mtisi S, Nicol A (2003) Caught in the act: New stakeholders, decentralisation and water management process in Zimbabwe. Sustainable livelihoods in Southern Africa, Institute of Development Studies, Research paper no. 14
Nyabeze WR (2004) Estimating and interpreting hydrological drought indices using a selected catchment in Zimbabwe. Phys Chem Earth Parts a/b/c 29(15–18):1173–1180
Shumba BT, Hlatywayo DJ, Midzi V (2009) Focal mechanism solution of the 15th March 2008, Nyamandlovu earthquake. S Afr J Geol 112(3–4):381–386
The Herald (2018) Tugwi-Mukosi feasibility studies complete. 18 August 2018.
The Herald (2020) Gwayi-Shangani dam construction. 15 May 2020
Tom T, Munemo E (2015) Republic of Zimbabwe national water policy: a desk review of the gaps between the policy and its implementation. Int J Publ Policy Admin Res 2(3):60–72

Chapter 3
Sustainable Water Treatment and Management Approaches in the Urban Environments of Developing Countries

Divya Vinod and Harikaranahalli Puttaiah Shivaraju

Abstract In the current scenario, more than half of the population is settled in urban areas, which serve as hubs of development, prosperity, and manufacturing. However, managing water resources in these developing regions poses a significant challenge in providing hygienic and efficient solutions. The most common challenges faced by developing countries include flooding and extreme weather events. This chapter aims to provide a critical assessment of emerging approaches to sustainable water treatment and management. We will discuss various types of membrane science techniques and advanced water treatment technologies within the context of sustainable water treatment. We will highlight the concept of the circular economy, the implementation of Sponge city initiatives, and decentralized solutions as key components of sustainable water management approaches. We will also explore various opinions on these factors and describe why the transformational agenda for sustainable urban water management perspectives has not yet been fully achieved.

Acronyms and Abbreviations

MBfRs	Membrane Biofilm Reactors
ZLD	Zero liquid discharge
POME	Palm oil mill effluent
MD	Membrane distillation

D. Vinod · H. P. Shivaraju (✉)
Department of Environmental Sciences, JSS Academy of Higher Education and Research, Mysuru 570015, India
e-mail: shivarajuenvi@gmail.com

D. Vinod
Center for Water, Food and Energy, Dombaranahalli Post, GREENS Trust, Turuvekere Taluka, Tumkur District, Harikaranahalli, Karnataka 572215, India

S. Suriyanarayanan et al. (eds.), *Water Management in Developing Countries and Sustainable Development*, Water Resources Development and Management,
https://doi.org/10.1007/978-981-99-8639-2_3

MDS	Molecular dynamic simulations
DSS	Decision support systems
ANN	Artificial neural network
GI	Green infrastructure
GR	Green roofs

3.1 Introduction

Rapid industrialization and rural migration are major contributors to the water crisis. The increasing urbanization has resulted in a growing demand for water. Furthermore, the contamination of groundwater and surface water is frequently reported, making it challenging for society to access clean water (Pauli 2020). Climate change also exacerbates this situation. These factors combined have led to a depletion of freshwater resources. Relying on a single distribution unit for water supply poses a significant challenge to sustainable water resources (Carrard et al. 2019). Given the current global population, governments and decision-makers are now exploring alternative options to address the existing water supply issues and promote efficient water reuse. One of the main reasons for the world's water scarcity is that less than 1% of total freshwater is available, and even that limited resource is often managed improperly (Roshan and Kumar 2020).

The shortage of water and sanitation poses a significant challenge in both urbanized and developing regions. Developed countries often manage to address these challenges by allocating expenses for operation and maintenance, but they still face resource depletion due to rapid urbanization, resulting in inadequate service supply to societies (Machete and Marques 2021). On the other hand, poor regions in developing countries lack proper access to facilities, leaving communities without access to water, sanitation, and hygiene. As a result, meeting sustainable development goals becomes unattainable in these areas (Emile et al. 2022).

The food and agriculture sectors are the leading consumers of water globally, utilizing approximately 100 times more water than other sectors. The scarcity of freshwater, extensive water usage in industries, and the corresponding water demands for agricultural purposes contribute to this challenge (van Vliet et al. 2021).To address this issue, the utilization of seas and oceans as a cost-effective source for surface water recovery has gained attention. Researchers have conducted numerous studies in the field of water desalination, aiming to find sustainable solutions (Dhakal et al. 2022).

Only limited investments have been made in sewage water treatment facilities, with greater emphasis placed on water treatment and supply rather than wastewater collection and management practices (Hoffmann et al. 2020). However, with ongoing development trends and population growth in developing areas, there is a growing recognition of the importance of wastewater treatment. There is a rising awareness of the detrimental effects of sewage water (Su et al. 2020).

Many research studies have demonstrated that if treated wastewater is effectively managed, it can serve as a significant source of water supply to meet the needs of the growing population. The Mediterranean region has embraced a new strategy of utilizing treated wastewater extensively for irrigation purposes, which has become a major water resource in those areas. Progress in water security and providing a safe water supply in the Mediterranean region is crucial for achieving social, economic, and political stability (Rocha et al. 2020). Embracing these techniques will fulfill the requirement for sustainable utilization of treated wastewater. However, a major challenge lies in implementing low-cost wastewater treatment technologies that can maximize the efficient utilization of limited water resources while ensuring health and safety standards associated with the use of treated wastewater (Rashid et al. 2021).

Beginning with the Brundtland Report of 1987 and continuing with the New Urban Agenda of 2016, sustainable development has been recognized as a crucial concept. Within this overarching concept, significant attention has been given to "sustainable urban development," specifically in the realm of "sustainable urban water management". Urban water management encompasses various aspects, such as urban drainage, wastewater treatment and collection, water supply in cities, and the preservation of freshwater resources. However, despite its importance, there have been numerous issues and concerns raised surrounding this term (Hajian and Kashani 2021).

In this chapter, we explore several techniques for water treatment and management in urban areas, including membrane science techniques and advanced water treatment technologies. Additionally, we delve into various approaches to urban water management.

3.2 Sustainable Water Management Challenges

Among all the planets we are aware of, ours is unique due to its abundance of water. However, its geographic and temporal patterns are not uniform, despite the fact that the world's freshwater supply is sufficient to meet all present and anticipated water demands. In numerous regions, our freshwater supplies are often insufficient to satisfy domestic, economic growth, and ecological demands. In these regions, the scarcity of safe water that can adequately meet the needs of both sanitation and drinking poses a significant barrier to economic development, human health, and the preservation of environmentally sustainable ecosystems. It is imperative that we work together to find solutions to these challenges. However, we face numerous obstacles in achieving that goal, particularly considering the unpredictable and changing nature of future climate patterns, as well as the rapidly growing population that fuels economic and social expansion, globalization, and urbanization. Comprehensive research is crucial in all aspects of water management to identify the most efficient approaches to tackle these challenges. Water Resources Research, a prominent journal since 1965, has played a vital role in documenting and disseminating recent discoveries regarding the management of water quantity, quality, and cost. In order to ensure the long-term accessibility and potability of water resources, it is imperative that sustainable water management overcome a number of challenges. Some of the major challenges include the following:

1. Water scarcity- Water scarcity is a problem in many parts of the globe due to factors such as growing populations, urbanization, and climate change. Sustainable water management is significantly hampered by the rising demand for water and depleted freshwater supplies (Sun et al. 2023).
2. Urbanization and Demographic Growth: Urbanization and significant population increases put enormous strain on water supplies. As the number of people living in cities grows, the consumption of water rises, exerting pressure on both natural systems and existing water supply systems (Sarkar and Lakshmana 2022).
3. Climate Change: Severe weather conditions such as droughts and floods are worsening due to climate change, which is also altering the distribution of precipitation. These changes make it challenging to manage water sustainably, as they affect both the distribution and accessibility of water resources (Abbass et al. 2022).
4. Water Pollution: Water quality is seriously threatened by contamination of water sources from industrial discharge, runoff from farms, and improper waste disposal. Pollution not only harms human well-being but also disrupts biodiversity and aquatic environments (haghi et al. 2022).
5. Inadequate Infrastructure: Many areas lack the storage, distribution, and treatment facilities necessary for effective water management. Insufficient infrastructure leads to water losses, inefficient water consumption, and limited access to sanitation facilities (Zulu et al. 2022).
6. Lack of Integrated Approaches: Many organizations, including governmental organizations, local governments, corporations, and agricultural sectors, are often involved in water management. However, efforts to manage water resources sustainably may be hampered by a lack of integrated approaches and cooperation among various stakeholders (Tejaswini et al. 2022).
7. Financial and Institutional Challenges:Adequate institutional structures and sufficient funding are needed to implement sustainable water management practices. The institutional capabilities for water governance may be insufficient, and financing for water management projects is often limited (Herrera 2019).
8. Equity and Access: It remains challenging to ensure equal access to water resources for all people and communities. Access to clean water and sanitary facilities is often limited for disadvantaged groups, especially marginalized communities and rural areas (Jiménez et al. 2020).
9. Water-Energy-Food Nexus: The infrastructures for energy, food, and water are interconnected, and addressing one without considering the others may have unforeseen repercussions. Achieving sustainable outcomes requires balancing the water-energy-food nexus through organized management and planning (Pahl-Wostl et al. 2021).

10. Public Awareness and Participation: It is critical to educate people about the value of sustainable water management and promote active engagement. Successfully addressing water management problems can be challenging when there is a lack of understanding, indifference, and limited involvement from the public (Enqvist et al. 2019).

To address these issues and ensure the long-term conservation and management of water resources for present and future generations, a comprehensive strategy is needed. This strategy should include technological advances, policy reforms, community participation, and international cooperation.

3.3 Sustainable Water Treatment Techniques

3.3.1 Membrane Science Techniques

Membrane science techniques are modern technologies that are widely used for separation and purification processes. Furthermore, research in the field of membrane technology has garnered significant attention from researchers due to its potential for sustainable water treatment techniques. Numerous hybrid techniques have emerged, incorporating membrane science as a crucial component (Apel et al. 2019).

The current trends in membrane science techniques encompass solutions rooted in circular economy and integration. Additionally, membrane synthesis techniques, such as additive manufacturing, are gaining prominence. Emerging modeling approaches, such as molecular dynamic simulations and artificial intelligence, are also being utilized. Furthermore, fabrication approaches, including 3D printing and 4D printing, are being incorporated in membrane science (Yusuf et al. 2020).

3.3.1.1 Circular Economy and Integration

a. **Unsustainable linear economy practices**

The linear economy in the process of wastewater treatment via membrane-based processes involves the disposal of materials used in the operation until their cycle is completed, as well as the waste generated during the operation, such as brine waste, spent cooling water, and sludge water (Morón-López et al. 2019). In contrast, a circular economy is an economic model that aims to keep existing resources within a closed-loop system by extending facility utilization and reducing waste. Currently, the most common and convenient practice for disposing of membranes is through landfilling or incineration, as evidenced by approximately 840,000 reports published on this topic. However, to provide sustainable solutions for the future, this technology must prioritize waste minimization. While membrane reuse has been identified as one of the fastest methods, it does come with certain challenges (Senán-Salinas et al. 2019). One such challenge is biofouling, which can affect the feasibility of reusing membranes. A recommended solution with higher environmental effects is the use of oxidizing agents like $KMnO_4$ or NaOCl to remove the polyamide layer and convert it into UF or NF membranes. However, it should be noted that not all membranes passing through the polyamide layer will yield positive results, as the outcome depends on the specific type of polyamide layer (Moradi et al. 2019).

Brine, which is discharged from seawater or brackish water plants, contains chemicals, precipitates, and salts. The discharge of brine into the environment poses two significant challenges. Firstly, it reduces the availability of freshwater during the recovery process, and secondly, it leads to the uncontrolled release of pollutants contained in the brine. The temperature of the brine plume upon release can reach up to 57 °C at the point of discharge. This elevated temperature can have adverse effects on the growth and reproduction of marine ecosystems (Fernández-Torquemada et al. 2019).

Reports have highlighted the severe environmental consequences resulting from the unsustainable discharge of brine in Spain, the Red Sea, and the Middle East. Spain, which established the first desalination plant in Europe approximately 40 years ago, is known for its high reliance on desalination technology within Europe. Researchers have observed a reduction in the densities of echinoderms in seagrass meadows during brine discharge, specifically originating from the Alicante seawater desalination plant. In a study conducted by (Gacia et al. 2007), it was found that the exposure to brine and increased nutrient availability led to increased necrosis and reduced carbohydrate storage in the tissues of Posidonia oceanica meadows.

The Red Sea, with a depth of 490 m, acts as a geographical barrier separating the African continent from the Arab region. A survey conducted by Mabrook (1994) reported that the increased salinity in the Mediterranean Sea and the Red Sea could lead to the complete disappearance of planktonic organisms, corals, and fish species. Several studies have indicated that brine discharge may contribute to an increase in salinity rates in both the Mediterranean Sea and the Red Sea by the year 2050 (Bashitialshaaer et al. 2011). In a study by Robert et al., it was observed that brine discharge containing high levels of metal constituents can impact biological communities, and biomonitoring studies have shown the accumulation of heavy metals in benthic sediments (Roberts et al. 2010).

b. **Zero liquid discharge techniques in wastewater treatment**

Zero liquid discharge (ZLD) has recently gained attention as a method for water recovery from saline water sources such as groundwater and surface water. However, one type of wastewater that has not received sufficient focus for reuse or recycling in manufacturing industries is boiler blowdown, which is often directly discharged into the environment (Yadav et al. 2022).

The evaporation-crystallization process is a conventional ZLD treatment method employed to minimize the amount of blowdown discharged. Initially, reverse osmosis (RO) is used to extract fresh water from the blowdown, followed by evaporation to concentrate the RO brine, and finally, crystallization is utilized to separate the dissolved salts (Yaqub and Lee 2019). This process has proven to be effective for treating boiler blowdown, but it may not be as efficient for other types of wastewater containing dissolved salts, such as those released from integrated gasification and wet flue gas desulfurization cycle plants (Moreira et al. 2022). As mentioned above, zero liquid discharge (ZLD) techniques, along with other techniques, are utilized for achieving ZLD in wastewater treatment, including the complete recovery of water from palm oil mill effluent (POME) and the removal of high amounts of wastewater generated during zeolite synthesis (Lee et al. 2020).

In ZLD technology for wastewater recovery in the textile industry, combined membranes consisting of ultrafiltration (UF), nanofiltration (NF), reverse osmosis (RO), and membrane distillation (MD) are employed. It should be noted that the cost of freshwater recovery using ZLD techniques is higher compared to conventional wastewater treatment methods (Bahadur et al. 2019). ZLD technologies can be broadly categorized into membrane-based and non-membrane-based methods. In recent studies, membrane-based methods have gained popularity and are widely preferred. Non-membrane methods play a role as pre-treatment before the membrane processes. Furthermore, membrane-based techniques have been utilized for the recovery of valuable metals and chemicals, enhanced water recovery, and solute recovery (Panagopoulos et al. 2022).

c. **Re-use and recycling of membrane**

Numerous researchers have been striving to develop circular economy solutions for the reuse of reverse osmosis (RO) membranes. Reusing RO membranes has been recognized as an appealing wastewater treatment technique and an alternative means of meeting the necessary reuse standards through "membrane autopsy". Lawler et al. (2015) explored various approaches for reusing, recycling, and sustainably discarding these membranes, including an assessment of the feasibility of each proposed option.

The first step involves reusing used RO membranes at a lower cost compared to manufacturing new ones. Other methods of reuse encompass recycling options, conventional disposal methods, and calorific recovery. For fouled RO membranes that cannot be reused, an emerging technology called Membrane Biofilm Reactors (MBfRs) has been developed, which is effective in treating poor-quality water. MBfRs employ a gas-permeable membrane to capture pollutants from the water. The operation functions in a way that gaseous molecules move across the membrane as donors or acceptors of electrons, catalyzing the biological activity of biofilms present on the external layer of the RO membrane (Morón-López et al. 2019).

Another solution is the indirect reuse of damaged RO membranes through their recycling as supporter materials for anion-exchange membranes, as demonstrated by Lejarazu-Larrañaga et al. (2020). The study presented two circular economy approaches for fouled RO membranes: (1) utilizing them as support materials in the fabrication of anion exchange membranes, and (2) extracting the polypropylene components for assembling electrodialysis (ED) stacks. Experimental results showed that the anion exchange membranes fabricated from used RO membranes achieved

87% of permselectivity after necessary testing. Furthermore, ED stacks utilizing RO membranes achieved an 84.5% salt rejection rate (Suwaileh et al. 2020).

Another option within the circular economy framework is recycling, although it is not as widely preferred compared to the reuse option. This is because recycled membranes may go through a production stage where the materials are in an expanded state. The efficiency of the recycling method is generally lower compared to the reuse method (Arikan et al. 2020).

3.3.1.2 Emerging Modeling Approaches

With the increasing demand for membrane-based techniques in water treatment applications, researchers have turned their attention to understanding how these techniques respond to membrane and water properties at the molecular level. Molecular dynamic simulations (MDS) are used to study the dynamic behavior and interactions of molecules over time. In membrane-based MDS, computer simulations are employed to determine the movement of atoms or molecules surrounding the membrane (He et al. 2023). The interactions among these constituents are measured at the nanometer scale. However, this process is complex and requires advanced experimental and analytical techniques. Two mechanisms are commonly used to describe molecule transport through membranes: pore flow and solution diffusion (Manikandan et al. 2022). Pore flow is specific to porous membranes, while solution diffusion applies to nonporous membranes. MDS has been used to simulate the properties of various types of membranes, including polymeric membranes, inorganic membranes, metal–organic frameworks, mixed matrix membranes (MMMs), and others (Guo et al. 2014).

Using MDS, researchers have focused on simulating the dynamic properties and water structure across polyamide reverse osmosis (RO) membranes. This simulation primarily focuses on the behavior of oxygen and hydrogen atoms within the carboxylic acid and amide groups on the membrane. Studies have shown that around 90% of hydrogen bonds around the polyamide membrane are trapped (Lock et al. 2020).

MDS has been utilized to demonstrate the void fraction created by hydration, which results in an increase in the volume of the membrane. As water transport increases, the free volume of the polyamide RO membrane expands. MDS can also be used to simulate the impact of absorption on fouled membranes using fouling model mechanisms (Song et al. 2022).

Decision support systems (DSS) are computerized information structures that support decision-making in wastewater treatment plants. These systems allow users to search and analyze data and results, which can be used to generate well-informed reports (Doufene et al. 2020). DSS is primarily used for the effective and efficient planning and allocation of resources in membrane-based plants. It is often employed to optimize the selection of locations for constructing various types of water treatment plants (Sucu et al. 2021).

Artificial neural network (ANN) is a widely used modeling technique that simulates the biological neural networks found in the human brain, effectively solving technical problems (Zhang et al. 2019). ANN consists of a collection of processing elements called neurons, organized in parallel layers with weighted connections. The layers include input, output, and hidden layers, with at least one hidden layer required, and additional layers depending on the complexity of the model's structure (Asteris and Mokos 2020). The input layer receives the input data, and the output layer provides the results. Each input is connected to a neuron through a bias connection. Artificial neurons process the inputs, incorporating internal states and optional thresholds using activation functions, and produce outputs using output functions. The parameters of an ANN include the number of hidden layers, the number of neurons in each hidden layer, the activation function, and the training algorithm. By optimizing these parameters, optimal results can be achieved using the ANN technique (Jawad et al. 2020).

3.3.2 Advanced Water Treatment Technologies

3.3.2.1 Bioremediation

Bioremediation is one of the most efficient methods for sustainable advanced water treatment, as it degrades organic compounds in an environmentally friendly and economically sustainable manner, enhancing feasibility. This process utilizes microbial techniques to address emerging environmental pollutants (Kour et al. 2021). In the presence of these harmful wastes, microbes adapt and naturally modified microbial strains grow, transforming toxic pollutants into non-toxic forms. The degradation of xenobiotics depends on enzymatic activities. Different techniques based on biological enzymatic activities include rhizoremediation, vermicomposting, and phytoremediation (Patel et al. 2022).

Phytoremediation involves the use of plants to reduce the concentration of environmental pollutants at contaminated locations. Rhizoremediation, on the other hand, involves the removal of specific pollutants from contaminated sites through the combined efforts of microbial flora and plant roots (Ihsanullah et al. 2020). Vermifiltration and vermicomposting utilize worms to treat organic wastes, serving as natural waste management techniques that produce soil-enriching compounds as by-products. Wastewater discharged from households and industries shows promise as a sustainable treatment method. As a result, complete pathogen reduction is achieved at the end-product level (Singh et al. 2020).

This technique utilizes microorganisms, enzymatic reactions, and plants to detoxify and degrade environmental contaminants in water. Numerous reports have demonstrated the efficacy of fungi and bacterial species in degrading organic pollutants such as PAHs (Verma and Kuila 2019). A study showed that *Aeromonas hydrophila*, *Raoultella ornithinolytica*, *Bacillus megaterium*, and *Serratia marcescens* exhibited approximately 90% degradation efficiency for acenaphthene and fluorene. However, certain parameters such as temperature and pH can influence this process. Optimal temperatures for PAH degradation were found to be above 30 °C, with 38 °C being considered optimal. It should be noted that bioremediation is not an inflexible or robust technique. With increased temperature, the solubility of oxygen decreases, thus reducing the availability of dissolved oxygen for microbial growth (Alegbeleye et al. 2017).

3.3.2.2 Photocatalytic Degradation

Nanotechnology plays a crucial role in the environmental remediation process, particularly in photocatalysis, which treats wastewater in the presence of sunlight. Heterogeneous photocatalysis is an advanced process that utilizes semiconductors as the main source of energy. Direct discharge of pollutants from industries such as paint, textile, leather, etc., into water bodies poses serious health risks to both humans and aquatic life. Many conventional treatment methods are ineffective or inefficient in removing toxic organic pollutants (Berkani et al. 2022).

Semiconductors are capable of absorbing light and generating electron pairs. These charge carrier species reside on the surface of the semiconductor, where they react with adsorbed species through chemical processes (Ouyang et al. 2021). Photocatalysis is widely used for degrading organic pollutants, generating hydrogen through water splitting or reforming with the assistance of biomolecules, and transforming CO2 into valuable by-products. While several semiconductors have shown their effectiveness in photocatalysis, TiO2 and ZnO have emerged as universal photocatalytic semiconductors (Qin et al. 2021).

When considering universal photocatalytic semiconductors, it is essential to take into account various characteristics such as intensity, wavelength, pH, etc., in both gas and liquid phase applications mentioned above. Materials of photoactive semiconductors require enhancements for industrial or practical applications (Rafiq et al. 2021). A common approach to synthesizing photocatalytic materials, often referred to as the practical approach, involves a variety of different procedures and serves as a major foundation for many studies on photocatalytic processes. In some cases, the synthetic pathway for producing photocatalytic solids focuses on hierarchical structures to control the porosity, defect structure, and bulk characteristics of single-phase or composite materials. Determining whether a material exhibits surface or bulk-phase physical properties can sometimes be a complex task, blurring the line between single-phase and composite materials (Chen et al. 2020).

In the case of composite structures, additional efforts are made to synthesize materials with control over the electronic and structural properties of their constituents, further enhancing their photoactivity (Koe et al. 2020).

There are various types of organic pollutants present in water, including phenolic compounds, dyes, surfactants, hydrocarbons, and organohalides. These pollutants are toxic, carcinogenic, chemically stable, and resistant to decomposition in water (Ren et al. 2021). The photodecomposition of TiO2 in water on cyanide was first identified by Frank and Bard (1977). Subsequently, Carey and colleagues investigated the catalytic degradation of polychlorinated biphenyls under UV light by TiO2, which laid the research foundation for photocatalysis (Carey et al. 1976). Ollies et al. further studied the photocatalytic oxidative breakdown of halogenated organic compounds, such as trichloroethylene and trichloromethane, using a TiO2 sensitized system. This area of research has become a major focus in recent years (Pruden and Ollis 1983).

3.4 Sustainable Urban Water Management Approaches

Managing water demand is a critical factor in achieving urban water sustainability. In many developing countries, water is supplied through networks of buried pipes that connect to societies for water supply (Zubaidi et al. 2020). Urban areas rely heavily on water supply systems, but developing countries have faced unprecedented drought situations in their history. Overpopulation in these countries has resulted in water management issues and a decline in water supply (Capodaglio and Olsson 2019).

Sustainable urban water management (SUWM) aims to manage the water cycle in urban areas to achieve more favorable outcomes compared to traditional management approaches. The exact definition of SUWM may vary. Developing cities worldwide are taking initiatives to enhance their water sustainability through technological and management innovations, while other underdeveloped regions continue to face severe water crises (Oral et al. 2020). Urban water stress is expected to persist due to increased water demand and other hydrological factors, posing significant challenges in the coming years. Therefore, understanding the concept of SUWM and implementing appropriate initiatives play a crucial role in the development of these countries (Stefanakis 2019).

In this chapter, we aim to discuss several major initiatives that are commonly practiced in sustainable water management. These include the circular economy concept, smart urban water management approaches, and the implementation of Sponge City. The initiatives encompass the construction of green infrastructure, urban permeable pavement, rainwater harvesting, and ecological water management. Additionally, other approaches such as artificial intelligence techniques, triangular model initiatives, and triple-loop learning are explored. Furthermore, we provide an overview of the challenges faced in developing countries.

3.4.1 The Concept of Circular Economy in the Urban Water Sector

A circular economy (CE) is a sustainable strategy that aims to increase economic benefits while reducing stress on natural resources. The concept of CE was first introduced by Pearce and Turner in 1990 (Pearce and Turner 1990). The main objective of the CE is to shift focus towards the recovery of natural resources from waste, leading to social, economic, and environmental benefits. In the context of water management, the concept of CE involves the intricate interrelations between various components of the urban water cycle, which are influenced by approaches toward managing water resources (Velasco-Muñoz et al. 2021).

Regulators and decision-makers play a vital role in promoting the implementation of water reuse, and their efforts can be observed in developing countries through innovative decentralized wastewater reuse schemes (Gravagnuolo et al. 2019). In these countries, various framework directives provide a stimulus for the implementation of water reuse within an integrated water management scheme. The principles of CE are based on a closed-loop concept, which promotes the multiple reuse of energy, materials, and resources with minimal operational requirements for each use (Ezeudu et al. 2019).

Reports indicate that numerous technologies have been developed to reclaim and capture valuable resources from water. For example, some technologies have been designed to recover cooling water and valuable resources in industrial applications, such as ferrite, ferric hydroxide, and rare earth metals (Joensuu et al. 2020).

New policies in the CE concept can bring about a paradigm shift, proposing an efficient and sustainable model in water management with enhanced resource recovery. Over the course of history, hygienic systems for wastewater treatment have exhibited a cyclical tendency. In the early twentieth century, decentralized systems were initially adopted due to population distribution and the absence of collection systems. However, by the mid-nineteenth century, centralized systems became more popular, offering greater benefits for human health and the environment (Guerra-Rodríguez et al. 2020). Currently, decentralized systems that rely on point source segregation are commonly adopted by developing countries and are considered more sustainable than centralized systems (Rodríguez et al. 2020).

The concept of the 6Rs plays an important role in CE for sustainable urban water management practices. Each of the 6Rs is relevant to achieving sustainability in urban areas, and they are discussed as follows:

The first strategy of the 6Rs is the reduction of freshwater consumption, which leads to a decrease in wastewater generation. Sewage reduction can be achieved by implementing awareness programs in neighborhoods, industries, and other organizations. Additionally, water consumption can be reduced by installing advanced appliances. For example, dual flush systems can be used instead of conventional flushing systems. In urban areas, leakages in the distribution system contribute to approximately 30–40% of water loss, so reducing leakages can also help decrease freshwater demand. Properly implementing these measures can alleviate stress on wastewater treatment facilities.

Reuse involves cascading the utilization of used water for various purposes, starting from its crude form, in order to reduce the demand for freshwater consumption. Wastewater reuse is an alternative approach to addressing water scarcity in water-scarce areas. For instance, the wastewater generated from urban areas can be used for agricultural activities, providing essential nutrients required for the growth and development of crops. Rivers receive partially treated wastewater that is intended for indirect reuse. Therefore, direct reuse poses potential environmental and health effects, and a comprehensive assessment is necessary before initiating such practices.

Recycling involves the use of treated sewage water, reusing it for the same purpose for which it was originally generated. Efficient recycling plays a crucial role in significantly reducing the demand for freshwater. In developing countries, due to increasing freshwater demand, municipalities are compelled to treat greywater and utilize it for secondary purposes.

Next is the reclamation of wastewater, which involves treating wastewater and using it outside of the original process. With advancements in treatment technologies, wastewater can now be treated to a quality equivalent to drinking water. The extent of wastewater reclamation depends on the pollution level and the intended purpose for which the reclaimed water will be used.

Recovery is the process of extracting valuable resources from wastewater. Wastewater recovery plays a crucial role in promoting sustainability in the present era. According to the concept of Circular Economy (CE), the waste obtained from wastewater is considered a valuable resource. It can provide resources such as water, energy, and materials like fertilizer, bioplastics, metals, protein-based food, building materials, and other consumer products, depending on the pollutants present.

The strategy of restoration involves artificially restoring or replenishing water sources to enhance water security during dry seasons, aiming to maintain the water balance in the area. Therefore, the concept of Circular Economy (CE) in urban areas contributes significant benefits by providing various alternative resources, ultimately reducing the demand for freshwater (Kakwani and Kalbar 2020).

3.4.2 Implementation of Sponge City

Sponge cities have evolved from conventional models of urban water management. Urban stormwater management has been practiced since 3000 BCE as a means to collect rainwater and prevent floods. The development of sponge cities is based on four major principles.

The first principle of a sponge city is to enhance the city's surface to absorb and retain rainwater, providing a water supply for households and mitigating stormwater runoff that can lead to floods.

The second principle focuses on water management ecology, incorporating self-purification systems and eco-friendly waterfront designs.

The third principle involves the implementation of green infrastructure to restore, purify, and reuse stormwater, protecting cities from soil and water pollution. This approach also helps mitigate the urban "heat island" effect and supports sustainable urban development.

The fourth principle emphasizes the use of permeable pavements to further improve the effectiveness of sponge cities. The implementation of a sponge city is divided into two broad scenarios: macro-scale and micro-scale (Nguyen et al. 2019).

In the micro-scale scenario, the focus is primarily on designing solutions at the site level, which may include constructed wetlands, rain gardens, and stormwater bio-retention systems. Scaling up the micro-scale approach of the sponge city concept extends to the catchment level, providing enhanced hydrological and bio-ecological benefits. On the macro-scale, stormwater systems are integrated with natural hydrological systems to protect riparian corridors, which encompass trees, grass, shrubs, and buffer areas. A novel model is used that incorporates available spatial data, such as green infrastructure practices, land use, socio-economic factors, and hydrological conditions. Various types of sustainable implementation methods for sponge cities are explained below (Nguyen et al. 2020).

3.4.2.1 Construction of Green Infrastructure

Green infrastructure (GI) has been developed with the aim of protecting ecosystems and creating sustainable urban environments. The implementation of GI involves nature-based solutions, such as filter drains, infiltration systems, detention basins, swales, and filter steeples. GI can be broadly categorized into two main types: green roofs and bio-retention (Ying et al. 2021).

Green Roofs

Green roofs (GR), also known as rooftop gardens or living roofs, are designed to be covered with plants or vegetation. Documentation of green roofs can be found in Low Impact Development (LID) strategies and Sustainable Urban Drainage System (SUDS) techniques. Green roofs play a vital role in combating the urban heat island effect, improving air and water quality, reducing stormwater runoff, and enhancing recreational activities (Manso et al. 2021).

The concept of green roofs dates back to ancient times, but it gained more attention in the early 1960s in Germany as a means to reduce energy consumption in buildings (Wooster et al. 2022). The United States Environmental Protection Agency has documented construction, design, implementation, maintenance, and the benefits of green roofs (Liu et al. 2021).

A green roof (GR) typically consists of five components: a waterproof membrane, filter membrane, drainage system layer, soil, and plants. The selection of each layer depends on climatic and physical factors. There are primarily four types of green roofs: intensive, semi-intensive, single-course extensive, and multi-course extensive. Among these, the single-course and multi-course extensive green roofs are the most common due to their low weight, cost-effectiveness, ease of maintenance, and lack of irrigation requirements (Liberalesso et al. 2020).

Bioretention Systems (BRS)

Bioretention, also known as rain gardens or biofilters, is a widely used system in rainwater management. A bioretention system consists of five main components: drainage layer, detention layer, transition layer, filter media, and submerged zone (Yusuf and Hasia 2021). Each layer is constructed based on the specific characteristics of the area. Bioretention systems are considered the most sustainable and feasible solution in managing rainwater. They are designed to balance stormwater peak flow, improve groundwater recharge, and reduce runoff volume. Bioretention systems are effective in regulating water quality and controlling non-point source pollution. They are commonly used to mitigate urban flood disasters. Researchers have dedicated significant attention to the field of bioretention in hydrological systems (Baek 2020).

3.4.2.2 Urban Permeable Pavement

Permeable pavement is a technique used in Sponge cities that utilizes a permeable membrane in ground pavement construction. It offers benefits such as groundwater purification, rainwater infiltration, noise reduction, cooling, and humidification (Kubilay et al. 2021). Urban pavements, including parking lots, walkways, roads, and squares, can be constructed using permeable pavement. The type of pavement chosen depends on the load requirements and traffic volume (Vujovic et al. 2021). The four common types of permeable pavements used in urban areas are permeable concrete (PC), grass pavement (GP), permeable interlocking concrete pavers (PICP), and permeable asphalt (PA). Permeable concrete (PC) demonstrates superior performance compared to PICP and PA, excluding the potential effect of clogging due to particle deposition on the pavement surface. Many developing countries have embraced the use of permeable pavements as part of their urban water management practices to mitigate flood disasters (Liu et al. 2020).

Permeable pavement parking offers numerous benefits, primarily its ability to facilitate rapid infiltration of stormwater, resulting in reduced water runoff. The construction of these pavements is adaptable to various conditions, but attention must be given to potential issues related to clogging and the life cycle of permeable pavements (Zhu et al. 2019).

3.4.2.3 Resourcing Rainwater

Due to rapid population growth, industrialization, and urbanization, the excessive use of water is placing a heavy burden on existing freshwater resources. Therefore, alternative water resources are needed to address the water scarcity issue. Sustainable water management is believed to play a crucial role in socio-economic development (Yannopoulos et al. 2019).

One ancient method for mitigating water shortage is stormwater harvesting. Rainwater harvesting (RWH) involves collecting and storing rainwater from building surfaces and rooftops for household and irrigation purposes (Prenner et al. 2021). Techniques such as tank systems and green roof infrastructure were introduced in the early 1960s in countries like Japan and Germany to construct RWH systems. Rainwater harvesting has been successfully implemented in countries like the United States, America, and Europe, and is now being adopted in many other developing countries (Darabi et al. 2021).

RWH offers two main advantages: meeting increased water demand and mitigating water runoff. However, the focus of RWH has primarily been on restoring and recycling stormwater, with less attention given to the numerous additional benefits it can provide (Huang et al. 2021; Słyś et al. 2020).

3.4.2.4 Ecological Water Management

The concept of ecological water management in a sponge city ensures the restoration of water through self-purification systems and waterfront design, creating healthy water landscapes for the ecosystem (Xing et al. 2021). Water self-purification involves physical, chemical, and biological processes, with biological purification considered an environmentally friendly technique. Several factors influence water self-purification systems (Anderson et al. 2019). Firstly, hydrodynamic forces contribute to the mixing of pollutants, their movement, and the dissolved oxygen content in water. Secondly, the soil acts as a pollutant remover through sedimentation, filtration, and absorption. Plants play a role in removing heavy metal pollutants, nitrogen, and phosphorus. Finally, microorganisms contribute to the degradation of contaminants. Ecological waterfront design, a form of ecological water management, plays a vital role in integrating cities and developing both macro and micro-urban environments (Xia et al. 2021).

3.4.3 Decentralized Solutions

Decentralized solutions refer to small and medium-sized water infrastructures that rely on readily available water resources such as stormwater runoff and greywater. These solutions can work independently or in combination with conventional water systems (Thakur et al. 2021). Decentralized solutions typically involve recycling wastewater, greywater, and harvesting stormwater. While size and scale are important, they are not the only characteristics that distinguish decentralized water infrastructures from traditional ones. The key distinction lies in the integrative capability of decentralized infrastructures when applied across previously segregated areas of urban water management (Tsegaye et al. 2020).

This characteristic allows for the expansion of internal water circulation and provides a variety of water supply options within urban water systems. Decentralized urban water systems offer advantages by addressing the primary drawbacks of current centralized systems, as they are integrated, sustainable, and resilient (Forrest et al. 2020). However, decentralized urban water systems can also be seen as an updated version of conventional centralized systems. Instead of completely replacing the conventional system, making some changes to it can achieve the resilience and sustainability desired in urban water systems. By enhancing the interconnections of the system, a decentralized system can come closer to achieving the Sustainable Urban Water Management (SUWM) paradigm (Herslund et al. 2019).

3.5 Conclusion

In this chapter, we have discussed the defining features of sustainable urban water solutions, including sustainable urban water systems and sustainable urban water management. The advantages of wastewater reuse for multiple purposes have been clearly highlighted, as it can effectively address water scarcity in urbanized regions. The rapid pace of urbanization has placed significant pressure on municipalities to manage the increasing demand for water supply. Consequently, alternative strategies are needed to enhance livelihoods in developing countries.

The literature review has explored various options to achieve sustainable water management challenges and treatment. The first approach toward wastewater treatment focuses on membrane science techniques, highlighting the importance of circular economy principles, integration, and emerging modeling approaches. The subsequent section discusses the selection of advanced water treatment technologies for developing countries, emphasizing the common characteristics of bioremediation and photocatalytic degradation.

The second part of the chapter explores sustainable water management approaches in urban areas. The first approach discusses the concept of circular economy in the urban water sector, emphasizing the importance of the 6Rs (Reduce, Reuse, Recycle, Recover, Redesign, and Remanufacture) in urban water management. The next approach focuses on the implementation of Sponge cities, which involves techniques such as green infrastructure, urban permeable pavement, rainwater harvesting, and ecological water management. Lastly, decentralized solutions for urban water management are briefly explained.

In light of rapid urban growth, it is crucial to implement eco-friendly sanitation systems that are sustainable and adaptable to society's water requirements. Decision-makers should consider various factors such as region, topography, climate, and socioeconomic status when determining appropriate treatment techniques. Further research is still needed to enhance existing wastewater treatment techniques and management approaches. Increased research in this field will contribute to the development of developing countries.

References

Abbass K, Qasim MZ, Song H, Murshed M, Mahmood H, Younis I (2022) A review of the global climate change impacts, adaptation, and sustainable mitigation measures. Environ Sci Pollut Res 29(28):42539–42559. https://doi.org/10.1007/S11356-022-19718-6

Alegbeleye OO, Opeolu BO, Jackson V (2017) Bioremediation of polycyclic aromatic hydrocarbon (PAH) compounds: (acenaphthene and fluorene) in water using indigenous bacterial species isolated from the Diep and Plankenburg rivers, Western Cape, South Africa. Brazilian J Microbiol 48(2):314–325. https://doi.org/10.1016/J.BJM.2016.07.027

Anderson EP, Jackson S, Tharme RE, Douglas M, Flotemersch JE, Zwarteveen M, Lokgariwar C, Montoya M, Wali A, Tipa GT, Jardine TD, Olden JD, Cheng L, Conallin J, Cosens B, Dickens C, Garrick D, Groenfeldt D, Kabogo J, Arthington AH (2019) Understanding rivers and their social relations: a critical step to advance environmental water management. Wiley Interdiscipl Rev Water 6(6). https://doi.org/10.1002/WAT2.1381

Apel PY, Bobreshova OV, Volkov AV, Volkov VV, Nikonenko VV, Stenina IA, Filippov AN, Yampolskii YP, Yaroslavtsev AB (2019) Prospects of membrane science development. Membr Membr Technol 1(2):45–63. https://doi.org/10.1134/S2517751619020021

Arikan EB, Ozay Y, Unal BO, Vatanpour V, Dizge N (2020) Treatability of household waste plastic garbage bag recycling industry wastewater with membrane. Membrane Water Treatment 11(6):375. http://www.techno-press.org/content/?page=article&journal=mwt&volume=11&num=6&ordernum=1. Accessed 2 June 2023

Asteris PG, Mokos VG (2020) Concrete compressive strength using artificial neural networks. Neural Comput Appl 32(15):11807–11826. https://doi.org/10.1007/S00521-019-04663-2/METRICS

Baek S (2020) Improvement of low impact development (LID) models coupled with watershed models: application, performance, and optimization. https://scholarworks.unist.ac.kr/handle/201301/31768

Bahadur N, Bhargava N (2019) Novel pilot scale photocatalytic treatment of textile & dyeing industry wastewater to achieve process water quality and enabling zero liquid discharge. Elsevier. https://www.sciencedirect.com/science/article/pii/S2214714419304891. Accessed 2 June 2023

Bashitialshaaer R, Persson K, of MA-IJ (2011) undefined. (n.d.) Estimated future salinity in the Arabian Gulf, the Mediterranean Sea and the Red Sea consequences of brine discharge from desalination. Tvrl.Se. Retrieved December 11, 2023, from http://www.tvrl.se/ma/Documents/ESTIMATED%20FUTURE%20SALINITY%20IN%20THE%20ARABIAN%20GULF,.pdf

Berkani M, Smaali A, Kadmi Y, Almomani F, Vasseghian Y, Lakhdari N, Alyane M (2022) Photocatalytic degradation of Penicillin G in aqueous solutions: kinetic, degradation pathway, and microbioassays assessment. J Hazard Mater 421:126719. https://www.sciencedirect.com/science/article/pii/S0304389421016848. Accessed 2 June 2023

Capodaglio AG, Olsson G (2019) Energy issues in sustainable urban wastewater management: use, demand reduction and recovery in the urban water cycle. Sustainability 12(1):266. https://doi.org/10.3390/SU12010266

Carey JH, Lawrence J, Tosine HM (1976) Photodechlorination of PCB's in the presence of titanium dioxide in aqueous suspensions. Bull Environ Contam Toxicol 16(6):697–701. https://doi.org/10.1007/BF01685575

Carrard N, Foster T, Willetts J (2019) Groundwater as a source of drinking water in southeast Asia and the Pacific: a multi-country review of current reliance and resource concerns. Water 11(8):1605. https://www.mdpi.com/508462. Accessed 1 June 2023

Chen D, Cheng Y, Zhou N, Chen P, Wang Y, Li K, Ruan R (2020) Photocatalytic degradation of organic pollutants using TiO2-based photocatalysts: a review. J Clean Prod https://www.sciencedirect.com/science/article/pii/S0959652620317728. Accessed 2 June 2023

Darabi H, Moradi E, Davudirad AA, Ehteram M, Cerda A, Haghighi AT (2021) Efficient rainwater harvesting planning using socio-environmental variables and data-driven geospatial techniques.

J Clean Prod https://www.sciencedirect.com/science/article/pii/S0959652621019247. Accessed 2 June 2023
Dhakal N, Salinas-Rodriguez SG, Hamdani J, Abushaban A, Sawalha H, Schippers JC, Kennedy MD (2022) Is desalination a solution to freshwater scarcity in developing countries? Membranes 12(4):381. https://doi.org/10.3390/MEMBRANES12040381
Doufene A, Sakhrani V, Alkhenani A, Yu B, Alsaati A (2020) Model based system design to support variability and flexibility, and early bidding phase of new procurement. http://oastats.mit.edu/handle/1721.1/125793
Emile R, Clammer JR, Jayaswal P, Sharma P (2022) Addressing water scarcity in developing country contexts: a socio-cultural approach. Human Soc Sci Commun 9(1):1–10. https://doi.org/10.1057/s41599-022-01140-5
Enqvist JP, Ziervogel G, Johan Enqvist CP, Climate and, A. (2019). Water governance and justice in cape town: an overview. Wiley Interdiscipl Rev Water https://doi.org/10.1002/wat2.1354
Es' haghi SR, Karimi H, Rezaei A, Ataei P (2022) Content analysis of the problems and challenges of agricultural water use: a case study of Lake Urmia Basin at Miandoab, Iran. J Sagepub.Com 12(2). https://doi.org/10.1177/21582440221091247
Ezeudu OB, Ezeudu TE (2019) Implementation of circular economy principles in industrial solid waste management: case studies from a developing economy (Nigeria). Mdpi.Com. https://www.mdpi.com/557334. Accessed 2 June 2023
Fernández-Torquemada Y, Carratalá A, Luis J, Lizaso S (2019) Impact of brine on the marine environment and how it can be reduced. https://doi.org/10.5004/dwt.2019.24615
Forrest N, Stein Z, Wiek A (2020) Transferability and scalability of sustainable urban water solutions–a case study from the Colorado River Basin. Resources, Conservation and Recycling https://www.sciencedirect.com/science/article/pii/S0921344920301117. Accessed 2 June 2023
Frank SN, Bard AJ (1977) Heterogeneous photocatalytic oxidation of cyanide and sulfite in aqueous solutions at semiconductor powders. J Phys Chem 81(15): 1484–1488. https://doi.org/10.1021/J100530A011/ASSET/J100530A011.FP.PNG_V03
Gacia E, Invers O, Manzanera M, Ballesteros E, Romero J (2007). Impact of the brine from a desalination plant on a shallow seagrass (Posidonia oceanica) meadow. Estuarine, Coastal and Shelf Science https://www.sciencedirect.com/science/article/pii/S0272771406005221. Accessed 3 June 2023
Gravagnuolo A, Angrisano M, Fusco Girard L (2019) Circular economy strategies in eight historic port cities: Criteria and indicators towards a circular city assessment framework. Sustainability 11(13):3512. https://www.mdpi.com/486486. Accessed 2 June 2023
Guerra-Rodríguez S, Oulego P, Rodríguez E, Singh DN, Rodríguez-Chueca J (2020) Towards the implementation of circular economy in the wastewater sector: Challenges and opportunities. Water https://www.mdpi.com/719028. Accessed 2 June 2023
Guo CX, Zhao D, Zhao Q, Wang P, Lu X (2014) Na+-functionalized carbon quantum dots: a new draw solute in forward osmosis for seawater desalination. Chem Commun 50(55):7318–7321. https://doi.org/10.1039/C4CC01603C
Hajian M, Kashani SJ (2021) Evolution of the concept of sustainability. From Brundtland Report to sustainable development goals. Sustain Resource Manag Modern Appr Contexts 1–24. https://doi.org/10.1016/B978-0-12-824342-8.00018-3
He J, McCutcheon JR, Li Y (2023) Effect of different manufacturing methods on polyamide reverse-osmosis membranes for desalination: insights from molecular dynamics simulations. Desalination 547:116204. https://doi.org/10.1016/J.DESAL.2022.116204
Herrera V (2019) Reconciling global aspirations and local realities: challenges facing the sustainable development goals for water and sanitation. World Dev 118:106–117. https://doi.org/10.1016/J.WORLDDEV.2019.02.009
Herslund L, Mguni P (2019) Examining urban water management practices–challenges and possibilities for transitions to sustainable urban water management in Sub-Saharan cities. Sustain Cities Soc 48:101573. https://www.sciencedirect.com/science/article/pii/S2210670719300721. Accessed 2 June 2023

Hoffmann S, Feldmann U, Bach PM, Binz C, Farrelly M, Frantzeskaki N, Hiessl H, Inauen J, Larsen TA, Lienert J, Londong J, Lüthi C, Maurer M, Mitchell C, Morgenroth E, Nelson KL, Scholten L, Truffer B, Udert KM (2020) A research agenda for the future of urban water management: exploring the potential of nongrid, small-grid, and hybrid solutions. Environ Sci Technol 54(9):5312–5322. https://doi.org/10.1021/ACS.EST.9B05222/ASSET/IMAGES/MEDIUM/ES9B05222_0002.GIF

Huang Z, Nya EL, Rahman MA, Mwamila TB, Cao V, Gwenzi W, Noubactep C (2021) Integrated water resource management: Rethinking the contribution of rainwater harvesting. Sustainability 13(15):8338. https://www.mdpi.com/2071-1050/13/15/8338. Accessed 2 June 2023

Ihsanullah I, Jamal A, Ilyas M, Zubair M, Khan G, Atieh MA (2020) Bioremediation of dyes: Current status and prospects. J Water Proc Eng 38:101680. https://www.sciencedirect.com/science/article/pii/S2214714420305584. Accessed 2 June 2023

Jawad J, Hawari AH, Zaidi S (2020) Modeling of forward osmosis process using artificial neural networks (ANN) to predict the permeate flux. Desalination 484:114427. https://doi.org/10.1016/J.DESAL.2020.114427

Jiménez A, Saikia P, Giné R, Avello P, Leten J, Lymer BL, Schneider K, Ward R (2020) Unpacking water governance: a framework for practitioners. Water 12(3):827. https://doi.org/10.3390/W12030827

Joensuu T, Edelman H, Saari A (2020) Circular economy practices in the built environment. J Clean Prod 276:124215. https://www.sciencedirect.com/science/article/pii/S0959652620342608. Accessed 2 June 2023

Kakwani N, Kalbar PP (2020) Review of circular economy in urban water sector: challenges and opportunities in India. J Environ Manag 271:111010. https://www.sciencedirect.com/science/article/pii/S0301479720309385. Accessed 2 June 2023

Koe WS, Lee JW, Chong WC, Pang YL, Sim LC (2020) An overview of photocatalytic degradation: photocatalysts, mechanisms, and development of photocatalytic membrane. Environ Sci Pollut Res 27(3):2522–2565. https://doi.org/10.1007/S11356-019-07193-5

Kour D, Kaur T, Devi R, Yadav A, Singh M, Joshi D, Singh J, Suyal DC, Kumar A, Rajput VD, Yadav AN, Singh K, Singh J, Sayyed RZ, Arora NK, Saxena AK (2021) Beneficial microbiomes for bioremediation of diverse contaminated environments for environmental sustainability: present status and future challenges. Environ Sci Pollut Res 28(20):24917–24939. https://doi.org/10.1007/S11356-021-13252-7

Kubilay A, Ferrari A, Derome D, Carmeliet J (2021) Smart wetting of permeable pavements as an evaporative-cooling measure for improving the urban climate during heat waves. J Building Phys 45(1):36–66. https://doi.org/10.1177/1744259120968586

Lawler W, Alvarez-Gaitan J, Leslie G, Le-Clech P (2015) Comparative life cycle assessment of end-of-life options for reverse osmosis membranes. Desalination 357:45–54. https://www.sciencedirect.com/science/article/pii/S0011916414005360. Accessed 3 June 2023

Lee ZS, Cheng CK, Chin SY (2020) Hydrothermal treatment of palm oil mill effluent (POME) under oxidative and non-oxidative conditions. IOP Conf Ser Mater Sci Eng 965(1). https://doi.org/10.1088/1757-899X/965/1/012002

Liberalesso T, Oliveira Cruz C, Matos Silva C, Manso M (2020) Green infrastructure and public policies: an international review of green roofs and green walls incentives. Land Use Policy 96:104693. https://doi.org/10.1016/J.LANDUSEPOL.2020.104693

Liu Y, Li T, Yu L (2020) Urban heat island mitigation and hydrology performance of innovative permeable pavement: a pilot-scale study. J Clean Prod 244:118938. https://www.sciencedirect.com/science/article/pii/S0959652619338089. Accessed 2 June 2023

Liu H, Kong F, Yin H, Middel A, Zheng X, Huang J, Wen Z (2021) Impacts of green roofs on water, temperature, and air quality: a bibliometric review. Build Environ 196:107794. https://www.sciencedirect.com/science/article/pii/S0360132321002018. Accessed 2 June 2023

Lock SSM, Lau KK, Jusoh N, Shariff AM, Gan CH, Yiin CL (2020) An atomistic simulation towards molecular design of silica polymorphs nanoparticles in polysulfone based mixed matrix membranes for CO2/CH4 gas separation. Polym Eng Sci 60(12):3197–3215. https://doi.org/10.1002/PEN.25547

Mabrook B (1994) Environmental impact of waste brine disposal of desalination plants, Red Sea, Egypt. Desalination 97(1–3):453–465. https://doi.org/10.1016/0011-9164(94)00108-1

Machete I, Marques R (2021) Financing the water and sanitation sectors: a hybrid literature review. Infrastructures 6(1):9. https://doi.org/10.3390/INFRASTRUCTURES6010009

Manikandan S, Subbaiya R, Saravanan M, Ponraj M, Selvam M, Pugazhendhi A (2022) A critical review of advanced nanotechnology and hybrid membrane based water recycling, reuse, and wastewater treatment processes. Chemosphere 289:132867. https://www.sciencedirect.com/science/article/pii/S0045653521033397. Accessed 2 June 2023

Manso M, Teotónio I, Silva CM, Cruz CO (2021) Green roof and green wall benefits and costs: a review of the quantitative evidence. Renew Sustain Energy Rev 135:110111. https://www.sciencedirect.com/science/article/pii/S1364032120304020. Accessed 2 June 2023

Moradi MR, Pihlajamäki A, Hesampour M, Ahlgren J, Mänttäri M (2019) End-of-life RO membranes recycling: reuse as NF membranes by polyelectrolyte layer-by-layer deposition. J Membr Sci 584:300–308. https://doi.org/10.1016/j.memsci.2019.04.060

Moreira V, Lebron YAR, Couto CF, Maia A, Moravia WG, Amaral MCS (2022) Process development for textile wastewater treatment towards zero liquid discharge: integrating membrane separation process and advanced oxidation techniques. Proc Saf Environ Protect 157:537–546. https://www.sciencedirect.com/science/article/pii/S0957582021005735. Accessed 2 June 2023

Morón-López J, Nieto-Reyes L, Senán-Salinas J, Molina S, El-Shehawy R (2019) Recycled desalination membranes as a support material for biofilm development: a new approach for microcystin removal during water treatment. Sci Total Environ 647:785–793. https://doi.org/10.1016/j.scitotenv.2018.07.435

Nguyen T, Ngo HH, Guo W, Wang XC, Ren N, Li G, Ding J, Liang H (2019) Implementation of a specific urban water management-Sponge City. Sci Total Environ 652:147–162. https://www.sciencedirect.com/science/article/pii/S0048969718340518. Accessed 2 June 2023

Nguyen T, Ngo HH, Guo W, Wang XC (2020) A new model framework for sponge city implementation: Emerging challenges and future developments. J Environ Manag 253:109689. https://www.sciencedirect.com/science/article/pii/S0301479719314070. Accessed 2 June 2023

Oral HV, Carvalho P, Gajewska M, Ursino N, Masi F, van Hullebusch ED, Kazak JK, Exposito A, Cipolletta G, Andersen TR, Finger DC, Simperler L, Regelsberger M, Rous V, Radinja M, Buttiglieri G, Krzeminski P, Rizzo A, Dehghanian K, Zimmermann M (2020) A review of nature-based solutions for urban water management in European circular cities: a critical assessment based on case studies and literature. Blue-Green Syst 2(1):112–136. https://doi.org/10.2166/BGS.2020.932

Ouyang Z, Yang Y, Zhang C, Zhu S, Qin L, Wang W, Qin F (2021) Recent advances in photocatalytic degradation of plastics and plastic-derived chemicals. J Mater Chem A 9(23):13402–13441. https://pubs.rsc.org/en/content/articlehtml/2021/ta/d0ta12465f. Accessed 2 June 2023

Pahl-Wostl C, Gorris P, Jager N, Koch L, Lebel L, Stein C, Venghaus S, Withanachchi S (2021) Scale-related governance challenges in the water–energy–food nexus: toward a diagnostic approach. Sustain Sci 16(2):615–629. https://doi.org/10.1007/S11625-020-00888-6

Panagopoulos A, Giannika V (2022) Decarbonized and circular brine management/valorization for water & valuable resource recovery via minimal/zero liquid discharge (MLD/ZLD) strategies. J Environ Manag 324:116239. https://www.sciencedirect.com/science/article/pii/S0301479722018126. Accessed 2 June 2023

Patel A, Singhania RR, Albarico FPJB, Pandey A, Chen CW, Dong CD (2022) Organic wastes bioremediation and its changing prospects. Sci Total Environ 824:153889. https://www.sciencedirect.com/science/article/pii/S0048969722009810. Accessed 2 June 2023

Pauli BJ (2020) The flint water crisis. Wiley Interdiscip Rev Water 7(3):e1420. https://doi.org/10.1002/WAT2.1420

Pearce DW, Turner RK (1990) Economics of natural resources and the environment. 378. https://books.google.com/books/about/Economics_of_Natural_Resources_and_the_E.html?id=ex8vaG6m4RMC

Prenner F, Pucher B, Zluwa I, Pitha U, Langergraber G (2021) Rainwater use for vertical greenery systems: development of a conceptual model for a better understanding of processes and influencing factors. Water 13(13):1860. https://www.mdpi.com/2073-4441/13/13/1860. Accessed 2 June 2023

Pruden AL, Ollis DF (1983) Degradation of chloroform by photoassisted heterogeneous catalysis in dilute aqueous suspensions of titanium dioxide. Environ Sci Technol 17(10):628–631. https://doi.org/10.1021/ES00116A013/ASSET/ES00116A013.FP.PNG_V03

Qin K, Zhao Q, Yu H, Xia X, Li J, He S, Wei L, An T (2021) A review of bismuth-based photocatalysts for antibiotic degradation: insight into the photocatalytic degradation performance, pathways and relevant mechanisms. Environ Res 199:111360. https://www.sciencedirect.com/science/article/pii/S001393512100654X. Accessed 2 June 2023

Rafiq A, Ikram M, Ali S, Niaz F, Khan M, Khan Q, Maqbool M (2021) Photocatalytic degradation of dyes using semiconductor photocatalysts to clean industrial water pollution. J Indus Eng Chem 97:111–128. https://www.sciencedirect.com/science/article/pii/S1226086X21000940. Accessed 2 June 2023

Rashid R, Shafiq I, Akhter P, Iqbal MJ, Hussain M (2021) A state-of-the-art review on wastewater treatment techniques: the effectiveness of adsorption method. Environ Sci Pollut Res 28(8):9050–9066. https://doi.org/10.1007/s11356-021-12395-x

Ren G, Han H, Wang Y, Liu S, Zhao J, Meng X, Li Z (2021) Recent advances of photocatalytic application in water treatment: A review. Nanomaterials 11(7). https://doi.org/10.3390/NANO11071804

Roberts D, Johnston E, research NK-W (2010) undefined. (n.d.) Impacts of desalination plant discharges on the marine environment: A critical review of published studies. Elsevier. Retrieved December 11, 2023, from https://www.sciencedirect.com/science/article/pii/S0043135410002903

Rocha J, Carvalho-Santos C, Diogo P, Beça P, Keizer JJ, Nunes JP (2020) Impacts of climate change on reservoir water availability, quality and irrigation needs in a water scarce Mediterranean region (southern Portugal). Sci Total Environ 736:139477. https://doi.org/10.1016/J.SCITOTENV.2020.139477

Rodríguez C, Florido C, Jacob M (2020) Circular economy contributions to the tourism sector: a critical literature review. Sustainability 12(11):4338. https://www.mdpi.com/725778. Accessed 2 June 2023

Roshan A, Kumar M (2020) Water end-use estimation can support the urban water crisis management: a critical review. J Environ Manage 268. https://doi.org/10.1016/J.JENVMAN.2020.110663

Sarkar R, Lakshmana CM (2022) Measuring urbanisation growth of urban agglomeration, urban growth sustainability and role of urban primacy in India. J Asian African Stud https://doi.org/10.1177/00219096221111360

Senán-Salinas J, García-Pacheco R, Landaburu-Aguirre J, García-Calvo E (2019) Recycling of end-of-life reverse osmosis membranes: comparative LCA and cost-effectiveness analysis at pilot scale. Resour Conserv Recycl 150. https://doi.org/10.1016/j.resconrec.2019.104423

Singh P, Singh VK, Singh R, Borthakur A, Madhav S, Ahamad A, Kumar A, Pal DB, Tiwary D, Mishra PK (2020) Bioremediation: a sustainable approach for management of environmental contaminants. Abatement Environ Pollut Trends Strat 1–23. https://doi.org/10.1016/B978-0-12-818095-2.00001-1

Słyś D, Stec A (2020) Centralized or decentralized rainwater harvesting systems: a case study. Resources 9(1):5. https://www.mdpi.com/615792. Accessed 2 June 2023

Song X, Teuler JM, Guiga W, Fargues C, Rousseau B (2022) Molecular simulation of a reverse osmosis polyamide membrane layer. In silico synthesis using different reactant concentration ratios. J Membrane Sci 643:120010. https://doi.org/10.1016/J.MEMSCI.2021.120010

Stefanakis AI (2019) the role of constructed wetlands as green infrastructure for sustainable urban water management. Sustainability 11(24):6981. https://doi.org/10.3390/SU11246981
Su Y, Gao W, Guan D, Zuo T (2020) Achieving urban water security: a review of water management approach from technology perspective. Water Resour Manage 34(13):4163–4179. https://doi.org/10.1007/S11269-020-02663-9/METRICS
Sucu S, van Schaik MO, Esmeli R, Ouelhadj D, Holloway T, Williams JB, Cappon HJ (2021) A conceptual framework for a multi-criteria decision support tool to select technologies for resource recovery from urban wastewater. J Environ Manag 300:113608. https://www.sciencedirect.com/science/article/pii/S0301479721016704. Accessed 2 June 2023
Sun S, Tang Q, Konar M, Fang C, Liu H, Liu X, Fu G (2023) Water transfer infrastructure buffers water scarcity risks to supply chains. Water Res 229:119442. https://www.sciencedirect.com/science/article/pii/S0043135422013872. Accessed 14 June 2023
Suwaileh W, Johnson D, Hilal N (2020). Membrane desalination and water re-use for agriculture: State of the art and future outlook. Desalination 491:114559. https://www.sciencedirect.com/science/article/pii/S0011916420310213. Accessed 2 June 2023
Tejaswini MSSR, Pathak P, Ramkrishna S, Ganesh PS (2022) A comprehensive review on integrative approach for sustainable management of plastic waste and its associated externalities. Sci Total Environ 825:153973. https://doi.org/10.1016/J.SCITOTENV.2022.153973
Thakur T, Mehra A, Hassija V, Chamola V, Srinivas R, Gupta KK, Singh AP (2021) Smart water conservation through a machine learning and blockchain-enabled decentralized edge computing network. Appl Soft Comp 106:107274. https://www.sciencedirect.com/science/article/pii/S1568494621001976. Accessed 2 June 2023
Tsegaye S, Missimer TM, Kim JY, Hock J (2020) A clustered, decentralized approach to urban water management. Water 12(1):185. https://www.mdpi.com/613702. Accessed 2 June 2023
Van Vliet MT, Jones ER, Flörke M, Franssen WH, Hanasaki N, Wada Y, Yearsley JR (2021) Global water scarcity including surface water quality and expansions of clean water technologies Global water scarcity including surface water quality and expansions of clean water technologies. Environ Res Lett 16(2):24020. https://doi.org/10.1088/1748-9326/abbfc3
Velasco-Muñoz J, Mendoza JMF, Aznar-Sánchez JA, Gallego-Schmid A (2021) Circular economy implementation in the agricultural sector: Definition, strategies and indicators. Resourc Conserv Recycl 170:105618. https://www.sciencedirect.com/science/article/pii/S0921344921002275. Accessed 2 June 2023
Verma S, Kuila A (2019) Bioremediation of heavy metals by microbial process. Environ Technol Innov 14:100369. https://www.sciencedirect.com/science/article/pii/S2352186418305911. Accessed 2 June 2023
Vujovic S, Haddad B, Karaky H, Sebaibi N, Boutouil M (2021) Urban heat island: causes, consequences, and mitigation measures with emphasis on reflective and permeable pavements. CivilEng 2(2):459-484. https://www.mdpi.com/2673-4109/2/2/26. Accessed 2 June 2023
Wooster EIF, Fleck R, Torpy F, Ramp D, Irga PJ (2022) Urban green roofs promote metropolitan biodiversity: a comparative case study. Build Environ 207:108458. https://www.sciencedirect.com/science/article/pii/S0360132321008544. Accessed 2 June 2023
Xia J, Li Z, Zeng S, Zou L, She D, Cheng D (2021) Perspectives on eco-water security and sustainable development in the Yangtze River Basin. Geosci Lett 8(1). https://doi.org/10.1186/S40562-021-00187-7
Xing Z, Huang H, Li Y, Liu S, Wang D, Yuan Y, Zhao Z, Bu L (2021) Management of sustainable ecological water levels of endorheic salt lakes in the inner mongolian plateau of China based on eco-hydrological processes. Hydrol Proc 35(5). https://doi.org/10.1002/HYP.14192
Yadav A, Labhasetwar PK, Shahi VK (2022) Membrane distillation crystallization technology for zero liquid discharge and resource recovery: opportunities, challenges and futuristic perspectives. Sci Total Environ 806:150692. https://www.sciencedirect.com/science/article/pii/S0048969721057703. Accessed 2 June 2023

Yannopoulos S, Giannopoulou I, Kaiafa-Saropoulou M (2019) Investigation of the current situation and prospects for the development of rainwater harvesting as a tool to confront water scarcity worldwide. Water https://doi.org/10.3390/w11102168

Yaqub M, Lee W (2019) Zero-liquid discharge (ZLD) technology for resource recovery from wastewater: a review. Sci Total Environ 681:551–563. https://www.sciencedirect.com/science/article/pii/S0048969719320741. Accessed 2 June 2023

Ying J, Zhang X, Zhang Y, Bilan S (2021) Green infrastructure: systematic literature review. 35(1):343–366. https://doi.org/10.1080/1331677X.2021.1893202. http://Www.Tandfonline.Com/Action/AuthorSubmission?JournalCode=rero20&page=instructions

Yusuf A, Sodiq A, Giwa A, Eke J, Pikuda O, De Luca G, Di Salvo JL, Chakraborty S (2020) A review of emerging trends in membrane science and technology for sustainable water treatment. J Clean Prod 266:121867. https://doi.org/10.1016/J.JCLEPRO.2020.121867

Yusuf B, Hasia D (2021) Integration of Eco-friendly design principles in the design of mixed-use building, Abuja, Nigeria. http://repository.futminna.edu.ng:8080/jspui/handle/123456789/13499

Zhang Q, Yu H, Barbiero M, Wang B, Gu M (2019) Artificial neural networks enabled by nanophotonics. Light Sci Appl 8(1):1–14. https://doi.org/10.1038/s41377-019-0151-0

Zhu H, Yu M, Zhu J, Lu H, Cao R (2019) Simulation study on effect of permeable pavement on reducing flood risk of urban runoff. Int J Transp Sci Technol 8(4):373–382. https://www.sciencedirect.com/science/article/pii/S2046043018301060. Accessed 2 June 2023

Zubaidi SL, Ortega-Martorell S, Al-Bugharbee H, Olier I, Hashim KS, Gharghan SK, Kot P, Al-Khaddar R (2020) Urban water demand prediction for a city that suffers from climate change and population growth: gauteng province case study. Water 12(7):1885. https://doi.org/10.3390/W12071885

Zulu E, Zulu S, Chabala M, Musonda I, Kavishe N, Chileshe N (2022) Challenges and advocated solutions for environmental protection legislation for building infrastructure projects in developing countries: evidence from Zambia. Project Leadership Soc 3:100056. https://doi.org/10.1016/J.PLAS.2022.100056

Chapter 4
Modern Water Treatment Methods: Exploring Public Acceptance and Socio-economic Factors Influencing Their Implementation

Sutapa Roy and Nitin Swamy

Abstract Since the beginning of human history, available water quality has been a primary concern. The development of the industrial revolution and the accompanying rise in population has contributed to a dramatic deterioration in the quality of the world's water supply, which has had disastrous effects on the planet's well-being. This dire condition has caused scientific brains to develop new technologies to address this issue. In the past two decades, various innovative water treatment technologies have been developed, evaluated, and introduced to the market. This chapter highlights advanced technologies like ultraviolet irradiation, ion exchange, nanotechnology, membrane filtration, aquaporin, solar filtration, photocatalyst, and adsorption, which would provide a promising outcome soon. The water sector has traditionally adopted new technologies at a glacial pace. They still have a long way to go, but they have made significant progress in proving their dependability and viability in large-scale municipal water treatment facilities. This chapter focuses on the viability and use of these technologies in light of the socio-economic conditions of developing nations.

Keywords Industrial revolution · Solar filtration · Nanotechnology · Socio-economic · Photocatalyst

S. Roy
Department of Chemistry, St. Aloysius College(Autonomous), Jabalpur, India

N. Swamy (✉)
Department of Biotechnology, St. Aloysius College(Autonomous), Jabalpur, India
e-mail: nitin.swamy21@gmail.com

S. Suriyanarayanan et al. (eds.), *Water Management in Developing Countries and Sustainable Development*, Water Resources Development and Management,
https://doi.org/10.1007/978-981-99-8639-2_4

4.1 Introduction

Water is a vital constituent for the survival of living beings. Hence it is of prime importance to maintain the quality of water. Access to pure and clean water is essential for maintaining the proper health of the biotic components. The world is facing a situation where water quality is constantly deteriorating due to an exponential increase in pollution caused by various anthropogenic activities. These natural and artificial contaminants find their way into the water bodies from different sources like domestic, industrial, municipal, sewer leakage, landfill leachates, wastewater treatment plant, etc. (Rasheed et al. 2019). Records reveal that in the last decade, approximately 4 billion of the population had no access to water stock, further estimating that the growth in the world population would hike to 8.6 billion by 2030, causing an increase in water demand to 400% by 2030(Islam and Karim 2019). This rise in water demand originated due to increased population, rapid industrial development, and climate change (Pearce 2008). The growth in the demand for water supply has gathered global attention for developing modern, cost-effective technologies to reuse wastewater to meet the expected need (Chung et al. 2012). This chapter is an effort to give an overview of the progressive change in water treatment techniques, various categories of water treatment technology, and the acceptability of such technologies by the public, keeping in mind the socio-economic status.

4.2 Journey of Water Treatment

Water treatment by purification and filtration can be traced back to ancient civilizations. However, in ancient history, the quantity of water overpowered its quality. Ancient records of Sanskrit and Greek writings reveal that treatment methods such as filtering through charcoal, exposing to sunlight, boiling, and straining were applied based on odor, appearance, and taste. It took a while to recognize that these were not the sole criteria for judging water quality. It was during early 1500 B.C. that the Egyptians used a chemical (alum) for the first time to settle down the suspended particles in drinking water. During the 1700s filtration method was introduced, which was found to be quite effective in removing particulate matter from water. Epidemiologist Dr John Snow linked the outbreak of Cholera in London to be due to the contamination caused by microbes present in the drinking water, and this brought about a change in the definition of clean water and the scenario of water treatment methods. Disinfectants like chlorine and ozone were used for water treatment during the early 1900s in the United States and Europe to control the growth of microbes in water. Gradually with passing time and Globalization, it was identified that the negative impact of industrial and agricultural developments also contributes to the waning quality of water (EAP report).

4.3 Categorization of Water Treatment Technologies

4.3.1 Primary Treatment

This procedure is considered the initial stage of the treatment process, which involves the elimination of suspended and floating solids from untreated raw water. It includes screening and sedimentation processes. Physical separation is mechanical, while chemicals can be used for sedimentation.

4.3.2 Secondary Treatment

The organic matter that escaped during the primary treatment is eliminated during the secondary treatment process. This process comprises the following steps- (i) Settling of suspended solids and carbonaceous pollutants, (ii) Treatment of water for removal of nitrogen and carbon, (iii) Passing of water through sedimentation basin for chlorine treatment, (iv) Further undergoing filtration followed by disinfection using sodium hypochlorite.

The first step forms biological sludge that can be used to produce biogas for generating heat and electricity. The second step involves the denitrification process to reduce eutrophication effects. After passing through the third step the water still contains microbial, chemical, and inorganic contaminants. After passing through the final step, the water is considered safe for irrigation and not for drinking.

4.3.3 Tertiary Treatment

The tertiary treatment assists in removing impurities up to 99%, for which the secondary treatment is unsuccessful. To make the water fit for consumption, diverse individual or combined advanced methods are adopted for deleting biological and chemical contaminants. The various tertiary treatment methods have been broadly classified into physical, chemical, and biological categories (Fig. 4.1).

(I) **Physical Treatment Technologies:**

(i) **Photodegradation**

Photodegradation is a technique applied for removing organic wastes from wastewater by using electromagnetic radiation, and it is considered a chemical-less process of purification. The source of radiation may be ultraviolet radiation or solar radiation. The absorption of radiation by a molecule leads to physical and chemical changes, forming new species. These radiation-based changes could be either direct or indirect

Fig. 4.1 Types of tertiary treatment

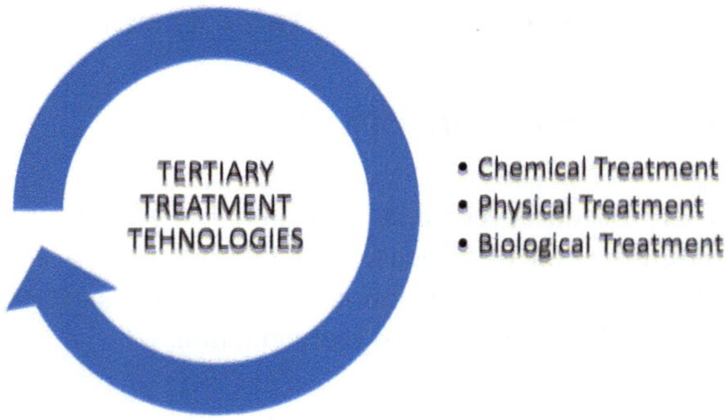

processes. The direct process involves the absorption of a photon by the pollutant molecule, which undergoes homolytic cleavage producing degradation products. On the contrary, the indirect process includes the absorption of a photon by a photosensitive molecule-producing species that interacts with the target pollutant, thus causing its degradation. The photodegradation process is classified into two types depending upon the initiation reaction for degradation, if the initiation is due to an oxidation reaction, then it is termed an advanced oxidation process (AOP) and when due to a reduction reaction, then it is known as an advanced reduction process (ARP) (Olatunde et al. 2020).

(ii) **Adsorption Techniques**

In the adsorption technique, the wastewater pollutants adhere to the surface of an adsorbent after coming in contact with it, thereby removing non-degradable organic compounds and other micro-contaminants. The most widely used adsorbent is activated carbon, however, there are numerous adsorbents with specific applications such as natural or synthetic zeolites, natural clay minerals, silicic acid, silica gel, and activated aluminum (Pillai 2020). This technique finds application in an industrial environment due to its effectiveness, low cost, ease of operation, reusable property, and resistance to noxious substances.

(iii) **Dissolved Air Floatation**

DAF technique involves the separation of solids from fluid, *i.e.*, wastewater, with the application of very fine bubbles. These air bubbles adhere to the suspended particles reducing their density. These particulates rise to the surface due to these buoyant bubbles. However, coagulants or flocculants are added before aeration to clump the suspended particles. This technique treats paper mill effluent and wastewater from oil refineries, industrial plants, chemical plants, and gas processing plants.

(iv) **Sonication Technique**

Ultrasonic technology is an innovative method for the removal of pollutants from wastewater. Sound is represented as mechanical energy passing through any form of matter by pressurized waves. Applying ultrasonic waves with or above 20,000 Hz of frequency is considered "sonication." In this process, ultrasonic's physical and chemical effects are applied to abolish organic and microbial contaminants in water. In the case of physical effects, powerful turbulences and shock waves carry out the disintegration of organic and microbial toxins. The chemical effects liberate highly reactive species, which are short-lived intermediates that cause chemical bond cleavage.

(v) **UV- Radiation Technique**

This method involves the disinfection of wastewater by the application of U.V. radiation. It is the UV-C zone out of the three U.V.- A, B, and C zones, which has the property to destroy disease-causing microorganisms. In this technique, wastewater can pass through an advanced pre-treatment component to remove suspended particles and organic matter that can interfere in the later stages of treatment. After that, the wastewater moves through the U.V. disinfection unit equipped with a U.V. light source, this interaction with the U.V. radiation destroys the genetic materials of the microbes, preventing them to reproduce. This process is very effective if the suspended particles, turbidity, and flow rate of the wastewater are kept at a low level.

(vi) **Solar Filtration**

Solar filtration is a low-technology solution which makes it widely popular for the purification and disinfection of water. In this process, solar energy is trapped through solar panels and stored in batteries in the form of electric energy that is used for the filtration and purification of water.

(vii) **Micro-Nanobubbles**

Nanobubbles are generated from gas-filled ozone or any other gas which is then passed through water. These nanobubbles remain in the water for a longer period, during which the ions present develop an electric double layer at the gas–water interface. When the bubble bursts, causing the disappearance of the interface and the release of energy accumulated by high concentrations of cations and anions which generate hydroxyl ion radicals. These free radicals degrade the organic matter present in the wastewater (Zhengguo et al. 2019).

(viii) **Photocatalytic Fuel Cell**

A photocatalytic fuel cell comprises a radiation source, cathode, photoanode containing photocatalyst, and organic matter-based wastewater as the chemical fuel. Exposure of the photocatalyst present on the photoanode with radiation causes the excitation of electrons thus creating electron–hole pair which generates electricity due to the electron movement from the photoanode to the cathode. The holes on the photoanode oxidize water and hydroxyl ions forming hydroxyl radicals

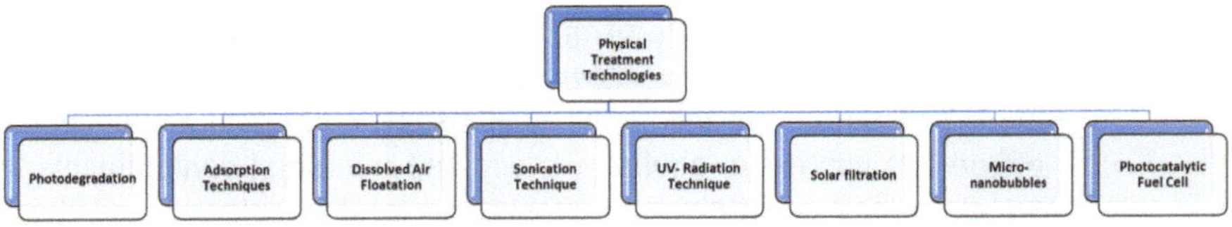

Fig. 4.2 Types of various physical treatment technologies

which degrade the organic matter in wastewater. Further, the organic matter present on the surface of the cathode is also degraded by the electrons that flow from the photoanode to the cathode (Nurhafizah and Meng 2022) (Fig. 4.2).

(II) **Chemical Treatment Technologies:**

(i) **Ozonation Process**

This process has been used since the 1950s to treat municipal and industrial wastewater. Ozone, a strong oxidant, generates hydroxyl radicals which initially destroy the cell wall and enter the cell through the cell membrane, causing the oxidation of the enzymes required for the decomposition of glucose, thus damaging the organelles and genetic materials. This process is more effective when combined with photocatalytic ozonation catalysts and U.V. treatment.

(ii) **Advanced Oxidation Process**

This process of cleaning is used in the treatment of wastewater containing organic, inorganic, and biological contaminants, those which conventional methods cannot remove. This procedure uses ozone, hydrogen peroxide, and U.V. radiation for purification, causing oxidation of the contaminants with the hydroxyl radicals formed. This process does not produce any hazardous byproducts thus not causing any harmful effect on the environment.

(iii) **Fenton and Photo-Fenton Processes**

The Fenton reaction is the originator of the advanced oxidation process (AOP), which uses hydrogen peroxide and Fe2 + ion to form hydroxyl radicals in acidic pH and ambient temperature. In the advanced Photo-Fenton process U.V. radiation and iron catalysts are used to form hydroxyl radicals, while the rest of the procedure remains the same. This modern method is much faster than the conventional Fenton process.

(iv) **Chlorination Process**

Chlorination is a quite effective and widely used method because of its modest application and low cost. This process is carried out in the presence of chlorine sources viz. elemental chlorine/ sodium hypochlorite or calcium hypochlorite. The free available chlorine inactivates the

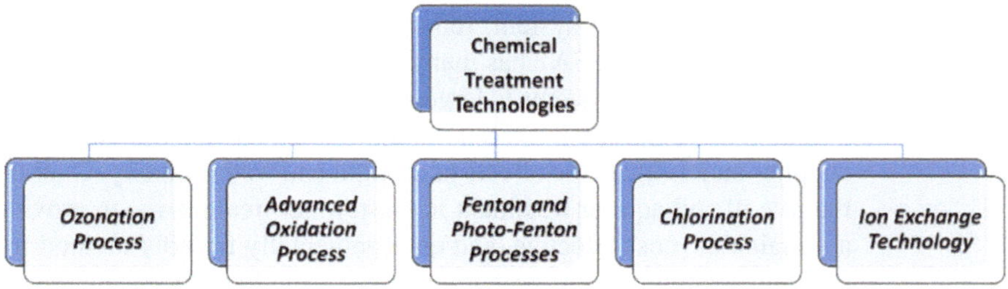

Fig. 4.3 Types of various chemical treatment technologies

microorganism by entering through the envelope into the cytoplasm and then damaging the DNA, RNA, and enzymes.

(v) **Ion Exchange Technology**

This technology involves the application of a fixed-bed synthetic resin through which wastewater is passed initiating the exchange of the contaminated ions with that of the resin ions, making the water free from contaminants. The removal of the adsorbed contaminant ions can regenerate the resin. The ion exchange resins are categorized into four types viz. strong acid cationic, weak acid cationic, strong base anionic, and weak base anionic resin depending upon the nature of the contaminant ions to be removed. This technology is efficient in the removal of anions and cations present as pollutants in wastewater (Fig. 4.3).

(III) Biological Treatment Technologies:

(i) **Soil Aquifer Treatment (SAT)**

Soil aquifer treatment (SAT) is a natural wastewater treatment technology that uses soil and groundwater to treat and purify wastewater. In SAT, treated wastewater is applied to the soil's surface and allowed to percolate through the soil and into the underlying aquifer. The soil and aquifer operate as natural filters when the wastewater goes through, eliminating impurities and pollutants. The role of soil aquifer treatment in wastewater treatment is to provide a sustainable and cost-effective method for treating and reusing wastewater. SAT has several advantages over traditional treatment technologies, such as its low cost, low energy requirements, and minimal chemical use. SAT can also be used to provide a sustainable water source for irrigation and other non-potable uses. The treatment process in SAT occurs through several mechanisms. Initially, soil microbes digest wastewater organic matter as it percolates through the soil. These microorganisms break down the organic matter, converting it into carbon dioxide, water, and other harmless byproducts. Second, the soil acts as a physical filter, removing suspended solids, pathogens, and other contaminants from the wastewater as it passes through. Finally, the underlying aquifer acts as a natural storage and filtration system, removing contaminants and pollutants from the wastewater. SAT has many applications, ranging from small-scale decentralized systems to large-scale centralized facilities. In addition to treating and reusing wastewater, SAT can replenish groundwater supplies and mitigate the effects of drought and water scarcity. Overall, the role of soil aquifer treatment in wastewater treatment is to provide a sustainable, cost-effective, and environmentally friendly method for treating and reusing wastewater while also providing a water source for other uses (Essandoh et al. 2011).

(ii) **Membrane Bioreactors**

Membrane bioreactors (MBRs) are an advanced wastewater treatment system that combines biological treatment with membrane filtration. MBRs use a combination of bacteria and other microorganisms to break down organic matter and pollutants in the wastewater and then use a membrane to filter out the remaining solids, pathogens, and contaminants. The critical components of an MBR system include a bioreactor tank, which houses the microorganisms responsible for breaking down the organic matter in the wastewater, and a membrane filtration system, which separates the treated water from the remaining solids and contaminants. The membrane can be constructed using various materials, including polymeric, ceramic, metal, and external or submerged. The use of MBRs offers several advantages over traditional wastewater treatment systems. For example, MBRs provide a high treatment quality, as the membrane can filter out virtually all suspended solids, pathogens, and contaminants. This results in much higher quality treated water than other treatment systems, making it suitable for various reuse applications. Since the membrane filtering system obviates the need for settling tanks and other secondary treatment components, MBRs have a smaller footprint than conventional wastewater treatment systems. They are, therefore, ideal for densely populated urban areas with little space. Finally, MBRs are very flexible to changing conditions due to their ease of modification to meet changes in flow rate, influent quality, and treatment objectives. As a result, they are well-suited for a wide range of wastewater treatment applications, ranging from small-scale decentralized systems to large-scale centralized facilities. In summary, the role of MBRs in wastewater treatment is to provide a high-quality, compact, and adaptable treatment system that combines biological treatment with membrane filtration to produce treated water suitable for various reuse applications (Skouteris et al. 2012).

(iii) **Biosorption**

Biosorption uses living or non-living biological materials to remove pollutants and toxins from wastewater. The biological material, known as the biosorbent, binds to the pollutants and removes them from the wastewater, reducing their concentration in the effluent. Biosorption is an effective method for removing heavy metals, dyes, and other pollutants from wastewater. Several biosorbents, such as bacteria, fungi, algae, and plant matter, can carry out the process. These materials have a high affinity for specific pollutants and can selectively remove them from wastewater, making the process highly efficient. One of the main advantages of biosorption is that it is a low-cost and environmentally friendly wastewater treatment method. The biosorbents used in the process are natural materials that are readily available and do not require expensive manufacturing or treatment processes. Additionally, biosorption does not generate harmful byproducts, making it a safe and sustainable wastewater treatment method. Biosorption can also be used with other treatment processes, such as activated sludge systems and membrane filtration, to enhance the overall efficiency of the treatment process. For example, biosorption can be used to remove heavy metals from the wastewater before it is treated with an activated sludge system, reducing the load on the system and improving its performance. In summary, biosorption is essential in wastewater treatment by providing an effective and sustainable method for removing pollutants and contaminants from wastewater. The process is low-cost, environmentally friendly, and can be combined with other treatment processes to enhance efficiency (Lettinga et al. 1980).

(iv) **Biofilms**

Biofilms play an important role in wastewater treatment by facilitating the growth of microorganisms that break down organic matter and pollutants in the wastewater. A biofilm is a thin layer of microorganisms that adhere to surfaces, forming a complex community of bacteria, fungi, and other organisms. Biofilms are frequently employed in biological treatment systems, such as activated sludge systems, trickling filters, and biofilters, for wastewater treatment. In these systems, wastewater is transported over or through a biofilm-coated surface, which provides a surface for microbial growth. The microorganisms in the biofilm consume and break down organic matter and pollutants in the wastewater, reducing the number of contaminants in the effluent. Biofilms are particularly effective at treating wastewater containing high concentrations of organic matter, as the microorganisms in the biofilm can adapt to changing conditions and break down a wide range of organic compounds.

Additionally, biofilms are often more resistant to shock loads and fluctuations in wastewater quality than other treatment systems, as they can buffer changes in the wastewater environment by maintaining a stable community of microorganisms. Overall, using biofilms in wastewater treatment can improve the efficiency and sustainability of the treatment process by reducing the amount of energy and chemicals required for water purification and reducing the environmental impact of wastewater discharge. But the effectiveness of biofilm-based treatment systems can be affected by things like the composition of the wastewater, the temperature and pH of the system, and the way the treatment process is set up. Therefore, careful monitoring and optimization of biofilm-based treatment systems are necessary to ensure their effectiveness and reliability (Fernández et al. 2007).

(v) **Aquaporins**

Aquaporins are a family of membrane proteins facilitating water transport across cell membranes. Aquaporins can improve water purification efficiency in wastewater treatment by increasing water's flux across membranes and reducing energy consumption. One potential application of aquaporins in wastewater treatment is reverse osmosis (R.O.) membranes. R.O. is a widely used water purification method involving passing water through a semi-permeable membrane to remove impurities. However, the process is energy-intensive, and the membranes can become clogged by fouling, which reduces their effectiveness. Water can be transported more rapidly and efficiently across the membrane by incorporating aquaporins into the R.O. membranes, reducing the energy required for water purification.

Additionally, using aquaporins can reduce the tendency of the membranes to become fouled, further improving their efficiency and extending their useful life. Aquaporins can also be used in other water treatment applications, such as desalination and wastewater reclamation. In these applications, aquaporins can improve the efficiency of water transport across membranes, reducing the cost and energy required for water treatment. Overall, using aquaporins in wastewater treatment can improve the efficiency and sustainability of water purification, reducing energy consumption and improving water quality. However, further research is needed to optimize the use of aquaporins in different water treatment applications and to address the challenges associated with incorporating these proteins into existing treatment systems (Camilleri-Rumbau et al. 2019).

(vi) **Enzymes**

Enzymes play an essential role in wastewater treatment by accelerating the breakdown of organic matter, which helps improve the treatment process' overall efficiency. Enzymes are biological molecules acting as catalysts, speeding up chemical reactions without being consumed or altered. In wastewater treatment, enzymes are used to break down organic pollutants resistant to degradation, such as fats, oils, and greases. Enzymes such as lipases and proteases are added to the wastewater to break these pollutants into minor, more easily degradable compounds. Enzymes can also enhance the performance of biological treatment systems, such as activated sludge systems, by increasing the rate of biochemical reactions within these systems. By accelerating the degradation of organic matter, enzymes can help reduce the amount of time and resources needed for wastewater treatment and the number of pollutants discharged into the environment. In summary, enzymes play an important role in wastewater treatment by improving the efficiency of treatment processes and reducing the environmental impact of wastewater discharge (Recktenwald et al. 2008).

(vii) **Plants**

Plants can play a crucial role in wastewater treatment through a process known as phytoremediation. Phytoremediation is a natural, low-cost, and sustainable method of treating contaminated water using plants. There are several ways that plants can help in the treatment of wastewater:

- **Phytoextraction:** Plants can absorb and accumulate heavy metals, organic compounds, and nutrients from contaminated water. This process, known as phytoextraction, can remove pollutants such as lead, cadmium, and arsenic from wastewater.
- **Phytodegradation:** Plants can also break down and detoxify pollutants through phytodegradation. This process involves the breakdown of contaminants by plant enzymes and microorganisms in the plant's root zone.
- **Phytostabilization:** Plants can also help immobilize soil contaminants and prevent them from leaching into groundwater. This process is known as phytostabilization and is particularly effective in treating wastewater that contains heavy metals.
- **Oxygenation:** Finally, plants can help oxygenate and aerate water through photosynthesis, which can improve the overall quality of the water (Fig. 4.4).

Several plant species are used in water treatment processes. Some of the most commonly used plant species include:

1. **Typha (Cattail):***Typha* species are commonly used in constructed wetland systems for wastewater treatment. These plants have a high pollution tolerance and can effectively remove nitrogen and phosphorus from wastewater.
2. **Phragmites (Common Reed)**: *Phragmites* species are often used in constructed wetland systems to treat wastewater and stormwater. They have a high capacity for removing nitrogen, phosphorus, and heavy metals from water.

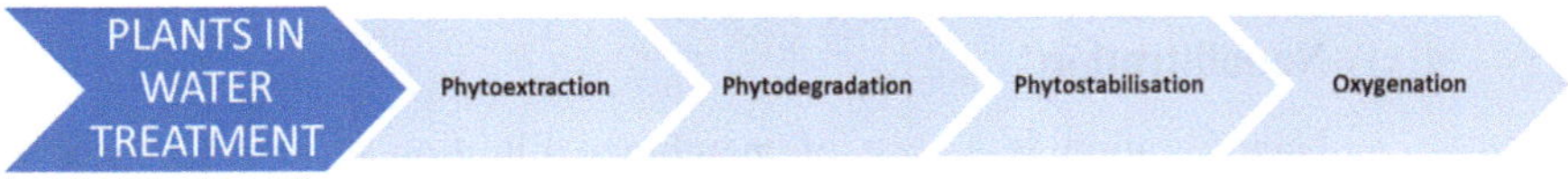

Fig. 4.4 Roles of plants in water treatment

3. **Lemna (Duckweed):***Lemna* species are small floating plants that are commonly used in wastewater treatment systems. They have a high rate of photosynthesis and can effectively remove nutrients and pollutants from water.
4. **Salvinia (Water Spangles):***Salvinia* species are floating ferns often used in wastewater treatment systems. They effectively remove nitrogen, phosphorus, and heavy metals from water.
5. **Reeds (Phragmites australis):***Phragmites australis* is a reed species often used in constructed wetland systems to treat wastewater and stormwater. They have a high capacity for removing nitrogen, phosphorus, and heavy metals from water.

These plant species are just a few examples of the many that can be used in water treatment. The choice of plant species depends on the specific treatment goals and the water's conditions. Using plants in wastewater treatment can be a cost-effective and sustainable way to treat contaminated water and protect the environment (Saba et al. 2019).

(viii) **Microorganisms**

Microorganisms play a crucial role in the biological treatment of water. This process removes pollutants and impurities from water, and it relies on the natural abilities of microorganisms to break down organic matter. In wastewater treatment, microorganisms break down organic matter, reducing the levels of organic pollutants, including nitrogen and phosphorus. This process is called biological nutrient removal. Activated sludge treatment, one of the most widely used wastewater treatment processes, is a biological treatment that uses microorganisms to break down pollutants in wastewater. In this process, wastewater is mixed with a population of microorganisms, typically bacteria, and aerated to promote the growth and activity of these microorganisms. The microorganisms consume the organic matter in the wastewater, converting it into biomass, carbon dioxide, and other byproducts. After that, the sedimentation process removes the biomass and any other solids that may be present in the water. The water that has been cleaned is then released back into the natural environment. Microorganisms are utilized not only for the treatment of wastewater but also for the treatment of drinking water. For example, biological filters that rely on microorganisms can remove contaminants from drinking water, such as bacteria, viruses, and toxic chemicals. Overall, the role of microorganisms in biological water treatment is to provide a natural, efficient, and cost-effective way to remove pollutants from water, protecting both human health and the environment (Tabatabaei et al. 2010).

(ix) **Nanofiltration**

Nanofiltration is a type of membrane filtration technology used in biological water treatment to remove pollutants from water. It is a pressure-driven process that utilizes a semi-permeable membrane to

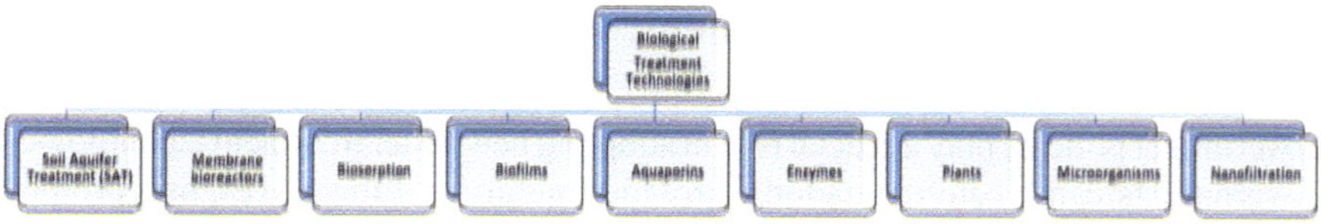

Fig. 4.5 Types of various biological treatment technologies

separate contaminants from the water based on their size and charge. In nanofiltration, water is pressurized and forced through the membrane, and contaminants too large to pass through the membrane's pores are trapped on one side while the treated water flows through to the other. This process can remove many contaminants, including organic compounds, salts, and inorganic contaminants. Nanofiltration is especially useful for eliminating compounds resistant to biological degradation, such as pesticides, heavy metals, and medicines. This makes it indispensable to biological water treatment methods that rely on microbes to decompose organic waste. In addition to its use in removing pollutants, nanofiltration can also soften water by removing hardness-causing minerals, such as calcium and magnesium. This makes nanofiltration a helpful technology in water treatment applications where the hardness of the water is a concern. Overall, nanofiltration is a valuable tool in biological water treatment, effectively removing pollutants from water and improving the treated water quality (Nataraj et al. 2006) (Fig. 4.5).

4.4 The Public Response Toward Accepting New Water Treatment Strategies:

The public's response toward new water treatment technologies can vary greatly depending on various factors such as the location, the specific technology, and the level of education and awareness about the water treatment issue. People are becoming more aware of the importance of clean and safe water and are open to new technologies that can provide this. Innovations in water treatment, such as advanced filtration systems, renewable energy-powered treatment plants, and decentralized treatment solutions, are becoming more popular and widely accepted. However, some people may be skeptical or resistant to new technologies due to concerns about cost, complexity, and impact on their communities. In such cases, water treatment providers and policymakers must engage with the public and address their concerns through education and transparency. Overall, the public's response to new water treatment technologies is dynamic and influenced by many factors, but there is growing support for innovative solutions that provide safe, clean, and sustainable water for all.

4.5 The Role of the Government:

Government policies play a crucial role in adopting new water treatment technologies. Governments can create regulations, provide funding and incentives, and promote public education and awareness about the importance of clean and safe water. In many countries, governments actively promote adopting new water treatment technologies by providing financial incentives and subsidies to companies and organizations investing in these solutions. This helps to reduce the financial barriers that can prevent the widespread adoption of these technologies. Governments also set regulations and standards for water quality, which drive innovation and development of new water treatment technologies. By setting high water safety and quality standards, governments can encourage the development and adoption of new technologies to meet these standards.

Additionally, governments can promote public education and awareness about the importance of clean and safe water. By educating the public about the benefits of new water treatment technologies and the risks associated with poor water quality, governments can help to build support for these solutions and encourage their adoption. Overall, government policies play a crucial role in promoting the adoption of new water treatment technologies. By providing financial support, setting regulations and standards, and promoting public education and awareness, governments can help to ensure that these technologies are widely adopted and help provide clean and safe water for all.

4.6 Socio-Economic Factors in Adopting a New Water Treatment

(i) **Lack of financial resources**:

A significant challenge developing countries face in adopting new water treatment methods is the lack of financial resources. Building and operating water treatment facilities can be prohibitively expensive, especially for countries with limited budgets and competing priorities. The cost includes not only the infrastructure but also the ongoing operational and maintenance expenses. Additionally, developing countries may have difficulty accessing loans and financing options due to their economic situation and creditworthiness. Limited financial resources can impede the implementation of advanced water treatment technologies, leading to continued reliance on traditional, often inadequate, methods. Developing countries can explore financing mechanisms such as public–private partnerships, concessional loans, and grants from international development organizations to address this challenge. Governments can prioritize water sector investments and allocate budgetary resources for water treatment infrastructure. Additionally, innovative financing models like performance-based contracts and output-based aid can incentivize private sector participation in the provision of water treatment services. Strengthening revenue collection mechanisms, improving cost recovery, and exploring alternative funding sources such as climate financing and impact investments can also contribute to overcoming financial barriers (World Bank 2019).

(ii) **Limited access to technology**:

Developing countries often face limited access to advanced water treatment technologies, which can impede their adoption. Access constraints may arise from multiple factors. Firstly, complex equipment and machinery required for modern water treatment may not be readily available or affordable in developing countries. Limited manufacturing capabilities and supply chains can further restrict access to technology. Secondly, the lack of qualified personnel with the necessary technical expertise to operate and maintain advanced water treatment systems can be a significant barrier. Training programs and capacity-building initiatives are essential to develop a skilled workforce capable of managing new technologies. Moreover, intellectual property rights, patents, and trade barriers can hinder technology transfer and collaboration between countries. These challenges require partnerships and collaborations with technology providers, research institutions, and international organizations. Technology transfer programs and knowledge-sharing platforms can facilitate access to advanced water treatment technologies. Governments can also incentivize technology providers to transfer their knowledge and equipment to developing countries by offering tax incentives or establishing technology hubs for research and development (United Nations 2021).

(iii) **Political instability**:

Political instability poses significant challenges to adopting modern water treatment systems in developing countries. It can lead to various issues, such as a lack of funds, poor governance, and corruption, which hinder the implementation and sustainability of water treatment projects. Political unrest can disrupt the continuity of government policies and programs, resulting in funding uncertainties and delays in project execution. Instability may also adversely affect the investment climate, discouraging private sector participation in the water sector. Building resilient institutions and improving governance structures are crucial to address political instability. Strengthening legal frameworks, regulatory frameworks, and accountability mechanisms can help ensure transparency and minimize corruption risks. Long-term planning and policy stability are essential to provide a conducive environment for investment in water treatment infrastructure. Engaging with international development partners and leveraging their support can provide funding and technical expertise stability. Fostering political stability through dialogue, conflict resolution, and good governance practices contributes to the sustained implementation of modern water treatment systems (Global Water Partnership 2020).

(iv) **Limited human resources**:

Developing countries often face challenges related to limited human resources when adopting new water treatment programs. These challenges encompass a lack of awareness among the population regarding the importance of clean water and the associated health risks. In many communities, there may be a lack of knowledge about the benefits of treated water and the potential dangers of consuming contaminated water sources. This lack of awareness can lead to resistance or apathy toward new water treatment initiatives. To overcome limited human resources, comprehensive public education campaigns are essential. These campaigns should raise awareness about the importance of clean water and the risks associated with consuming contaminated water. They can involve community outreach programs, school-based initiatives, and media campaigns to reach a broad audience. Engaging local community leaders, health workers, and educators can help disseminate accurate information and foster behavioral change. Capacity-building programs are also crucial to develop a skilled workforce capable of implementing and maintaining modern water treatment systems. Training programs should target water operators, technicians, and local authorities to enhance their understanding of water quality management and treatment processes. These programs can be conducted with international organizations, universities, or technical institutions to ensure knowledge transfer and best practices. Furthermore, community engagement is vital in garnering support for water treatment programs. Involving local communities in the decision-making process can address their concerns and preferences. Establishing community-based water committees or user groups can empower communities to take ownership of water treatment initiatives, leading to better acceptance and sustainability of the systems (United Nations Children's Fund and World Health Organization 2019).

(v) **Lack of public awareness**:

In many developing countries, a lack of general knowledge about the significance of clean water and the risks associated with consuming polluted water poses a significant challenge. This lack of awareness can make it difficult to gain support for additional water purification initiatives in the future. However, targeted awareness campaigns can address this issue. Education programs should focus on raising awareness about the health impacts of contaminated water, emphasizing the benefits of clean water, and promoting proper hygiene practices. These campaigns can utilize multiple communication channels such as radio, television, print media, and community gatherings to reach a broad audience. Additionally, leveraging local cultural practices and traditions can help deliver messages effectively. Collaboration with local influencers, community leaders, and religious institutions can enhance the credibility and reach of awareness campaigns. Engaging schools and educational institutions to include water hygiene and sanitation education in their curriculum can also have a long-lasting impact on future generations. It is crucial to design messages and materials that are culturally sensitive, engaging, and easy to understand for the target audience. Utilizing local languages and cultural references can enhance the effectiveness of the awareness campaigns (United Nations Children's Fund and World Health Organization 2019).

(vi) **Cultural barriers:**

Certain cultural behaviors and practices in underdeveloped nations can pose significant barriers to adopting modern water treatment methods. For example, using untreated water for irrigation or washing can contaminate water. Overcoming these cultural barriers requires a comprehensive approach that addresses cultural norms and beliefs. Engaging with local communities and conducting cultural assessments can provide insights into the prevailing practices and beliefs related to water. It is important to involve community members in designing and implementing water treatment interventions to ensure they are culturally appropriate and acceptable. Tailoring water treatment interventions to fit within the cultural context can help overcome resistance. This may involve adapting technologies to suit local conditions, considering traditional water management practices, and involving local communities in decision-making. Community-based participatory approaches, social marketing campaigns, and behavior change communication strategies can facilitate the adoption of new practices and promote the importance of clean water (World Health Organization 2019).

By addressing these socio-economic factors, developing countries can overcome barriers to adopting modern water treatment technologies and improve access to safe and clean water sources for their populations.

4.7 A Model for the Adoption of Modern Water Purification Technologies in Developing Countries

Access to clean and safe drinking water is a fundamental human right, yet many developing countries face significant challenges in meeting this basic need. Traditional water purification methods often fall short of providing reliable and sustainable solutions. This model proposes a systematic approach for developing countries to adopt modern water purification technologies, leveraging innovative solutions to improve access to safe drinking water.

- **Conduct Needs Assessment and Resource Evaluation:**

Before adopting modern water purification technologies, conducting a comprehensive needs assessment and evaluating available resources is crucial. This assessment should consider population size, water sources, water quality, existing infrastructure, budget constraints, and local preferences.

- **Establish Partnerships and Collaborations:**

Developing countries should foster partnerships and collaborations with international organizations, governmental agencies, non-governmental organizations

(NGOs), research institutions, and industry experts. These collaborations can provide technical expertise, funding, capacity-building programs, and knowledge-sharing opportunities to facilitate the adoption of modern water purification technologies.

- **Develop National Water Purification Strategy:**

  A well-defined national water purification strategy is essential to guide the adoption of modern technologies. This strategy should outline specific goals, targets, and timelines for improving water quality and access. It should also address policy and regulatory frameworks, institutional capacities, and financing mechanisms for sustainable implementation.

- **Technology Selection and Adaptation:**

  Selecting appropriate water purification technologies requires considering the context and challenges the developing country faces. Modern technologies such as membrane filtration, reverse osmosis, ultraviolet (UV) disinfection, and solar-powered systems offer efficient and reliable solutions. However, the selected technologies should be adaptable to local conditions, easy to operate and maintain, cost-effective, and scalable.

- **Pilot Projects and Demonstration Sites:**

  Pilot projects and demonstration sites are valuable for testing and evaluating the performance of modern water purification technologies in real-world conditions. These projects allow for fine-tuning of the technology, identification of potential challenges, and the generation of local acceptance and support. Lessons learned from pilot projects can inform the scaling-up process.

- **Capacity Building and Training:**

  Investing in capacity building and training programs is crucial to ensure successful adoption and long-term sustainability. This includes training water operators, technicians, and local communities on the operation, maintenance, and troubleshooting of modern water purification technologies. Training programs should be tailored to the local context, considering language barriers, literacy levels, and cultural sensitivities (Fig. 4.6).

- **Monitoring, Evaluation, and Knowledge Sharing:**

  Establishing a robust monitoring and evaluation system is vital to track the progress and impact of adopting modern water purification technologies. Regular water quality testing, data collection, and analysis are essential to ensure the technologies effectively remove contaminants and improve overall water quality. Sharing knowledge and best practices with other developing countries and global networks can foster innovation and promote successful implementation.

This model may help developing countries adopt modern water purification technologies, improving access to clean and safe drinking water. Collaboration, strategic planning, appropriate technology selection, capacity building, and monitoring are

Fig. 4.6 A systematic approach for developing countries to adopt modern water purification technologies

crucial in achieving sustainable and scalable solutions. Adopting modern water purification technologies will contribute to realizing the United Nations Sustainable Development Goal 6, ensuring access to clean water and sanitation for all.

4.8 Examples of Modern Water Treatment Technologies in Developing Nations:

The examples highlight the successful adoption of modern water treatment technologies in developing countries, showcasing their positive impact on improving water quality and access to safe drinking water.

Example 1: BioSand Filters in Cambodia

BioSand Filters (BSFs) have been widely adopted as an effective water treatment solution in Cambodia. BSFs are low-cost, low-maintenance, and sustainable water filtration devices that can remove contaminants such as bacteria, viruses, and suspended solids from water. One successful implementation of BSFs in Cambodia is the WaterSHED program. The program focuses on promoting the use of BSFs in rural communities, providing training on filter construction and maintenance, and raising awareness about the importance of clean water. As a result, thousands of households in Cambodia have gained access to safe drinking water through the use of BioSand Filters. (Source: WaterSHED.)

Example 2: Community Water Treatment Plants in India

India has implemented community water treatment plants to enhance access to clean water in rural areas. These plants employ modern treatment technologies like reverse osmosis (RO) and ultraviolet (UV) disinfection to eliminate contaminants from water sources. One notable project is the Ammachi Water Treatment Initiative, led by the Mata Amritanandamayi Math, a humanitarian organization. The initiative focuses on installing community water treatment plants in rural villages across India, providing safe drinking water to thousands of people. These plants incorporate advanced technologies to ensure the quality and safety of the treated water. (Source: Ammachi Labs.)

Example 3: Solar-Powered Desalination Plant in Morocco

Morocco has implemented a groundbreaking solar-powered desalination plant to address water scarcity in coastal regions. The plant in the city of Agadir utilizes solar energy to power the desalination process, converting seawater into clean drinking water. The Noor Ouarzazate Solar Complex, one of the world's largest solar power plants, provides the renewable energy needed for the desalination plant. This innovative project has significantly increased water availability in the region, benefiting the local population and agricultural activities. (Source: The World Bank. (2019)).

Example 4: Floating Treatment Wetlands in Bangladesh

In Bangladesh, floating treatment wetlands (FTWs) have gained attention as a nature-based solution for water treatment. FTWs consist of floating platforms with vegetation that enhance water quality by providing natural filtration and nutrient uptake. In collaboration with international partners, the Center for Environmental and Geographic Information Services (CEGIS) has implemented FTWs in water bodies across Bangladesh. These innovative systems help to purify contaminated water and mitigate pollution from industrial and domestic sources. FTWs improve water quality and provide habitat for aquatic species, contributing to biodiversity conservation. (Source: Center for Environmental and Geographic Information Services).

Example 5: Decentralized Water Treatment Systems in Peru

Peru has adopted decentralized water treatment systems to tackle water contamination problems in rural and peri-urban areas. These systems aim to treat water at the community level, ensuring households not connected to centralized water supply networks can access safe drinking water. One notable project is the "Sistema de Tratamiento de Agua para Consumo Humano" (Water Treatment System for Human Consumption) implemented by the Peruvian Ministry of Housing, Construction, and Sanitation. These decentralized systems incorporate various treatment technologies such as filtration, disinfection, and chlorination to remove contaminants and ensure water safety at the local level. (Source: Ministry of Housing, Construction, and Sanitation, Peru).

Adopting modern water treatment technologies in developing countries exemplifies the commitment to addressing water challenges and promoting sustainable access to clean water. These examples demonstrate the diverse range of modern water treatment technologies adopted in developing countries. Whether through nature-based solutions like floating treatment wetlands or community-level systems like decentralized water treatment, these initiatives showcase the ongoing efforts to improve water quality, enhance public health, and achieve sustainable water management in diverse contexts (Varis and Abu-Zeid 2009).

4.9 Conclusion

When evaluating water treatment methods for developing countries, it is critical to prioritize low-cost, simple-to-run, and maintain methods that are successful at treating a wide range of water sources. Contemporary water treatment systems can improve the efficiency and efficacy of water treatment, but they may necessitate more significant initial investments and specialized expertise for operation and maintenance. Some modern water treatment methods that may be suitable for developing countries include Reverse Osmosis (R.O.), Ultrafiltration (U.F.), Membrane Bioreactors (MBR), Electrocoagulation (E.C.), and Activated Carbon Filtration. These modern water treatment systems may suit underdeveloped countries depending on local water quality challenges, resource availability, and operator experience. To choose the best approach, evaluate the water treatment system's long-term viability and consult local populations.

References

Ammachi Labs (n.d.) Ammachi water treatment initiative. https://www.ammachilabs.org/awi

Camilleri-Rumbau MS, Soler-Cabezas JL, Christensen KV, Norddahl B, Mendoza-Roca JA, Vincent-Vela MC (2019) Application of aquaporin-based forward osmosis membranes for

processing of digestate liquid fractions. Chem Eng J 371:583–592. https://doi.org/10.1016/j.cej.2019.02.029
Center for Environmental and Geographic Information Services (n.d.) Floating treatment wetlands. https://www.cegisbd.com/tech_solution/floating-treatment-wetland-ftw/
Chung TS, Li X, Ong RC, Ge Q, Wang H, Han G (2012) Emerging forward osmosis (F.O.) technologies and challenges ahead for clean water of energy application. Curr Opin Chem Eng 1(3):246–257
Ministry of Housing, Construction, and Sanitation (Peru) (n.d.) Water treatment system for human consumption. http://www.vivienda.gob.pe/cuarta/portal/?page_id=3559
Essandoh HMK, Tizaoui C, Mohamed MHA, Amy G, Brdjanovic D (2011) Soil aquifer treatment of artificial wastewater under saturated conditions. Water Res 45(14):4211–4226. https://doi.org/10.1016/j.watres.2011.05.017
Fernández N, Díaz EE, Amils R, Sanz JL (2007) Analysis of microbial community during biofilm development in an anaerobic wastewater treatment reactor. Microb Ecol 56(1):121–132. https://doi.org/10.1007/s00248-007-9330-2
Islam SMF, Karim Z (2019) World's demand for food and water: the consequences of climate change. In: Desalination-challenges and opportunities. IntechOpen Limited, London, United Kingdom. https://doi.org/10.5772/intechopen.85919
Lettinga G, van Velsen AFM, Hobma SW, de Zeeuw W, Klapwijk A (1980) Use of the upflow sludge blanket (USB) reactor concept for biological wastewater treatment, especially for anaerobic treatment. Biotechnol Bioeng 22(4):699–734. https://doi.org/10.1002/bit.260220402
Nataraj SK, Hosamani KM, Aminabhavi TM (2006) Distillery wastewater treatment by the membrane-based nanofiltration and reverse osmosis processes. Water Res 40(12):2349–2356. https://doi.org/10.1016/j.watres.2006.04.022
Nurhafizah MdS, Meng GT (2022) A review on future wastewater treatment technologies: micro-nanobubbles, hybrid electro-Fenton processes, photocatalytic fuel cells, and microbial fuel cells. Water Sci Technol 85(1):319–341. https://doi.org/10.2166/wst.2021.618
Olatunde OC, Kuvarega AT, Onwudiwe DC (2020) Photo enhanced degradation of contaminants of emerging concern in waste water. Emerg Contam 6:283–302. https://doi.org/10.1016/j.emcon.2020.07.006
Pearce GK (2008) UF/MF pre-treatment to R.O. in seawater and wastewater reuse applications: a comparison of energy costs. Desalination 222(1–3):66–73. https://doi.org/10.1016/j.desal.2007.05.029
Pillai SB (2020) Adsorption in water and used water purification. In: Lahnsteiner J (ed) Handbook of water and used water purification. Springer, Cham, Switzerland. https://doi.org/10.1007/978-3-319-66382-1_4-1
Rasheed T, Bilal M, Nabeel F, Adeel M, Iqbal HMN (2019) Environmentally-related contaminants of high concern: potential sources and analytical modalities for detection, quantification, and treatment. Environ Int 122:52–66. https://doi.org/10.1016/j.envint.2018.11.038
Recktenwald M, Wawrzynczyk J, Dey ES, Norrlöw O (2008) Enhanced efficiency of industrial-scale anaerobic digestion by the addition of glycosidic enzymes. J Environ Sci Health Part A 43(13):1536–1540. https://doi.org/10.1080/10934520802293693
Saba B, Khan M, Christy AD, Kjellerup BV (2019) Microbial phyto-power systems—a sustainable integration of phytoremediation and microbial fuel cells. Bioelectrochemistry 127:1–11. https://doi.org/10.1016/j.bioelechem.2018.12.005
Skouteris G, Hermosilla D, López P, Negro C, Blanco Á (2012) Anaerobic membrane bioreactors for wastewater treatment: a review. Chem Eng J 198–199:138–148. https://doi.org/10.1016/j.cej.2012.05.070
Tabatabaei M, Rahim RA, Abdullah N, Wright A-DG, Shirai Y, Sakai K, Sulaiman A, Hassan MA (2010) Importance of the methanogenic archaea populations in anaerobic wastewater treatments. Process Biochem 45(8):1214–1225. https://doi.org/10.1016/j.procbio.2010.05.017

The Global Water Partnership (GWP) (2020) The role of water governance in achieving the sustainable development goals. https://www.gwp.org/globalassets/global/toolbox/publi-1/reports/research-series/The_Role_of_Water_Governance_in_Achieving_the_SDGs.pdf
The World Bank (2019) Solar-powered desalination in Morocco. https://www.worldbank.org/en/news/feature/2019/09/18/solar-powered-desalination-in-morocco
United Nations Children's Fund (UNICEF) and World Health Organization (WHO). (2019). Progress on household drinking water, sanitation and hygiene 2000–2017: special focus on inequalities. https://www.who.int/water_sanitation_health/publications/jmp-report-2019/en/
United Nations Development Programme (UNDP) (2021) Integrated water resources management for sustainable development. https://www.undp.org/content/undp/en/home/librarypage/environment-energy/water-governance/integrated-water-resources-management-for-sustainable-developme.html
US EPA (1999) 25 years of the safe drinking water act: history and trends. http://www.epa.gov/safewater/sdwa25/sdwa.html
Varis O, Abu-Zeid K (2009) Socio-economic and environmental aspects of water management in the 21st century: trends, challenges and prospects for the MENA region. Int J Water Resour Dev 25(3):507–522. https://doi.org/10.1080/07900620902972679
WaterSHED (n.d.) BioSand filters. https://www.watershedasia.org/biosand-filters
World Bank (2019) The human capital project: water quality and human capital–case study report. https://openknowledge.worldbank.org/bitstream/handle/10986/32645/Water-Quality-and-Human-Capital-Case-Study-Report.pdf
World Health Organization (WHO) (2019) Guidelines for drinking-water quality. https://www.who.int/water_sanitation_health/publications/drinking-water-quality-guidelines-4-including-1st-addendum/en/
Zhengguo X, Tallal BA, Li D (2019) Applications of micro–nano bubble technology in environmental pollution control. Micro Nano Lett 14(7):782–787

Chapter 5
Ensuring Sustainability of Groundwater Resources: A Review of Challenges and Initiatives by Southern African Arid and Semi-arid Countries

Nnenesi A. Kgabi and Hilma R. Amwele

Abstract Water research in most Southern African Development Community (SADC) countries is focused more on water resource management and modeling. Keywords such as "conservation, Southern Africa, Africa, climate change, rainfall, systems, and river-basin yield more results in most research" (Elema 2019). The limited focus by researchers on sustainable groundwater management is evident. Also, the lack of consideration of the interconnectedness of groundwater quality, quantity, climate change, and human activities at regional, national, and local levels is also receiving limited attention. This chapter aims to review the challenges and initiatives by Southern African arid and semi-arid countries.

Keywords Groundwater availability · Water quality monitoring · Groundwater sustainability · Groundwater management

5.1 Introduction

Inadequate information makes it difficult to manage and develop sustainable groundwater resources (Gaye and Tindimugaya 2019) needed to guide the planning and implementation of augmentation programmes in Africa.

N. A. Kgabi (✉)
Unit for Environmental Science and Management, North-West University, Potchefstroom, South Africa
e-mail: Nnenesi.Kgabi@nwu.ac.za

H. R. Amwele
Department of Agriculture and Natural Resources Sciences, Namibia University of Science and Technology, Windhoek, Namibia

S. Suriyanarayanan et al. (eds.), *Water Management in Developing Countries and Sustainable Development*, Water Resources Development and Management,
https://doi.org/10.1007/978-981-99-8639-2_5

Gaye and Tindimugaya (2019) noted the fact that, though the considerable amount of groundwater contributes to river flows, it is frequently excluded from river basin management plans.

In the SADC Region, groundwater is the main source of supply for residential water and is crucial for livestock and related agricultural activities. Overall, water use in the Region is anticipated to be 11.6% by volume, with household supply accounting for about 20% by volume (Molapo et al. 2000). Gaye and Tindimugaya (2019) also recognised the significance of groundwater as a resource for Sub-Saharan Africa, both ecologically and socio-economically. Therefore, an understanding of the gravities on groundwater resources, particularly on economic growth, and climate change (Nijsten et al. 2018) is crucial to ensure sustainability of the resource.

Southern Africa is characterised by unfavourable hydro-climatic conditions of low rainfall, high evaporation, and low recharge associated with the aridity levels (see Fig. 5.1) of the region. Matchaya et al. (2009) reported that the region, which has a high degree of climatic variability and unpredictable rainfall patterns, is 75% arid. Also, Turton et al. (2006) observed that low rainfall to runoff conversion rates in Southern Africa have an impact on surface water, river flows, and groundwater recharge. While the two studies reported the unfavourable conditions, Braune et al. (2008) confirmed that the prevalent drought conditions in the SADC Region impede achievement of sustainable development goals. Braune et al. (2008) also observed that, the semi-arid and arid conditions in the region contribute to lower recharge rates, particularly in western Zimbabwe, southern Angola, Botswana, Namibia, and western South Africa. Low rates of groundwater recharge are exacerbated by climate change due to higher temperatures, lower rainfall, and high evaporation.

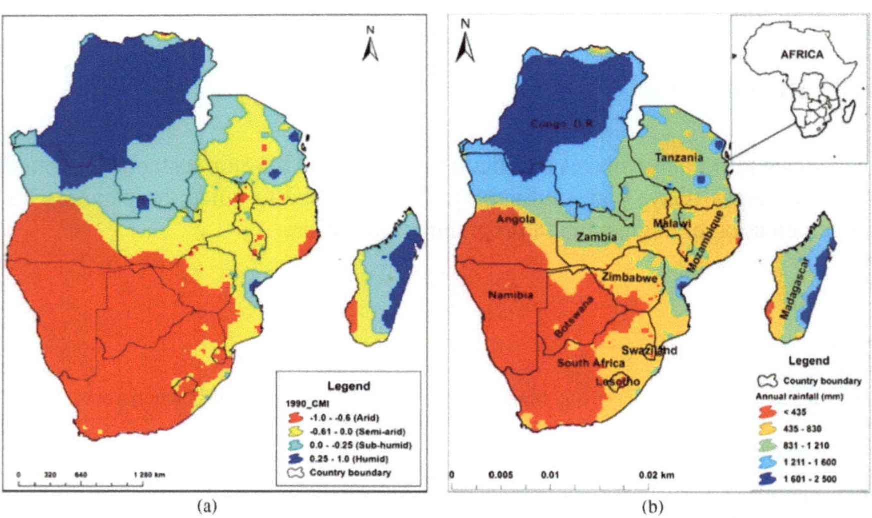

Fig. 5.1 **a** Distribution of water scarcity and aridity levels in Southern Africa **b** Yearly distribution of rainfall in Southern Africa. *Source* Matchaya et al. (2019)

It is evident that effective implementation of adaptation methods is necessary given the demand of climate variability on the region's water supplies. However, most arid and semi-arid countries consider drilling of boreholes as a common "solution" during drought periods (SADC 2009).

For a country such as Namibia, the driest country in the SADC, and without surface water sources, the primary source of water supply for all seasons is groundwater. Figure 5.2 shows the groundwater depended on supply scheme in Namibia. It should be noted that if continued, the practice can lead to depletion of the resource.

The groundwater supply scheme is based on the accessibility of groundwater, which is measured in terms of the abstraction potential. According to Nijsten et al. (2018), abstraction potential is represented as a proportion of total water use. It is a measurement or indicator of the degree to which communities rely on groundwater for household, industrial, and agronomic water consumption. "The yearly groundwater abstraction is divided by the annual recharge to determine the groundwater development stress", (Nijsten et al. 2018). The abstraction potential and groundwater development stress can be used as indicators of productivity (storage and yield) of groundwater aquifers.

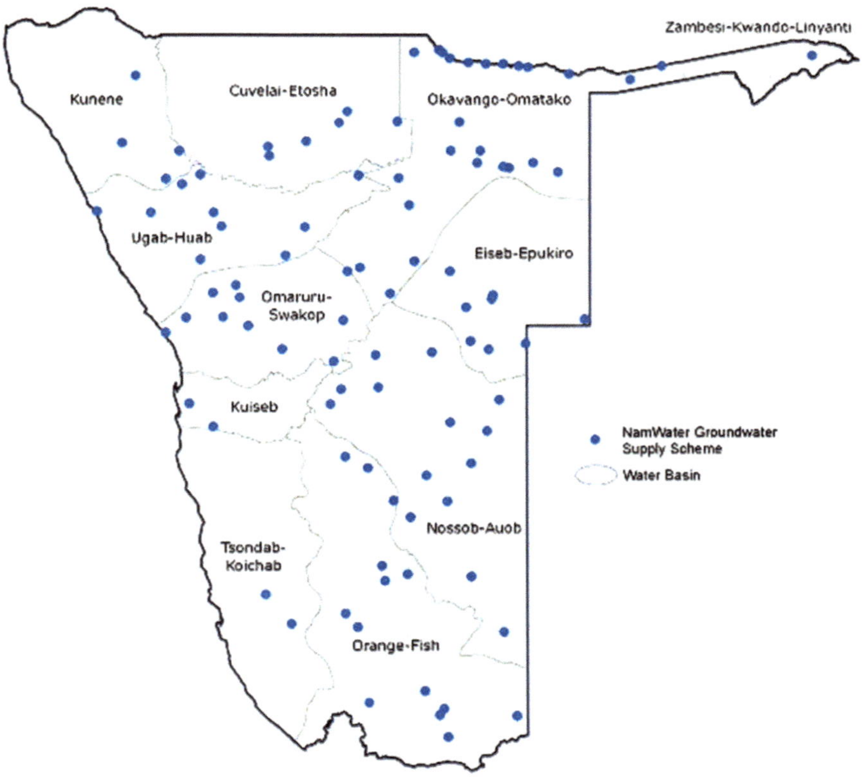

Fig. 5.2 Groundwater supply scheme in Namibia. *Source* Christelis and Struckmeier (2011)

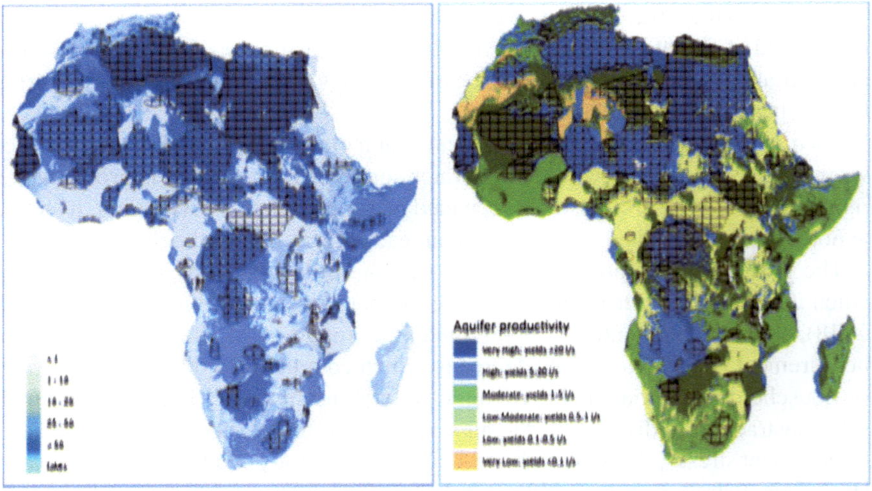

Fig. 5.3 Transboundary Aquifers in Africa joined with aquifer storage [m] (left) and aquifer yields [l/s] (right). *Sources* IGRAC & UNESCO-IHP (2015), BGS et al. (2017), Nijsten et al. (2018)

Figure 5.3 shows a map of transboundary aquifers in Africa joined with aquifer storage and yields.

In order to counter the possible depletion of the groundwater resource, there is a need to assess the possibility of implementing groundwater augmentation programmes such as managed artificial recharge of aquifers.

5.2 Methods

A review was conducted of articles from selected scientific and open-source literature. The studies selected were focused on different research aspects of groundwater resource management.

"Convenience sampling was used for selection of studies to be assessed. The researcher picked the studies that were top on the Google search engine output, and that were freely accessible (mainly open access). It should be noted that selection of the studies does not have merit or ranking implications", (Kgabi 2020). There is no intended classification of research conducted in the Region. Additionally, the information was accessed outside of organisations, primarily via websites, published research networks, and government databases.

Comparisons, looking for patterns, themes, and statistical trends, creating a consistent chain of substantiation, and creating conceptual hypothetical soundness were all part of the content analysis process for generating conclusions.

5.3 Results and Discussion

In the past two decades, the SADC made efforts to identify, monitor, and address groundwater management challenges of the Region, with the aim to ensure sustainability of the groundwater resources. The efforts made at regional level are grouped below according to groundwater availability, quality, and quantity monitoring, management, and sustainability initiatives.

5.3.1 Groundwater Availability

Groundwater availability is closely linked to water scarcity in arid regions. According to Matchaya (2009), "water stress, water shortage or surplus, and water crisis are all included in the idea of water scarcity." To avoid the water deficit and crisis situations, countries need to develop context-specific monitoring and augmentation programmes.

Nijsten et al. (2018) confirmed that continuous monitoring of groundwater levels, abstraction of water from boreholes, and groundwater chemistry are necessary to provide a sustainable water supply.

Internationally Shared Aquifer Resources Management (ISARM) is one of the groundwater assessment initiatives in the Region. The goal of ISARM is to improve the understanding of the scientific, institutional, legal, socio-economic, and environmental challenges related to the management of transboundary aquifers (Nijsten et al. 2018). The ISARM encourages comprehensive assessment of TBA including environmental, socio-economic, hydrogeological, and legal elements of the aquifer area.

The groundwater availability challenges differ from country to country based on the climatic conditions, water governance and management, and monitoring and reporting. The challenges of limited availability (both quantity and quality) of the groundwater resources (see Table 5.1) are however, common to most countries.

Some of the groundwater regional initiatives focused on groundwater availability include (1) Creating the SADC region's basic shared requirements for groundwater development; (2) Development of a regional information system for groundwater; (3) Formation of regional groundwater monitoring network; (4) Regional hydrogeological map and atlas compilation for the SADC region; (5) Karoo aquifers' groundwater resource assessment; (6) Evaluation of the Precambrian basement aquifers' regional groundwater resources; (7) Limpopo or Save Basin assessment of groundwater resources; and, (8) The regional transboundary water resources project.

One of the common standards for groundwater development is the shared water courses Protocol of the SADC (adopted 1995, revised 2000). The Protocol was drafted to develop the rules for cooperative management of regional water resources.

Table 5.1 Groundwater challenges in the SADC

Water basin(s)	Country	Challenges	Lessons learnt	Reference(s)
Not specified	Botswana	The need for detailed analysis keeps increasing because knowledge on the actual potential of groundwater is limited "Reservoir abstractions are likely to increase in future due to connection of more villages to water transfer schemes, and use of recently completed reservoirs", (Setlhogile et al. 2017) "Groundwater accounts for approximately 58% of total water abstraction, while the remainder is surface water", (Setlhogile et al. 2017)	The main source of water in rural areas is groundwater	Setlhogile et al. (2017)
Orange-Senqu	Botswana, Namibia and South Africa	In Namibia and Botswana groundwater is a limited but significant source of supply	Accounting for groundwater is handled separately from surface water in Namibia and Botswana	Lange et al. (2007)
	South Africa Gauteng Province	Water scarcity, pollution, land degradation and climate change		Earle et al. (2005)
		Increasing industrialisation and population growth High demand for water Massive water pollution issues		Conley and Van Niekerk (1998)
		Surface hydrology, basin-wide environmental factors, water quality, groundwater, legal requirements, and economic activity		UNDP GEF (2008)
Khan and Swakop Rivers	Namibia	Concern over possible contamination of groundwater that seeps into fractured rocks from tailing storage facilities, particularly radionuclides. reaching the fleeting Khan and Swakop rivers that are nearby		Barnard (2005); Kohrs (2008)

(continued)

Table 5.1 (continued)

Water basin(s)	Country	Challenges	Lessons learnt	Reference(s)
		The alluvial groundwater's uranium concentration ranges from 2 gl^{-1} to 528 gl^{-1}, with a mean of 39 gl^{-1}		Kringel et al. (2010)
		The Khan and Swakop rivers' groundwater is saline and unfit for human consumption. Members of the community use it just to irrigate crops		CSIR (1997)

The Protocol's overarching goal is to promote greater collaboration for the responsible, sustainable, and synchronised management, safeguard, and use of shared watercourses.

The Regional Groundwater Management Programme was also developed with the overarching goal of creating a background that would promote the cooperative management of shared aquifers by establishing a framework and particular tools to enable efficient management of water resources.

The regional transboundary water resources project reported by SADC (2009) focused on: "(i) development forecasts, climate change scenarios, and the overall situation; (ii) socioeconomic and physical effects; and (iii) the adaptive capacity, along with hydrology, demand, and infrastructure evaluation."

5.3.2 *Quantity and Quality Monitoring*

"SADC has made tremendous strides toward addressing groundwater with its all-inclusive regional water resource policies and objectives, and the launch of its Groundwater Management Program," Braune et al. (2008). However, the monitoring component has almost always presented challenges. The transboundary nature of water sources together with diverse levels of socio-economic development of the member states is also adding to the complex situation.

The transboundary aquifers (TBAs) in the region are often exposed to pollution. Several major challenges in joint monitoring within a transboundary context like the SADC were identified by Nijsten et al. (2018) as the synchronisation of systems, methods, and data formats across the participating member states.

The SADC HYCOS Programme is a good example of one of the monitoring initiatives by the region. The Hydrological Cycle Observing System (HYCOS) consists of a network of monitoring stations (see Fig. 5.4) generating data on the transboundary river basins in the SADC. The SADC environmental management and SADC-HYCOS were strengthened to monitor SADC freshwater resources,

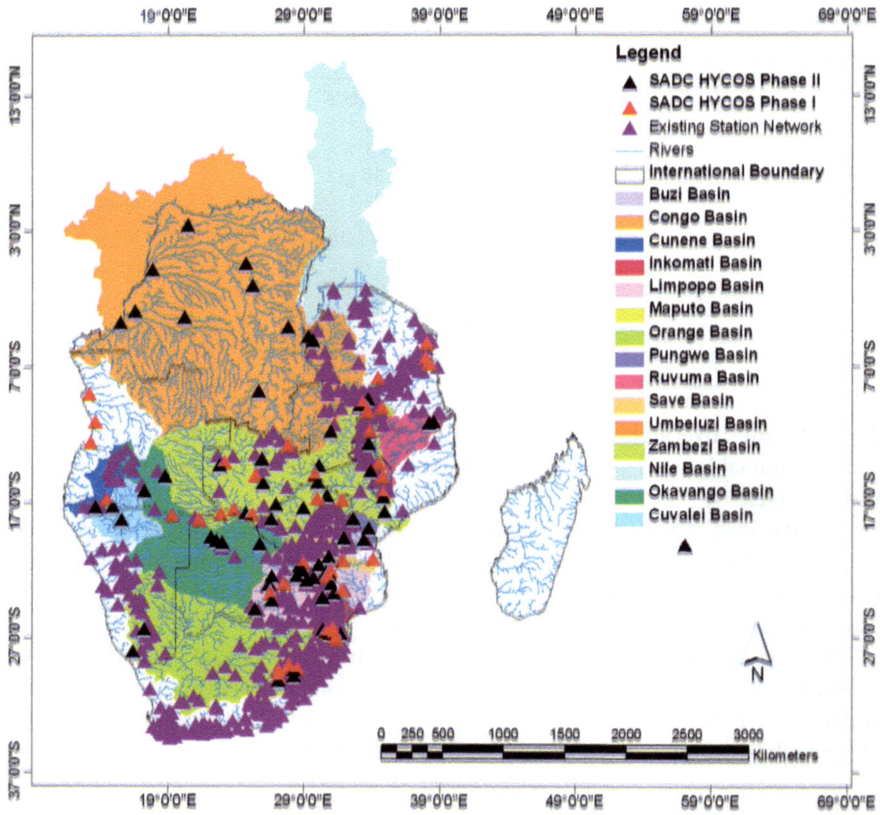

Fig. 5.4 SADC HYCOS monitoring stations. *Source* Nijsten et al. (2018)

member nations' adherence to Protocol legislation, and general water resources (SADC-HYCOS 1996).

The SADC HYCOS started with fewer stations in Phase 1 and added more during Phase 2. The project experienced problems with stations that could not be maintained properly due to lack of technical capacity in some of the participating countries. Some of the countries accumulated the data but did not have the know-how to use the data in the planning and decision-making.

Other challenges experienced both in the HYCOS and the region at large is the limited collaboration and/or communication. The collaboration helps to close gaps created by distance—e.g., monitoring of availability and quality water resources across countries sharing a specific water source. Communication is often hampered by language barriers—e.g., Cuvelai Watercourse Commission (CUVECOM) with two riparian states, Angola and Namibia, one using Portuguese and one English as an official language.

Other initiatives reported by Altchenko et al. (2017) include cooperation by the Stampriet TBA countries and the Ramotswa TBA countries which might succeed to form joint groundwater monitoring programmes for the Region.

5.3.3 Groundwater Management

"Some hydrogeological characterisation has been done across Africa, particularly in rural regions that are prone to drought, but more often for groundwater resource evaluation and development than for groundwater management" (Nijsten et al. 2018).

SADC (2010) also noted that, despite the reliance on groundwater by roughly 80% of Africa's rural population, when it comes to management, groundwater as a resource frequently receives less consideration than surface water.

A variety of uncertainties including socio-economic development and climate change, and unsuccessful groundwater governance structures (Knüppe 2011) could be obstructing the sustainable groundwater management in Southern Africa. The observation was made by Turton et al. (2006) that water resource management is almost always transboundary is crucial and should be considered when making decisions at local and national level.

According to the United Nations (1992), the water must be managed more efficiently than it has been managed in the past because the water will have a negative impact on ecosystems, industrial development, human health and welfare, and food security. Braune and Xu (2008) reported that groundwater is a primary supply for livestock consumption and small-scale crop irrigation in several SADC Member States. Braune and Xu (2008) also estimated that around 14% of the entire agricultural irrigated area uses groundwater in Namibia, compared to 18% in South Africa and 56% in Botswana. The chance to enhance the lives of the poor community members in Africa using groundwater for small-scale agriculture irrigation is enormous, and it may grow in the near future (FAO 2003). However, it is also important to take into account the groundwater-dependent industries including rural development, healthcare, agriculture, and the environment (Braune et al. 2008). Table 5.2 gives summary of the groundwater dependency for SADC member states.

Initiatives by the African Ministers' Council on Water (AMCOW) "to encourage river basin organisations to institutionalise groundwater management in order to ensure regional ownership of the new development" are crucial to creating synergies within the southern African region.

It should be noted that, creating synergies within the region for effective management of groundwater will also depend on availability and accessibility of groundwater management tools including data and policies; regional cooperation, networks, and governance structures such as transboundary water commissions; and capacity building institutions and networks.

Table 5.2 Summary of groundwater dependency in the SADC Region

Member state	Rural	Urban	Agriculture	Industry	Overall dependency
Angola	Moderate	Moderate	Moderate	Minor	Moderate
Botswana	Major	Moderate	Major	Major	Major
D R Congo	Minor	Minor	Minor	Minor	Minor
Lesotho	Moderate	Moderate	Minor	Minor	Minor
Malawi	Major	Minor	Moderate	Minor	Moderate
Mauritius	Moderate	Moderate	Moderate	Moderate	Moderate
Mozambique	Moderate	Moderate	Minor	Minor	Moderate
Namibia	Major	Major	Major	Major	Major
Seychelles	Moderate	Moderate	Minor	Minor	Minor
South Africa	Major	Moderate	Moderate	Moderate	Moderate
Swaziland	Major	Minor	Moderate	Minor	Moderate
Tanzania	Major	Moderate	Moderate	Minor	Moderate
Zambia	Moderate	Moderate	Minor	Moderate	Moderate
Zimbabwe	Major	Moderate	Major	Moderate	Major

Source Adjusted from Wellfield Consulting Services and British Geological Survey (2003)

5.3.3.1 Availability and Accessibility of Groundwater Management Tools

Groundwater Data

Gaye and Tindimugaya (2019) reported that "the existing state of knowledge and the lack of data on groundwater systems provide substantial obstacles to the sustainable exploitation of these resources." SADC (2009) noted that "more data should be gathered and presented on groundwater issues in ways that decision-makers, policymakers, and politicians can easily comprehend." Braune et al. (2008) also noted that the relatively low significance accorded to groundwater is linked to insufficient groundwater monitoring, exploration, and data gathering.

Though crucial for effective management of the groundwater resource, the monitoring data including flow regimes, recharge rates, and quality controls is patchy (Gaye and Tindimugaya 2019) and fragmentary (Nijsten et al. 2018), thus making effective management almost impossible.

Initiatives by the Global Water Partnership to develop the IWRM Toolbox are appreciated. According to Braune et al. (2008), the toolbox can be used to assess the enabling environment, institutional development, and management measures within the Integrated Water Resources Management (IWRM) framework.

Other initiatives to ensure availability and accessibility of data/information include the SADC Water Sector International Cooperating Partner Collaboration Portal, SADC Groundwater Grey Literature Archive, SADC Groundwater and Drought Management Portal, SADC Economic Accounting for Water Use Project, and the SADC GeoNetwork Portal–Water Datasets. All these can be accessed from https://www.sadc.int/sadc-secretariat/.

Policies

The SADC has a comprehensive policy framework for ensuring sustainable water resource management. According to Matchaya (2009), the SADC Regional Water Policy and the SADC Protocol on Shared Watercourses are included in the framework. The Regional Water Policy (RWP), the Regional Water Strategy (RWS), the Protocol, and other subregional and national policy and strategy processes (SADC 2005) are synchronised (see Fig. 5.5).

The main challenge faced by the SADC revolves around ensuring implementation by the member states. However, most SADC member states have not considered the groundwater resource in the national policies and strategies.

Even though the SADC convention on joined watercourses is intended to support regional collaboration for the utilisation, safeguard, sustainable management, and coordinated use of joined watercourses (Matchaya 2009), it is evident that, the implementation is not easily realised. Most transboundary shared watercourse commissions still grapple with monitoring and capacity-related limitations.

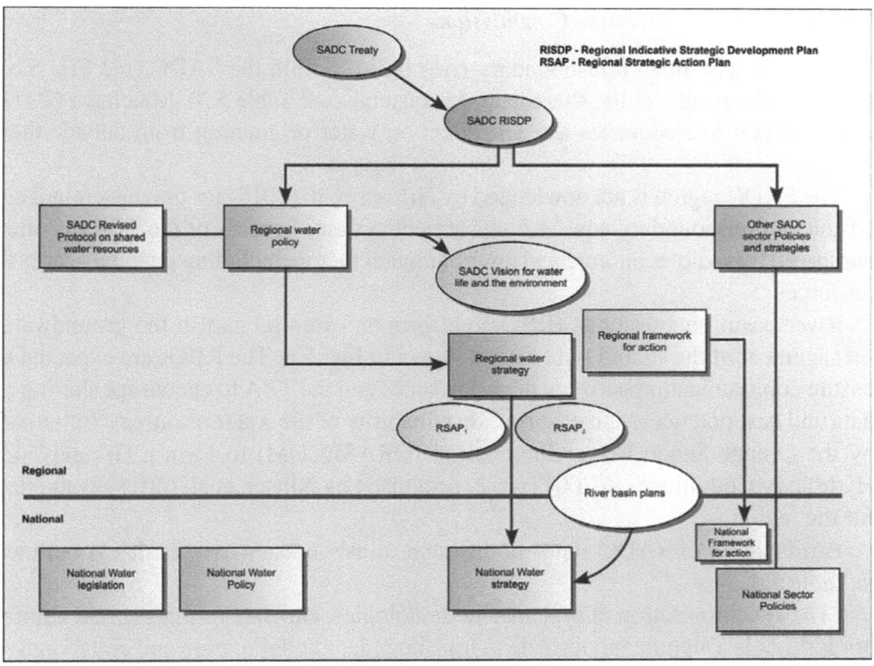

Fig. 5.5 The relationship between regional water policies and strategies. *Source* SADC (2005)

5.3.3.2 Regional Cooperation, Networks, and Governance Structures

Knüppe (2011) confirmed that "supporting concepts, like tools for collaboration, participation, and information networks, have been created to make it easier to apply adaptive water management strategies and, as a result, bring about institutional change in the groundwater management political arena." The SADC relies on transboundary shared watercourse commissions and transboundary aquifer systems to ensure implementation of the SADC protocol on jointed watercourses, the Regional Water Policy (RWP), the Regional Water Strategy (RWS), Protocol and related policies and strategies.

The policies and strategies mentioned above are crucial in guiding the decisions that help to protect quantity and quality of water courses. However, effective implementation of the SADC transboundary shared watercourse strategies depends on the socio-economic status of the river basin communities. Collaborations also play crucial role. Thus, establishment of River Basin Organisations (RBOs) helps to ensure effective implementation of the strategies. The lack or limited financial resources always hamper the implementation. To ensure effective management of shared watercourses, RBOs made arrangements, and developed frameworks, instruments, and tools that guide decision-making for effective implementation of strategies.

Transboundary Watercourse Commissions

There are fifteen main transboundary river basins within the SADC (see Fig. 5.6). These are the results of the shared aquifer systems (see Table 5.3). Matchaya (2009) observed that five countries are dependent on water originating from outside their borders.

The SADC region is acknowledged by Nijsten et al. (2018) for having a relatively advanced transboundary aquifer management system. Success of the TBAs is often hampered by socio-economic and environmental factors including unequal access to resources.

River basin organisations (RBOs) perform an essential part in the groundwater management of the shared river basins shown in Fig. 5.6. The RBOs are expected to ensure communication between the basin states and the TBA to encourage sharing of data and best practices; and ensuring sustainability of the water resources. Initiatives by the Orange-Senqu River Commission (ORASECOM) to form a Groundwater Hydrology Committee (in 2007) were recognised by Nijsten et al. (2018) as the first for the region.

Participation of SADC states in different transboundary river basins is outlined in Table 5.4.

"The synchronisation of systems, methodologies, and data formats across aquifer borderlines is a significant obstacle to transboundary collaboration and collaborative monitoring" (Nijsten et al. 2018). The condition is further confounded by partaking in RBOs outside the main development community. For example, because of the geographic location, Angola, Tanzania and the Democratic Republic of Congo (DRC) also have a footprint in the RBOs whose administrations are largely outside the SADC. These include the Nile Basin Initiative (NBI) for Tanzania and the DRC;

Fig. 5.6 River basins within the SADC. *Source* SADC (2010)

and the Commission Internationale du Bassin Congo-Oubangui-Sangha (CICOS)—Angola and the DRC.

5.3.3.3 Capacity Building Institutions and Networks

There are several groundwater associated networks and building capacity institutions in Africa. Institutions/networks perform a vital role in the improvement of

Table 5.3 Known transboundary aquifer systems shared by SADC countries

Aquifer Riparian State	Cunene Coastal	Cuvelai	Congo Coastal	Congo Intra-Catonic	Gariep Coastal	Incomati Coastal	Kagera	Kalahari	Karoo Sedimentary	Kenya-Tanzania Coastal	Kilimanjaro
Angola	X	X	X	X							
Botswana								X			
Congo (DRC)			X	X							
Lesotho									X		
Madagascar											
Malawi											
Mauritius											
Mozambique						X					
Namibia	X	X			X			X			
South Africa					X	X		X	X		
Swaziland						X					
Tanzania							X			X	X
Zambia				X							
Zimbabwe											
States sharing	2	2	2	3	2	3	1	3	2	1	1

Aquifer Riparian State	Limpopo Granulite-Gneiss Belt	Nata-Gwaai	Okavango	Okavanga-Epuriko	Pafuri Alluvial	Pomfret-Vergelegen Dolomic	Ramotswa Dolomite	Rovuma Coastal	Shire Valley Alluvial	Tull Shashe	Upper Rovuma	Shared Aquifers
Angola			X									5
Botswana	X	X	X	X		X	X			X		8

(continued)

Table 5.3 (continued)

Aquifer Riparian State	Limpopo Granulite-Gneiss Belt	Nata-Gwaai	Okavango	Okavanga-Epuriko	Pafuri Alluvial	Pomfret-Vergelegen Dolomic	Ramotswa Dolomite	Rovuma Coastal	Shire Valley Alluvial	Tull Shashe	Upper Rovuma	Shared Aquifers
Congo (DRC)												2
Lesotho												1
Madagascar												0
Malawi									X			1
Mauritius												0
Mozambique					X			X	X		X	5
Namibia			X	X								6
South Africa	X				X	X	X			X		9
Swaziland												1
Tanzania								X			X	5
Zambia		X	X									3
Zimbabwe	X	X			X					X		4
States sharing	3	3	4	2	3	2	2	2	2	3	3	

Source Turton et al. (2006)

Table 5.4 Transboundary river basin organisations with membership of SADC states

River basin	Member states
Limpopo Watercourse Commission–LIMCOM	Botswana Mozambique South Africa Zimbabwe
Cuvelai Watercourse Commission–CUVECOM	Angola Namibia
Orange-Senqu Commission–ORASECOM	Botswana Lesotho Namibia South Africa
Zambezi Watercourse Commission–ZAMCOM	Angola Botswana Malawi Mozambique Namibia Tanzania Zambia Zimbabwe
Okavango River Basin Commission–OKACOM	Angola Botswana Namibia
Commission Internationale du Bassin Congo-Oubangui-Sangha–CICOS	Angola Gabon Cameroon Democratic Republic of Congo Republic of the Congo Central African Republic
Nile Basin Initiative–NBI	Burundi Democratic Republic of Congo Egypt Ethiopia Kenya Rwanda South Sudan Sudan Tanzania Uganda

human capacity and additional groundwater related research and technical knowledge and skills to grow and manage African resources. These were identified by Gaye and Tindimugaya (2019) as Southern Africa Development Community (SADC), Global Water Partnership-Southern Africa (GWP-SA), Africa Groundwater Network (AGW-Net), Capacity Building Center on Integrated Water Resources Management (WaterNet), Capacity Building Network for Integrated Water Resources Management (Cap-Net) and the new SADC Groundwater Management Centre (see Table 5.5). It

Table 5.5 Participation in capacity building institutions and networks

Member state	SADC WaterNet	GWP-SA	AGW-Net	Cap-Net	SADC GMC	SADC GMI
Angola	Y	Y	Y	Y	Y	Y
Botswana	Y	Y	Y	Y	Y	Y
Comoros	Y		Y	Y	Y	Y
Democratic Republic of Congo	Y	Y		Y	Y	Y
Eswatini	Y			Y	Y	Y
Lesotho	Y	Y	Y	Y	Y	Y
Madagascar	Y		Y	Y	Y	Y
Malawi	Y	Y	Y	Y	Y	Y
Mauritius	Y		Y	Y	Y	Y
Mozambique	Y	Y	Y	Y	Y	Y
Namibia	Y	Y	Y	Y	Y	Y
Seychelles	Y		Y	Y	Y	Y
South Africa	Y	Y	Y	Y	Y	Y
Swaziland	Y	Y	Y	Y	Y	Y
United Republic of Tanzania	Y		Y	Y	Y	Y
Zambia	Y	Y	Y	Y	Y	Y
Zimbabwe	Y	Y	Y	Y	Y	Y

'Y' signifies YES

is important for SADC member states to also use the networks to share best practice and to improve their groundwater management profiles.

The lack of capacity in terms of managing, regulating, and protecting groundwater resources also observed by SADC (2009) is evident in most SADC countries. However, Knüppe (2011) identified a stakeholder responsive and comprehensive program and strategy to capacity building by SADC and AMCOW. The initiative was aimed to incorporate the groundwater resource management into the region's IWRM plan.

5.3.4 *Groundwater Sustainability Initiatives*

Several groundwater sustainability initiatives have been taken by different projects, programmes, and networks to address the groundwater quality and quantity challenges in the region. Participation by SADC member states in the programmes/

initiatives is summarised in Table 5.6. These include the production of hydrogeological data by IGRAC and the development of a web-based system for the SADC region's TBA information storage and display. The ISARM's experience was used to compile with the guiding principles for the evaluation of TBAs across disciplines (IGRAC & UNESCO-IHP 2015).

The AMCOW serves the whole continent of Africa. The AMCOW is the initiator of the resolution on the institutionalisation of groundwater in the continent (Braune et al. 2008).

Currently, South Africa, Namibia, and Botswana are cooperating on the Stampriet TBA, in addition, the Ramotswa TBA shared by South Africa and Botswana might be the region's first cooperative groundwater monitoring programme (Altchenko et al. 2017; UNESCO-IHP & IGRAC 2015).

The SADC Dialogue in 2009 was meant to analyse and review how best the Integrated Water Resources Management (IWRM) strategy for managing water resources

Table 5.6 Participation by SADC member states

Member state	AMCOW	SADC WIN	ISARM	IGRAC	Regional transboundary water assessment	Stampriet TBA
Angola	Y	Y				
Botswana	Y	Y	Y	Y	Y	Y
Comoros	Y	Y				
Democratic Republic of Congo	Y	Y				
Eswatini	Y	Y				
Lesotho	Y	Y				
Madagascar	Y	Y				
Malawi	Y	Y				
Mauritius	Y	Y				
Mozambique	Y	Y				
Namibia	Y	Y	Y	Y	Y	Y
Seychelles	Y	Y				
South Africa	Y	Y	Y	Y	Y	Y
Swaziland	Y	Y				
United Republic of Tanzania	Y	Y				
Zambia	Y	Y				

'Y' signifies YES

might help persuade different sectors to develop, utilise, and maintain groundwater resources sustainably (SADC 2009).

5.4 Conclusion

This study highlighted the initiatives to address the challenges linked to climate change impacts, governance of groundwater resources, groundwater resource management, quantity and quality, limited capacity (particularly technical capacity), availability and user-friendliness of data, and groundwater management tools (including policies) in Southern Africa.

The sustainability status of the groundwater resource was assessed in terms of groundwater availability, quality and quantity monitoring, availability and accessibility of groundwater management tools, regional cooperation, networks and governance structures, capacity building institutions and networks, and groundwater sustainability initiatives. The study highlighted initiatives to address the challenges linked to climate change impacts, governance of groundwater resources, groundwater resource management, quantity and quality, limited capacity (particularly technical capacity), availability and accessibility of data, and groundwater management tools in the region.

Therefore, it is necessary to utilise different methods or management strategies to ensure the sustainable management and development of groundwater resources. These include "establishing groundwater monitoring systems, managing transboundary aquifers, addressing the effects of climate change on groundwater, determining how increased abstraction from various types of aquifers will affect the sustainability of groundwater, and developing groundwater management capacity", (Gaye and Tindimugaya 2019) (Fig. 5.7).

Implementation of the following recommendations (adapted from Beetlestone 2009) is crucial:

- A good knowledge base: understanding of groundwater-climate interactions, vulnerability, groundwater flow, amount and location of recharge, prediction on impact of abstraction.
- The sub-committees within river basin organisations that generate a network of institutions which create interaction between local, regional, national, and global levels.
- A link of experts from countries are part of the network as custodians of the technical knowledge.
- Instruments or tools such as code of good practice, harmonised procedures, models, regional hydrogeological map, etc.
- Training and capacity building that include community members, to develop national groundwater management structures.

Fig. 5.7 Groundwater resource governance. *Source* Foster (2006)

Acknowledgements The authors acknowledge in-kind support by the North-West University and the UNESCO Chair on Sustainable Water Research for Climate Adaptation in Arid Environment at the Namibia University of Science and Technology.

References

Altchenko Y, Genco A, Pierce K, Woolf R, Nijsten G, Ansems N, Magombeyi M, Ebrahim G, Lautze J, Villholth KG, et al (2017) Resilience inthe Limpopo basin: the potential role of the transboundary raotswa aquifer–final draft. [ResearchReport] USAID Southern Africa. hal-02329681

Barnard M (2005) Uranium in groundwater 'not serious': rössing. https://www.namibian.com.na/index.php?id=14669&page=archive-read

Beetlestone, P. (2009) Challenges to transboundary aquifer management in the SADC region. http://archive.iwlearn.net/sadc-groundwater.org/upload/file_372.pdf. Accessed 22 December 2021

Braune E, Xu Y (2008) Groundwater management issues in Southern Africa—an IWRM perspective. IWRM special edition. Water SA 34(6), 699–706. [Online]. http://www.wrc.org.za

Braune E, Hollingworth B, Xu Y, Nel M, Mahed G, Solomon H (2008) Protocol for the assessment of the status of sustainable utilization and management of groundwater resources with special reference to Southern Africa. Water Research Commission Report TT 318/08, July, ISBN 978–1–77005–712–8

BGS, DFID, NERC, ESRC (2017) Africa groundwater atlas. Online. https://www.bgs.ac.uk/africa groundwateratlas/. Accessed 25 June 2017

Christelis G, Struckmeier W (2011) Groundwater in Namibia: an explanation to the hydrogeological Map, 2nd Edn Cape Town: Department of Water Affairs, Ministry of Agriculture, Water and Rural Development; The Geological Survey of Namibia, Ministry of Mines and Energy: The Namibia Water Corporation and the Federal Institute for Geoscience and Natural Resources on behalf of the German Ministry of Economic Cooperation (22)

(PDF) Sustainability of Groundwater for Irrigation Purposes in Semi-arid Parts of Namibia. Available from: https://www.researchgate.net/publication/356809708_Sustainability_of_Groundwater_for_Irrigation_Purposes_in_Semi-arid_Parts_of_Namibia [accessed Dec 17 2023]
Conley A, Van Niekerk P (1998) Sustainable management of international waters: the orange river case. In: Savenje H, van der Zaag P (eds) The management of shared river basins: experiences from the SADC and EU. Ministry of Foreign Affairs, The Hague, The Netherlands
CSIR (1997) An assessment of the potential environmental impacts of the proposed aquifer recharge scheme on the Khan River, Namibia. Final report by CSIR Division of Water, Environment & Forest Technology to Rössing Uranium Limited, Swakopmund, Namibia
Development Bank of Southern Africa (DBSA) (2016) Knowledge management and innovation (KMI) UNIT. DBSA Research Agenda, Issues and Opportunities in Southern African Water Security
Earle A, Malzbender D, Turton AR, Mazungu E (2005) A preliminary basin profile of the orange-senqu river. AWIRU, University of Pretoria, Pretoria.
Elema NM (2019) The contribution of The water research fund for Southern Africa (WARFSA) to knowledge production and policy in the SADC water sector, PhD in Science and Technology Studies. Dissertation, Stellenbosch University
Food and Agricultural Organization of the United Nations (2003) groundwater management—the search for practical approaches. Water Reports 25. FAO, Rome
Foster S (2006) Groundwater for life and livelihoods—the framework for sustainable use. Presentation at 4th World Water Forum, Mexico. www.iah.org
Gaye CB, Tindimugaya C (2019) Review: challenges and opportunities for sustainable groundwater management in Africa. Hydrogeol J 27:1099–1110. https://doi.org/10.1007/s10040-018-1892-1
IGRAC, UNESCO-IHP (2015) Transboundary aquifers of the world [map]. 15 Scale 1: 50 000 000, edition 2015. IGRAC, Delft, Netherlands
Kgabi NA (ed) (2020) Integrated transboundary water–climate management tools in water security and climate adaptation in Southern Africa, vol 1. AOSIS, Cape Town, pp i–224
Kohrs B (2008) Uranium—a blessing or a curse? What you need toknow about the uranium industry in Namibia (A factsheet by earthlife Namibia)
Knüppe K (2011) The challenges facing sustainable and adaptive groundwater management in South Africa. Water SA 37(1):67–79. http://www.wrc.org.za
Kringel R, Wagner F, Klinge H (2010) Groundwater quality assessment in the Khan and Swakop River catchment with respect to geogenic background concentrations of dissolved uranium. BMZ No: 2008.2007.6. Hannover, Germany
Lange G, Mungatana E, Hassan R (2007) Water accounting for the Orange River Basin: an economic perspective on managing a transboundary resource. Ecol Econ 61:660–670
Matchaya G (2009) Land ownership security in Malawi. Afr J Agric Res 4(1):001–013
Matchaya G, Nhamo L, Nhlengethwa S, Nhemachena C (2019) An overview of water markets in Southern Africa: an option for water management in times of scarcity. Water 11:1006. https://doi.org/10.3390/w11051006
Molapo P, Pandey SK, Puyoo S (2000) Groundwater resource management in the SADC region: a field of regional cooperation. In: IAH 2000 Conference, Cape Town
Nijsten G, Christelis G, Villholth KG, Braune E, Gaye CB (2018) Transboundary aquifers of Africa: review of the current state of knowledge and progress towards sustainable development and management. J Hydrol Reg Stud 20:21–34
SADC (2005) Regional strategic action plan for integrated water resource management
SADC (2009) SADC: multi—stakeholder water dialogue report. May, Johannesburg, South Africa
SADC (2010) Regional water infrastructure programme "Watering life together, forever". Brochure. https://www.sadc.int/files/7213/5293/3540/Regional_Water_Infrastructure_Programme_brochure.pdf
SADC (2000) Revised protocol on shared watercourses in the Southern African Development Community. https://www.sadc.int/files/3413/6698/6218/Revised_Protocol_on_Shared_Watercourses_-_2000_-_English.pdf

SADC-HYCOS (1996) SADC-HYCOS project document. Geneva

Setlhogile T, Arntzen J, Pule OB (2017) Economic accounting of water: the Botswana experience. Phys Chem Earth 100:287e295

Turton A, Patrick M, Cobbing J, Julien F (2006) The challenges of groundwater in Southern Africa. Navigating Peace, No. 2, August. Woodrow Wilson International Centre for Scholars

UNDP GEF (2008) Preliminary transboundary diagnostic analysis for the Orange-Senqu River Basin. UNDP GEF

UNITED NATIONS (1992) Agenda 21, Chapter 18. United Nations, conference on environment and development. 3–14 June 1992. Rio de Janeiro, Brazil

Wellfield Consulting Services Pty Ltd & British Geological Survey (2003) SADC: Regional Situation Analysis. Report no. RFP # WB 1861–571/02

Chapter 6
Microplastics Contamination in Urban Water System: A Risk Assessment and Mitigation

Jijoe Samuel Prabagar and Shivaraju Harikaranahalli Puttaiah

Abstract Microplastics are considered to be pollutants of emerging concern. A systematic method is required since it is a complex issue to attain the overview and negate management solutions that have peripheral outcomes. An important query regarding microplastics is the number of risks they can cause to the urban water system. No risk assessment is available that considers the diverse properties of microplastics versus the natural particles. Herein, current trends of microplastic risk assessment are illustrated along with the impact of microplastics on chemical risks in the water system. Risk assessment frameworks were also summarized which include consistent risk characterization by aligning exposure and effect thresholds. Further in this review, various mitigation strategies were summarized to limit microplastic contamination in the urban water system. Numerous promising solutions are present for microplastic mitigation, irrespective of preventative actions and treatment depending on the source. The responsibility aimed at respective measures and influence on the entire urban water system should be considered when developing a management plan for microplastics.

6.1 Introduction

An increase in anthropogenic activities in recent years has led to several environmental-related pollutions that are mostly concerted in urban areas through stormwater and wastewater as pathways (Fältström and Anderberg 2020). In recent years, microplastics have surfaced as a new class of pollutants occurring in different

J. S. Prabagar · S. Harikaranahalli Puttaiah (✉)
Department of Environmental Sciences, JSS Academy of Higher Education and Research, Mysuru 570015, India
e-mail: shivarajuhp@jssuni.edu.in

S. Harikaranahalli Puttaiah
Center for Water, Food and Energy, GREENS Trust, Harikaranahalli, Dombaranahalli Post, Turuvekere Taluka, Tumkur District, Karnataka 572215, India

S. Suriyanarayanan et al. (eds.), *Water Management in Developing Countries and Sustainable Development*, Water Resources Development and Management,
https://doi.org/10.1007/978-981-99-8639-2_6

types of water bodies. Microplastics are particles that are known to have a size below 5 mm down to 1 μm (Hartmann et al. 2019). Microplastics are referred to as "primary" if they are produced with intent at the microscale and are commonly found in personal care products (PCPs) as well as scrubbers used in industries (Wang et al. 2019). Larger plastics, like fishing nets, plastic bags or tires wear undergo weathering to form secondary microplastics and are shaped irregularly (Lehtiniemi et al. 2018). Rapid research regarding microplastics has increased in recent days rapidly with theoretical and pragmatic advancements. But, research on microplastics is still in the early stages with indeterminate and inconsistent results. Due to this reason, it is challenging to structure an adequately consistent assessment of the source and pathways of microplastics which can be equipped for widespread and effectual reduction approaches.

Marine and land sources are the main originating foundations of microplastics. Inadequate wastewater management, city traffic, and manufacturing spills are considered to be the major land-based microplastic sources in an urban environment (Jan Kole et al. 2017; Sun et al. 2019). Similarly, laundry of synthetic substances from textile fibres and microbeads found in PCPs are a few other sources of microplastics (Sun et al. 2020). Without the aid of advanced techniques, current conventional wastewater treatment plants are capable of retaining upto 99% microplastics (Conley et al. 2019; Tang and Hadibarata 2021). The conveyance and dispersal of microplastics in urban water systems depend on their physical properties and their environmental circumstances (Patil et al. 2020). Plastics, which are the main source of microplastics, are sustainable if recycled and managed suitably. Though plastics are recycled and are managed adequately, a significant amount of them ends up in the environment (Gregory 1991). In a review, it is stated that approximately out of 8300 million Mt of plastics that are globally produced, 6300 million Mt are categorized as waste, of which 9% were recycled, 12% were incinerated, and the rest 79% were disposed of as landfills (Geyer et al. 2017). Inefficient waste management due to various factors such as lack of manpower, economic assistance, and resource necessities are the reason for the wide gap resulting in the discharge of microplastics in urban waters.

Assessing the risk of microplastics can be challenging. Studies have shown that microplastics are found in the atmosphere, soil, urban water system, aquatic biota, edible goods, and the human body (He et al. 2018; Leslie et al. 2022; Venghaus and Barjenbruch 2017b). Microplastics can be assorted and nebulous. Due to their diverse exposure properties and techniques used by researchers regarding impact assessment, makes it indeterminate to determine the risk of microplastics. Challenges regarding the same have been addressed in the various review of the literature (Jan Kole et al. 2017; Lebreton et al. 2018; Leslie et al. 2022; Wu et al. 2021). Still, researchers are missing some critical aspects when studying the risk factors of microplastics. In general, microplastics are termed as environmental contaminants containing distinctive features, like unequalled complexity (He et al. 2021). But when associated with their natural equivalents, the properties of microplastic particles do not vary much. Microplastic researchers are now motivated by perseverance profiles, fate progressions, and the mechanism of these particles. Reflecting on the intricacy of microplastics and their natural counterparts, a clear picture of microplastic risks posed to the urban environment can be determined. Developing mitigation strategies for microplastic removal in urban waters makes it difficult due to the limited and ambiguous information as well as a lacking framework (Pettipas et al. 2016). However, handling microplastics is of utmost priority and various mitigation methods have been suggested and executed. But assessments regarding the same were not studied from a wider perception. It is vital to develop a pioneering outlook with ambition to structure inclusive and effectual mitigation strategies. To structure such an outlook, microplastic analysis can benefit through a systematic approach. A systematic perspective can avoid the movement of pollutants within the urban area. Herein, we discuss risk assessment frameworks and mitigation strategies that can further help in assessing the microplastics in urban waters.

6.2 Risk Assessment of Microplastics

6.2.1 Existing Trends of Risk Assessment

Though the literature regarding the exposure of microplastics to human exists, risk assessment regarding the same is lacking as well for urban waters (Cox et al. 2019; Zhang et al. 2020a). But, the risk assessment of microplastics in urban waters was implemented based on intensities of exposure and threshold effect concentration (Everaert et al. 2018; Zhang et al. 2020b). A literature survey or data modelling is equipped to understand the exposure concentration (Everaert et al. 2018). The exposure data obtained are often of inadequate quality and are challenging to compare due to the implementation of various analytical methods (Bailey et al. 2021; Koelmans et al. 2020; Mattsson et al. 2021). Similarly, comparability and quality inadequacies are present, and when combined the consistency of the risk characterization is comprised.

Species sensitivity distributions (SSDs) are employed to analyze the microplastics present in urban waters by determining their concentration in a set of species (Posthuma et al. 2001). At the community level, SSDs enhance the significance of microplastic risk assessment and are engaged to achieve the hazard concentration for 5% of species (HC_5). In general, for a particular precise category of stressor, SSDs utilize the effect threshold value including the end point for species greater than ten whereas the environmental variables are kept constant. But for microplastics, such data are not accessible yet (Koelmans et al. 2022). In reality, microplastics characterize a range of particle variety, but on a laboratory scale, monodisperse units are ingested resulting in interference of intake hence restraining bioavailability. Due to the action of diverse means of action, microplastic action triggers signify that HC_5 values attained are indefinite with regard to determining the appropriate particles and related outcomes (De Ruijter et al. 2020).

6.2.2 Approaches to Developing Risk Assessment

The inadequate comparability and quality of microplastic risk assessment can be resolved by coordinating the structures among studies, sternly following the quality standards, and employing accurate microplastic concentration in laboratory studies. Microplastic probability density functions (PDFs) are utilized as a substitute for repeating the mentioned tests from the beginning which can rescale and support the now accessible data regarding the exposure and threshold concentration, including that of SSDs (Koelmans et al. 2020; Kooi et al. 2021; Mohamed Nor et al. 2021). The uncertainty of various causatives can be delivered by initial recognition of the effect mechanism, and later affiliating and enumerating the appropriate exposure and effect threshold data. For example, when 'food dilution' is recognized as a particular effect mechanism of microplastics, the equivalent dose metric and the exposure and effect threshold data can be correlated and enumerated (Koelmans et al. 2020). The explained method is well used in toxicology and can be implemented in microplastic assessment in urban waters, which can include assessing the physicochemical effects. The elaborated theory can be used for probable risk assessments established on food-web modelling.

6.2.3 Chemical Risks

In urban waters, including the risks of microplastics and their physical impacts, they can also concurrently trigger chemical impacts given if the exposure to chemicals is elevated. It is very common to find microplastics with chemicals like additives and other pollutants in urban waters which are stated as plastic chemicals (Burns and Boxall 2018; Kögel et al. 2020; Wiesinger et al. 2021). However, the term used is indefinite since phase distribution confirms the presence of identical chemicals that are as well found in further environmental substances, like, residue, soil, water, algae, and so on (Birdwell et al. 2007; Burns and Boxall 2018; Gassel and Rochman 2019; Liping et al. 2014). Relying on the concentrations of exposure, the biota stands at risk of contact with these chemicals originating from any medium. For example, the major source of perilous chemicals is microplastics where urban waters can act as a medium, which can result in mortality in pertinent environmental circumstances.

For a long time, it was theorized that during the process of ingestion, the desorption of chemicals via microplastics can enhance chemical exposure resulting in higher risks of chemicals surrounding the urban biota which is termed a microplastic vector effect (Koelmans et al. 2022). But to date, there has been no risk reported and the microplastic vector effect is considered to play no vital role in major habitats (Wang et al. 2020). This conclusion can be due to a couple of reasons. First, chemical exposure is reflected in major experimental studies through the ingestion of microplastics, ignoring the equivalent uptake route. The comparative significance of microplastics as a vector is overvalued aimed at uptake and transport, but underrates the overall exposure to chemicals (Gouin 2021). Considering the overall uptake courses, uptake via microplastic ingestion is reckoned to be diminutive (Bakir et al. 2016). It can also indicate that toxicity testing of microplastic chemicals by neglecting the concurrent exposure of the same chemicals with different media can be of no use when imitating the environmental conditions in the laboratory. Second, the chemical concentration gradient among plastic and urban biotic receptors is improbable in most reported studies. For example, desorption related to plastic chemicals was investigated on animals with a hygienic system at unpretentiously elevated excessive plastic–water proportions and moderately low extract volumes (Capolupo et al. 2020). Since substantial dilution is due to the additives in urban areas, any perceived toxicity will not unambiguously specify if the risk occurs or not. It is evident that at hotspot sites enhanced concentration of microplastics ingesting can occur. But considering the standard chemical dilution in an urban environment, additional media uptake, minimal assimilation of plastic, food dilution effect, and no proper chemical concentration gradient in urban conditions, there is a possibility that before surpassing the toxicity threshold, the organism might starve.

It does not always mean that when ingested, the chemical effects are low when associated with physical element consequences. The effects of chemicals and their probable risks are greatly dependent on context that is manipulated by chemical mixture alignment, the concentration of chemicals, the path in which the chemical is transferred, or contact with water (Koelmans et al. 2022). It is best to assess the risk of microplastic hazards concerning physical and chemical toxicity, after the norm of discrete dose metrics as well as effective procedures, which are combined during the final stage of risk assessment.

6.2.4 Risk Assessment Framework of Microplastics

Risk assessment frameworks for microplastic contaminants were proposed by various researchers (Bucci and Rochman 2022; Gouin et al. 2019; Liu et al. 2022). General approaches and identification of challenges concerning microplastic complexity are given by these frameworks. The definite range of microplastics is streamlined by utilizing classes and bins intended for the content of the chemical, hence ignoring the discrepancies in the groups (De Ruijter et al. 2020; Gouin et al. 2019). Subsequently, the inadequate availability, as well as the non-comparability of data, was not considered when executing a strategy. Finally, a framework that was proposed lacked the effect threshold component, dose dependency was ignored as well as additives and environmental chemicals were imprecisely differentiated. But with rapid development in science over the past few years, challenges regarding the mentioned problems were addressed, as a result, a broad risk assessment framework of microplastics is now defined which applies to the urban environment. The framework is governed by three innovative elements: exemplifying toxicity appropriate to the distinctive particle by equipping PDFs; assessing if the exposure data obtained as well as the effect data is valuable by employing quality assurance and control screening method, and finally determining the plastic-associated chemical exposure via probable routes by using framework calculations. Herein, a brief insight into the new risk assessment frameworks for microplastics is provided.

Three classes of data are necessary for the risk assessment. Initially, in the process stage, problems are defined, exposure planning, assessment of effect, characterization of risks, and quality assurance are employed (Besseling et al. 2019). Next, looking at the definition of the problem, the detection of adversative effects mechanism on urban biota is necessary since microplastic contaminants are miscellaneous causing effects concurrent with altered effect mechanism (Kooi and Koelmans 2019). The mechanisms are related to an ecological or toxicologically pertinent endpoint which is measured by an ecologically or toxicologically relevant dose metric (ERM or TRM) (Kooi et al. 2021). Finally, since for a portion of the range, the dose metrics are appropriate, it can be obtained from the full range exposure data (Koelmans et al. 2020).

The defined problem, which can be referred to as population protection otherwise a leading species, originates from the regulatory domain. Next, exposure and effect assessment that depends on quality assurance and control, analytical methods management, and effect test method are carried out (De Ruijter et al. 2020). Fourier transform infrared analysis helps in enhancing the exposure and effect assessment and the cost of such techniques is expected to reduce once they are conventional. Both exposure and effect evaluation is crucial for assessing the risk of microplastics. Testing the miscellaneous microplastic mixture with consistency is considered to be a standard procedure, however, the variances in existing data can be bridged by data rectification and alignment techniques (Kooi and Koelmans 2019). For surveying risk assessment, tools such as quantitative quality assurance and criteria valuation are established (De Ruijter et al. 2020; Koelmans et al. 2019). The one who is assessing the risk can equip these tools to select the data relevant for assessment, which is vital when assessing the risk for regulatory policies.

The probable effects of microplastics are food dilution, oxidative stress, and internal damage (Bucci et al. 2020). For such mechanisms, appropriate dose metrics are necessary. For example, to represent the food dilution mechanism, the volume would be an appropriate dose metric whereas fibre toxicity represents internal damage, and the aspect ratio is pertinent (Braakhuis et al. 2014; Gualtieri 2021). Specific surface area can also be one such relevant metric for oxidative stress (Schmid and Stoeger 2016). Chemical-related risks tend to adhere to the plastic surface or can be absorbed as a whole. Considering the environmental scenario, the microplastic mixture may lead to multiple concurrent mechanisms, suggesting that various relevant dose metrics are present (Koelmans et al. 2020). It is observed that few metrics possess limitations for bioavailability justification. For illustration, when the fragment size is big to be consumed, food dilution applies to bioavailable segments that are ingestible for a range of microplastics (De Ruijter et al. 2020). The concept of bioavailability is a general topic related to the risk assessment of conventional chemicals and is well-relevant for microplastics too.

After the identification of the relevant effect mechanism and dose metrics, the exposure data must be aligned (Besseling et al. 2019). To assess the risk of chemicals, a standard protocol is accessible and has been further broadened to microplastic contaminants (Gouin et al. 2011; Koelmans et al. 2016). Aligning the effect of the microplastics on exposure data, first factors such as volume, mass, specific area, or aspect ratio must be used to express exposure (Kooi et al. 2021). A rational approximation for volume can be provided by mass if there is proportionality among the two since the density of microplastic contaminants found in the environment is almost 1 as well as limited variables other than microplastic properties such as shape and size (Gouin et al. 2011). By using the existing data, several relevant doses and exposure metrics can be enumerated by employing PDFs of microplastics (Mohamed Nor et al. 2021). Since microplastics are very diverse, it is a must to consider multiple effect mechanisms and therefore considering dose metrics is a necessity rather than exclusion, systematic microplastic range characterization is essential. A novel focal plane array Fourier transforms infrared analysis delivers the required information to attain such PDFs. However, the present PDFs related to microplastics found in the environment can be equipped for risk assessment (Koelmans et al. 2022). For example, comprehensive PDFs intended for volume, mass, area, aspect ratio, and surface area are delivered by characterization of microplastics experimented with from various environmental sectors. This illustrates a favourable technique for associating the dose of microplastics and characteristics to influence threshold and mechanism. For each effect mechanism, risk assessment is conducted independently, where the risk assessor after this will provide a precautionary method after going through the conclusion on the risk assessed.

6.3 Mitigation of Microplastics in Urban Water System

With the categorization of microplastics as new emerging contaminants, some rules and control strategies were proposed and implemented. Though the pathway, source, and endpoint of microplastics are undefined, certain sources leading to urban waters were identified which can provide vital data. Stormwater, wastewater, and combined sewage overflow are noted to be significant pathways to attaining surface waters (Bollmann et al. 2019). Further, the enormous volume of microplastics is either found in sewage sludge or storm sediments (Olesen et al. 2019). Eventhough the flow of microplastics is uncertain, various actions and approaches are reported. Table 6.1 illustrates the source, recommended strategies, and approaches to prevent microplastic contamination in urban waters.

Table 6.1 Insight into microplastic source, recommended strategies, occurrence, and responsible authorities

Source	Pathway	Recommended strategies	Preventive or decentralized	Responsible authorities	Movement of pollution
Microbeads	Wastewater	Alternate material/Ban	Preventive	Producers and authorities for assessment and accordance	Nil
Washing	Wastewater	• Change in textile construction • Change in laundry and consumption behaviour • Pre-washing	• Preventive • Decentralized • Decentralized	• Manufacturers • Residents • Producer and authority for guaranteeing compliance	• Nil • Nil • Wastewater to solid
Pre-production of plastics	Wastewater or stormwater	• Implementation of rules • Drain filters	• Preventive • Decentralized	• Authorities • Equipped in factories and authorities for monitoring	• Nil • Wastewater to solid
Aggregated	Stormwater	• Stormwater retention pond • Sweeping of streets	• Decentralized • Preventive	• Regional authorities or water and wastewater services • Regional authorities	• Stormwater to sediments • Stormwater to solid waste
Traffic	Stormwater	• Filters • Enhancing resilience and resistance	• Decentralized • Preventive	• Responsible for driving and concerned personnel to assure compliance • Producers	• Stormwater to solid waste • Nil
Artificial turfs	Stormwater	Alternation in maintenance	Preventive	Owner of the field and regional authorities to assuring compliance	Nil
Aggregated	Wastewater	Advanced treatment methods	Centralized	Water and wastewater utilities	Distributing at WWTP

6.3.1 *Proposed Strategies for Microplastics Mitigation in the Urban Water System*

Banning microbeads in cosmetic commodities has been proposed in various countries to control microplastic pollution (Ju et al. 2021). Since it addressed just a small portion of the source and not cosmetics as a whole, this ban received criticism. Synthetic materials such as fibres are released into urban waters due to washing (De Falco et al. 2019). By altering our consumption behaviours or washing practices, such microplastics released into urban waters can be prevented. In washing machines, a decentralized treatment option has been provided which is a filter to prevent the release of synthetic materials into the water system (Fältström and Anderberg 2020). The textile factories upstream are prewashing the fabrics to prevent contamination which is a common technique used in factories (Carney Almroth et al. 2018). It has been reported that during the preliminary washes, the release of fibres has reduced but a large number of fibres are released in the first washes (Cesa et al. 2020). No reduction in the release of fibres during initial washes was also reported. Hence, prewashing does not guarantee the elimination of fibres that are released from domestic waters. To reduce the release of fibre, advanced knitting techniques and prevention strategies at the manufacturing level must beemployed and a thread that reduces fibre loss must be used (Carney Almroth et al. 2018).

Plastic precursor factories have been reported to discharge an enormous quantity of plastic pellets into the environment (Karlsson et al. 2018). Drain filters are fixed in a few factories to curb the discharge of pellets. Even though laws exist to limit the release of pellets from production sources, they should be implemented with greater filtration efficiency. In urban areas, microplastic sources can be many that can end up in stormwater as well as in retention pond sediments signifying that the ponds surrounded by the urban area act as a sink (Olesen et al. 2019). In the upstream, microplastics are reported to be present in street dust including other contaminants and this can be prevented by sweeping the streets which can be restricted from entering into the water phase (Polukarova et al. 2020).

Microplastic pollution in urban areas can be due to road traffics. To prevent the wear and tear of the tyres, a preventive approach exists, but tyres properties are correlated. But increasing the resistance of the tyres can lead to rolling resistance which can alter the fuel efficiency and slip resistance which correlates to safety (Jan Kole et al. 2017). Installing filters can help limit the microplastic pollution that is caused by traffic found in stormwater (Venghaus and Barjenbruch 2017a). But installing filters has received criticism due to its intervention with traffic. So they were suggested to be fitted in stormwater wells, gas stations, and parking lots. Artificial turfs are constructed by stuffing them with old tyres. This can lead to microplastic pollution which eventually ends up in the urban water system and this can be prevented by altering the filling material and modifying the field management (Fältström and Anderberg 2020). This can comprise dumping snow and scrubbing clothes before exiting the field. Granulate traps can be equipped in stormwater wells present within the field or the surrounding drainage system.

Curbing the existence of macroplastics has been the main task for a long time since it breaks down to form microplastics, and then nanoplastics, and hence strategies to limit macroplastics are more compared to microplastics (Pettipas et al. 2016). In the case of urban waters, this mainly involves littering. The use of plastic bags in urban areas can be decreased by posing a ban, tax or fee which can reduce plastic littering (Knoblauch and Mederake 2021). Enhancing the urban waste management system, volunteering for behavioural change and economic incentives can help limit the consumption and littering of plastics (Ogunola et al. 2018). Though it can be challenging to assess the efficacy of the behavioural measures, reports have shown positive results on coastal wastes (Willis et al. 2018). In wastewater treatment plants (WWTP), additional treatment phases can help limit the release of microplastics in urban water systems (Carney Almroth et al. 2018). Though most of the microplastics are retained in conventional WWTP, they are still persistent in effluents and this can be limited by additional advanced treatment techniques in WWTP (Ngo et al. 2019). Some of the advanced treatment techniques that are used in WWTPs are disc filters, dissolved air filters, rapid sand filters, and membrane bioreactors (Esfandiari and Mowla 2021; Simon et al. 2019; Talvitie et al. 2017). Even though most of these techniques can retain higher microplastic contaminants, it is necessary to develop a novel technology exclusively meant for microplastics.

6.3.2 Understanding Microplastic Management in Urban Waters

From the three substance cases, a few practices can be noted that are vital for microplastic mitigation. One such characteristic is primary influence attempts varied within the substance. The control strategies of cadmium were preventively resulting in the curbing of cadmium residues but in various departments the copper values were limited. Treatment was water was focused downstream to limit the entry of pharmaceuticals discharge. The different mitigation methods for cadmium and copper are because copper is an essential material with high toxicity, but cadmium has no positive effects. Cadmium is rather predominantly found in essential materials as byproducts like zinc and phosphate rock (Van Der Voet et al. 1999). There are two motives for focusing on the centralized treatment of pharmaceuticals. First, imposing a ban or finding an alternative material can be very tough since restricting medicines can lead to ethical issues (Eggen et al. 2014). Secondly, if the collection system is installed previously and treatment is high-priced, centralized treatment can be considered since it is economically feasible (Libralato et al. 2012).

Most of the recommended strategies and approaches for microplastics are preventive because, unlike cadmium, microplastics are not necessities. But imposing a ban or finding an alternative material can be difficult since microplastics have different sources. Installing an additional treatment step in WWTPs is also not feasible since the difference between microplastics and pharmaceuticals are numerous. Similar to that of copper and cadmium, the conventional treatment techniques can retain most microplastics and a vast amount of microplastics enter the WWTPs due to large influents. Since microplastics have several sources, they all usually do not end up in WWTPs, unlike pharmaceuticals. It is evident that regulations regarding stormwater emission are widely not present and thus it is difficult to manage the pathway of microplastics which have no proper monitoring and structure.

Communal-level decisions are made for implementing preventive measures; however, real-world responsibilities are divided into two categories, producers and citizens. Special level authorities have the duty to practical control that producers assist and citizens are notified. Some responsibilities were also placed on producers regarding cadmium mitigation by replacing or curbing the production of cadmium. Microplastics entering the advanced centralized treatment processes are generally circulated in the WWTP or displaced to sludge where the wastewater utilities take charge. But for decentralized treatment, the responsibility is dependent on the source and various performers may have a duty. For instance, in a washing machine, fixing a filter is the producer's responsibility but accurate usage is the citizen's responsibility. Additionally, the responsibility of the authority is to control the compliance of the producer. To centralize waste management, the suggested mitigating strategy is to move microplastics from water to solid waste. A change in responsibility and coordination is required to manage such movements.

As reported in a study, the degree of flow and probable toxicity is not sufficient for a successful mitigation process (Sedlak 2017). Action must be taken if the solution is very expensive which can be inferred by financial costs as well as daily living costs. Banning microbeads cannot interrupt daily life drastically, but imposing a ban on synthetic textiles and driving can impact drastically. Currently, there are no values set for microplastic limits. But management and responsibility can be impacted if restrictions on microplastics are placed. Introducing firm necessities on WWTP outlet water may favour centralized methods, and reliability will be on wastewater efficacies. Including microplastics in sewage sludge directive, rather than the treatment can produce incentives or enhanced sewage sludge incineration. Incorporating water framework regulation for microplastics can influence land-related sources and various urban sectors can share responsibility.

6.4 Conclusion

Particles are universal in the environment and microplastics are one such particle pollutant emerging in recent times. Microplastic contaminants in the urban water system are expected to increase due to anthropogenic activities and excessive production of plastics. Steadily, a shift towards smaller particles is expected in regard to size distribution. Currently, the risk regarding microplastics is not considered to be severe, but researchers have stated that gradually the impact of microplastics on the environment and humans will be evident. For a better understanding of the timeline regarding risks, enhancing the risk assessment technique can be beneficial. Considering microplastics as a range of characteristics rather than a definite entity is a novel concept. Though the latest frameworks address the assortment of microplastics regarding analytical detection, assessment of exposure, and risk assessment, they are not implemented in real-life applications. The hypothesized frameworks must be examined, certified, and experimented on the different environmental surroundings and human health risk assessment difficulties. The new frameworks help to assess microplastic risks as well as merge and align research methods which have been previously disintegrated and unrelated. The traditional knowledge concerning risk assessment is aimed at whichever stressor, a framework is a must.

The concerns of microplastics are still in development and many challenges such as cadmium, copper, and pharmaceutical contaminants are yet to be tackled. An explicit outline of sources, pathways, and control strategies can be established by these challenges. Sources will be abundant, dispersed, and tough to mitigate if we consider the previous assessment of microplastics. Since microplastic pollutants are very diverse, it makes them difficult to manage. Preventive measures have the advantage of curbing microplastic pollutants in the urban water system. But it is very difficult to implement or merge with other techniques to considerably influence the load, particularly if they are voluntary action-based. A discrete set of responsibilities, as well as the success rate concerning compartment management, is necessary for decentralized treatment. Constant knowledge regarding the movement of microplastics in urban waters as well as flow management is vital. Merging the flow and actor assessment with a systematic approach can help achieve detailed insight into microplastic flow in urban waters. Herein, the frameworks regarding the risk assessment and mitigation strategies of microplastics in urban waters are described and hope that they will be able to further support and provide efficacy in this complex research area.

Acknowledgements The authors would like to extend their gratitude to, JSS Academy of Higher Education and Research, Mysuru and NAM S&T Centre, New Delhi for providing an opportunity to publish this review.

References

Bailey ND, Basu S, Onink V, Jongedijk CE, Hoffman MJ, Kaandorp MLA, Dijkstra HA, Van Sebille E (2021) Modelling size distributions of marine plastics under the influence of continuous cascading fragmentation. Environ Res Lett 16(5):054075. https://doi.org/10.1088/1748-9326/ABE9EA

Bakir A, O'Connor IA, Rowland SJ, Hendriks AJ, Thompson RC (2016) Relative importance of microplastics as a pathway for the transfer of hydrophobic organic chemicals to marine life. Environ Pollut 219:56–65. https://doi.org/10.1016/j.envpol.2016.09.046

Besseling E, Redondo-Hasselerharm P, Foekema EM, Koelmans AA (2019) Quantifying ecological risks of aquatic micro- and nanoplastic. Crit Rev Environ Sci Technol 49(1):32–80. https://doi.org/10.1080/10643389.2018.1531688

Birdwell J, Cook RL, Thibodeaux LJ (2007) Desorption kinetics of hydrophobic organic chemicals from sediment to water: a review of data and models. Environ Toxicol Chem 26(3):424–434. https://doi.org/10.1897/06-104R.1

Bollmann UE, Simon M, Vollertsen J, Bester K (2019) Assessment of input of organic micropollutants and microplastics into the Baltic Sea by urban waters. Mar Pollut Bull 148:149–155. https://doi.org/10.1016/j.marpolbul.2019.07.014

Braakhuis HM, Park MVDZ, Gosens I, De Jong WH, Cassee FR (2014) Physicochemical characteristics of nanomaterials that affect pulmonary inflammation. In Particle and fibre toxicology, vol 11, no 1, p 18. BioMed Central. https://doi.org/10.1186/1743-8977-11-18

Bucci K, Tulio M, Rochman CM (2020) What is known and unknown about the effects of plastic pollution: a meta-analysis and systematic review. Ecol Appl 30(2). https://doi.org/10.1002/eap.2044

Bucci K, Rochman CM (2022) Microplastics: a multidimensional contaminant requires a multidimensional framework for assessing risk. Microplast Nanoplast 2(1):1–9. https://doi.org/10.1186/s43591-022-00028-0

Burns EE, Boxall ABA (2018) Microplastics in the aquatic environment: Evidence for or against adverse impacts and major knowledge gaps. In Environmental toxicology and chemistry, vol 37, no 11, pp 2776–2796. John Wiley & Sons, Ltd. https://doi.org/10.1002/etc.4268

Capolupo M, Sørensen L, Jayasena KDR, Booth AM, Fabbri E (2020) Chemical composition and ecotoxicity of plastic and car tire rubber leachates to aquatic organisms. Water Res 169.https://doi.org/10.1016/j.watres.2019.115270

Carney Almroth BM, Åström L, Roslund S, Petersson H, Johansson M, Persson NK (2018) Quantifying shedding of synthetic fibers from textiles; a source of microplastics released into the environment. Environ Sci Pollut Res 25(2):1191–1199. https://doi.org/10.1007/s11356-017-0528-7

Cesa FS, Turra A, Checon HH, Leonardi B, Baruque-Ramos J (2020) Laundering and textile parameters influence fibers release in household washings. Environ Pollut 257.https://doi.org/10.1016/j.envpol.2019.113553

Conley K, Clum A, Deepe J, Lane H, Beckingham B (2019) Wastewater treatment plants as a source of microplastics to an urban estuary: removal efficiencies and loading per capita over one year. Water Res X 3:100030. https://doi.org/10.1016/J.WROA.2019.100030

Cox KD, Covernton GA, Davies HL, Dower JF, Juanes F, Dudas SE (2019) Human consumption of microplastics. Environ Sci Technol 53(12):7068–7074. https://doi.org/10.1021/ACS.EST.9B01517/SUPPL_FILE/ES9B01517_SI_001.PDF

De Falco F, Di Pace E, Cocca M, Avella M (2019) The contribution of washing processes of synthetic clothes to microplastic pollution. Sci Rep 9(1):1–11. https://doi.org/10.1038/s41598-019-43023-x

De Ruijter VN, Redondo-Hasselerharm PE, Gouin T, Koelmans AA (2020) Quality criteria for microplastic effect studies in the context of risk assessment: a critical review. Environ Sci Technol 54(19):11692–11705. https://doi.org/10.1021/acs.est.0c03057

Eggen RIL, Hollender J, Joss A, Schärer M, Stamm C (2014) Reducing the discharge of micropollutants in the aquatic environment: the benefits of upgrading wastewater treatment plants. Environ Sci Technol 48(14):7683–7689. https://doi.org/10.1021/es500907n

Esfandiari A, Mowla D (2021) Investigation of microplastic removal from greywater by coagulation and dissolved air flotation. Process Saf Environ Prot 151:341–354. https://doi.org/10.1016/j.psep.2021.05.027

Everaert G, Van Cauwenberghe L, De Rijcke M, Koelmans AA, Mees J, Vandegehuchte M, Janssen CR (2018) Risk assessment of microplastics in the ocean: modelling approach and first conclusions. Environ Pollut 242:1930–1938. https://doi.org/10.1016/j.envpol.2018.07.069

Fältström E, Anderberg S (2020) Towards control strategies for microplastics in urban water. Environ Sci Pollut Res 27(32):40421–40433. https://doi.org/10.1007/s11356-020-10064-z

Gassel M, Rochman CM (2019) The complex issue of chemicals and microplastic pollution: a case study in North Pacific lanternfish. Environ Pollut 248:1000–1009. https://doi.org/10.1016/j.envpol.2019.03.002

Geyer R, Jambeck JR, Law KL (2017) Production, use, and fate of all plastics ever made. Sci Adv 3(7):e1700782. https://doi.org/10.1126/sciadv.1700782

Gouin T (2021) Addressing the importance of microplastic particles as vectors for long-range transport of chemical contaminants: perspective in relation to prioritizing research and regulatory actions. Microplast Nanoplast 1(1):1–19. https://doi.org/10.1186/s43591-021-00016-w

Gouin T, Roche N, Lohmann R, Hodges G (2011) A thermodynamic approach for assessing the environmental exposure of chemicals absorbed to microplastic. Environ Sci Technol 45(4):1466–1472. https://doi.org/10.1021/es1032025

Gouin T, Becker RA, Collot AG, Davis JW, Howard B, Inawaka K, Lampi M, Ramon BS, Shi J, Hopp PW (2019) Toward the development and application of an environmental risk assessment framework for microplastic. Environ Toxicol Chem 38(10):2087–2100. https://doi.org/10.1002/ETC.4529

Gregory MR (1991) The hazards of persistent marine pollution: drift plastics and conservation islands. J R Soc N Z 21(2):83–100. https://doi.org/10.1080/03036758.1991.10431398

Gualtieri AF (2021) Bridging the gap between toxicity and carcinogenicity of mineral fibres by connecting the fibre crystal-chemical and physical parameters to the key characteristics of cancer. Curr Res Toxicol 2:42–52. https://doi.org/10.1016/j.crtox.2021.01.005

Hartmann NB, Hüffer T, Thompson RC, Hassellöv M, Verschoor A, Daugaard AE, Rist S, Karlsson T, Brennholt N, Cole M, Herrling MP, Hess MC, Ivleva NP, Lusher AL, Wagner M (2019) Are we speaking the same language? Recommendations for a definition and categorization framework for plastic debris. Environ Sci Technol 53(3):1039–1047. https://doi.org/10.1021/ACS.EST.8B05297/ASSET/IMAGES/MEDIUM/ES-2018-05297K_0006.GIF

He D, Zhang X, Hu J (2021) Methods for separating microplastics from complex solid matrices: comparative analysis. J Hazard Mater 409:124640. https://doi.org/10.1016/j.jhazmat.2020.124640

He D, Luo Y, Lu S, Liu M, Song Y, Lei L (2018) Microplastics in soils: analytical methods, pollution characteristics and ecological risks. In TrAC-Trends in analytical chemistry, vol 109, pp 163–172. Elsevier. https://doi.org/10.1016/j.trac.2018.10.006

Jan Kole P, Löhr AJ, Van Belleghem FGAJ, Ragas AMJ (2017) Wear and tear of tyres: a stealthy source of microplastics in the environment. In International journal of environmental research and public health, vol 14, no 10. Multidisciplinary Digital Publishing Institute (MDPI). https://doi.org/10.3390/ijerph14101265

Ju S, Shin G, Lee M, Koo JM, Jeon H, Ok YS, Hwang DS, Hwang SY, Oh DX, Park J (2021) Biodegradable chito-beads replacing non-biodegradable microplastics for cosmetics. Green Chem 23(18):6953–6965. https://doi.org/10.1039/d1gc01588e

Karlsson TM, Arneborg L, Broström G, Almroth BC, Gipperth L, Hassellöv M (2018) The unaccountability case of plastic pellet pollution. Mar Pollut Bull 129(1):52–60. https://doi.org/10.1016/j.marpolbul.2018.01.041

Knoblauch D, Mederake L (2021) Government policies combatting plastic pollution. In Current opinion in toxicology, vol 28, pp 87–96. Elsevier. https://doi.org/10.1016/j.cotox.2021.10.003

Koelmans AA, Redondo-Hasselerharm PE, Mohamed Nor NH, Kooi M (2020) Solving the nonalignment of methods and approaches used in microplastic research to consistently characterize risk. Environ Sci Technol 54(19):12307–12315. https://doi.org/10.1021/ACS.EST.0C02982/SUPPL_FILE/ES0C02982_SI_001.PDF

Koelmans AA, Bakir A, Burton GA, Janssen CR (2016) Microplastic as a vector for chemicals in the aquatic environment: critical review and model-supported reinterpretation of empirical studies. In Environmental science and technology, vol 50, no 7, pp 3315–3326. American Chemical Society. https://doi.org/10.1021/acs.est.5b06069

Koelmans AA, Mohamed Nor NH, Hermsen E, Kooi M, Mintenig SM, De France J (2019) Microplastics in freshwaters and drinking water: critical review and assessment of data quality. In Water research, vol 155, pp 410–422. Pergamon. https://doi.org/10.1016/j.watres.2019.02.054

Koelmans AA, Diepens NJ, Mohamed Nor NH (2022) Weight of evidence for the microplastic vector effect in the context of chemical risk assessment. Springer, Cham, pp 155–197. https://doi.org/10.1007/978-3-030-78627-4_6

Koelmans AA, Redondo-Hasselerharm PE, Nor NHM, de Ruijter VN, Mintenig SM, Kooi M (2022). Risk assessment of microplastic particles. In Nature reviews materials, vol 7, no 2, pp 138–152. Nature Publishing Group. https://doi.org/10.1038/s41578-021-00411-y

Kögel T, Bjorøy Ø, Toto B, Bienfait AM, Sanden M (2020) Micro- and nanoplastic toxicity on aquatic life: determining factors. In Science of the total environment, vol 709, p 136050. Elsevier. https://doi.org/10.1016/j.scitotenv.2019.136050

Kooi M, Koelmans AA (2019) Simplifying microplastic via continuous probability distributions for size, shape, and density. Environ Sci Technol Lett 6(9):551–557. https://doi.org/10.1021/acs.estlett.9b00379

Kooi M, Primpke S, Mintenig SM, Lorenz C, Gerdts G, Koelmans AA (2021) Characterizing the multidimensionality of microplastics across environmental compartments. Water Res 202:117429. https://doi.org/10.1016/J.WATRES.2021.117429

Lebreton L, Slat B, Ferrari F, Sainte-Rose B, Aitken J, Marthouse R, Hajbane S, Cunsolo S, Schwarz A, Levivier A, Noble K, Debeljak P, Maral H, Schoeneich-Argent R, Brambini R, Reisser J (2018) Evidence that the great pacific garbage patch is rapidly accumulating plastic. Sci Rep 8(1):4666. https://doi.org/10.1038/s41598-018-22939-w

Lehtiniemi M, Hartikainen S, Näkki P, Engström-Öst J, Koistinen A, Setälä O (2018) Size matters more than shape: ingestion of primary and secondary microplastics by small predators. Food Webs 17:e00097. https://doi.org/10.1016/J.FOOWEB.2018.E00097

Leslie HA, van Velzen MJM, Brandsma SH, Vethaak AD, Garcia-Vallejo JJ, Lamoree MH (2022) Discovery and quantification of plastic particle pollution in human blood. Environ Int 163:107199. https://doi.org/10.1016/j.envint.2022.107199

Libralato G, Volpi Ghirardini A, Avezzù F (2012) To centralise or to decentralise: an overview of the most recent trends in wastewater treatment management. J Environ Manage 94(1):61–68. https://doi.org/10.1016/j.jenvman.2011.07.010

Liping L, Guanghuan C, Jingyou D, Mingyang S, Huanyu C, Qiang Y, Xinhua X (2014) Mechanism of and relation between the sorption and desorption of nonylphenol on black carbon-inclusive sediment. Environ Pollut 190:101–108. https://doi.org/10.1016/j.envpol.2014.03.027

Liu Z, Huang Q, Wang H, Zhang S (2022) An enhanced risk assessment framework for microplastics occurring in the Westerscheldt estuary. Sci Total Environ 817:153006. https://doi.org/10.1016/j.scitotenv.2022.153006

Mattsson K, Björkroth F, Karlsson T, Hassellöv M (2021) Nanofragmentation of expanded polystyrene under simulated environmental weathering (thermooxidative degradation and hydrodynamic turbulence). Front Mar Sci 7:1252. https://doi.org/10.3389/FMARS.2020.578178/BIBTEX

Mohamed Nor NH, Kooi M, Diepens NJ, Koelmans AA (2021) Lifetime accumulation of microplastic in children and adults. Environ Sci Technol 55(8):5084–5096. https://doi.org/10.1021/ACS.EST.0C07384/SUPPL_FILE/ES0C07384_LIVESLIDES.MP4

Ngo PL, Pramanik BK, Shah K, Roychand R (2019) Pathway, classification and removal efficiency of microplastics in wastewater treatment plants. In Environmental pollution, vol 255, no 2. Environ Pollut. https://doi.org/10.1016/j.envpol.2019.113326

Ogunola OS, Onada OA, Falaye AE (2018) Mitigation measures to avert the impacts of plastics and microplastics in the marine environment (a review). In Environmental science and pollution research, vol 25, no 10, pp 9293–9310. Environ Sci Pollut Res Int. https://doi.org/10.1007/s11356-018-1499-z

Olesen KB, Stephansen DA, van Alst N, Vollertsen J (2019) Microplastics in a stormwater pond. Water (switzerland) 11(7):1466. https://doi.org/10.3390/w11071466

Patil S, Bafana A, Naoghare PK, Krishnamurthi K, Sivanesan S (2020) Environmental prevalence, fate, impacts, and mitigation of microplastics—a critical review on present understanding and future research scope. Environ Sci Pollut Res 28(5):4951–4974. https://doi.org/10.1007/S11356-020-11700-4

Pettipas S, Bernier M, Walker TR (2016) A Canadian policy framework to mitigate plastic marine pollution. Mar Policy 68:117–122. https://doi.org/10.1016/J.MARPOL.2016.02.025

Polukarova M, Markiewicz A, Björklund K, Strömvall AM, Galfi H, Andersson Sköld Y, Gustafsson M, Järlskog I, Aronsson M (2020) Organic pollutants, nano- and microparticles in street sweeping road dust and washwater. Environ Int 135:105337. https://doi.org/10.1016/j.envint.2019.105337

Posthuma L, Suter GW, Traas TP (2001) Species sensitivity distributions in ecotoxicology. In Species Sensitivity Distributions in Ecotoxicology. https://doi.org/10.1201/9781420032314

Schmid O, Stoeger T (2016) Surface area is the biologically most effective dose metric for acute nanoparticle toxicity in the lung. J Aerosol Sci 99:133–143. https://doi.org/10.1016/j.jaerosci.2015.12.006

Sedlak D (2017) Three lessons for the microplastics voyage. In Environmental science and technology, vol 51, no 14, pp 7747–7748. American Chemical Society. https://doi.org/10.1021/acs.est.7b03340

Simon M, Vianello A, Vollertsen J (2019) Removal of > 10 μm microplastic particles from treated wastewater by a disc filter. Water (switzerland) 11(9):1935. https://doi.org/10.3390/w11091935

Sun Q, Ren SY, Ni HG (2020) Incidence of microplastics in personal care products: an appreciable part of plastic pollution. Sci Total Environ 742:140218. https://doi.org/10.1016/J.SCITOTENV.2020.140218

Sun J, Dai X, Wang Q, van Loosdrecht MCM, Ni BJ (2019) Microplastics in wastewater treatment plants: detection, occurrence and removal. In Water research, vol 152, pp 21–37. Pergamon. https://doi.org/10.1016/j.watres.2018.12.050

Talvitie J, Mikola A, Koistinen A, Setälä O (2017) Solutions to microplastic pollution–removal of microplastics from wastewater effluent with advanced wastewater treatment technologies. Water Res 123:401–407. https://doi.org/10.1016/j.watres.2017.07.005

Tang KHD, Hadibarata T (2021) Microplastics removal through water treatment plants: its feasibility, efficiency, future prospects and enhancement by proper waste management. Environ Challenge 5:100264. https://doi.org/10.1016/J.ENVC.2021.100264

Van Der Voet E, Van Oers L, Guinée JB, Udo De Haes HA (1999) Using SFA indicators to support environmental policy. Environ Sci Pollut Res 6(1):49–58. https://doi.org/10.1007/BF02987121

Venghaus DV, Barjenbruch M (2017a) Microplastics in urban water management. Czasopismo Techniczne 1.https://doi.org/10.4467/2353737xct.17.011.6108

Venghaus DV, Barjenbruch M (2017b) Microplastics in urban water management. Czasopismo Techniczne 1(1):137–146. https://doi.org/10.4467/2353737xct.17.011.6108

Wang T, Li B, Zou X, Wang Y, Li Y, Xu Y, Mao L, Zhang C, Yu W (2019) Emission of primary microplastics in mainland China: invisible but not negligible. Water Res 162:214–224. https://doi.org/10.1016/J.WATRES.2019.06.042

Wang J, Coffin S, Schlenk D, Gan J (2020) Accumulation of HOCs via precontaminated microplastics by Earthworm Eisenia fetida in soil. Environ Sci Technol 54(18):11220–11229. https://doi.org/10.1021/acs.est.0c02922

Wiesinger H, Wang Z, Hellweg S (2021) Deep dive into plastic monomers, additives, and processing aids. Environ Sci Technol 55(13):9339–9351. https://doi.org/10.1021/acs.est.1c00976

Willis K, Maureaud C, Wilcox C, Hardesty BD (2018) How successful are waste abatement campaigns and government policies at reducing plastic waste into the marine environment? Mar Policy 96:243–249. https://doi.org/10.1016/j.marpol.2017.11.037

Wu J, Jiang Z, Liu Y, Zhao X, Liang Y, Lu W, Song J (2021) Microplastic contamination assessment in water and economic fishes in different trophic guilds from an urban water supply reservoir after flooding. J Environ Manage 299:113667. https://doi.org/10.1016/j.jenvman.2021.113667

Zhang Q, Xu EG, Li J, Chen Q, Ma L, Zeng EY, Shi H (2020a) A review of microplastics in table salt, drinking water, and air: direct human exposure. Environ Sci Technol 54(7):3740–3751. https://doi.org/10.1021/ACS.EST.9B04535/SUPPL_FILE/ES9B04535_SI_001.PDF
Zhang X, Leng Y, Liu X, Huang K, Wang J (2020b) Microplastics' pollution and risk assessment in an Urban River: a case study in the Yongjiang River, Nanning City, South China. Exposure Health 12(2):141–151. https://doi.org/10.1007/s12403-018-00296-3

Chapter 7
Discovery of Novel Butyrylcholinesterase Inhibitors as Potential Candidates for the Treatment of Alzheimer's Disease Caused Due to the Presence of Aluminium in Drinking Water

Sushma Pradeep, C. P. Kavana, M. R. Sai Chakith, Pruthvish Reddy, Pallavi K. Shekar, Aishwarya H. Keerthi, Shiva Prasad Kollur, and Chandan Shivamallu

Abstract Alzheimer's disease (AD) is a neurodegenerative disorder that is associated with abnormal cognition. The main signs of AD include memory loss that worsens with time and a reduction in cognitive function. When misfolded proteins accumulate in the aged brain, they cause inflammation, oxidative damage, synaptic dysfunction, and energy loss. These molecular lesions are what make AD detectable.

S. Pradeep · C. P. Kavana · C. Shivamallu (✉)
Department of Biotechnology and Bioinformatics, School of Life Sciences, JSS Academy of Higher Education and Research, Mysuru, Karnataka 570 015, India
e-mail: chandans@jssuni.edu.in

S. Pradeep
e-mail: sushmap@jssuni.edu.in

M. R. Sai Chakith
Department of Pharmacology, JSS Medical College, JSS Academy of Higher Education and Research, Mysuru, Karnataka 570 015, India
e-mail: saichakithmr@jssuni.edu.in

P. Reddy
Department of Biotechnology, Acharya Institute of Technology, Bengaluru, Karnataka, India
e-mail: pruthvi.19@gmail.com

P. K. Shekar
Department of Biochemistry, JSS Medical College, JSS Academy of Higher Education and Research, Mysuru, Karnataka 570 015, India
e-mail: pallavishekar169@jssuni.edu.in

A. H. Keerthi
Department of Microbiology, School of Life Sciences, JSS Academy of Higher Education and Research, Mysuru, Karnataka 570 015, India
e-mail: aishwaryahk@jssuni.edu.in

S. Suriyanarayanan et al. (eds.), *Water Management in Developing Countries and Sustainable Development*, Water Resources Development and Management,
https://doi.org/10.1007/978-981-99-8639-2_7

Hereditary and environmental factors contribute to the development of AD. A neurotoxic substance called Aluminium (Al) induces oxidative stress, which is connected to the development of AD prominently in women. A new pharmacological entity that is efficient is constantly needed to treat cognitive impairment like AD. Natural products like medicinal herbs are mined to find new medication possibilities. To treat cognitive dysfunction, *Centella asiatica* (CA), a perennial herbaceous plant is frequently utilized in traditional medicine. Using chemoinformatics, we were able to discover the bioactive components of this plant. To determine the potential molecular roles of the phytoconstituents of CA in AD, we conducted several bioinformatics investigations. Before predicting their possible protein targets, we first discovered the bioactive phytoconstituents from CA that may be involved in the etiology of AD. The current study shows that Furfural has a higher affinity for protein binding based on the RMSD, RMSF, and H-bond trajectories.

Keywords Alzheimer's disease · Aluminium · Centella Asiatica

7.1 Introduction

Alzheimer's disease (AD) is a neurodegenerative disorder that affects the brain and is age-related, progressive, and irreversible. It is characterized by cognitive impairment, memory loss, and such other behavioural difficulties that intrude with day-to-day activities. Amyloid beta (Aβ) plaques, cholinergic impairments, and neurofibrillary tangles (NFT) are extracellular deposits that are present in AD. From a clinical perspective, AD significantly impacts patients' lifestyles, which are marked by a prodromal phase followed by a progressive decline in cognitive abilities and memory, necessitating ongoing medical care. The main clinical characteristics of AD include early memory loss and memory impairment, subsequent topographical difficulties, and loss of attention, judgement, and confidence (Pradeep et al. 2021). The cognitive symptoms of AD are now treated with acetylcholinesterase (AChE) inhibitors such as rivastigmine (Exelon), tacrine (Cognex), donepezil (Aricept), galantamine (Razadyne), and N-methyl-D-aspartate (NMDA) receptor antagonist memantine. However, because donepezil operates on the memantine and AChEenzyme on the NMDA receptor, these medications become less effective as the condition worsens. The negative effects of these include bradycardia, high respiratory secretion, hypotension, reduced intra-ocular pressure, and vomiting, among many others (TarunMinocha et al. 2022).

Aging, a family history of AD, and the presence of the genes for -amyloid precursor protein (APP), presenilin 1, and apolipoprotein E (APOE) are some of the factors that can contribute to the development of AD. Being exposed to some

S. P. Kollur
School of Physical Sciences, Amrita Vishwa Vidyapeetham, Mysuru Campus, Mysuru, Karnataka 570026, India

environmental toxins, such as pesticides, industrial chemicals, and toxic metals like lead and aluminum, as well as suffering severe or repetitive traumatic brain traumas (TBI), may raise the chance of developing AD. Most European reports indicate that two-thirds of AD cases with clinical diagnoses clinical diagnosis are female. Since the likelihood of getting AD rises with age, women's longer lifespan is cited as the main explanation for this gender disparity (Pradeep et al. 2022a).

By altering several biomolecules linked to neurotoxicity, Aluminum (Al) contributes to the etiology of several neurodegenerative illnesses. Al has been linked to oxidative stress, plaque development in the cortex and hippocampus, oligomer cross-linking, and deposition. Al exposure is practically inevitable due to its prevalence in the environment, daily activities, food, and drinking water (Pradeep et al. 2022b). According to several number of studies, consuming aluminum causes an increase in the expression of amyloid protein in mouse tissues, which may be a key step in the progression of AD. According to ecological research, drinking water with Al concentrations of 0.1–0.2 mg/L may increase the chance of developing AD, with relative risks or odds ratios ranging from 1.35 to 2.67 (Chandrashekar Srinivasa et al. 2022).

Centella asiatica (CA) is an essential herbal remedy that is always steady in Chinese, Ayurvedic, and Indonesian medicine. This medicinal plant, which is categorized as both a nutraceutical and a cosmeceutical, is used to improve cognition, treat skin conditions, and foster liver and kidney health. Because it has been demonstrated to increase insights and enhance memory, CA is recognized as a herb that repairs nerve and brain cells. Due to the presence of bioactive substances such as rutin, quercetin, kaempferol, gallic acid, naringin, luteolin, catechin, and stigmasterol. AChE activities were also reduced, suggesting a potential role for CA in the prevention and treatment of AD (Prasad et al. 2021a). Widespread used in medicine, CA is capable of suppressing H2O2-induced cell passing, lowering free radical concentrations, preventing-amyloid cell passing, and inhibiting cell passing in addition to inhibiting free radical concentrations, according to in vitro studies using various CA plant extracts. The symptoms of mild to severe AD are treated with donepezil. It is a selective and temporary AChE inhibitor that increases the availability of acetylcholine.

Several proteins may have a key role in the pathogenesis of AD, which plays a crucial role in AD, butyrylcholinesterase (BChE) is one among them. BChE serves as AChE's backup and a scavenger for toxins that can prevent AChE from doing its job. The AChE protein may hasten the synthesis of A in the brain and result in AD12. The study aimed to demonstrate how CA bioactive substances interact with an in-silico analysis of the Alzheimer's protein BChE (Sushma et al. 2020).

The computer-aided drug design is a powerful method in development of potential drugs for prevention or treatment of AD. The primary method in drug discovery research has typically been the design and synthesis of selective ligands pointing to single disease targets, however due to the multifactorial pathophysiology of many diseases, such as AD, this strategy has not always been successful. *In-silico* molecular docking (MD) techniques aim to predict a ligand's ideal way of binding to a macromolecular protein (Kollur et al. 2021).

7.2 Materials and Methods

7.2.1 Protein Retrieval and Validation

The Protein Data Bank tool (https://www.rcsb.org/) is used to retrieve the Butyryl-cholinesterase (PDB ID: 4QAD) structure. After modelling the structure, the PDB coordinate file was submitted to the Procheck server. The most favoured portions of the server plot, which is a graphical depiction of protein 3D structure, were confirmed. The protein model was confirmed, and the sections are identified by distinct colours (Uppar et al. 2021).

7.2.2 Ligand's Selection and Preparation

The phytochemical compounds of CAplant (Myrcene, Hexanal, Beta-bisabolene, Furfural, Toluene, Farnesol, Humulene, Linalool, Nerol, and Limonene) three-dimensional structure files were obtained from the IMPPAT database in 3D PDB format (https://cb.imsc.res.in/imppat/) (Ankegowda et al. 2020) (Table 7.1).

7.2.3 Protein Preparation

Protein (PDB Id 4AQD) was downloaded from the PDB database in PDB format. Obtained protein structures were cleaned using PyMol software (https://pymol.org/2/). All mammalian tissues have the serine hydrolase BChE, which functions as the enzyme that breaks down the neurotransmitter acetylcholine in the brain (Prasad et al. 2020a; Patil et al. 2022; Shivalingaiah et al. 2022a).

7.2.4 Molecular Docking

To understand the interaction between the phytochemical compounds of the *Centella asiatica* plant with the Human butyrylcholinesterase (4AQD) PyRx v0.8 (https://pyrx.sourceforge.io/) open-access docking software was used to study molecular interaction and docking study (Shivalingaiah et al. 2022b; Mariwamy et al. 2022; Srinivasa et al. 2021; HeggadadevanakoteKendaganna et al. 2021). The grid box was designed for the selected binding site residues to provide the best trajectory with the heist binding affinity values, shielding the active binding site residues for the selected binding site residues (Prasad et al. 2020b; Kumar et al. 2021).

Table 7.1 Phytochemical compounds with scientific names, molecular formulas, compound names, and 2D structures

Sl. no	Scientific name	Molecular formula	Compound name	2D structure of compounds
1	Centella asiatica	$C_{10}H_{16}$	Myrcene	
2	Centella asiatica	$C_6H_{12}O$	Hexanal	
3	Centella asiatica	$C_{15}H_{24}$	Beta-Bisabolene	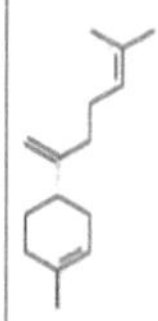
4	Centella asiatica	$C_5H_4O_2$	Furfural	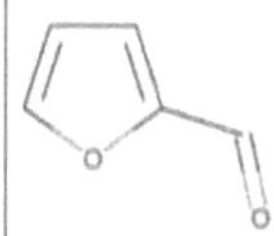
5	Centella asiatica	C_7H_8	Toluene	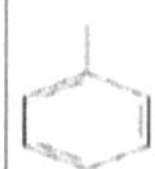
6	Centella asiatica	$C_{15}H_{26}O$	Farnesol	
7	Centella asiatica	$C_{15}H_{24}$	Humulene	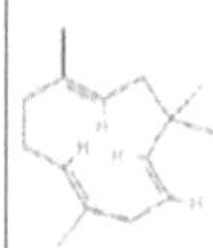
8	Centella asiatica	$C_{10}H_{18}O$	Linalool	
9	Centella asiatica	$C_{10}H_{18}O$	Nerol	
10	Centella asiatica	$C_{10}H_{16}$	Limonene	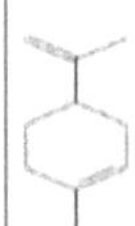

7.2.5 Molecular Docking Visualization

The hydrogen bonds, bond length, and hydrophobic interactions between the 4AQD protein and the docked conformations obtained from the ligand against the specific protein were visualized using Discovery Studio (https://discover.3ds.com/discovery-studio-visualizer-download) (Patil et al. 2021a; Santosh Kumar et al. 2021). Discovery studio was used to create a 3D and 2D interaction between the targeted protein and the CAplant phytochemical compounds (Shankar et al. 2021; Vinay et al. 2021; Prasad et al. 2021b; Avinash et al. 2021).

7.2.6 ADME Properties

The docked phytocompound investigations revealed ligand interaction with 3 compounds, which were then submitted to in-silico ADME characteristic analysis using the application system Swiss-ADME http://www.swissadme.ch/. The traditional approach to pharmacokinetics divides the goal into different factors, which may then be analysed separately by specific methods it has been demonstrated that early prediction of ADME in the investigation stage drastically reduces the fraction of pharmacokinetics-related failure inside the clinical phases, so further research by Lipinski et al. analysed orally active substances to define physiochemical properties for the high chance of being an oral drug, in which the 'Rule-of-Five' described the correlation among pharmacokinetic and physicochemical characteristics (Pruthvish et al. 2021). So, bioavailability is provided by radar for a rapid evaluation of drug-likeness, and the six syntheses and characterization of lipophilicity, size, polarization, solubility, flexibility, and water content are considered that a drug intended for parenteral administration must be extremely soluble in water to produce a suitable amount of active substance in the small quantity of such pharmacological dosage to predict Solubility are considered in Swiss ADME analysation with prediction (Prasad et al. 2020c; Navyashree et al. 2021).

7.3 Results

7.3.1 Protein Retrieval and Validation

The 3-dimensional structure of human butyrylcholinesterase was downloaded using Protein Data Bank in the PDB format. The 3D structure was visualized under Discovery studio. To forecast the quality of protein structure, use the Ramachandran plot. A high-quality Ramachandran plot comprises 88.5% of the torsional angles in the region being followed (Patil et al. 2020) (Fig. 7.1).

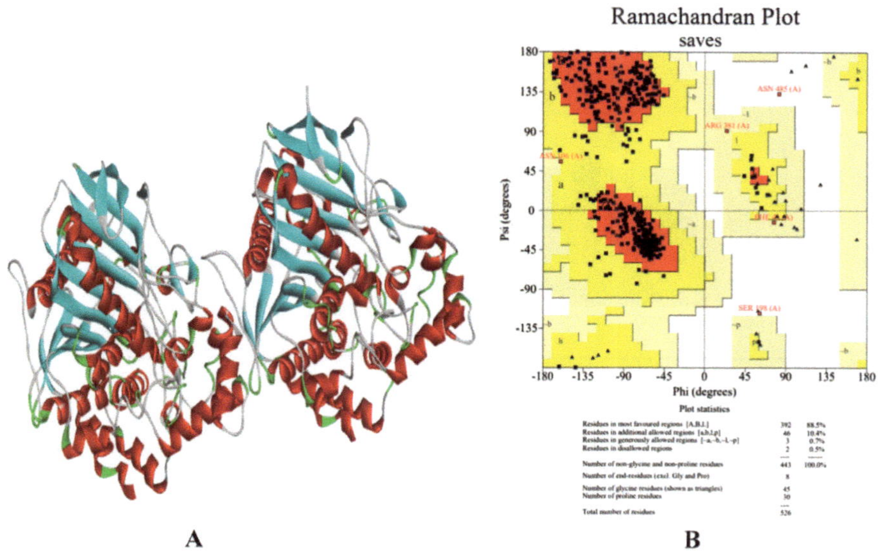

Fig. 7.1 **A** Crystal structure of Human butyrylcholinesterase and **B** Validation using Ramachandran plot

7.3.2 *Molecular Docking*

The outcomes of the docking experiment demonstrated that all the phytocompounds produced from the extract interacted effectively with the protein targets. Out of 10 docked postures for each protein–ligand interaction, the one with the least binding energy was selected as the best docked pose. The 4 ligand shows the good interaction with targeted protein (Patil et al. 2021b). The Furfural has 3 hydrogen bond interactions with SER 198, GLY 117, GLY 116, and it exhibits a binding affinity of −4.4 kcal/mol. Hexanal form binding affinity of −4.3 kcal/mol by forming 3 hydrogen bonds with TRP 82, TRP 430, and TRP 440. The Nerol has 2 hydrogen bond interaction with TRP 28, GLY 78, and TRL 440. And it exhibits binding affinity of −6.0 kcal/mol. The Linalool has 3 hydrogen bond interactions with GLY 116 and GLU 197 and it exhibits binding affinity of −5.1 kcal/mol (Fig. 7.2).

7.3.3 *ADME Properties*

To study general characteristics, physicochemical properties, lipophilicity, water solubility characteristics, pharmacokinetic parameters, drug-likeness rule and bioavailability score, and medicinal chemistry properties, 3 potent phytocompond of *Centella asiatica* were analysed using the Swiss ADME web tool. The estimation of drug absorption, distribution, metabolism, and excretion (ADME) is currently

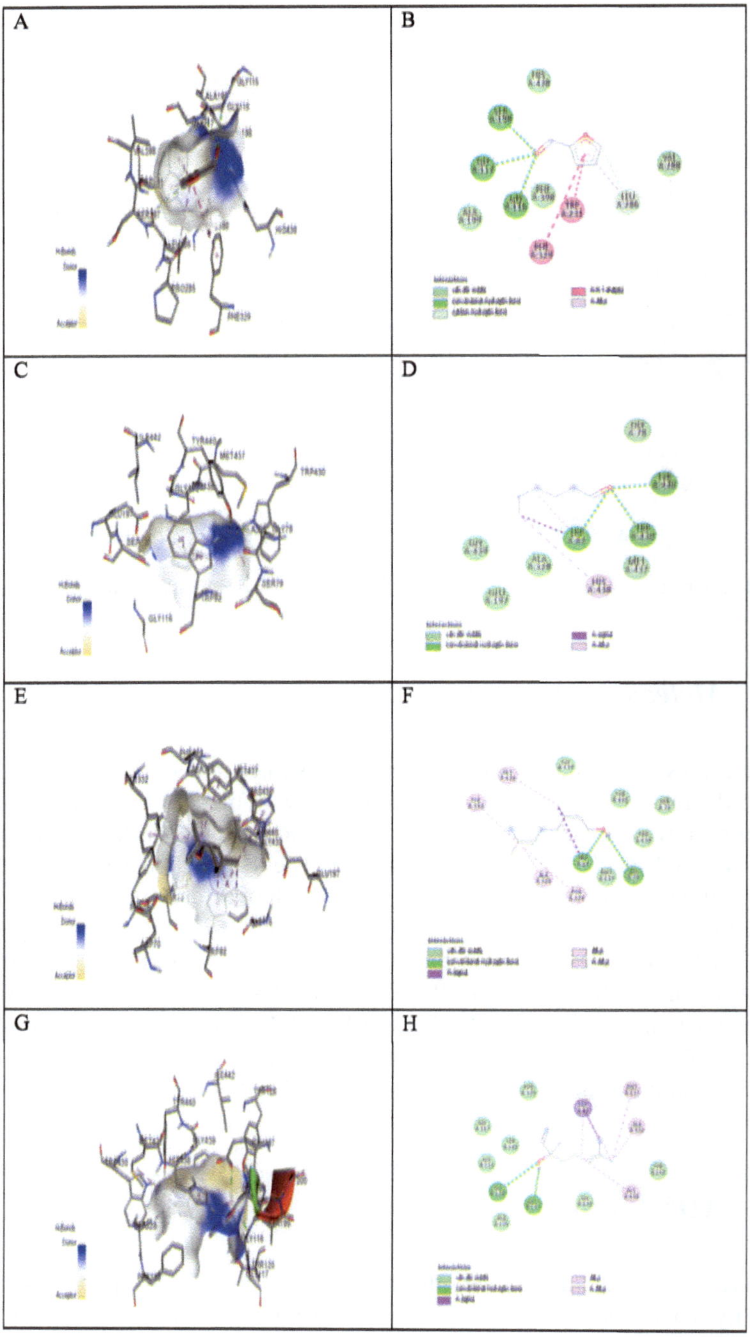

Fig. 7.2 Insilco docking studies representing the two-dimensional and three-dimensional targeted protein (4AQD) complex with ligand interaction. **A** Furfural, **B** Hexanal, **C** Nerol, **D** Linalool

supported by computer-aided drug development. These methods presume reliable and anticipatory data fast and support experimental methods. The percentage of pharmacokinetics-related failures in the clinical phase has been found to be significantly reduced by the first assessment of ADME properties in the discovery phase (BramhadevPattnaik et al. 2021).

A high level of GI absorption was seen with FURFURAL, HEXANAL, NEROL. Except for a few, the majority of the chemicals in *Centella asiatica* are not P-gp substrates. The Swiss ADME model outputs "Yes" or "No" depending on whether the substance under investigation is more likely to be a P-gp substrate, inhibitor, or non-inhibitor (CYP1A2, CYP2c9, CYP2C19, CYP2D6, and CYP3A4). All 3 phytocompounds were "No" for P-gp substrate. And also FURFURAL, HEXANAL, NEROL for CYP1A2. The higher negative the log Kp (with Kp in cm/s), the less skin permeant the molecule is, according to a multiple linear regression. NEROL (−4.67 cm/s) is the least permeant chemical among the phytocompound of *Centella asiatica* (Fig. 7.3 and Table 7.2).

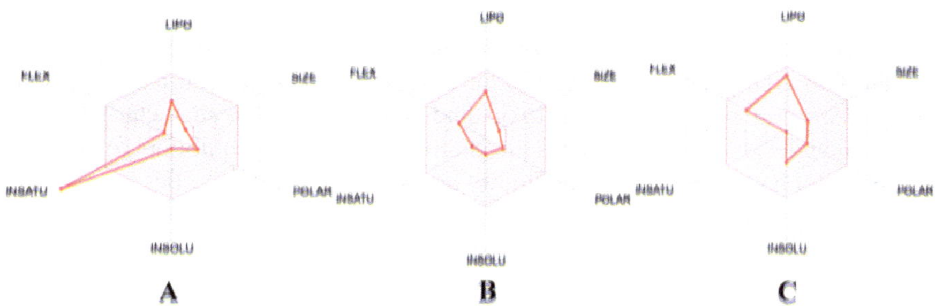

Fig. 7.3 ADME studies of Insilico computational phytocompound of **A** Furfural **B** Hexanal **C** Nerol

Table 7.2 This *Insilco* approach further studied with the pharmacokinetic absorption, distribution, metabolism, and excretion results of Furfural, Hexanal, and Nerolby SWISSADME

ADMET entry	FURFURAL	HEXANAL	NEROL
Lipinski	Yes	Yes	Yes
Bioavailability score	0.55	0.55	0.55
GI absorption	High	High	High
BBB permeant	Yes	Yes	Yes
P-gp substrate	No	No	No
CYP1A2 inhibitor	No	No	No
CYP2C19 inhibitor	No	No	No
CYP2C9 inhibitor	No	No	No
CYP2D6 inhibitor	No	No	No
CYP3A4 inhibitor	No	No	No
Log K_p (skin permeation)	−6.60 cm/s	−5.65 cm/s	−4.67 cm/s

7.4 Conclusion

Apparently, aluminium (Al) is a potent neurotoxin. Al may contribute to the pathophysiology or aetiology of Alzheimer's disease (AD), however it is still unclear if this association is causal. Although Al is relatively weakly absorbed in the digestive system, it cannot currently be ruled out that some Al fractions found in drinking water may be highly bioavailable. The data relating Al and AD as a whole justifies intensive study efforts. Clarification of the cellular and molecular mechanisms underlying Al toxicity as well as the fundamental metabolism and kinetics of Al in the human body should be the main objectives of such studies. The present study suggests that aluminium may be one of the factors associated with dementia and especially Alzheimer's disease and must now be investigated further, since simple measures could be taken to reduce such exposure in the elderly. Also, the study was to find phytoconstituents that can bind to the critical targets of amyloid hypothesis of AD using a computational approach and to check the AChE inhibition activity of the extract. In silico molecular docking approach revealed that most of the compounds derived from *Centella asiatica* extract can bind to the selected targets, according to the binding scores and analysis of the interactions of the compounds. Further we studied with ADME studies its prediction also shows that furfural has good solubility, absorption, permeation, non-toxic, and non-carcinogenic characteristics. Our study concludes that the furfural has the potential to be strong inhibitors for targeted protein and it has the high binding affinity and good ADME properties.

Author Contributions Conceptualization, SP and CS; methodology, PR, HK and SP; software, SP, CP and KS; validation, ES, VS, NM, RRA, SPK, SP, and CS; formal analysis, ES, VS, NA, RRA, and SMS; investigation, SP; writing—original draft preparation, SP, SMR and SCP, and DG-M; writing—review and editing, CS, SP, SMR, AAS, MYA, SEIE; visualization, SP, CP. All authors have read and agreed to the published version of the manuscript.

Conflict of Interest: There is no conflict of interest.

References

Ankegowda VM, Kollur SP, Prasad SK, Pradeep S, Dhramashekara C, Jain AS, Prasad A, Srinivasa C, SridharaSetty PB, Gopinath SM, RP S, Bahkali AH, Syed and Shivamallu C (2020) Phyto-mediated synthesis of silver nanoparticles using terminalia chebula fruit extract and evaluation of its cytotoxic and antimicrobial potential. Molecules 25(21):5042

Avinash KO, Pradeep S, Shivamallu C, Gopenath TS, Kiran Kumar MN, Kanthesh BM, Kumar JR (2021) In silico screened flavanoids of glycyrrhiza glabra inhibit CPLA2 and SPLA2 in LPS stimulated macrophages. Bull Env Pharmacol Life Sci 10(5):72–80

BramhadevPattnaik SS, Patil CS, Amachawadi RG, Dash AP, Yadav MP, Prasad KS, Sushma P, Jain AS, Shivamallu C (2021) COVID-19 pandemic: a systematic review on the coronaviruses of animals and SARS-CoV-2. J Exp Biol Agricult Sci 9(2):117–130

Chandrashekar Srinivasa SR, Kumar S, Pradeep S, Prasad SK, Veerapur R, Ansari MA, Alomary MN, Alghamdi S, Almehmadi M, Kavitha GC, Daphedar AB, Kakkalameli SB, Shivamallu

C, Kollur SP (2022) Eco-friendly synthesis of MnO2 nanorods using *Gmelina arborea Fruit* extract and its anticancer potency against MCF-7 breast cancer cell line. Int J Nanomed

HeggadadevanakoteKendaganna P, Shivamallu C, Shruthi G, NagabushanChitagudigi M, Pradeep S, Karunakar P, Prasad Kollur S, Amachawadi RG, Patil SS, Syed A, Elgorban AM, Bahkali AH, Veerapur R, Kollur SP (2021) In silico screening and validation of KPHS_00890 protein of Klebsiella pneumoniae proteome: an application to bacterial resistance and pathogenesis. J King Saud Univ Sci 33(6):101537

Kollur SP, Prasad SK, Pradeep S, Veerapur R, Patil SS, Amachawadi RG, Lamraoui G, Al-Kheraif AA, Elgorban AM, Syed A, Shivamallu C (2021) Luteolin-fabricated ZnO nanostructures showed PLK-1 mediated anti-breast cancer activity. Biomolecules 11(3):385

Kumar V, Ramu R, Shirahatti PS, Chandana Kumari VB, Sushma P, Mandal SP, Patil SM (2021) α-Glucosidase, α-Amylase Inhibition, kinetics and docking studies of novel (2-Chloro-6-(trifluoromethyl)benzyloxy)arylidene) based rhodanine and rhodanine acetic acid derivatives. Chem Select 6(36):9637–9644

Mariwamy VH, Kollur SP, Bindya S, Begum M, Shivamallu C, Dharmashekara C, Pradeep S, Jain AS, Prasad SK, Syed A, Elgorban AM, Al-Rejaie S, Ortega-Castro J, Frau J, Flores-Holguín N, Glossman-Mitnik D (2022) N-((1H-pyrrol-2-yl)methylene)-6-methoxypyridin-3-amine and its Co(II) and Cu(II) complexes as antimicrobial agents: chemical preparation, in vitro antimicrobial evaluation, in silico analysis and computational and theoretical chemistry investigations. Molecules

Minocha T, Birla H, Obaid AA, Vipin Rai PS, Shivamallu C, Moustafa M, Al-Shehri M, Al-Emam A, Tikhonova MA, Yadav SK, Poeggeler B, Singh D, Singh SK (2022) Flavonoids as promising neuroprotectants and their therapeutic potential against Alzheimer's disease. Oxid Med Cell Longev 13

Navyashree B, Jain AS, Sushma P, Chandan D, Patil SS, Kollur SP, Latha KC, Chandan S (2021) Computational based approach in discovering the phytocompound based complex inhibition against Ralstonia Solanacearum and Root-knot nematode Meloidogyne Incognita. Bull Env Pharmacol Life Sci 10(11):23–34

Patil SM, Sujay S, Chandana Kumari VB, Tejaswini M, Sushma P, Shirahatti PS, Ramu R (2020) Bioactive peptides: its production and potential role on health. Int J Innovat Sci Eng Technol 7(1)

Patil SS, Suresh KP, Shinduja R, Amachawadi RG, Chandrashekar S, Sushma P, Kollur SP, Syed A, Sood R, Roy P, Shivamallu C (2021) Prevalence of methicillin-resistant Staphylococcus aureus in India: a systematic review and meta-analysis. Oman Med J

Patil SS, Shivamallu C, Dharmashekara C, Pradeep S, Suresh KP, Ashwini P, Kollur SP, Yadav MP, Chandrashekar S, Pattnaik B (2021) Diversity due to mutations in circulating virus strains of SARS- CoV-2 may delay control of COVID-19. J Appl Biol Biotechnol 1–8

Patil SS, Shivamallu C, Dharmashekara C, Pradeep S, Suresh KP, Ashwini P, Kollur SP, Yadav MP, Chandrashekar S, Pattnaik B (2022) Diversity due to mutations in circulating virusstrains of SARS-CoV-2 may delay control of COVID-19. J Appl Biol Biotech

Pradeep S, Jain AS, Dharmashekara C, Prasad SK, Akshatha N, Pruthvish R, Amachawadi RG, Srinivasa C, Syed A, Elgorban AM, Al Kheraif AA, Ortega-Castro J, Frau J, Flores-Holguín N, Shivamallu C, Kollur SP, Glossman-Mitnik D (2021) Synthesis, computational pharmacokinetics report, conceptual DFT-based calculations and anti-acetylcholinesterase activity of hydroxyapatite nanoparticles derived from *Acorus Calamus* plant extract. Front Chem 9:741037

Pradeep S, Patil SM, Dharmashekara C, Jain AS, Ramu R, Shirahatti PS, Mandal SP, Reddy P, Srinivasa C, Patil SS, Ortega-Castro J, Frau J, Flores-Holgúın N, Shivamallu C, Kollur SP, Glossman-Mitnik D (2022) Molecular insights into the *in silico* discovery of corilagin from *Terminalia chebula* as a potential dual inhibitor of SARS-CoV-2 structural proteins. J Biomol Struct Dyn

Pradeep S, Prabhuswaminath SC, Reddy P, Srinivasa SM, Shati AA, Alfaifi MY, Eldin I, Elbehairi S, Achar RR, Silina E, Stupin V, Manturova N, Glossman-Mitnik D, Shivamallu C, Kollur SP (2022) Anticholinesteraseactivity of *Areca catechu*: In Vitro and *in silico* green synthesis approach insearch for therapeutic agents against Alzheimer's disease. Front Pharmacol 13:1044248

Prasad KS, Pillai RR, Shivamallu C, Prasad SK, Jain AS, Pradeep S, Armaković S, Armaković SJ, Srinivasa C, Kallimani S, Amachawadi RG, Ankegowda VM, Marraiki N, Elgorban AM, Syed A (2020a) Tumoricidal potential of novel Amino-1,10-phenanthroline derived imine ligands: chemical preparation, structure, and biological investigations. Molecules 25(12):2865

Prasad A, Shruthi G, Sushma P, Jain AS, Chandan D, Nagendra Prasad MN, Kollur SP, Chandrashekar S, Shivamallu C (2020c) Helicobacter pylori infection: a bioinformatic approach. Int J Pharm Sci Res 11:5469–5483

Prasad SK, Pradeep S, Shimavallu C, Kollur SP, Syed A, Marraiki N, Egbuna C, Gaman M-A, Kosakowska O, Cho WC, Patrick-Iwuanyanwu KC, Ortega Castro J, Frau J, Flores-Holguín N, Glossman-Mitnik D (2021a) Evaluation of *Annona muricata* Acetogenins as potential anti-SARS-CoV-2 agents through computational approaches. Front Chem 8:624716. https://doi.org/10.3389/fchem.2020.624716

Prasad KS, Shivamallu C, Gopinath SM, Srinivasa C, Dharmashekara C, Sushma P, Jain AS, Ashwini P (2021b) Schiff base ligands derived from 4-chloro-6-methylpyrimidin-2-amine: chemical synthesis, bactericidal activity and molecular docking studies against targeted microbial pathogen. Int J Health Allied Sci 10:157–166

Prasad KS, Pillai RR, Ghimire MP, Ray R, Richter M, Shivamallu C, Jain AS, Prasad SK, Sushma P, Armakovic S, Armakovic SJ, Amachawadi RG (2020) Indole moiety induced biological potency in pseudo-peptides derived from 2-amino-2-(1H-indole-2-yl) based acetamides: chemical synthesis, in vitro anticancer activity and theoretical studies. J Mol Struct 128445

Pruthvish R, Gopinath SM, Sushma P, Jain AS, Kollur SP, Shivmallu C (2021) In-silico evaluation of anti-cancerous activity of herbal plant extracts. Bull Env Pharmacol Life Sci 10(4):105–117

Santosh Kumar SR, Srinivasa C, Shivamallu C, Prasad KS, Pradeep S, Syed A, Bhakali A, Shankar A, Patil SS, Ashwini P, Chandan D, Triveni R, Melappa G, Krishna V (2021) Callus induction and shoot regeneration from the immature flower bud of Caesalpinia bonducella and its antileptospiral potential by in vitro and in silico analysis. Phcog Mag 17(Suppl S1):38–44

Shankar A, Gopinath SM, Kollur SP, Sushma P, Jain AS, Patil SS, Srinivasa CS, Shivamallu C (2021) Structural diversity and role of phytochemicals against P38-a mitogen-activated protein kinase and epidermal growth factor receptor kinase domain: a privileged computational approach. J Pure Appl Microbiol15(4):2263–2269

Shivalingaiah T, Sai Chakith MR, Pradeep S, Pallavi KS, Kavana CP, Navyashree B, Guru Kumar D, Girish MS, Srinivas C, Kollur SP, Shivamallu C (2022) In-silico evaluation of anti-acne property of *Syzygium(S.) aromaticum.* Int J Food Nutr Sci

Shivalingaiah T, Shekar PK, Pradeep S, Sai Chakith MR, Navyashree B, Guru Kumar D, Chandrashekar Srinivas GMS, Kollur SP, Shivamallu C (2022) Computational approach in search of therapeutic for colorectal cancer. Int J Food Nutri Sci

Srinivasa C, Umesha S, Pradeep S, Ramu R, Ansari MA, Kollur SP, Shivamallu C (2021) Salicylic acid-mediated enhancement of resistance in tomato plants against Xanthomonas perforans. Saudi J Biol Sci. ISSN 1319–562X

Sushma P, Jain AS, Chandan D, S Prasad SK, Prasad KS, Asad S, Shivamallu C (2020) Herbal combination therapy against Alzheimer's disease. J Alzheimer's Dis Rep 4(1):417–429.https://doi.org/10.3233/ADR-200228

Uppar V, Chandrashekharappa S, Shivamallu C, Sushma P, Kollur SP, Ortega-Castro J, Frau J, Flores-Holguín N, Basarikatti AI, Chougala M, Mohan MM, Banuprakash G, Jayadev VKN, Nandeshwarappa BP, Veerapur R, Al-Kheraif AA, Elgorban AM, Syed A, Mudnakudu-Nagaraju KK, Padmashali B, Glossman-Mitnik D (2021) Investigation of antifungal properties of synthetic Dimethyl-4-Bromo-1-(Substituted Benzoyl) Pyrrolo[1,2-a] Quinoline-2,3-dicarboxylates analogues: molecular docking studies and conceptual DFT-based chemical reactivity descriptors and pharmacokinetics evaluation. Molecules 26(9):2722

Vinay S, Dharmashekar C, Sushma P, Jain AS, Kollur SP, Bindya S, Srinivasa C, Patil SS, Prasad A, Shivamallu C (2021) In-Silico elucidation of Abrusprecatorius phytochemical against Diabetes mellitus. Bull Env Pharmacol Life Sci 10(4):118–128

Chapter 8
Combatting Plastic Contamination: Strategies for Removing Microplastic from Water Resources

Athulya Thankachan, Gajendran Chellaiah, P. P. Sivan, U. Surendran, Mahendran Chellaiah, Ramamoorthy Ayyamperumal, Adwaitha Sunilkumar, and Dhayanand Jayaseelan

Abstract The rise in the amount of microplastics present in aquatic habitats has developed into a major ecological concern in recent times. Microplastics are widely distributed in the environment due to inadequate rates of plastic waste disposal, a lack of standard detection equipment, and specific cleanup techniques. Studies indicate that microplastics, by absorbing heavy metals, pathogens, and chemicals used in polymer production, can spread pollutants. They are consumed by fish and other

A. Thankachan
Department of Agricultural Engineering, SriShakthi Institute of Engineering and Technology, Coimbatore, Tamil Nadu, India

G. Chellaiah (✉)
Department of Civil Engineering, SNS College of Technology, Coimbatore, Tamil Nadu, India
e-mail: cgajendran@gmail.com

P. P. Sivan
Department of Civil Engineering, Government Engineering College, Thrissur, Kerala, India

U. Surendran
Centre for Water Resources Development and Management (CWRDM), Kunnamangalam, Kerala, India

M. Chellaiah
Department of Electronics and Communication Engineering, Alagappa Chettiar Government College of Engineering and Technology, Karaikudi, Tamil Nadu, India

R. Ayyamperumal
MOE Key Laboratory of Western China's Environmental System, College of Earth and Environmental Sciences, Lanzhou University, Lanzhou, P.R. China
e-mail: ramamoorthy@lzu.edu.cn

A. Sunilkumar
Gujarat Environmental Institute of Management, Gandhinagar, Gujarat, India

D. Jayaseelan
Jayanagar, 3Rd Street Vadavalli, Coimbatore, Tamil Nadu, India

S. Suriyanarayanan et al. (eds.), *Water Management in Developing Countries and Sustainable Development*, Water Resources Development and Management, https://doi.org/10.1007/978-981-99-8639-2_8

aquatic life and eventually reach humans at the top of the food chain. This leads to clogged digestive tracts, disrupted digestion, and reduced reproductive growth in living organisms. These consequences have increased the concern for microplastics as a potentially harmful new issue, which demands their regulation in aquatic media. The last difficulties in microplastic treatment systems are the detection and separation of microplastics from discharged effluent. This work provides a critical examination of current and newly created methodologies. The aim of this article is to investigate methods for detecting and eliminating microplastics, as well as the challenges they pose. By implementing a waste discharge standard that minimizes their effects on aquatic ecosystems, we can work towards reducing the overall impact of microplastics on the environment.

Keywords Microplastic contamination · Polymerization · Aquatic environment · Detection analysis · Removal characteristics

8.1 Introduction

Microplastics (MP) are becoming more prevalent in aquatic habitats as annual plastic production increases. Currently, only around 71% of plastic trash is recycled, leading to an increase in microplastic contamination. Instead, the remaining 35% are directly absorbed by the environment. Plastic goods were manufactured in a little more than 335 million metric tons in 2016 and 348 million metric tons in 2017. Therefore, in just one year, plastic manufacturing worldwide rose by almost 4%. By 2050, it is expected that the amount of plastic waste that is currently directly incorporated into the marine ecosystem would have quadrupled (Derraik 2002). Richard Thompson, a marine biologist, classified 20 μm plastic particles as MPs (Thompson et al. 2004). In 2009, the National Oceanic and Atmospheric Administration (NOAA) categorized plastic debris with a diameter of less than 5 mm as Microplastics (MPs). The degradation and pealing of various plastic wastes produce these plastic particles, which pollute the environment (Hirai et al. 2011; Law et al. 2014; Van Cauwenberghe et al. 2013). The toxicity of microplastics (MPs) is determined by their chemical composition and the presence of harmful chemicals during polymerization that increase crosslinking (Meeker et al. 2009).

Studies have revealed that the build up of microplastics (MPs) in water bodies can harm ecosystems and subject wildlife to plastic waste exposure (Sun et al. 2019). Studies show that MPs can move more than 100 kms on wind currents while carrying pathogens. Even in the midst of the current pandemic, the virus-like new coronavirus (SARS-CoV-2) is infecting people in different places without the involvement of any affected patients (Ebere et al. 2020). Because MPs may facilitate pandemic transmission due to their larger adsorption surfaces and prolonged microbe retention, more research on their impacts (up to 5 days). Santos recent study examined various analytical methods for locating MPs and emphasized the need for additional research on the development of their detection and removal techniques, allowing the construction of practical strategies for reducing microplastic pollution (Rocha-Santos and Duarte 2015).

Furthermore, Akiyama recently reviewed the removal of microplastics from wastewater using acoustic waves (Poerio et al. 2019) focus was on removing microplastics using traditional filtration systems, while Padervand discussed coagulation methods. However, the data from their investigations on membrane filtering technologies and microbial degradation were scant and Sun investigated MPs' existence, detection, and removal from wastewater treatment facilities. The authors carried out extensive research on MPs and used FTIR, Raman spectroscopy, and stereomicroscopy to characterize them. Although research on MP removal technology is still in its early stages of initial, basic, advanced and final treatments in wastewater treatment plants and achieved 35–59%, 50–98%, 0.2–14% and 0.2–2% microplastic removal, respectively. Even though it could be utilized in advanced treatment procedures, microbial deterioration in biological treatment systems was frequently not well-explained. Innovative techniques for eliminating microplastics from wastewater have also been devised.

The main objective of this manuscript is to thoroughly examine the implementation of various technologies for detecting and removing MPs in wastewater, through comprehensive testing. Currently, the data included in a single report provide a clear understanding of the state of microplastic detection, facilitating a quick comprehension of the current situation.

8.2 Microplastic Detection Methods

8.2.1 Microplastic Detection

Microplastics (MPs) are ubiquitous in the environment and can be found as solid particles of various densities, colours, sizes, polymer forms and kinds, as well as when plastic particles interact with other contaminants. These characteristics are solely responsible for MP transportation in the ecosystem. Small plastic particles (PPs) are commonly found in various bodies of water. For example, small PPs have been found in various estuaries in China (Zhao et al. 2014, 2015), including Lake Taihu in the Yangtze Delta and Wuhan's urban water systems (Wang et al. 2017). The weathering process, on the other hand, converts large plastics into small PPs (Lambert and Wagner 2016). Researchers are concerned about the lack of information on the distribution of polypropylene sizes in the natural environment, which has led to the need for the development of appropriate methods (Zhang et al. 2018).

According to morphological research, microplastics (MPs) can be divided into six categories: fibres, spheres, foam, sheets, fragments and films. Among these categories, fibres are the most commonly found in natural environments (Zhang et al. 2018). Microfibres have been found in water in concentrations ranging from 0.02 to 25.8 fibres/L and in sizes ranging from 0.09 to 27.06 mm (Suaria et al.

2020). The main source of microfibre pollution is the garment industry, specifically from the washing of clothing during the manufacturing process (Salvador Cesaet al. 2017). Transparent microplastic particles often originate from disposable and short-lived plastics such as containers, bags and bottles while colored polypropylene (PP) microplastics are typically produced from long-lasting plastic packaging materials (Zhang et al. 2018).

8.3 Methods of Detection

Identifying microplastics (MPs) can be difficult due to size, shape, colour and polymer type variations. Therefore, multiple detection methods are often used. The main techniques for identifying microplastics (MPs) are physical methods such as colour and size, and chemical methods like composition and structure (Shim et al. 2017). Visual analysis, which involves using your unaided eyes and forceps to distinguish larger and some smaller coloured MPs, is used (Heo et al. 2013). The majority of tiny particles, however, are likely to be missed during analysis. As a result, the initial identification step is quick and simple (Hidalgo-Ruz and Thiel 2013) and the additional advancement in supplying in-depth information about MPs is necessary, which has led to an evolution in MPs' detection.

MP counting and morphological analysis are the stereomicroscope's two primary uses. However, its reduced magnification capacity limits its size, and how effectively it detects depends on the user's skill. It is believed that throughout the test, a measurement error of up to 70% is conceivable, with this percentage increasing as particle size lowers (Hidalgo-Ruz et al. 2012). It is challenging to utilize this technique to distinguish between synthetic fibres made of cotton and fibres that are readily available in nature (Jamal et al. 2004). Before being inspected under a microscope, the sample might be dyed to increase accuracy. To distinguish between natural and synthetic fibre, employ a pink stain (such as Rose Bengal staining solution) (Ziajahromi et al. 2017). The inaccuracy of stereomicroscopic detection can be removed using this kind of sample pretreatment. However, this approach has some limitations, including the need for more manpower and time, and the incapacity to distinguish between various types of polymers (Sun et al. 2019). However, employing this method along with a suggested pre-treatment for detecting relatively larger particles (>5 mm) will be the optimal strategy to decrease testing errors.

Some MPs are analysed from the sample using a compound microscope. Samples are collected from wastewater and now this sample has been poured into Whatman 41 filter paper of pour size 20 μm. This filter paper was placed in the hot air oven for 2 h for drying. And now this dried filter paper was kept under the microscope for analysis of microplastic. 10X, 40X magnification was given to the microscope for analysis of microplastic from dried filter paper which contains samples. By analysing this process with the help of a microscope we found that there are nanofibres present in the filter paper which is confirmed as MP (Fig. 8.1).

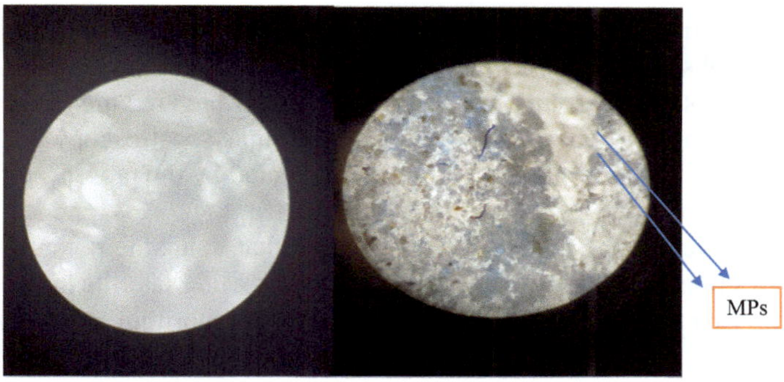

Fig. 8.1 Analysis of MPs using a microscope (40X, 10X)

PET, PP and PE-based MPs can also be examined using Polarized Light Microscopy and other modern and advanced microscopic techniques. On the basis of the source polymers and manufacturing techniques, plastic crystallinity can be categorized. Because polarized light may flow through plastics and is affected by the crystallinity of the plastic, it can be used to identify microplastic. Incontrast, PLM cannot identify thick or opaque MPs (Sierra et al. 2020). An additional assessment could be conducted to determine the typical size and colour distribution of microplastics (MPs) in order to minimize errors in the particle-level morphological process.

Chemical analysis of microplastics can provide insight into their chemical composition, which is crucial for enhancing detection accuracy. Both destructive and non-destructive methods can be used to determine the chemical properties of a substance. Destructive methods for analysing microplastics include High-Performance Liquid Chromatography (HPLC), Gas Chromatography-Electron Impact Mass Spectrometry (GC-EI-MS) with pyrolysis and thermal desorption. Non-destructive techniques include Raman spectroscopy and Fourier Transform Infrared (FTIR) spectroscope (Araujo et al. 2018; Browne et al. 2011). One of the most popular analytical methods for this purpose is spectroscopy (Hidalgo-Ruz et al. 2012).

Popular non-destructive methods for finding MPs include FTIR. This method involves exposing plastic particles to infrared light, which creates a spectrum that is ideal for the vibration of an atomic chemical bond. To accurately identify the makeup of plastic particles, a library of reference spectra that includes a variety of plastic sources, including those that are not typical, must be established. This library will be used to match the spectrum obtained from the plastic particle under analysis (Murphy et al. 2016). The procedure requires a lot of labour because the specimen must be examined and aligned under a microscope with optical capabilities before performing spectrum analysis on individual microplastics, despite the wide range of equipment available for identifying microplastics (Harrison et al. 2012). Sample correction with these methods is still difficult, though. The load on the process would be reduced while the data accuracy would be improved by integrating auto-particle detection with an image analyser.

Another common technique for identifying microplastics is Raman spectroscopy. This vibrational spectroscopy technique uses inelastic light scattering and is known for its improved spatial resolution compared to FTIR, which requires fewer pixels to create a digital image (Schymanski et al. 2018). Raman spectroscopy is also highly responsive to non-polar groups and remains unaffected by the presence of CO2 and water in the surrounding environment (Li et al. 2018).

A new remote sensing technique has emerged that holds the potential to detect microplastics floating on the surface of coastal waters using satellite images at a sub-pixel scale. This technique employs the Naive Bayes algorithm to process sensing data, resulting in an 86% detection accuracy (Biermann et al. 2020). Raman spectroscopy, automated particle identification, and image analysis can be used together to detect microplastics in various samples. In this technique, individual plastic particles are arranged on a single glass plate and individually evaluated through spectroscopy and image analysis. This approach makes it easy to identify the characteristics of microplastics (Shim et al. 2017).

8.4 Microplastic Removal Methods

8.4.1 Microorganism

Microplastic levels in biota species are rising quickly (Cole et al. 2013; Ivar Do Sul and Costa 2014). Plastic particles are present everywhere, even in the most distant regions of the planet such as the Antarctic Islands or the deep oceans (Desforges et al. 2014; Woodall et al. 2014). Therefore, biodegradation of MP might be an effective way to get rid of this plastic waste. Microbes decompose plastic waste using enzymes, transforming it into water, methane, biomass, carbon dioxide and a variety of inorganic substances. The speed and extent of microplastic biodegradation depend on the physical and chemical characteristics of the polymer (Shah et al. 2008).

8.4.2 Adsorption

Biosorption is a technique that utilizes adsorbent materials to extract microplastics from water. The adsorbate attaches to the adsorbent surface through a basic physical and chemical interaction, which could be Van der Waals attraction or ionic exchange (Mrvčić et al. 2012). Recent research suggests that the Bladderwrack, commonly found in marine environments, has the ability to absorb microplastics due to the presence of alginic acid in its cell wall (Sundbk et al. 2018). The presence of carboxylic functional groups on the alginate polymer of brown algae is found to be related to the microplastics adsorption ability of the algae. A higher number of these functional groups on the surface of the algae, the more efficient it is in adsorbing microplastics.

This is because the carboxylic groups are responsible for the binding of the plastic particles to the surface of the algae.

A biodegradable, dense and spongy material infused with chitin and graphene oxide (ChGO) was found to be able to absorb various kinds of plastic particles in a study on the removal of MPs using adsorption. For the adsorption and desorption of microplastics, this sponge can be used three times in a row. In this case, the Hydrogen-bonding force, Pi-pi stacking and electrostatic force dominate the surface microplastic adsorption of the graphene-modified adsorbent. Three days after being buried to test the biodegradability of the graphene-coated sponge, hyphae were discovered on its surface. Due to soil bacteria, the sponge completely disintegrated after 28 days, demonstrating its biodegradability. Researchers have developed a three-dimensional graphene oxide adsorbent to separate PS-based microplastics. The strong benzene moiety of polystyrene and the carbon atoms of three-dimensional reduced graphene oxide work together efficiently in this process to remove microplastics from water.

In order to prevent secondary contamination into the environment, studies should also be done to establish the reusability of this nano adsorbent. The process of adsorption is therefore time- and labour-intensive. Commercialization is therefore a challenge. Additionally, it causes secondary contamination and requires particular precautions while being dumped or renewed for future use. Before using more sophisticated approaches, such as membrane systems based on nanomaterials or advanced oxidation systems, this method can be utilized as a pretreatment.

8.4.3 Magnetic Extraction

The study found that magnetic extraction is an effective method for separating microplastics from wastewater. This technique uses external magnetic attraction, oxalic acid as a Fe di-sorbent, and magnetic seeds made of Fe nanoparticles. Iron-based nanoparticles are chosen for their ferromagnetic properties, low cost, and larger specific surface area. To increase the hydrophobicity of the nanoparticle surfaces, hexadecyltrimethoxysilane is applied, which causes the plastic particles to clump together. This method can remove almost all PS and PE beads that are between 10 and 20 μm in size.

Smaller MPs (PET, PVC, and PP) (less than 1 mm) have, on the other hand, been reduced by up to 93%. Additionally, medium MPs (200 m–1 mm) were extracted from the sediment and the waters in proportions of 78 and 84, respectively. Consequently, this technique is efficient for removing small plastic particles (less than 10 m). Because these tiny particles cannot be broken down by natural means and cannot be reused, they result in subsequent pollution. Furthermore, the amount of plastic removed by harmful nanoparticles may be lessened by the existence of soil particles and lipophilic substances. Wang recently introduced TiO2-based photocatalytic micromotors to remove MPs in aqueous media. The researchers have developed two new techniques for removing microplastics. The first method, called Pheretic interaction, is used to remove primary microplastics, such as those found in personal care products. The second method involves using chains of magnetic micromotors to keep microplastics mobile under a magnetic field for improved segregation. However, it is still challenging to confirm the specificity and identification of plastic particles using micromotors (Wang et al. 2019).

8.4.4 Biofiltration

Research suggests that traditional wastewater treatment plants may not be as effective in removing certain nutrients and organic substances as they are for microplastics. However, these facilities can still remove more than 90% of microplastics. However, a significant proportion of microplastics still escape into the environment (Mason et al. 2016; Nizzetto et al. 2016; Simon et al. 2018). In this situation, Fan Liu and associates Liu created a biofilter to eliminate MPs from the effluents of secondary water treatment plants. In this investigation, MPs were removed from personal care items, pharmaceutical waste, and wastewater from treatment plants using a prototype biofilter plant. This biofilter comprises multiple layers, with the top layer receiving secondary sewage and the bottom layer receiving treated wastewater. There were a total of six separate thick layers of stone wool inside the biofilter (1.1 m in total height). A biofilter treatment was found to reduce particulate matter to 197 particles/m^3 (79%) with a mass concentration of 2.8 g/m^3, compared to 917 particles/m^3 with a mass concentration level of 24.8 g/m^3 in the effluent (89%). The biofilter had a higher efficiency in removing larger microplastics (>100 μm) with higher mass concentrations, than smaller ones. However, due to this limitation it is challenging to achieve full elimination of microplastics.

8.4.5 Electrocoagulation

Electrocoagulation is a different type of therapy from chemical and microbiological coagulation in which metal electrodes produce cations in an electrical field. It creates coagulants that feature metal electrodes that are more susceptible to come into contact with MPs and have a variety of advantages, including the elimination of sludge, automatic treatment and efficient and reasonably priced operation. Also removed with this technique were PE-based MPs (300–355 m). Studies at pH 7.5 revealed a higher percentage of clearance (99.24%) due to a stronger coagulant production due to this neutral pH. Additionally, it was found that the peak efficiency of the system was attained after 40 min of operation and that gradually increasing the current density did not enhance the removal efficiency. Moreover, the cost of removing microplastics has been evaluated and found that electrocoagulation offers a more efficient process at a lower cost than other methods of pollutant removal (Perren et al. 2018). Therefore, this method can be applied commercially to remove microplastics before they are discharged into the aquatic environment.

8.4.6 Acoustic Waves

Using acoustic sound waves, the MPs were removed from wastewater (Akiyama et al. 2020). In this study, the author took laundry wastewater as a sample. Laundry wastewater, which contains MP fibres, is one of the most significant sources of MPs (Salvador Cesaet al. 2017). Filters in laundry machines and wastewater treatment plants would be nearly impossible to recover all MPs after they were released into the natural environment, and it would be preferable to remove MP fibres before they exited laundry machines. The research focused on enhancing the quality of wastewater from laundry machines, but it does not aim to extract microparticles. There are already many microfluidic techniques available for separating microparticles using acoustic, electric, magnetic, optical or inertial forces (Browne et al. 2011). These tools have mostly been used in the biomedical field (Lenshof and Laurell 2010). This study recommends using acoustic focusing to collect MPs and MP fibres. The author chose to employ acoustophoretic force with BAW for controlling MPs because it allows for the application of a controlled force without the need for labels or microelectrodes (Leibacher et al. 2015). A BAW device was constructed to collect MPs from the middle branch of a trifurcated microchannel using acoustic concentration at the center. The results demonstrate that MPs can be extracted from wastewater from laundry machines and analysed using BAW acoustic focusing (Fig. 8.2).

PMMA stands for poly (methyl methacrylate), PET stands for polyethylene terephthalate, PP stands for polypropylene, PS stands for polystyrene and PTFE stands for polytetrafluoroethylene. PE stands for polyethylene. The compressibility was computed by taking the reciprocal of the bulk modulus.

Microparticles with either a positive or negative ACF move towards the center or the walls of the microchannel because of the acoustophoretic forces, with the microchannel width set to half the wavelength of water. The direction of the acoustophoretic force is determined by the acoustic contrast factor (ACF). Only particles with ACF greater than one at their center can be focused by the BAW device. Table 8.1 shows that all the polymers used had a positive ACF. It's important to note that polymers with lower ACF values, such as low-density polyethylene, require higher pressure amplitudes for manipulation because the ACF affects the acoustophoretic force.

8.4.6.1 Acoustic Focusing of PS Microparticles

The acoustophoretic force directed polystyrene (PS) microparticles passing through the microchannel to its center. Photos of the three overlapping triple branches are shown in Figure. Single frames from Movie S1 are used in the images. When the piezo

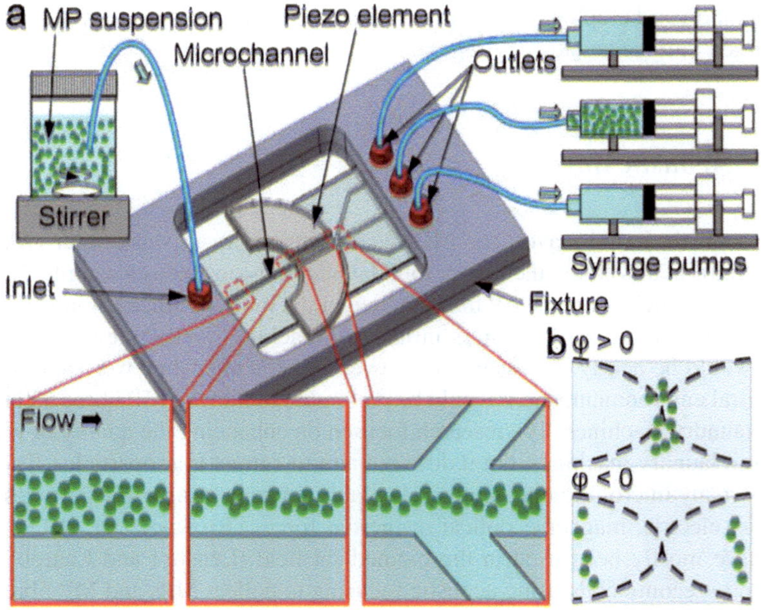

Fig. 8.2 A diagram showing the acoustic focusing of microparticles in the study. **a** A diagram of the BAW device setup and its associated equipment. The microparticles suspension was conveyed through the microchannel with the help of suction from three syringe pumps located at the outlets. The concentrated microparticles were eventually gathered from the central branch of the trifurcation (Akiyama et al. 2020). **b** A diagram illustrating the relationship between the acoustophoretic force (ACF) and the location of the focusing within the cross-section of the microchannel. Particles with a positive ACF are attracted to the nodes of the standing wave, while those with a negative ACF are drawn to the antinodes. The standing acoustic wave is depicted by the dotted lines (Akiyama et al. 2020)

Table 8.1 Physical characteristics and ACF values of illustrative polymers (Van Krevelen and Te Nijenhuis 2009)

	Density ρ[g·mL − 1]	Bulk modulus K [GPa]	Compressibility βc [GPa − 1]	ACF φ
PMMA	1.17	5.0	0.22	0.73
PS	1.02	4.0	0.20	0.51
PE ld Md hd	0.93 0.95 0.97	1.4 4.5 5.0	0.39 0.22 0.20	0.26 0.46 0.53
PET	1.38	4.0	0.25	0.75
Nylon 6	1.14	4.5	0.30	0.59
PP	0.90	3.5	0.28	0.26
Nylon 66	1.14	7.2	0.15	0.85
PTFE	2.2	2.5	0.30	0.82

components were inactive, PS microparticles were dispersed across the three trifurcation branches. However, when activated, the microparticles moved towards the middle branch but dispersed at the junction because of an increase in flow. Figure 8.3b shows the results of manual counting of the PS microparticles and indicates that 99% of them entered the center channel (Akiyama et al. 2020).

The smallest size of PS microparticles that this BAW gadget can gather from the channel side was then determined. Initially, image analysis of the PS microparticles was used to calculate the focusing time. A few minutes after the syringe pumps were switched off and the piezo components were turned on, the flow almost completely halted. Microparticles swiftly moved towards the center of the microchannel and concentrated there. Since the acoustophoretic force is proportional to the ACF, the smallest collection size for low-density polyethylene would be 6.6 μm (Fig. 8.4).

The related bar graphs are displayed in the bottom row; scale bar = 500 m; n = 5; error bars reflect one standard deviation (Akiyama et al. 2020).

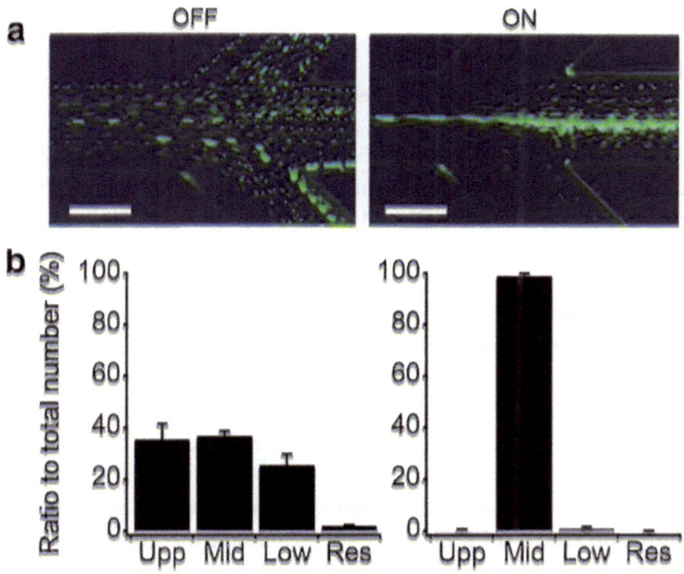

Fig. 8.3 PS microparticles are focused utilizing acoustics. **a** Combining ten-second frame-by-frame images resulted in overlaid fluorescent images of the trifurcation in the microchannel. Without the acoustic force, the flow was evenly distributed among the three branches. But when the acoustic force was applied, all PS microparticles gathered in the center of the microchannel and flowed into the central branch (Akiyama et al. 2020). **b** Graph displaying the percentage of microparticles in each microchannel and remaining microparticles on the left and right before and after the acoustic force is activated, respectively. MPs are labelled as Upp (upper), Mid (middle), and Low (lower). "Res" stands for residual MPs in the image microchannel. The scale bars measure 500 m. The sample size is five, and error bars denote one standard deviation (Akiyama et al. 2020)

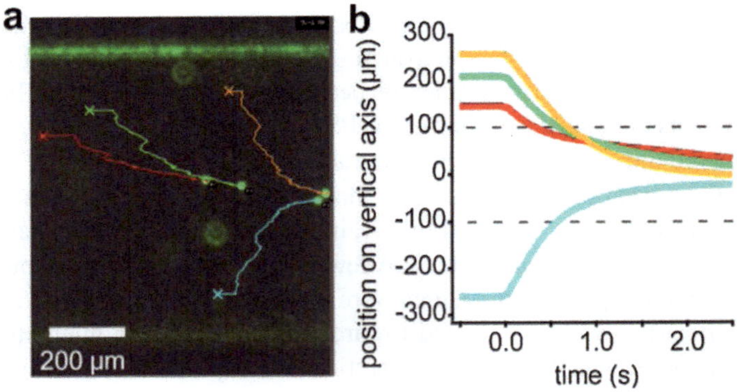

Fig. 8.4 Measurement of focus time by image analysis. At 0.0 s, the piezo components were activated, and the MP particles began to move towards the microchannel's center. **a** Four MP particle trajectories on a microscope picture. Crosses indicate the position at 0.5 s. **b** Graph of the positions of the microparticles in relation to the center of the microchannel over time in the vertical direction (Akiyama et al. 2020).

8.4.6.2 Acoustic Focus of MP Fibres

Movies S2 and S3 showed that Nylon 6 and PET fibre MPs were also centered in the microchannel, mostly collected from the side channels. Figure 8.5 displays the ratio of manually counted MP fibres and photo overlays. A small fraction of the MP fibres stuck to branch walls and edges. 6% of both types of MP fibres remained in the channel. Collection rates of 95% and 99% were achieved for PET and Nylon 6 fibres, respectively. The hydrodynamic force alone kept most of the fast-moving MP fibres flowing parallel to the channel without the need for acoustic focusing. The BAW device avoids clogging with MP fibres at the trifurcation because it aligns them with both acoustophoretic and hydrodynamic forces, resulting in high collection. Rod-shaped microparticles can also be directed to the center by hydrodynamic forces acting as inertia concentrators (Hur et al. 2011).

Usually, nylon mesh is used in commercial laundry machines to eliminate bigger fibres. A previous study (Pirc et al. 2016) examined the size of the MP fibre in the effluent of such devices (between 20 and 200 m). These fibres are more or less straight and have a diameter of more than 10 m. As a result, we hypothetically looked at straight fibres that may be thought of as cylindrical particles in the 20–200 m range. Additionally, Stoke's flow was used to calculate the focusing times for such fibres. The hydrodynamic force caused fibres to align parallel to the flow direction. The acoustophoretic force was supposed to push the fibres horizontally without altering their orientation.

Based on the focusing time of 0.7 s for the PS particles in Fig. 5b, it was estimated that the focusing times for PS fibres with lengths of 20 μm and 200 μm would be 0.82 s and 0.20 s, respectively. In the event of worst case scenario, the 2.8 s dwell time in straight channel at 300 L/min is shorter than the 1.8 s focus time for a 20 μm PP fibre with lowest ACF in Table 8.1. The results indicate that the BAW device has a

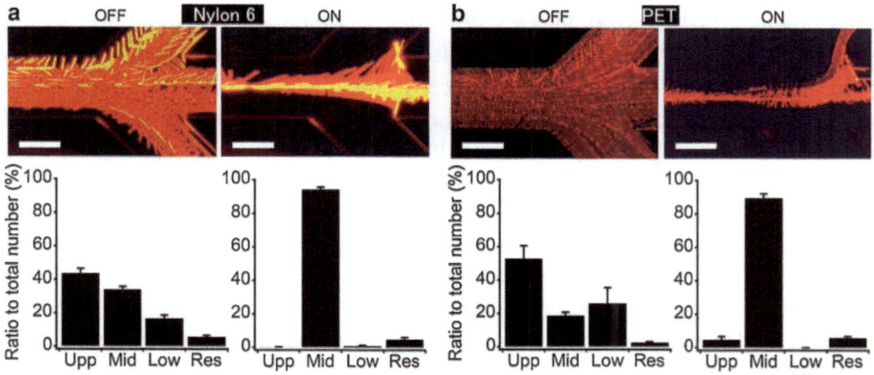

Fig. 8.5 Nylon 6 (**a**) and PET MP fibres' acoustic focusing (**b**) Fluorescence photos that have been superimposed in the top row depict MP fibres with and without acoustic focusing

strong enough acoustophoretic force to capture MP fibres from commercial washing facilities' filtered effluent. However, some MP fibres moved slowly or adhered to the microchannel walls, which could lead to clogging after extended use.

Despite moving faster than PS microparticle suspensions, MP fibres can settle at the bottom due to their large size and high density. The Hagen-Poiseuille flow reduces velocity near the channel walls. Additionally, in high aspect ratio microchannels, the acoustophoretic force at the top and bottom center drives towards the center of the corresponding wall (Muller et al. 2012). The adhesion of MP fibres to the walls may also be influenced by force. MP fibres, which are less dense than water, would naturally float to the top of the channel. Orienting the microchannel with gravity will assist in addressing sedimentation and levitation issues. Furthermore, the attachment of MP fibres need to be made easier by the rough surfaces produced by diamond-tool machining, such as frosted glass. Smoother surfaces can be produced using several processing techniques, including glass reflow, reactive ion etching, or wet etching with hydrofluoric acid. Placing the device vertically and smoothing the channel surface can effectively solve the issue. Laundry detergent often contains a surface-active ingredient that increases MP attachment by neutralizing surface charges in laundry machine effluent (Akiyama et al. 2020).

8.4.6.3 Towards the Practical Continuous Collection of MPs

Nearly all MPs fibres and particles from the branch's center were captured by the BAW device. PS microparticle and MP fibre blood-cell separation collection rates were comparable to the best results from prior studies (Wu et al. 2019). Previous studies showed that impedance-matched buffer flow achieved high removal rates for blood cells (nearly quantitative) and over 85% collection rate for platelets at flow rates up to 10 ml/min. However, using a buffer flow would dilute MP-containing samples, making it ineffective for MP collection.

A practical MP-removal device will require a combination of serial and parallel separation processes. Each washing cycle in a typical washing machine generates more than 100 L of effluent, which typically contains more than 100 fibres. Based on the assumption of a 99% collection rate for PS particles, the final collection rate after 10 sequential separations would be 90%, which is satisfactory for environmental studies. However, using a microfluidic device to treat an entire load of wastewater would take some time. The processing time of parallel processes can simply be decreased by raising the total flow rate. To create complex microchannels for serial and parallel processes, such as stacked glass-based microfluidic devices, semiconductor processing technologies can be used.

There is currently no consensus on the number of lawmakers who are dangerous to the environment and to human health. Most research on MP dispersal in the environment focuses on MPs larger than 10 m in diameter (Mai et al. 2018). The BAW device described in this research can, in theory, collect MPs with diameters of up to 5 μm, as previously stated.

Environmental sample sets in reality may contain significant amounts of organic matter. Unlike MPs, lightweight and soft debris in such materials would be separated from MPs since they would only experience minimal impact from the acoustophoretic force. However, heavy and hard debris such as crustacean shells or degraded wood, if present, would have a positive ACF and would be collected along with MPs. Initial tests showed that the BAW device only slightly focused on cellulose debris, indicating challenges in using acoustic focusing for MP isolation from environmental samples. Unseparated debris can be disregarded for analysis by labelling MPs with dyes like Nile Red, a widely used method in MP analysis.

8.4.6.4 Conventional Treatment Techniques

Traditional water treatment plants are not intended to remove microplastics from water. Certain treatment procedures employed in traditional plants, on the other hand, can help to reduce microplastic concentrations to some extent. Here are a few examples of procedures typically used in existing treatment plants:

1. Coagulation and Flocculation: Coagulation is the process of adding chemicals that cause microplastic particles to aggregate and create bigger flocs. Flocculation then aids in the settling of these flocs, allowing them to be removed via sedimentation or filtration.
2. Sedimentation: Sedimentation tanks are used in water treatment plants to allow suspended particles, including microplastics, to settle to the bottom. Sludge is formed from the settling particles.
3. Filtration: To remove suspended particles, conventional treatment plants frequently use various types of filters, such as sand filters or multimedia filters. These filters can also assist in the capture of bigger microplastics.

4. Activated Carbon Adsorption: In water treatment plants, activated carbon is often employed to remove organic pollutants. Because of its porous structure, it can also absorb some microplastics. However, its ability to remove tiny microplastics may be limited.
5. Disinfection: Microplastics present in water as free-floating particles may be removed by disinfection techniques such as chlorination or ultraviolet (UV) irradiation. However, rather than targeting microplastics, the primary goal of disinfection is to eradicate hazardous bacteria.
6. Advanced Filtration Technologies: Sand filters, for example, can remove bigger microplastics, but tiny particles may pass through. Because of their tiny pore sizes, advanced filtration technologies such as membrane filtration (including ultrafiltration and nanofiltration) have shown promise in absorbing smaller microplastics. Membrane filtration may successfully remove particles as small as a submicron or a nanometer.
7. Granular Activated Carbon (GAC) Filtration: While activated carbon is generally employed to remove organic compounds, granular activated carbon (GAC) filters have demonstrated some potential to absorb microplastics. GAC filters having microplastic removal-specific features, such as increased surface area or changed surface chemistry, may be more successful.
8. Ozonation and Advanced Oxidation Processes (AOPs): Ozonation, as well as other advanced oxidation methods such as hydrogen peroxide/UV or ozone/UV, are largely utilized for organic pollutant disinfection and oxidation. These methods have also shown some ability to break down microplastics into smaller bits, which can aid in their removal during following treatment phases.
9. Coagulation with Polymer Additives: Traditional coagulation procedures can be enhanced by the inclusion of specialized polymer additives that improve microplastic aggregation and settling. These polymers have the ability to enhance particle size and improve particle removal efficiency during sedimentation or filtration.
10. Biological Treatment: While conventional treatment plants are not specifically intended for the removal of microplastics, biological treatment procedures such as activated sludge systems have shown promise in capturing microplastics. Microorganisms in the treatment process can help remove microplastics by forming biofilms and then sedimenting them.
11. Research and Development: There is continuing research and development focusing on developing specific microplastic removal technology. This includes investigating novel methods such as electrocoagulation, electrostatic separation, and improved adsorbents specifically intended to target microplastics.
12. Activated Carbon: Because of its high surface area and adsorption capability, activated carbon (AC) is a flexible adsorbent commonly utilized in water treatment. Because of its physical features, activated carbon can effectively absorb bigger microplastic particles when it comes to microplastic removal. Because of the porous structure of activated carbon, there is plenty of surface area for microplastic adsorption, minimizing their presence in the treated water.

13. However, activated carbon is limited in its ability to remove tiny microplastics. Finer particles may be more likely to pass through the porous structure of the carbon, lowering overall efficiency. Furthermore, activated carbon's adsorption capacity might become saturated over time, necessitating regular replacement or regeneration to maintain its efficiency.
14. Reverse Osmosis (RO): Reverse osmosis is a highly effective membrane-based water treatment technology for removing microplastics. RO systems make use of a semipermeable membrane with extremely small pores, which typically range from 0.0001 to 0.001 μm. Contaminants, including microplastics, are rejected and eliminated as water is driven through the membrane under pressure, resulting in clean, purified water.
15. Microplastics of all sizes, including larger and smaller particles, can be successfully removed by reverse osmosis. The membrane's tiny pore size enables a high clearance effectiveness, frequently surpassing 99%. It is crucial to remember, however, that reverse osmosis is a relatively sophisticated and energy-intensive process that necessitates regular monitoring and maintenance to assure peak performance.
16. Biological Removal: Biological treatment procedures can help remove microplastics from water through a variety of mechanisms. Microorganisms in activated sludge systems create flocs that can entrap microplastic particles, assisting in their removal during the settling phase. Microorganisms can also breakdown and assimilate certain forms of microplastics, resulting in their partial or complete elimination from water.

Sand filters and trickling filters, for example, provide additional chances for microplastic removal. The biofilm that forms on the filter medium can operate as a physical barrier, trapping and preventing microplastics from passing through. Furthermore, the filtration process, in conjunction with microbial activity, can aid in the removal of microplastics. While biological removal methods can reduce microplastics to some extent, their performance can vary depending on parameters such as microplastic composition and size, microbial activity, and process design. More research is being performed to optimize biological treatment procedures for the removal of microplastics.

In general, a combination of activated carbon, reverse osmosis, and biological removal technologies can aid in the reduction of microplastic pollution in water treatment operations. To ensure effective microplastic removal and conserve water resources, it is critical to examine the individual properties of microplastics, process design parameters, and the necessity for continual monitoring and optimization. It is crucial to note that the efficiency of various approaches can vary depending on parameters such as microplastic size, shape, concentration, and treatment plant configuration.

8.5 Conclusion

The standardization of procedures for detecting and eliminating microplastics is a challenging task due to the diverse properties and uncertain nature of their complete elimination from the environment. This paper evaluates and compares the performance of both traditional and newly implemented methods for detecting and removing microplastics. At first, the detection of microplastics was carried out using manual techniques such as forceps and visual inspections. However, the use of a stereomicroscope was found to improve efficiency and accuracy. While central laboratory techniques such as SEM–EDX and PLM are widely recognized for their precision, they are limited by size restrictions and potential errors in identifying the color and composition of microplastics. In contrast, FTIR has become a commonly used method due to its affordability, portability and ease of use.

However, FPAFTIR, TGA-FTIR, and Raman spectroscopy are being used to expand the identification of microplastics (MPs) in the environment. The most widely used technique for removing MPs is through the use of ultrasonic waves. The study demonstrated that it is feasible to collect MPs using a bulk acoustic wave (BAW) device and acoustic focusing. The BAW device collected all MPs of various shapes from the central branch of a trifurcated microchannel. However, for practical applications, multiple similar processes would have to be integrated both serially and simultaneously. This research has significant potential for the removal of MPs from aqueous solutions such as the effluent from commercial laundry machines. The removal process is being developed to integrate adsorption, magnetic extraction, and electrocoagulation methods, however, their practical use is limited by high energy consumption, secondary MP contamination, and prolonged time duration. The environment contains a majority of microplastics made from polystyrene (PS) and polyethylene (PE) mostly found in clothing and household items. As a potential area for future study that can aid in the removal of MP pollution from water, membrane-based antifouling technology should also be considered.

References

Akiyama Y, Egawa T, Koyano K, Moriwaki H (2020) Acoustic focusing of microplastics in microchannels: A promising continuous collection approach. Sens Actuat B Chem 304. https://doi.org/10.1016/j.snb.2019.127328

Araujo CF, Nolasco MM, Ribeiro AMP, Ribeiro-Claro PJA (2018) Identification of microplastics using Raman spectroscopy: Latest developments and future prospects. Water Res 142:426–440. Elsevier Ltd. https://doi.org/10.1016/j.watres.2018.05.060

Biermann L, Clewley D, Martinez-Vicente V, Topouzelis K (2020) Finding plastic patches in coastal waters using optical satellite data. Sci Rep 10(1). https://doi.org/10.1038/s41598-020-62298-z

Blair RM, Waldron S, Phoenix VR, Gauchotte-Lindsay C (2019) Microscopy and elemental analysis characterisation of microplastics in sediment of a freshwater urban river in Scotland, UK. Environ Sci Pollut Res. https://doi.org/10.1007/s11356-019-04678-1

Browne MA et al (2011) Accumulation of microplastic on shorelines woldwide: sources and sinks. Environ Sci Technol 45(21):9175–9179. https://doi.org/10.1021/es201811s
Browne MA et al (2011) Accumulation of microplastic on shorelines woldwide: sources and sinks. Environ Sci Technol 45(21):9175–9179. https://doi.org/10.1021/es201811s
Cole M et al (2013) Microplastic ingestion by zooplankton. Environ Sci Technol 47(12):6646–6655. https://doi.org/10.1021/es400663f

Dekiff JH, Remy D, Klasmeier J, Fries E (2014) Occurrence and spatial distribution of microplastics in sediments from Norderney. Environ Pollut 186:248–256. https://doi.org/10.1016/j.envpol.2013.11.019

Derraik B (2002) The pollution of the marine environment by plastic debris: a review Jos e e G. www.elsevier.com/locate/marpolbul

Desforges JPW, Galbraith M, Dangerfield N, Ross PS (2014) Widespread distribution of microplastics in subsurface seawater in the NE Pacific Ocean. Mar Pollut Bull 79(1–2):94–99. https://doi.org/10.1016/j.marpolbul.2013.12.035

Dümichen E, Eisentraut P, Bannick CG, Barthel AK, Senz R, Braun U (2017) Fast identification of microplastics in complex environmental samples by a thermal degradation method. Chemosphere 174:572–584. https://doi.org/10.1016/j.chemosphere.2017.02.010

Ebere EC et al. (2020) Novel coronavirus (SARS-CoV-2) and airborne microplastics chemical speciation and risk implications of some heavy metals in sediment profiles of urban stream stretches, Yenagoa City, Bayelsa State, Nigeria view project personal skill, professional development and other Certificates view project Novel Coronavirus (SARS-CoV-2) and Airborne Microplastics. J Mater Environ Sci 2020(9):1454–1461, 1454. http://www.jmaterenvironsci.com

Elert AM et al (2017) Comparison of different methods for MP detection: what can we learn from them, and why asking the right question before measurements matters? Environ Pollut 231:1256–1264. https://doi.org/10.1016/j.envpol.2017.08.074

Erni-Cassola G, Gibson MI, Thompson RC, Christie-Oleza JA (2017) Lost, but found with nile red: a novel method for detecting and quantifying small microplastics (1 mm to 20 μm) in environmental samples. Environ Sci Technol 51(23):13641–13648. https://doi.org/10.1021/acs.est.7b04512

Fries E, Dekiff JH, Willmeyer J, Nuelle MT, Ebert M, Remy D (2013) Identification of polymer types and additives in marine microplastic particles using pyrolysis-GC/MS and scanning electron microscopy. Environ Sci Process Impacts 15(10):1949–1956. https://doi.org/10.1039/c3em00214d

Harrison JP, Ojeda JJ, Romero-González ME (2012) The applicability of reflectance micro-Fourier-transform infrared spectroscopy for the detection of synthetic microplastics in marine sediments. Sci Total Environ 416:455–463. https://doi.org/10.1016/j.scitotenv.2011.11.078

Heo NW et al (2013) Distribution of small plastic debris in cross-section and high strandline on Heungnam beach, South Korea. Ocean Sci J 48(2):225–233. https://doi.org/10.1007/s12601-013-0019-9

Hidalgo-Ruz V, Thiel M (2013) Distribution and abundance of small plastic debris on beaches in the SE Pacific (Chile): a study supported by a citizen science project. Mar Environ Res 87–88:12–18. https://doi.org/10.1016/j.marenvres.2013.02.015

Hidalgo-Ruz V, Gutow L, Thompson RC, Thiel M (2012) Microplastics in the marine environment: a review of the methods used for identification and quantification. Environ Sci Technol 46(6):3060–3075. https://doi.org/10.1021/es2031505
Hidalgo-Ruz V, Gutow L, Thompson RC, Thiel M (2012) Microplastics in the marine environment: a review of the methods used for identification and quantification. Environ Sci Technol 46(6):3060–3075. https://doi.org/10.1021/es2031505
Hirai H et al (2011) Organic micropollutants in marine plastics debris from the open ocean and remote and urban beaches. Marine Pollut Bull 62(8):1683–1692. https://doi.org/10.1016/j.marpolbul.2011.06.004

Hur SC, Choi SE, Kwon S, di Carlo D (2011) Inertial focusing of non-spherical microparticles. Appl Phys Lett 99(4). https://doi.org/10.1063/1.3608115

Ivar Do Sul JA, Costa MF (2014) The present and future of microplastic pollution in the marine environment. Environ Pollut 185:352–364. https://doi.org/10.1016/j.envpol.2013.10.036

Jamal M, Sarac AS, Magner E (2004) Conductive copolymer-modified carbon fibre microelectrodes: electrode characterisation and electrochemical detection of p-aminophenol. Sens Actuat B Chem 97(1):59–66. https://doi.org/10.1016/S0925-4005(03)00647-6

Lambert S, Wagner M (2016) Formation of microscopic particles during the degradation of different polymers. Chemosphere 161:510–517. https://doi.org/10.1016/j.chemosphere.2016.07.042

Law KL et al (2014) Distribution of surface plastic debris in the eastern pacific ocean from an 11-year data set. Environ Sci Technol 48(9):4732–4738. https://doi.org/10.1021/es4053076

Leibacher I, Reichert P, Dual J (2015) Microfluidic droplet handling by bulk acoustic wave (BAW) acoustophoresis. Lab Chip 15(13):2896–2905. https://doi.org/10.1039/c5lc00083a

Lenshofand A, Laurell T (2010) Continuous separation of cells and particles in microfluidic systems. Chem Soc Rev 39(3):1203–1217. https://doi.org/10.1039/b915999c

Li J, Liu H, Paul Chen J (2018) Microplastics in freshwater systems: a review on occurrence, environmental effects, and methods for microplastics detection. Water Res 137:362–374. Elsevier Ltd. https://doi.org/10.1016/j.watres.2017.12.056

Liu F, Nord NB, Bester K, Vollertsen J (2020) Microplastics removal from treated wastewater by a biofilter. Water (Switzerland) 12(4). https://doi.org/10.3390/W12041085

Löder MGJ, Kuczera M, Mintenig S, Lorenz C, Gerdts G (2015) Focal plane array detector-based micro-Fourier-transform infrared imaging for the analysis of microplastics in environmental samples. Environ Chem 12(5):563–581. https://doi.org/10.1071/EN14205

Mai L, Bao LJ, Shi L, Wong CS, Zeng EY (2018) A review of methods for measuring microplastics in aquatic environments. Environ Sci Pollut Res 25(12):11319–11332. Springer Verlag. https://doi.org/10.1007/s11356-018-1692-0

Mason SA et al (2016) Microplastic pollution is widely detected in US municipal wastewater treatment plant effluent. Environ Pollut 218:1045–1054. https://doi.org/10.1016/j.envpol.2016.08.056

Meeker JD, Sathyanarayana S, Swan SH (2009) Phthalates and other additives in plastics: human exposure and associated health outcomes. Philos Trans Royal Soc b Biol Sci 364(1526):2097–2113. https://doi.org/10.1098/rstb.2008.0268

Mrvčić J, Stanzer D, Šolić E, Stehlik-Tomas V (2012) Interaction of lactic acid bacteria with metal ions: opportunities for improving food safety and quality. World J Microbiol Biotechnol 28(9):2771–2782. https://doi.org/10.1007/s11274-012-1094-2

Muller PB, Barnkob R, Jensen MJH, Bruus H (2012) A numerical study of microparticle acoustophoresis driven by acoustic radiation forces and streaming-induced drag forces. Lab Chip 12(22):4617–4627. https://doi.org/10.1039/c2lc40612h

Murphy F, Ewins C, Carbonnier F, Quinn B (2016) Wastewater treatment works (WwTW) as a in the Aquatic

Nizzetto L, Futter M, Langaas S (2016) Are agricultural soils dumps for microplastics of Urban origin? Environ Sci Technol 50(20):10777–10779. American Chemical Society. https://doi.org/10.1021/acs.est.6b04140

Perren W, Wojtasik A, Cai Q (2018) Removal of microbeads from wastewater using electrocoagulation. ACS Omega 3(3):3357–3364. https://doi.org/10.1021/acsomega.7b02037

Pirc U, Vidmar M, Mozer A, Kržan A (2016) Emissions of microplastic fibers from microfiber fleece during domestic washing. Environ Sci Pollut Res 23(21):22206–22211. https://doi.org/10.1007/s11356-016-7703-0

Poerio T, Piacentini E, Mazzei R (2019) Membrane processes for microplastic removal. Molecules 24(22). MDPI AG. https://doi.org/10.3390/molecules24224148

Proceedings of the International Research Workshop on the Occurrence, Effects, and Fate of Microplastic Marine Debris (2009). www.MarineDebris.noaa.gov

Rocha-Santos T, Duarte AC (2015) A critical overview of the analytical approaches to the occurrence, the fate and the behavior of microplastics in the environment. TrAC Trends Anal Chem 65:47–53. Elsevier B.V. https://doi.org/10.1016/j.trac.2014.10.011

Salvador Cesa F, Turra A, Baruque-Ramos J (2017) Synthetic fibers as microplastics in the marine environment: a review from textile perspective with a focus on domestic washings. Sci Total Environ 598:1116–1129. Elsevier B.V. https://doi.org/10.1016/j.scitotenv.2017.04.172
Salvador Cesa F, Turra A, Baruque-Ramos J (2017) Synthetic fibers as microplastics in the marine environment: a review from textile perspective with a focus on domestic washings. Sci Total Environ 598:1116–1129. Elsevier B.V. https://doi.org/10.1016/j.scitotenv.2017.04.172
Schymanski D, Goldbeck C, Humpf HU, Fürst P (2018) Analysis of microplastics in water by micro-Raman spectroscopy: release of plastic particles from different packaging into mineral water. Water Res 129:154–162. https://doi.org/10.1016/j.watres.2017.11.011

Shah AA, Hasan F, Hameed A, Ahmed S (2008) Biological degradation of plastics: a comprehensive review. Biotechnol Adv 26(3):246–265. https://doi.org/10.1016/j.biotechadv.2007.12.005

Shim WJ, Hong SH, Eo SE (2017) Identification methods in microplastic analysis: a review. Anal Method 9(9):1384–1391. Royal Society of Chemistry. https://doi.org/10.1039/c6ay02558g

Sierra I, Chialanza MR, Faccio R, Carrizo D, Fornaro L, Pérez-Parada A (2020) Identification of microplastics in wastewater samples by means of polarized light optical microscopy. Environ Sci Pollut Res 27(7):7409–7419. https://doi.org/10.1007/s11356-019-07011-y

Simon M, van Alst N, Vollertsen J (2018) Quantification of microplastic mass and removal rates at wastewater treatment plants applying Focal Plane Array (FPA)-based fourier transform infrared (FT-IR) imaging. Water Res 142:1–9. https://doi.org/10.1016/j.watres.2018.05.019

Suaria G et al (2020) O C E A N O G R A P H Y microfibers in oceanic surface waters: a global characterization. https://www.science.org

Sun J, Dai X, Wang Q, van Loosdrecht MCM, Ni BJ (2019) Microplastics in wastewater treatment plants: detection, occurrence and removal. Water Res 152:21–37. Elsevier Ltd. https://doi.org/10.1016/j.watres.2018.12.050

Sun C, Wang Z, Chen L, Li F (2020) Fabrication of robust and compressive chitin and graphene oxide sponges for removal of microplastics with different functional groups. Chem Eng J 393. https://doi.org/10.1016/j.cej.2020.124796

Sundbæk KB, Koch IDW, Villaro CG, Rasmussen NS, Holdt SL, Hartmann NB (2018) Sorption of fluorescent polystyrene microplastic particles to edible seaweed Fucus vesiculosus. J Appl Phycol 30(5):2923–2927. https://doi.org/10.1007/s10811-018-1472-8

Thompson R et al (2004) Lost at sea: where is all the plastic? www.sciencemag.org/cgi/content/full/304/5672/838/

van Krevelen DW, te Nijenhuis K (2009) Mechanical properties of solid polymers. In Properties of polymers. Elsevier, pp 383–503. https://doi.org/10.1016/b978-0-08-054819-7.00013-3

van Cauwenberghe L, Vanreusel A, Mees J, Janssen CR (2013) Microplastic pollution in deep-sea sediments. Environ Pollut 182:495–499. https://doi.org/10.1016/j.envpol.2013.08.013

Wang W, Ndungu AW, Li Z, Wang J (2017) Microplastics pollution in inland freshwaters of China: a case study in urban surface waters of Wuhan, China. Sci Total Environ 575:1369–1374. https://doi.org/10.1016/j.scitotenv.2016.09.213

Wang L, Käppler A, Fischer D, Simmchen J (2019) Photocatalytic TiO 2 micromotors for removal of microplastics and suspended matter

Woodall LC et al (2014) The deep sea is a major sink for microplastic debris. Royal Soc Open Sci 1(4). https://doi.org/10.1098/rsos.140317

Wu M, Ozcelik A, Rufo J, Wang Z, Fang R, Jun Huang T (2019) Acoustofluidic separation of cells and particles. Microsyst Nanoeng 5(1). Nature Publishing Group. https://doi.org/10.1038/s41378-019-0064-3

Zhang K, Su J, Xiong X, Wu X, Wu C, Liu J (2016) Microplastic pollution of lakeshore sediments from remote lakes in Tibet plateau, China. Environ Pollut 219:450–455. https://doi.org/10.1016/j.envpol.2016.05.048

Zhang K et al (2018) Microplastic pollution in China's inland water systems: a review of findings, methods, characteristics, effects, and management. Sci Total Environ 630:1641–1653. Elsevier B.V. https://doi.org/10.1016/j.scitotenv.2018.02.300

Zhao S, Zhu L, Wang T, Li D (2014) Suspended microplastics in the surface water of the Yangtze Estuary System, China: first observations on occurrence, distribution. Mar Pollut Bull 86(1–2):562–568. https://doi.org/10.1016/j.marpolbul.2014.06.032
Zhao S, Zhu L, Li D (2015) Microplastic in three urban estuaries, China. Environ Pollut 206:597–604. https://doi.org/10.1016/j.envpol.2015.08.027
ZiajahromiBEng S (2007) Identification and quantification of microplastics in wastewater treatment plant effluent: investigation of the fate and biological effects

Chapter 9
Low-Cost Sustainable Nanotechnology Approaches for Water Quality Monitoring in Developing Countries

Seemesh Bhaskar and Sai Sathish Ramamurthy

Abstract Conventionally, techniques based on electrochemistry, optical feedback (including absorbance, fluorescence and scattering), electronic response (pH and conductivity) are being explored and revised in addition to the use of exorbitant methodologies such as GC–MS and Mass spectroscopy to analyze the quality of water from various water bodies including river, pond and lake to name a few. Predominantly, these methodologies are time and labor-intensive in addition to being not amenable for resource-limited settings. In this background, recently, there is a growing demand to develop technologies for water quality monitoring using advanced biophysicochemical strategies that predominantly involve the use of novel nanomaterials and interfaces. Importantly, development of methodologies that aid in rapid, cost-effective, real-time identification and monitoring of water quality for existing or emerging pollutants (ions, molecules and other analytes) is the need of the hour in order to establish and implement effective pollution control regulations, with appropriate surveys and surveillance. Although methodologies involving nanotechnology render newer and simplified ways of detecting biohazards and analyzing the same, the research focus toward the development of sustainable routes to address the current challenges of human and environmental health is seldom discussed. In this perspective, this chapter summarizes the newer developments specifically in

S. Bhaskar (✉)
Carl R. Woese Institute for Genomic Biology (IGB), Department of Electrical and Computer Engineering, Holonyak Micro and Nanotechnology Laboratory, University of Illinois, Urbana–Champaign, IL 61801, USA
e-mail: seemeshb@illinois.edu

S. S. Ramamurthy
Surface Plasmon based Translational and Advanced Research (STAR) Laboratory, Central Research Instruments Facility (CRIF), Department of Chemistry, Sri Sathya Sai Institute of Higher Learning, Prasanthi Nilayam, Puttaparthi, Anantapur, Andhra Pradesh 515134, India
e-mail: rsaisathish@sssihl.edu.in

S. Suriyanarayanan et al. (eds.), *Water Management in Developing Countries and Sustainable Development*, Water Resources Development and Management,
https://doi.org/10.1007/978-981-99-8639-2_9

nanotechnology domain with a focus toward advancement of bio-inspired nanointerfaces for detection of hazardous analytes in water samples. The transdisciplinary onset of synergizing the merits of fluorescence spectroscopy and nano-engineering has ameliorated the biosensing and water quality monitoring routes with drastically augmented sensitivity. This is majorly on account of the ability of materials with functional properties at nano-dimensions to boost light-matter interactions pertaining to extremely low concentrations of hazardous molecules and ions in water samples. While different methodologies are explored in this direction, this chapter deliberates the robust hybridization of bludgeoning bio-inspired photonic sensor platforms such as surface plasmon-coupled emission (SPCE) and photonic crystal-coupled emission (PCCE) along with ubiquitously used and cost-effective smartphone-based detection techniques. The understanding of the associated strategies and execution routes in realizing cutting-edge ultra-high sensitivity (attomolar and zeptomolar limit of detection), thereby presents unifying principles from different scientific disciplines such as materials chemistry, nanotechnology and photonics for addressing the current challenges of water quality monitoring in low-and-middle-income countries.

Keywords Nano-engineering · Smartphone sensing · Photo-plasmonics · Surface plasmon-coupled emission · Hybrid nanointerfaces · Point-of-care diagnostics · Water quality monitoring

9.1 Introduction

Undoubtedly, water is an important entity for sustaining terrestrial, aerial as well as aquatic lifeforms. The utility of water toward several human activities includes, but not limited to, agriculture, hydropower generation, flood prevention, recreation, industrial functionality, domestic purposes, fishing, medication and tourism has been exponentially increasing in the past few decades (Stillinger 1980; Gray 2017; Howard and Bartram 2003). While the water constitutes about ~60% of the normal human body, the cells and organs utilize water to essentially perform functionalities including perspiration, bowel moments, lubrication, urination and temperature regulation to name a few (Moss 2007). Importantly, adequate supply of water ensures protection to the spinal cord, tissues and joints, and hence remains an indispensable part of human life. As rightly said by noble laureate (1937) Albert von Szent-GyörgyiNagyrápolt, '*Water is life's matter and matrix, mother and medium. There is no life without water*', the necessary role of water in maintaining life on earth has become more clearer in the recent days, especially with the onset of population explosion and industrialization (Abel 2014; Goel 2006; Warren 1971).

The global population is expected to see an upsurge from 9.7 billion in 2050 to 11 billion in 2100, thereby increasing the tremendous need for freshwater sources (Chawla et al. 2020). Consequently, the issues pertaining to water scarcity are anticipated to grow exponentially. Moreover, in the last 100 years, the per capita utilization of water resources has significantly increased from 500 to 3800 km^3/

year (Chawla et al. 2020). While this is the scenario at global level, considering a developing country like India to analyze the population (~18% of the world's population) versus the available freshwater (~4% of the global resources), presents serious concerns for water management for the upcoming generation. Interestingly, it has been documented that the highly populated countries such as India and China are enhancing the industrial technologies encompassing petrochemical, pharmaceutical, semiconductor, agrochemical and self-driving to name a few, which has concurrently resulted in high usage of water resources in addition to contamination of water bodies via uncontrolled industrial waste (Nagar and Pradeep 2020). The inadequacy of regulated governmental policies for managing such wastes and appropriate water reuse and recycling has posed major challenges in developing countries. In this regard, the need for holistic balance between rapid industrialization (with extensive over-exploitation) and water resource management has raised several environmental concerns with regard to groundwater reserves, lakes, rivers and wetlands. While such issues related to industrialization indicate the need for developing revised water management protocols, the depreciation in the quality of water has raised serious health implications.

The quality of water determines whether or not it would be the cause of well-being or a cause of diseases and deaths. It has been noted that, more than 99% of the deaths that are associated with the water-related issues are emerging from developing countries, wherein the >80% of such population live in remote areas with negligible access to safe drinking water (Sitecontrol Progress on Drinking Water and Sanitation 2014; Chouler and Lorenzo 2015). Additionally, developing countries such as the Philippines depend on the traditional methodologies for obtaining water as well as for water quality assessment. Furnishing safe water resources with adequate management of water resources is hence crucial for a sustained economic growth and maintenance of livelihood safety. This assists in dwindling the risks involved with health-related ailments and diminishes the vulnerability of social communities to diseases. In this regard, an important Millennium Developed Goal discusses the prospects for making drinking water globally accessible in order to decrease the morbidity across the world (World Bank World Development Report 2008; Jackson et al. 2001; Development and for I., Development, E.C.C.D.-G. 2002). Such a realization attained with refinement of the drinking water via appropriate incorporation of sanitation, hygiene and other important parameters of resource management is anticipated to lessen the disease susceptibility and burden by 10%. The debate with respect to the individual contribution and the collective community impact on water misuse has remained an inevitable challenge with a clear exhibition of collective wisdom for sustainable water resource management (Nagar and Pradeep 2020).In addition to this, the uncertainty and gaps in the governmental regulations at local, national and international scenarios have undoubtedly worsened the current state-of-affairs especially with regard to water pollution and relevance of management in developing countries. In order to understand the quality of the water, appropriate sensing methodologies need to be developed. This chapter majorly discusses the relevance of biosensing using advanced technologies based on nanomaterials and photoplasmonics. To begin with, we discuss the major causes and effects of water pollution in order to comprehend the relevance of sensing hazardous molecules and ions in water to determine the quality of water bodies. The major portion of the chapter is further devoted to presenting the latest technologies that are developed for detection of ions and molecules of interest in water bodies. In this regard, the performance of surface plasmon-coupled emission (SPCE) and photonic crystal-coupled emission (PCCE) platforms are elaborately presented to the broad audience of interdisciplinary scientific disciplines (Bhaskar et al. 2020a, 2023a; Lakowicz 2004; Rai et al. 2021a).

9.2 Water Pollution: An Overview

Prior to the discussion of different methodologies being developed for the detection of molecules and ions that are relevant in water pollution, it is informative to briefly present the various routes that contribute to the causes of water pollution and the effects that are observed consequently. By and large, the point sources and nonpoint sources constitute the major two categories of water pollution (Peirce et al. 1998). As the name suggests, the point source driven water pollution is caused by a single location source such as channels in industrial waste and discharge pipes. On the other hand, the nonpoint source driven water pollution is associated with the scenarios where the pollutants enter the water resource from different points simultaneously. The waste runoffs observed in the construction sites and associated land troubles result in a huge contribution of nonpoint source for water pollution (Richardson and Ternes 2018; Dosi and Tomasi 1994; Xie et al. 2022). The world water distribution and the available sources of freshwater are pictorially presented in Fig. 9.1. The figure also summarizes the causes, effects and types of water pollution. In brief, the major causes of water pollution are urbanization, unjustifiable water, and land exploitation and uncontrolled population growth. Moreover, intensive farming with the use of hazardous herbicides, pesticides, fungicides and fertilizers has contributed significantly toward water pollution. Industrial contributions include oil pollution, nuclear/radioactive waste, pharmaceutical by-products, soaps, detergents, food toxins and sewers with heavy metals (mercury, lead, aluminum and zinc) (Goel 2006; Owa 2013).

The harmful effects of water pollution are numerous. While the aquatic lifeforms are directly affected reflecting on the health and growth of fishes and other aquatic organisms, the water pollution also substantially affects the terrestrial lifeforms by contaminating the associated lifeforms. The highest level of hazard is introduced especially when the sources of drinking water are contaminated by one means or other (Verma and Dwivedi 2013; Qu et al. 2013). The different sources of water pollution and the related effects are also presented in Fig. 9.1 to present a broad overview to the broad audience of the book. In this background, it is important to realize the seriousness of the situation where water pollution results in different health disorders including but not limited to: cholera, diarrhea, giardiasis, amoebiasis, gastroenteritis, encephalitis, dysentery, respiratory infections ascariasis, hepatitis, typhoid and polio

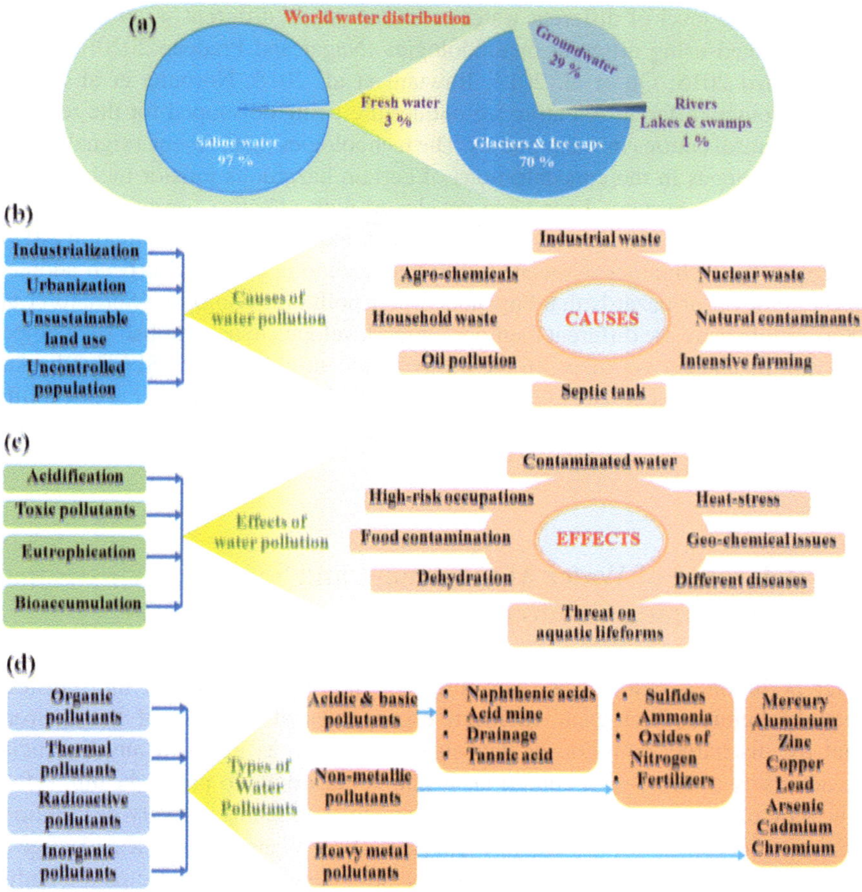

Fig. 9.1 Schematic representation of **a** earth's total water distribution emphasizing on the low availability of freshwater resources, **b** causes of water pollution, **c** effects of water pollution and **d** types of water pollution

(Goel 2006; Jackson et al. 2001; Rawtani et al. 2018). Additionally, reproductive, endocrine and neurological problems as well as certain types of cancers have been recorded on account of population exposure to water pollution. While the direct causes include the exposure to polluted water via consumption or bathing, the indirect causes constitute using vegetables, fruits and animals exposed to contaminated water. While these effects are well monitored in developed countries, such diligent monitoring needs substantial improvement in developing and under-developed countries (Nagar and Pradeep 2020; Noronha et al. 2019).

Therefore, understanding the purity levels of water is of large interest. In this context, effective strategies that measure the extent of water pollution need to be deployed in order to maintain adequate alarming systems. With this as the primary motivation, the level of different water contaminants in different water bodies has been examined using different methodologies (Nagar and Pradeep 2020; Chouler and Lorenzo 2015; Qu et al. 2013; Rawtani et al. 2018; Noronha et al. 2019). Simultaneously different scientific techniques are being developed for the removal of such harmful entities in water bodies. The technologies that present a signal output for small changes in the concentrations of certain hazardous ions or molecules or analytes in water systems have been developed with effective use of nanoscience and nanotechnology (Sitecon#trol Progress on Drinking Water and Sanitation 2014). Moreover, the techniques associated with nanotechnology have also been adopted for the removal of hazards that engender water pollution (Gonzalez-Serrano et al. 2004). While there are different methodologies developed to accomplish such objectives, in this chapter we focus on the utility of recently developed SPCE (Bhaskar et al. 2022a, b; Cao et al. 2014, 2019) and PCCE technologies and their applicability in the developing country (by considering the related research carried out in Indian scenario) (Bhaskar et al. 2020, 2023).

9.3 Relevance of Water Management and Sensing in Developing Countries

Water pollution is defined, as 'Any direct or indirect alteration of the physical, thermal, chemical, biological, radioactive properties of any part of the environment by, discharge, emission or deposit of wastes so as to affect any beneficial use adversely or to cause a condition, which is hazardous to public health, safety or welfare of animals, birds, wildlife, aquatic life or to plants of every description' (Environment Act of 1980) (Abel 2014; Warren 1971; Wanger 2019). Rapid surge in industrialization and associated progressions has overseen and neglected the undue byproducts in long-term technology development, as a result of which the hazards of water pollution have exponentially increased in the past decade. It is to be noted that the way water treatment, distribution, use and reuse has developed has emphasized the fact that the aspects related to water pollution have become greatly sophisticated demanding fast recovery. More importantly, the relevance of water management as well as detection/sensing of hazards in water needs utmost attention (Treacy 2019).

In order to obtain the relevance of water management and sensing in water research we present a comprehensive analysis of the number of documents published in the past ~25 years (1995–2022), as obtained from the Scopus database that is publicly available (Fig. 9.2). The literature search is primarily performed to comprehend the importance that the research community has devoted to different important aspects of water pollution. To begin with, from Fig. 9.2, we note that the importance of research with regard to water, water management, pollution and sensing has grown exponentially in the past 25 years. Notably, we see that the importance of sensing analytes of interest in water bodies has grown exponentially, although the number of documents is significantly lower as compared to that with water management.

Moreover, the number of documents constituting water pollution and sensing in developing countries is lesser as compared to rest of the world. These observations re-emphasize the need for considering adequate focus on awareness and water research toward developing countries and advance active water quality monitoring systems to detect/sense harmful analytes or ions in different water bodies.

The study of proper water management techniques, especially in developing countries, is of utmost importance to maintain a balance between the technological development and over-exploitation of water resources. Such over-exploitation and inappropriate reuse strategies are anticipated to worsen the current scenario of inadequate water management on account of the onset of critical climate changes. The drastic changes in temperature, rise in sea water level, unpredictability in local and continental rainfall, effect of greenhouse gases and the associated climate fluctuations are expected to effect the water management systems in an uninformed manner globally (Chawla et al. 2020). As presented in Fig. 9.2, water management is an important tool that is being increasingly researched in this perspective. Suggestively, the IPCC (2019) has predicted that with a 1.5 °C rise in temperature, approximately 178 million population are susceptible to scarcity of water resources, especially in dryland areas, and this number is evaluated to rise from 220 to 270 million vulnerable people with a rise in temperature from 2 °C to 3 °C (Chawla et al. 2020; Cooper et al. 2002). Research studies have predicted that the unexpected changes in the temperature may induce melting of polar ice, sea level rise and infiltration of saltwater into groundwater bodies, thereby making the situation extremely difficult for developing countries to recover from the impact. Such drastic changes are expected to be accompanied with unreliable possibilities for snowfall, snowmelt, drought, floods, cyclones and diminution of different surface water bodies with a long-term disastrous effect. Another important component of the water management domain comprises the close monitoring of hydrologic cycle, and strategic plans to alleviate the above-mentioned damaging effects on developing countries with effective synergy of environmentalists, researchers and governance officials (Chawla et al. 2020; Nagar and Pradeep 2020).

One of the Sustainable Development Goals (SDGs; SDG 6) as presented by the United Nations constitutes the importance of water security at a global level which demands holistic management of different resources of water (Chawla et al. 2020; Developments in Hydrometric Network Design 2009; Martins et al. 2016). Such an agenda requires constant monitoring of water availability and integrated routes to address the current challenges of water security and vulnerabilities globally. The details of soil moisture content, streamflow and precipitation to name a few that assist in determining the hydro-meteorological information have been monitored poorly on account of old or less prevalent socioeconomic data, especially in developing countries (Chawla et al. 2020). This has portrayed serious risks in clear prediction of accessibility of water resources, possible droughts and flood conditions, on account of under-developed technologies to manage hydro-meteorological information. While this is one part of the spectrum that demands immediate recovery, the implementation of novel technologies for water quality monitoring in developing countries is facing challenges on account of a smaller number of cost-effective techniques for on-field water health monitoring (Chowdhury et al. 2016). In this regard, nanomaterials are being extensively researched to overcome the caveats associated with detecting hazardous chemicals at trace concentrations, so as to administer appropriate remediation.

9.4 Utility of Nanomaterials for Monitoring Water Quality

Monitoring the quality of water at early stages assists in taking appropriate steps toward regulation and treatment. Among the several methodologies that are being adopted for early detection of pollutants in water bodies, (i) chemical and (ii) biological indicators are extensively used to accomplish the objective (Owa 2013; Zhang et al. 2020). With the adoption of chemical indicators, the researchers often measure the concentration levels of different chemical moieties in a variety of water bodies. If the study suggests analyte concentrations higher than the recommended levels, appropriate measures are taken to purify the water while it is being considered as polluted chemically. On the other hand, biological indicators involve specific studies where the living systems in water bodies such as fishes, insects and different invertebrates are analyzed for the exposure to chemicals (hazardous ions and molecules). This provides indirect information about the purity of the water systems that are under consideration (from which the living systems are collected).

Monitoring water quality can also be categorized as qualitative and quantitative. While the former involves strategies that assist decision-makers to observe and confirm the existence of a particular analyte of interest in a water body that is suspected to be contaminated with, the latter provides the information about the concentration levels of such analytes. Generally, the qualitative water monitoring systems are accompanied with sensors that have high sensitivity and selectivity. Importantly, public health which is predominantly dependent on water quality needs to be prioritized and both the qualitative and quantitative testing to provide better societal welfare. In the past few decades, the selective, sensitive and real-time analysis of different types of water contaminants (organic and inorganic) with cost-effective strategies are gaining increased attention (Soni et al. 2020; Teodoro et al. 2021; Walekar et al. 2017), especially to make the sensing system amenable for resource-limited settings. The use of nanoscience and nanotechnology has ameliorated the conventional biosensing technologies for monitoring air and water. Expedited novel strategies for water health monitoring with reliable and reproducible outputs are on the rise with the use of different nano-engineering strategies (Walekar et al. 2017; Nanomaterials for Sensing and Destroying Pesticides 2023).

Different types of nanomaterials such as cellulose nanofibrils (CNF), cellulose nanocrystals (CNC) as well as functionalized metal nanoparticles, semiconductor quantum dots, carbon dots, nanotubes, and a variety of nanocrystals, have been explored for detection of hazardous analytes in different water systems (Soni et al. 2020; Teodoro et al. 2021; Walekar et al. 2017; Nanomaterials for Sensing and Destroying Pesticides 2023), using different elegant methodologies. By and large, the nanosensors that are utilized for monitoring the quality of water can be categorized into (i) electrochemical, (ii) optical, (iii) magnetic, (iv) biomimetic and hybrid sensors (Gupte and Pradeep 2022). For instance, lead (Pb^{2+}) is a hazardous element found in water bodies and this has been detected using all the four techniques. Magar et al., demonstrated an electronical sensor using carbon nanotubes with 0.0083 ppb limit of detection (LOD) (Magar et al. 2017). Nguyen et al., demonstrated the same using nano-colorimetry (an optical method) (Nguyen and Shih 2019). Bagheri et al., detected the same using a magnetic nanocomposite of Schiff base/silica/magnetite (Bagheri et al. 2012). Chu et al. demonstrated the detection of lead ions in water samples using a biomimetic approach (Chu et al. 2015). Similarly, different metal ions and molecules that pose the risk of water pollution at altered concentrations have been detected using different techniques in which the nanomaterials are used to enhance the performance of the sensor. Fundamentally, these classifications are based on the kind of signal transduction elements used in the technology for the detection of analytes of interest.

9.5 Importance of Advanced Water Health Monitoring Through Sustainable Nanotechnology Approaches

The generally observed water contaminants can be categorized as follows: (i) harmful chemicals from industrial wastes, (ii) disease causing pathogens that include viruses and bacteria, (iii) hazardous chemicals and minerals that contain high concentration of specific ions (like heavy metal ions) leached via natural environmental effects, and (iv) certain contaminants that are explicitly not hazardous, but impact strange color, smell and taste to the water rendering it not acceptable for consumption. These four major aspects might work independently and/or in different combinations at different locations, therefore complicating the methods needed for water health monitoring. Predominantly, the first aspect pertaining to the harmful chemicals is given utmost importance as it has the potential to alter several parameters related to the toxicology of water content including charge, volatility, hydrophilicity, boiling point, reactivity, hydrophobicity, biodegradability, solubility, portability to name a few. In light of this, different methodologies based on advanced nano-engineering tools are being developed for the early detection of water contaminants, so as to advocate appropriate remediation steps.

Advanced nano-engineering techniques in the broad arena of water research are being employed for (i) detection of water pollutants, (ii) development of effective method for purifying the polluted water body, and (iii) to prevent water pollution in the first place (Yunus et al. 2012). There are different review articles that summarize the utility of different types of such nano-engineering methodologies for water quality monitoring applications in these three domains (Qu et al. 2013; Zhao et al. 2020; Thakkar et al. 2021). Furthermore, the emphasis on the aspect of water health monitoring using different nanotechnology approaches via development of active sensors/detection techniques has been extensively captured in different books and book chapters (Noronha et al. 2019; Pooja et al. 2020, Hu and Apblett 2014; Street 2014; Hussain and Mishra 2018; Diallo et al. 2009; Sarkar et al. 2021, Mustafa et al. 2021), and we encourage the readers to refer to the same for more details. Additionally, these book documents also provide the latest technologies that are being developed for purification of water resources using different nanotechnology approaches. In this background, we emphasize that the need for development of novel tools for prevention, detection and remediation of water pollutants has been gaining great interest.

While the first five sections of this chapter provides a brief overview of the different aspects of water pollution (causes & effects) in addition to the vast variety of techniques being developed for sensing hazardous analytes in different water bodies, the next few sections are devoted to presenting the latest photo-plasmonic technologies that are being developed for detection of analytes of interest in drinking water. Photo-plasmonics, as the name suggests is a field that involves the study of light (photon)—matter interactions where the plasmons (charge density oscillations) play a key role in modulating the light signal intensity (Bhaskar et al. 2021a, 2023). While there are numerous technologies that are developed using such photo-plasmonic platforms for the qualitative and quantitative analysis of water contaminants, in this chapter we focus on SPCE (Lakowicz 2004; Gryczynski et al. 2004; Cao et al. 2012) and PCCE (Bhaskar et al. 2020, 2020, 2022c) platforms on account of certain factors that may be listed as follows: (i) high sensitivity, (ii) high selectivity, (iii) high robustness of the platform, (iv) low-cost nano-engineering techniques, (v) ability to integrate with portable technologies and (vi) easy handling with rapid and facile read-out systems, (vii) ever-advancing design principles for augmenting the performance of the associated platform (Cao et al. 2019; Bhaskar and Ramamurthy 2021a; Mulpur et al. 2013; Chowdhury et al. 2008). The high sensitivity of these platforms arises on account of their ability to respond to mild variations in the refractive index of the local environment in the dielectric layer present immediately above the metallic thin film (in the case of SPCE platform) and photonic crystal (in the case of PCCE platform).

9.6 Latest Technologies with High Sensitivity for Water Monitoring: SPCE &PCCE

Among the several biosensing techniques that are available for detection of hazardous chemicals in water bodies, fluorescence spectroscopy has emerged with several advantages. In the past few decades, the versatility of nanotechnology and fluorescence spectroscopy has been uniquely synergized into a judicious platform termed plasmon-enhance fluorescence (PEF) which is also known as metal-enhanced fluorescence (MEF) technology (Bauch et al. 2014; Li et al. 2017). The emission from the fluorescent moieties is significantly enhanced or quenched in the presence of metallic nanomaterials depending upon the nature of interaction between them. The plasmonic nanomaterials sustain localized surface plasmon resonances (LSPRs) on account of which the associated sensing methodologies have observed a leapfrogging improvement in terms of sensitivity and selectivity of the biosensing platform (Bhaskar et al. 2021b, 2022b; Rathnakumar et al. 2021). Such PEF technologies are being incorporated into different spectroscopic techniques in order to obtain the information about the fluorescence signal intensity as well the vibrational stretching bands of the molecules involved. This sub-field of PEF has developed as well-known 'plasmon-enhanced spectroscopy' field (Cao et al. 2019; Lakowicz et al. 2008).

Although PEF and associated spectroscopic methodologies presented several advantages over the conventional fluorescence detection systems, limitations with regard to omnidirectional emission attributes, high background noise, low spectral resolution always remained a major challenge. In 2004, Lakowicz and co-workers innovated an effective platform to overcome such caveats in the cornerstone work on development of 'surface plasmon-coupled emission (SPCE)' technology (Lakowicz 2004; Gryczynski et al. 2004; Dutta Choudhury et al. 2015, 2013). This technology presented > 50% exceptionally directional and remarkably polarized emission with high reduction in background noise and excellent spectral resolution. In comparison to the conventional fluorescence spectrophotometer, the SPCE platform yielded tenfold enhancement in the fluorescence. Further, Chowdhury et al., demonstrated effective nano-engineering strategy for further improving the fluorescence signal intensity, where a spacer nanolayer of silver nanoparticles was adopted to achieve 60-fold SPCE enhancements (Chowdhury et al. 2008). Ever since the inception of SPCE substrate nano-engineering, different types of nanomaterials including metal, dielectrics, metallo-dielectric hybrids, magnetic nanohybrids, hollow and sharp nanoarchitectures have been explored in the SPCE platform presenting interesting physicochemical insights about the photoplasmonic coupling efficiency (Cao et al. 2014; Xie et al. 2019a, 2021; Chen et al. 2020; Xu et al. 2022). Interesting optical phenomena including Rabi splitting, Fabry–Perot photonic mode-coupling, Casimir force, Fano resonance and Purcell effect (Bhaskar et al. 2020b, 2021, 2023b) as well as processes such as CNT-assisted amplified coupling (Xie et al. 2019a), metal-dielectric plasmon coupling (Bhaskar and Ramamurthy 2021b), fluorescent polymer brushes (Weng et al. 2019), ultra-high sensitivity (Bhaskar et al. 2020a; Rai et al. 2021b, 2022a), interfacial molecular beacon (Cao et al. 2019), cavity-void plasmon (Bhaskar et al. 2021b; Rai et al. 2021c), heterometallic hybrid coupling (Rai et al. 2022b; Bhaskar and Ramamurthy 2019), graphene π-plasmon hybrid coupling (Bhaskar et al. 2022a; Mulpur et al. 2013; Xie et al. 2019b; Rai et al. 2022c), adsorption–desorption analysis (Pan et al. 2020), mesoporous carbon florets for photon cascading in nanocavity (Bhaskar et al. 2023a) and PLED (Wang et al. 2019) to name a few other cost-effective biosensing technologies.

Since the onset of different SPCE sensing frameworks developed for quantification of different analytes of interest, STAR laboratory has demonstrated the several functionalities of SPCE platform for detection of hazardous molecules in drinking water samples at extremely lower concentrations. Fundamentally, the SPCE substrates are made up of metallic thin film where the surface plasmon polaritons couple with the incident electromagnetic field to generate evanescent field, which in turn aid effective surface analysis. However, it has been observed that the related technologies that depend on metallic thin films present inevitable Ohmic losses that drastically reduce the overall performance of the sensing device. For instance, such Ohmic losses and surface induced quenching effects have been observed in SPCE, metal-dielectric-metal (MDM) and Tamm state-coupled emission (TSCE) technologies, thereby limiting their wide-spread use in the development of versatile biosensing frameworks (Dutta Choudhury et al. 2013; Badugu et al. 2014).

In an attempt to overcome the long-standing limitation with the SPCE platform, we developed an effective alternative platform with augmented performance using the dielectric substrates as opposed to metallic thin films used in SPCE technology. The alternative high and low refractive index nanolayers yielded one-dimensional photonic crystal (1DPhC) with a quarter wave-plate thickness dimensions presenting a photonic band gap (Bhaskar et al. 2020). Consequently, the 1DPhC assisted in supporting sharp resonant dip in the reflectance profile when appropriate phase matching conditions are satisfied (defined by physics equations). The extensive details of the fabrication and characterization of the 1DPhC and the optical response are presented in our recent pioneering works (Bhaskar et al. 2020, 2020). The directional and highly polarized emission observed using the 1DPhC substrates are termed as photonic crystal-coupled emission (PCCE).

The developments in the fluorescence spectroscopy toward the SPCE and PCCE technologies are pictorially captured in Fig. 9.3. In the case of a conventional fluorescence spectrophotometer, the excitation source and emission detector are placed exactly at 90°, in a bulky experimental setup (Fig. 9.3a). Since the emission from the luminophores in the cuvette is omnidirectional, only a portion of the emitted photons are collected. While the radiating dipoles are situated at the water–glass interface (Fig. 9.3b), the emitted photons are funneled into the HRI region (glass, $n_{g} = 1.52$) with a partial development of directional emission, which occurs due to the effect of critical angle (θ_C) (Bhaskar et al. 2021a; Lakowicz et al. 2008). Furthermore, while the same fluorophores are interfaced with a SPCE and PCCE platforms (Fig. 9.3b), the out-coupled photons from the prism side exhibit sharp angularity (or directionality) and high polarization (a property rendered by the substrate effect) with augmented signal collection efficiency. In this chapter, the recently explored sensing applications using SPCE and PCCE where the analytes (molecules and ions) that pose threats toward water pollution are effectively sensed are discussed. SPCE and PCCE studies can be performed in Kretschmann and reverse Kretschmann (RK) configurations. In order to maintain consistency, we have chosen explorations that have been carried out using RK configuration, in this chapter. In this regard, the laser excitation is performed from the free space side on to the plasmonic (SPCE) and photonic (PCCE) substrate and the prism coupled emission is collected from the curved surface of the prism (or distal part of the prism). The emission is passed through the appropriate emission filter and polarizer before the directional emission is captured using the optic fiber. The USB4000 fiber optic spectrometer connected to Ocean Optics SpectraSuite software is used as the detector in all the experiments

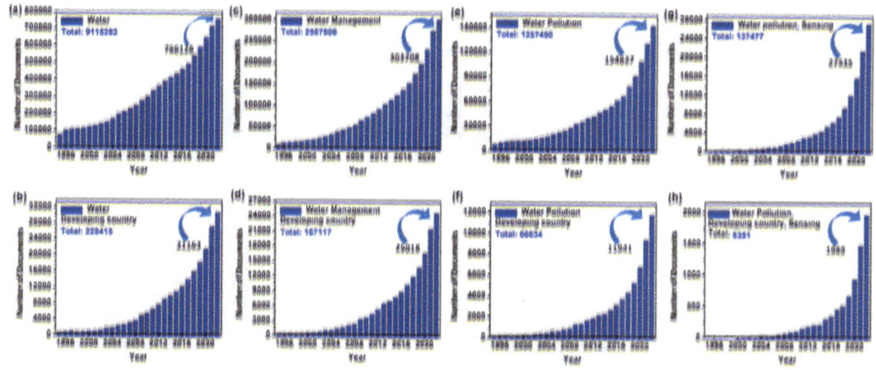

Fig. 9.2 The number of documents that are published from 1995 to 2022, as obtained from SCOPUSdatabase under the keyword sear constituting **a** water, **b** water, developing country, **c** water management, **d** water management, developing country, **e** water pollution, **f** water pollution, developing country, **g** water pollution, sensing, **h** water pollution, developing country, sensing

discussed in this chapter. A typical polymer matrix such as polyvinyl alcohol (PVA) is used for doping the nanomaterial of interest over the SPCE and PCCE substrate.

While the SPCE substrate is made up of the 50 nm metal thin film vapor deposited over the Pyrex glass slide, the PCCE substrate is made up of 1DPhC and is presented in Figs. 9.3d and e. The coupling of the emitted photons with the surface plasmon polaritons of metal thin film in SPCE substrate and with the different modes of the PCCE substrate is shown on the right side of Fig. 9.3d and e respectively. The modes sustained by the PCCE substrate are of two types: (i) the surface electromagnetic waves which are also known as Bloch surface waves (BSWs) which have highest field intensity at the surface of the 1DPhC. (ii) internal optical modes (IOMs) that have high photonic mode density in the interior of the 1DPhC (Bhaskar et al. 2020, 2020). Different nanomaterials of interest have been explored over both the substrates in different nanointerfaces which are briefly discussed in the succeeding section.

9.7 SPCE and PCCE Technology: Effective Nano-engineering Strategies

The majorly investigated nanointerfaces in SPCE and PCCE platforms are presented schematically in Fig. 9.4. The photoplasmonic coupling efficiency and the related fabrication details are presented in detail in our earlier publications (Rai et al. 2021b, 2022b). In the spacer nanointerface, the nanomaterial is interfaced between the radiating dipole and the metallic thin film. Consequently, the coupling of the propagating plasmon polaritons with the radiating dipoles is mediated via the strong field intensities generated by the localized plasmons of the plasmonic nanomaterial.

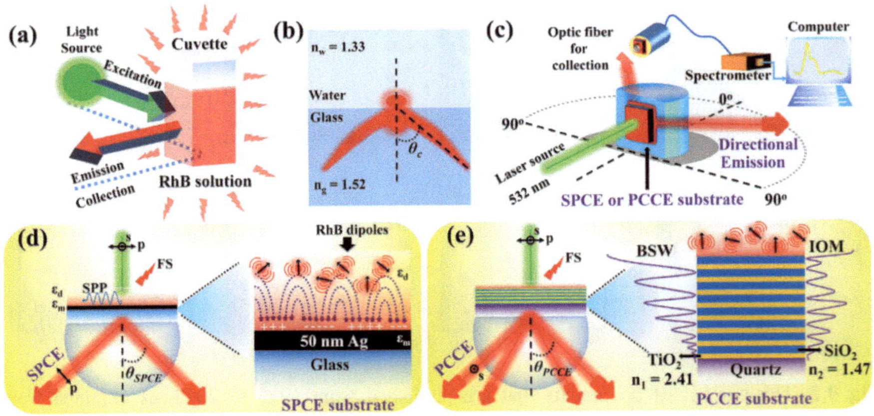

Fig. 9.3 Conceptual schematic of **a** fluorescence emission (of RhB) recorded in conventional fluorescence spectrophotometer, **b** the angular dispersion of fluorescence emission as observed in water–glass interface. The angle shown is a critical angle (θ_C) of emission. **c** Optical setup used for SPCE and PCCE experimental work with reverse Kretschmann (RK) configuration. Conceptual schematic of coupling of emission with **d** propagating surface plasmon polaritons (SPPs) of metal thin film (with θ_{SPCE} as the angle of emission) and **e** surface-trapped modes of 1DPhC (with θ_{PCCE} as angle of emission). The magnified images on the right side of (**d**) and **e** display the coupling at the nanointerface. Reprinted with permission from Reference (Bhaskar et al. 2021a), Copyright 2021, De Gruyter, Nanophotonics

Depending on the efficacy of the quenching and enhancement effects of the nanomaterial adopted, the SPCE enhancements are modulated. In the case of the cavity nanointerface, the nanomaterial of interest and the radiating dipoles are admixed and interfaced as a single coating over the SPCE and PCCE substrates. In this case, the radiating dipoles that are present in the infinitesimal nanogaps between the metallic thin film and the plasmonic nanomaterial are essentially present in the cavity plasmonic hotspots that sustain extremely high electromagnetic field intensity (Bhaskar et al. 2020a). However, since the radiating dipoles and the nanomaterial are in the same nanolayer, the effect of surface induced quenching is dominant as compared to similar effects observed in the spacer nanointerface. In order to present the best performance of both the spacer and cavity nanointerface, in 2020 we developed extended (ext.) cavity nanointerface which combines the benefits of both the interfaces (Bhaskar et al. 2020a). Here the cavity nano-region is extended to a specified distance as a result of which the surface induced quenching effects are dramatically reduced. Further since the nanomaterial and the radiating dipoles are coated as separate nanolayers, the architecture resembles the spacer interface with an opposite configuration (with respect to the metal surface). Depending on the target analyte (ions and molecules) of interest, different nanomaterials are used in different nano interfaces for achieving high fluorescence signal intensity in the SPCE and PCCE platforms. The rationale behind the use of a specific nanomaterial as compared to others is based on factors such as (i) interfacial interaction at the nano-junctions,

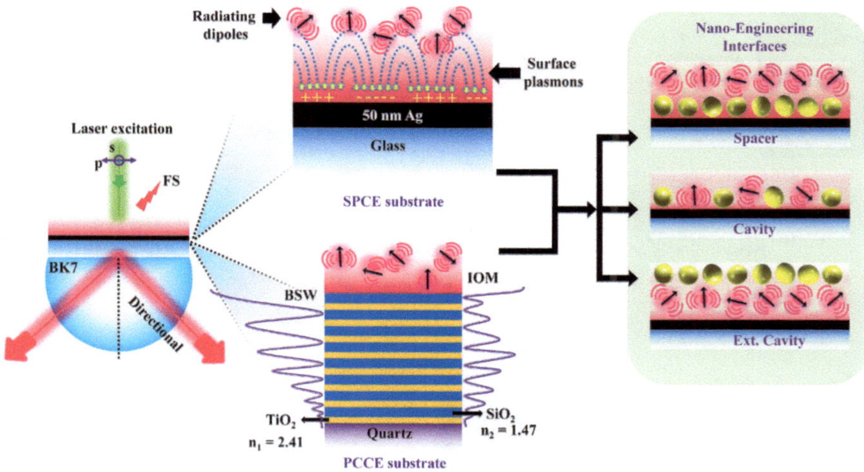

Fig. 9.4 Schematic representation of coupling of emission from radiating dipoles with the photo-plasmonic (SPCE and PCCE) interfaces (left and middle) in spacer, cavity and ext. cavity nanointerfaces (on the right). Reprinted with permission from Reference (Bhaskar et al. 2021a; Rai et al. 2022a), Copyright 2021, De Gruyter, Nanophotonics and Elsevier

(ii) size, (iii) shape, (iv) local environment, (v) surface roughness and (vi) surface chemistry.

9.8 Case Studies of Early Water Quality Monitoring in Developing Countries

9.8.1 Sensing Hazardous Molecules in Drinking Water Using SPCE Technology

Prior to discussing different sensing platforms achieved with the SPCE technology, it is informative to briefly state the associated nano-engineering technique adopted to realize high fluorescence intensity. This is because the high signal intensity plays a central role in developing the associated sensor of interest (Rai et al. 2021b; Rai et al. 2022). In the past decade, different methodologies have been developed for obtaining nano-assemblies on account of their ability to sustain delocalized Bragg plasmons and localized Mie plasmons that assist in realization of high EM field intensity. In this regard, Rai et al., innovated the cryosoret nano-engineering technology for realization of fourth-generation plasmonic hotspots that aid high sensitivity of analytes (Rai et al. 2022). Figure 9.5a presents the schematic representation of obtaining precise nano-assemblies using this technology. Briefly, this technique is an external stimulus driven method, where the homogeneous nanoparticle solution

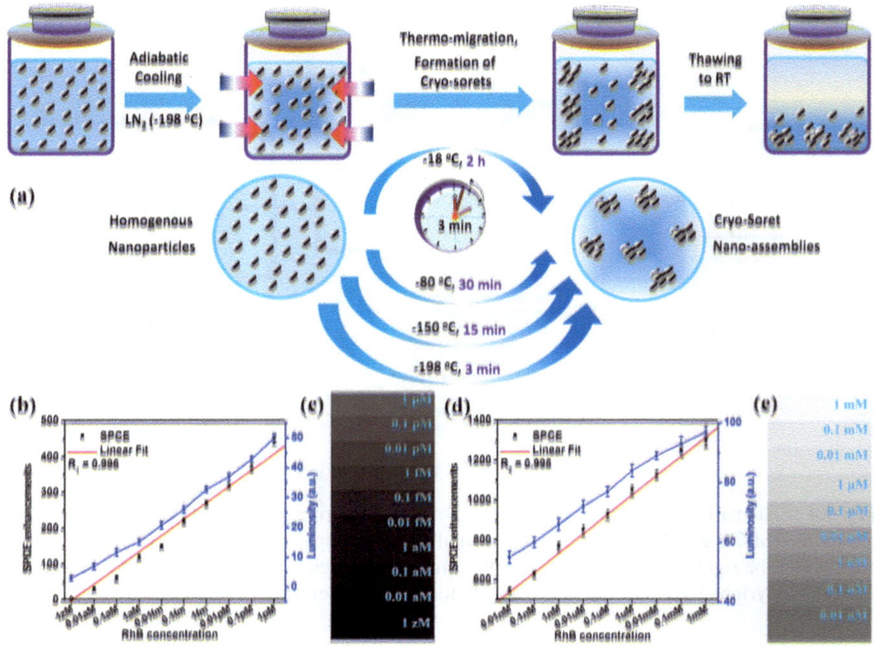

Fig. 9.5 **a** Conceptual schematic of the formation of cryosorets (CSs) using the adiabatic-cooling technology. The steps involve adiabatic cooling, thermomigration, and thawing the nano-assembly to room temperature. The decrease in the synthesis time to obtain soret nano-assemblies at different cooling temperatures (−18, −80, −150, and −196 °C) is also depicted. **b, d** Sensing of RhB presents a single-molecular limit of detection (zeptomolar). The respective luminosity values are plotted in the right y-axis along with the corresponding shade cards shown in (**c**) and (**e**), respectively. Reprinted with permission from Reference (Rai et al. 2022), Copyright 2022, American Chemical Society

is subjected to the low temperature gradient to observe the soret effect. Soret effect is the phenomena where the materials in a particular phase (generally liquid and gas) are subjected to movement from hot end to the cold end, when they are exposed to external temperature gradient (Moronshing and Subramaniam 2018). The report presented different optimization protocols using different low temperature conditions (−18 °C, −80 °C, −150 °C and −198 °C). Interestingly, it was observed that with lowering the temperature, the time required to obtain nano-assemblies of a particular configuration drastically reduced as captured in Fig. 9.5a, with the LN2 (−198 °C) driven synthesis yielding the fastest route to obtain the nano-assemblies (in less than 3 min). In this regard, the cryosorets obtained using this temperature exposure were utilized for sensing studies.

The hazardous molecule, rhodamine B in drinking water samples were quantified using hybrid cryosorets and the results are presented in Fig. 9.5b. Since the $AgTiO_2$cryosorets yielded the maximum enhancements while studied over graphene oxide (GO) pre-coated slides (with TiO_2 being nanorods), this particular nano-engineered material was utilized for performing the sensing of RhB (Rai et al. 2022). From Fig. 9.5b and d we see that the linear fits can be observed for sensing RhB in two different ranges with the experimentally observable limit of detection being zeptomolar concentration (single molecule limit). The high sensitivity achieved is on account of the hottest hotspots realized with the quintessential hybridization of different materials including (i) metal plasmonics with the use of AgNPs and metallic thin film, sustaining localized and propagating plasmons, (ii) HRI TiO_2 nanorods supporting the dielectric plasmons of lossless hotspots, (ii) graphene π-plasmons supported by the GO underlying layer present as a sandwich between the $AgTiO_2$cryosoret and the metallic thin film. Therefore, it is to be emphasized that the high sensitivity of the device can be accomplished using integrated hotspots from metal-dielectric-graphene hybrid photo-plasmonic coupling, assisting effective monitoring of analytes in water samples at extremely low concentrations, thereby aiding early water quality specialist care.

Different types of nanomaterials have been adopted for enhancing the SPCE signal intensity and HRI dielectric nanomaterials with nanorod (NR) morphology have gained noteworthy consideration. In this context, the Nd_2O_3 yielded 118-fold SPCE enhancements on account of lightning rod effect that resulted in high EM field intensity (Fig. 9.6a) (Bhaskar and Ramamurthy 2019). Moreover, these nanostructures with HRI also assist in substantial forward scattering of light as a result of which they perform as resourceful Huygen sources. The high fluorescence enhancements were employed for detection of environmentally harmful pollutant molecule, tannic acid in drinking water sample (Bhaskar and Ramamurthy 2019). The strong interaction of the Nd_2O_3NRs and the TA molecules results in changes in the surface properties of the Nd_2O_3 and the assemblies of them disintegrate into more discrete NRs in the presence of TA. Therefore, the uniform dispersion of the NRs is significantly increased which in turn assists in coupling of high number of NRs for a given concentration of tannic acid added. While lesser concentrations of TA were added (1fM to 100fM) there were no observable changes in the SPCE intensity with the use of Nd_2O_3NRs, because these concentrations do not alter the physicochemical properties of NRs. With higher concentrations of NRs (1 pM to 100 pM) the SPCE enhancements increased due to enhanced coupling between the more dispersed NRs and the radiating dipoles (Fig. 9.6a). Further increase in the concentrations of TA resulted in strong interaction with the PVA polymer matrix (resulting in agglomeration) adopted in the methodology, and hence presented lessened SPCE enhancements. In summary, the limit of detection of the sensor was constrained to 10 pM. The performance of the sensor was validated in drinking water samples and this presented reliable and reproducible results (Bhaskar and Ramamurthy 2019). Furthermore, in order to use cost-effective detection systems for sensing application, the expensive detector was replaced with the mobile phone-based detector system and the obtained luminosity values are plotted against the SPCE enhancements with an exceptional correlation between them. Such studies were performed by collecting the directional SPCE emission using mobile phone cameras and processing the same with freely available ColorGrab app (downloaded from the PlayStore.)

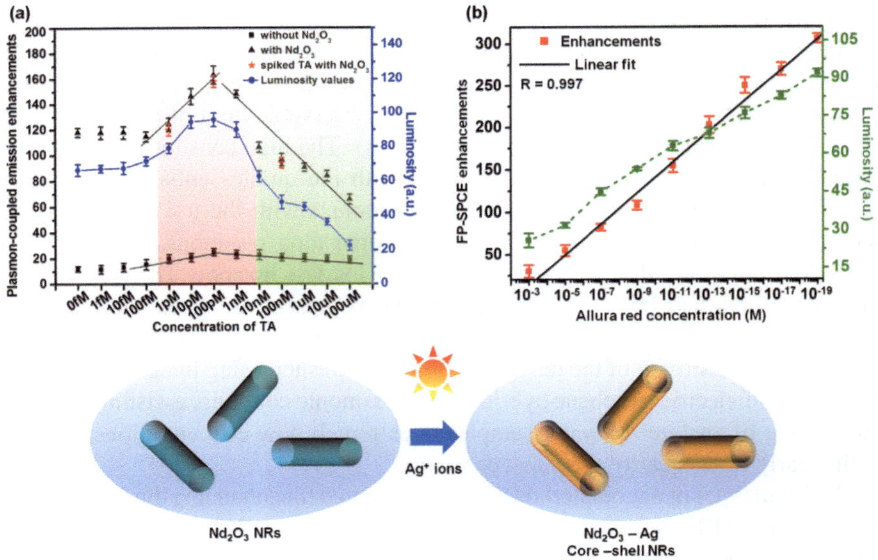

Fig. 9.6 **a** Left y-axis: modulation in plasmon-coupled emission enhancements with increasing TA concentration (red stars indicate spiked samples). Right y-axis: luminosity plot presenting alteration in the emission intensity from Rh6G with increase in TA concentration. Reprinted with permission from Reference (Bhaskar and Ramamurthy 2019) Copyright 2019, American Chemical Society. **b.** Attomolar sensing of allura red with excellent correlation between the SPCE enhancements and the luminosity values. Reprinted with permission from Reference (Bhaskar et al. 2023b), Copyright 2023, American Chemical Society

Furthermore, in an attempt to enhance the fluorescence enhancements observed with Nd_2O_3NRs, frugal nano-engineering strategy was adopted, where a mixture of Ag^+ ions and the NRs were exposed to direct sunlight. On account of the photo-induced electron transfer mechanism, Nd_2O_3-Ag core shell nanostructures were obtained with tunable optical and magnetic properties based on the concentration of Ag^+ ion utilized in the chemical nanosynthesis (Bhaskar et al. 2023b). Interestingly, we observed room-temperature ferromagnetism in such core–shell nanostructures while the starting materials are not ferromagnetic (Nd_2O_3NRs are paramagnetic and AgNPs are diamagnetic). These materials presented augmented SPCE enhancements on account of the effect of electric as well as magnetic plasmonic hotspots. Additionally, the metal-dielectric hybrid resonances assisted in the generation of high EM field intensity in the nano-junctions generated between the core–shell nanohybrids and the metallic thin film. On account of their unique magnetic response with a stable LSPR response for Ag shell, the hybrids are termed as ferroplasmons and the associated SPCE platform is addressed as FP-SPCE (Bhaskar et al. 2023b). The multifunctional capabilities of the ferroplasmons as active antibacterial as well as anticancer agents in biological and toxicology research and widespread applications are discussed in earlier work (Bhaskar et al. 2023b). These nanostructures were utilized for detection of hazardous allura red hazardous chemical. The direction interaction between the rhodamine moieties and the allura red was utilized to achieve the sensor where the higher concentrations of allura red resulted in drastic quenching of the rhodamine emission (Fig. 9.6b) (Bhaskar et al. 2023b). The sensing has been carried out in drinking water samples and the performance of the sensor is validated using a cost-effective smartphone-based detection system. A wide linear range was observed with an attomolar limit of detection, and an excellent correlation between the luminosity values and the SPCE enhancements is documented. While this section discusses the detection of a few hazardous in drinking water samples encompassing an industrial dye (rhodamine B), wastewater toxin (tannic acid) and food and beverage hazard (allura red) using the recent developments in SPCE technology, in the subsequent we present the utility of SPCE platform for the quantitative analysis of hazardous ions in drinking water samples.

9.8.2 Metal Ion Sensing in Drinking Water Using SPCE Technology

An important step toward monitoring water quality at early stages of contamination constitutes determination and quantification of hazardous ions in the water samples. In this section, we present recent advancements in the development of functional plasmonic nanostructures for the detection of certain hazardous ions in drinking water samples at extremely low concentrations. The relevance and need for sensing Cu^{2+}, Zn^{2+}, Hg^{2+}, Al^{3+} and I^{-}ions in drinking water samples is elaborated in associated recent works (Bhaskar et al. 2020; Bhaskar et al. 2020; Rathnakumar et al. 2021; Bhaskar et al. 2021b; Rai et al. 2021b). In an effective nano-engineering technique termed 'soret colloid (SC) nano-engineering' plasmonic nano-assemblies were synthesized using adiabatic cooling protocol (Bhaskar et al. 2020b, 2021b). The hybrid assembly of platinum and silver NPs in to sorets, namely AgPtSC yielded > 600-fold enhancement in the fluorescence signal intensity. This high emission intensity was utilized for strategic sensor development for copper ion detection and the details are presented in Fig. 9.7a using a smartphone-based detector system. In this methodology, a 5.5 pH Britton–Robinson (BR) buffer [Composition: 0.04 M phosphoric acid, 0.04 M boric acid,0.04 M acetic acid and 0.2 M sodium hydroxide] was used to prepare the alizarin red S (ARS) fluorophore. The simple complexation between the ARS and Cu^{2+} ions was adopted for the detection of latter. The absorption and emission maxima of ARS is ~400 nm and ~560 nm. The SPCE experiments were performed using the 532 nm laser source and no discernible emission was observed with the use of 0.5 μM ARS in the absence of any NPs. This is because the ARS molecules exhibited negligible absorption coefficient in this region of EM spectrum. Upon complexation with the Cu^{2+} ions (1 mM) the absorption of the ARS shifted to ~510 nm (which is closer to the laser excitation source of 532 nm). Consequently, ~40-fold increase in the out-coupled emission was observed. Further, with the incorporation of AgPtSCin the spacer nanointerface, the SPCE enhancements increased to ~610-fold and such high enhancements were utilized for indirect detection of Cu^{2+} ion at low concentrations (Bhaskar et al. 2021b). With the reduction in the concentration of Cu^{2+} ion, the SPCE enhancements decreased and a linear rangefrom 1 mM to 1 pM has been experimentally recorded, with the limit of detection being 1 pM. Further reducing the concentration of Cu^{2+} ions did not yield any observable changes in the fluorescence intensity and the emission enhancements remained the same as observed in the case of bare AgPtSC and ARS (i.e., ~35-fold).

In another approach to realize highest SPCE enhancements, heterometallic nanohybrids were synthesized by Rai et al., using plasmonic Ag and Au (Rai et al. 2021b). While the Au renders high chemical stability, Ag assists in providing less plasmonic structures with boosted EM field intensity, while it is hybridized with Au. In this background, AgAu nanohybrids were synthesized using a biocompatible, user eco-friendly approach and interfaced over the SPCE platform to realize ~1000-fold enhancement in fluorescence. The details of the synthesis, optimization and sensing strategy are elaborately presented in earlier work (Rai et al. 2021b). This augmented signal was utilized for detection of environmentally hazardous Zn^{2+} ions in drinking water samples. As elaborated in earlier section, the ARS molecules did not show fluorescence signal under 532 nm excitation in SPCE platform. With the addition of AgAu nanohybrids, a 54-fold enhancement in fluorescence was recorded with ARS, with all the other conditions remaining the same. This is termed as blank B1 in Fig. 9.7b and the observed enhancement is on account of the abundant plasmonic hotspots generated with anisotropic AgAu nanohybrids. Further, the addition of 0.1 mM Zn^{2+} ion yielded ~30-fold SPCE enhancements without the use of any nanomaterial for substrate engineering. This blank measurement is given the label B2, and the observed enhancement is attributed to the shift in the absorbance maximum of ARS towards higher wavelengths. While the AgAu nanohybrids and the Zn^{2+} ion was incorporated in the system, as high as 1000-fold enhancements were recorded. Such high enhancement assisted in indirectly sensing the Zn^{2+} ion, by serial dilution technique (with 10 times dilution at a time). Two linear ranges in the SPCE sensor were observed from 1 aM to 0.1 nM ($R = 0.987$) and from 0.1 nM to 0.1 mM ($R = 0.988$) with excellent reliability and reproducibility for triplicate measurements. The correlation experiments with the mobile phone-based sensing exhibited promising outcome where the luminosity values yielded good correspondence with the SPCE enhancements. Further particulars relating to percentage recovery, interference studies (selectivity) as well as recovery with regard to these explorations are discussed in detail in earlier works (Rai et al. 2021b).

In another approach, Rathnakumar et al., demonstrated the utility of Ag nanocubes for realization of high SPCE enhancements (>900-fold) (Rathnakumar et al. 2021). Ag nanocubes assist in accomplishing nano-regimes of extremely high EM field intensity on account of sharp nano-architectures (nano-edges and nano-vertices). Importantly, such structure is synthesized using bio-friendly methodology adopting biocompatible Lycoat polymer as the only reducing and capping agent, thereby satisfying the fundamental principles of green nano-synthesis (Rathnakumar et al. 2021). The high SPCE enhancements were utilized for attomolar detection of Hg^{2+} ions by incorporating the interaction between the sharp boundaries of the AgNPs and

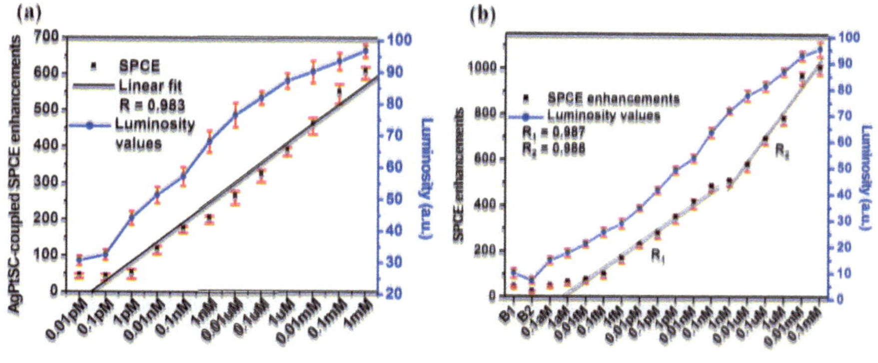

Fig. 9.7 **a** AgPtSC-coupled emission enhancements [left y-axis] and luminosity values [right y-axis] obtained for different concentrations of Cu^{2+} ion under study. **b** Overlap of SPCE enhancements (left y-axis) obtained from an Ocean Optics detector and the luminosity values (right y-axis) extracted from the mobile phone-based detector for the detection of Zn^{2+} ions using AgAu nanohybrids. Reprinted with permission from Reference (Bhaskar et al. 2021b), Copyright 2021, Elsevier. Reprinted with permission from Reference (Rai et al. 2021b), Copyright 2021, American Chemical Society

the mercury ions (Fig. 9.8), with the mechanism of sensing schematically presented in Fig. 9.8a. The addition of Hg^{2+} ions (1 mM) to the AgNPs resulted in instantaneous modification in the UV–Vis absorbance spectra as observed in Fig. 9.8b, with a distinguishing alteration in the color of the solution (Fig. 9.8c). This is on account of the capability of mercury ions to release the silver ions from the AgNPs, as a result of which the characteristic plasmonic resonance of AgNPs diminishes. It was seen that lower concentrations of Hg^{2+} ions occasioned in lessening of AgNPs LSPR peak intensity. While these samples were analyzed in the SPCE platform, improved sensitivity was accomplished on account of the generation of hybridized plasmons with the coupling of LSPR of AgNPs and SPPs of metallic thin film (Rathnakumar et al. 2021). With the decrease in the Hg^{2+} ions concentration, we observed that the SPCE enhancements were restored on account of lesser interaction between the AgNPs and the Hg^{2+} ions. The SPCE enhancements are correlated with the luminosity values obtained from mobile phone-based detector system. It was observed that an outstanding correlation exists between the two, thus qualifying mobile phone-based analysis of water samples for Hg^{2+} ions contamination. Accordingly, a systematic sensor platform was developed with high reliability, presenting a 10 aM limit of detection (Rathnakumar et al. 2021).

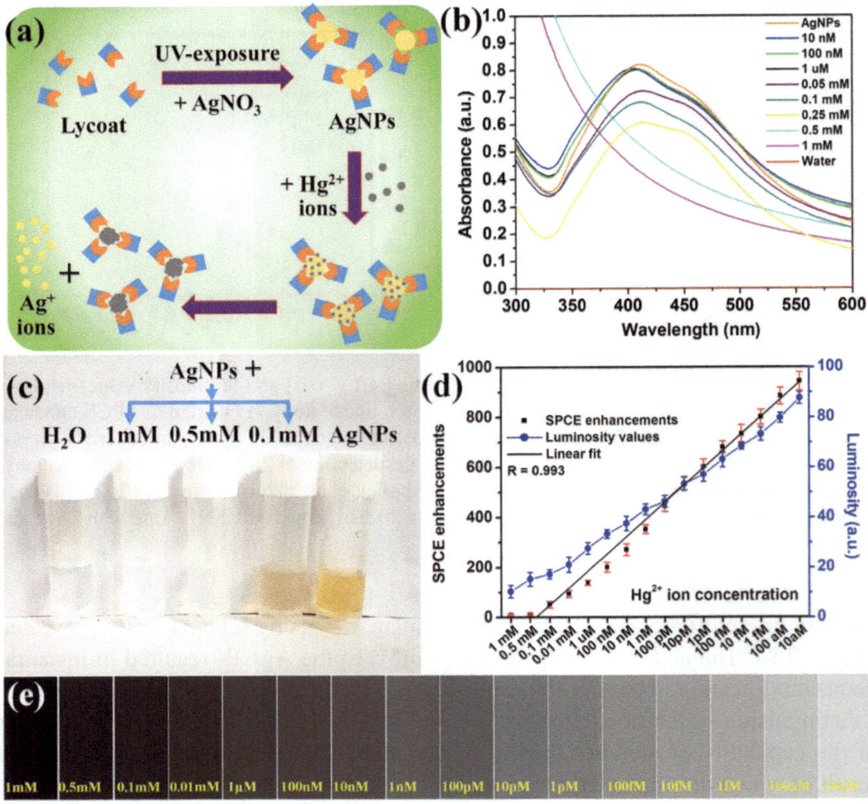

Fig. 9.8 **a** Schematic illustration of the effect of addition of Hg^{2+} ions to AgNPs capped with LYCOAT. **b** UV – vis absorbance pattern obtained upon the addition of different concentrations of Hg^{2+} to AgNPs. **c** Visual change in the color of the solution with and without Hg^{2+} ions. **d** SPCE enhancements and luminosity values were obtained using different concentrations of Hg^{2+} ions in the SPCE platform. **e** Shade cards corresponding to the luminosity values at different Hg^{2+} ion concentrations. Reprinted with permission from Reference (Rathnakumar et al. 2021), Copyright 2021, American Chemical Society

9.8.3 Metal Ion Sensing in Drinking Water Using PCCE Technology

In the preceding section we discussed the utility of SPCE platform for detection of ions and molecules of interest in drinking water. However, the presence of underlying metallic thin film poses certain challenges on account of the Ohmic losses sustained by them as well as limited substrate stability and less versatility for reuse. In order to overcome the Ohmic losses, certain sensing studies are carried out using PCCE platform and the results gotten therefrom are deliberated in this section. The utility of the PCCE platform for the detection of Al^{3+}ions and the I^{-}ions are presented in Figs. 9.9ab and cd respectively. The cavity nanointerface was utilized to monitor the

Al^{3+}ion concentration with the use of ARS as the radiating dipoles. It is important to briefly mention the importance of quantifying the concentrations of Al^{3+} ions in drinking water. These ions are extensively used in industries such as transportation, food additives and pharmaceutical industries. Altered concentrations of Al^{3+}ions pose serious risks to the human being causing Parkinson's disease, hampered child growth, Alzheimer's disease, breast cancer and dialysis encephalopathy to name a few. In this context, development of rapid, facile and highly sensitive detection method for sensing Al^{3+}ions in different water bodies is of utmost relevance. The incorporation of AgNPs with ARS radiating dipoles yielded ~17-fold PCCE enhancements on account of the generation of cavity photo-plasmonic hotspots (presented as blank measurement in Fig. 9.9a). With the addition of 1 mM Al^{3+} ions to the above solution ~ 230-fold enhancement in the fluorescence signal was observed on account of the shift in the absorbance of ARS toward the laser excitation source (Bhaskar et al. 2020). The high PCCE enhancements were utilized for detection of Al^{3+} ions by decreasing the concentration of the same in order to obtain a calibration plot. A liner sensing range was observed from 1 mM to 1 fM (for triplicate measurements) and the change in the intensity of the emission spectra observed with altering the Al^{3+} ion concentration (with 10 times dilution at every step) is shown in Fig. 9.9b. Furthermore, the spiking studies and the recovery analysis of Al^{3+} ions in the drinking water samples are elaborately discussed in the related work (Bhaskar et al. 2020).

In a unique approach for detection of I^- ions sharp-edged silver nanoprisms (AgNPrs) were utilized. AgNPrs of different structural morphologies were explored over the PCCE platform and conditions to realize optimum photo-plasmonic coupling were drawn. The AgNPrs with particular thickness and dimensions provided as high as 947-fold fluorescence enhancement on account of the intense plasmonic hotspots rendered by the AgNPrs, which support high Purcell effect (Bhaskar et al. 2020). From Fig. 9.9c, we note that with increasing concentration of I^-ions, the PCCE enhancements gradually decreased. This is on account of the ability of I^-ions to etch the sharp edges of the AgNPrs, thereby altering their structural morphology toward spherical geometries. Consequently, the intensity of the plasmonic hotspots decreases, which in turn results in dwindled PCCE enhancements. A linear relationship was observed from 0.01 mM to 1 fMwith an R value of 0.997, thereby presenting a promising calibration plot for the detection of I^-ions in water samples. It was observed that concentrations lower than 1 fM did not yield any observable changes in the PCCE enhancements thereby limiting the detection to femtomolar concentrations. For I^-ion concentration more than 0.01 mM, the PCCE enhancements remained as a constant of ~200-fold, which is similar to that observed with spherical AgNPs. The recommendations for iodine intake by human beings as provided by the World Health Organization range from 90 to 200 μg/mL [or from 0.1 to 1.96 μM]. In this regard, it is worth noting that the developed PCCE sensing platform aids in successful distinguishment of benign zone (from 0.1 to 2 μM) as well as a detrimental zone (2 μM) providing insights into the tendency toward hypo- and hyperthyroidism. In summary, so far, we have discussed the applicability of SPCE and PCCE substrates elaborately using certain case studies where these photoplasmonic substrates a utilized for detection of certain hazards in drinking water samples.

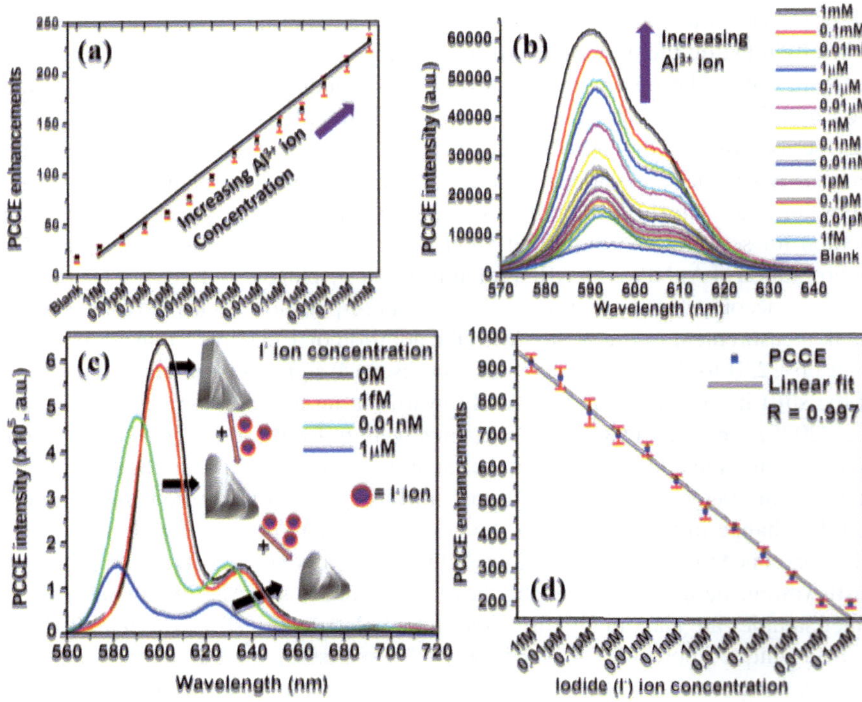

Fig. 9.9 **a** PCCE enhancements obtained with different concentrations of Al^{3+} ions from 1 mM to 1 fM with 10 times dilution at each step, with the use of 1 mM ARS in a 60 nm PVA matrix. Blank indicates 17-fold enhancements obtained with ARS and AgNPs. **b** PCCE spectra obtained using different concentrations of Al^{3+} ions from 1 mM to 1 fM with 10 times dilution for each spectrum. **c** Representative PCCE intensity plot displaying a correlation between the effect of iodide concentration, the shape of AgNPr, and the emission profile. **d** Variation in the emission intensity with the increasing concentration of iodide ions. Reprinted with permission from Reference (Bhaskar et al. 2020a; Rai et al. 2021a), Copyright 2020,American Chemical Society

9.9 Water Purification and Its Prospects

While we discussed the latest photo-plasmonics techniques based on SPCE and PCCE for the detection of certain analytes of interest in drinking water samples in the earlier sections, in this penultimate section the reader is introduced to certain technologies that have been developed for the effective removal of pollutants from drinking water samples. Different hazardous analytes have myriad health implications and are still being studied using advanced technologies. While certain hazards are classified as those which should never be consumed by human beings, exposure to certain other hazards needs to have a balanced approach. For instance, certain analytes are required by the human body for optimum performance of different faculties of the human body. However, while the concentration of the same analyte is drastically altered (increased or decreased), different health implications have been documented. While there are different methodologies being developed for removal of different analytes (molecules and ions) of interest, in this section we focus on removal of one such analyte, fluoride, due to its peculiar dual functionalities. Moreover, recently several techniques are being developed for efficient removal of fluoride ions as it is one of the major pollutants in developing country such as India and China (Ayoob and Gupta 2006). The lack of access to pure drinking water has been a growing issue in several countries and is more evident in underdeveloped and developing countries.

The use of water resources contaminated with high concentrations of different analytes like fluoride has dangerous effects on the health of consumers. Considering the case of Indian subcontinent, it has been observed that the quality of the groundwater (80% in rural areas and 50% in urban areas) that is being utilized for various domestic purposes is contaminated with fluoride menace (Roy 2003). Considering the world health organization (WHO) health standards, 1.0–1.5 mg/L is the range of permissible fluoride ion concentration in the drinking water samples (Fluorides and Oral Health 1994). While the adequate consumption of fluoride ion via different water sources performs as the tonic to healthy well-being, the altered concentrations (higher and lower) have emerged with major toxic effects. The adequate consumption of fluoride ions (as recommended for specific age groups) assists in building healthy teeth (supporting calcification of dental enamel in newborns), thereby preventing dental issues. Excess concentration of fluoride ions results in disorders related to dental, skeletal as well as long-term effects pertaining to tissue fluorosis, and at times death (Abe et al. 2004; Sathish et al. 2007; Sathish et al. 2008).

While the geogenic percolation of fluoride into the groundwater is one of the uncontrollable effects of high fluoride content in certain parts of the Indian subcontinent, the anthropogenic causes worsen the situation. The latter includes fertilizer manufacturing, semiconductor industry, coal power plants, ceramic and glass production, uranium enrichment as well as electroplating industry that result in generation of fluoride discharges. Further, while these are the effects of fluoride ions as such, the fluoride ions combine with metal ions with myriad complexation routes and result in varied levels of toxicity, thereby resulting in organelle disruption, oxidative stress, cell cycle arrest and apoptosis to name a few. Habitants from different regions of Indian subcontinent grieve from the unexpected health hazards of fluoride ions. Moreover, certain states that including Andhra Pradesh, Gujarat and Rajasthan are harshly affected where high fluoride contents (0.5 to 48 mg/L) in water samples lead to several health disorders (Janardhana et al. 2006; Ramamurthy et al. 2011). While there are certain municipal regulatory bodies in some countries for monitoring the fluoride ion concentration in water samples, the exact levels of fluoride in water are little understood in developing countries on account of unknown sources of fluoride contamination. In this regard, there is a growing demand to efficiently detect high concentrations of fluoride in water, as well as to develop novel, simple and economical methodologies for effective removal of excess fluoride (Sathish et al. 2008, 2006; Kamble et al. 2007).

The three major techniques utilized for the defluoridation of drinking water are precipitation, membrane-based and adsorption process. It is instructive to discuss the advantages and challenges of these techniques by considering defluoridation as a case study here. In the precipitation methodology, the salts of calcium and aluminum are used for the effective formation of precipitate (and co-precipitate of fluoride) and subsequently separate using appropriate methods. However, this method introduces the use of additional chemical moieties for effective precipitation which poses the challenge of undue contamination of drinking water with added chemicals. While the membrane-based method includes the use of nanofiltration and reverse osmosis, as well as electrodialysis, these technologies are often less economical as it involve the usage of high budget for maintenance purposes (Sathish et al. 2008, 2006; Kamble et al. 2007). In light of these observations, recently the focus is being devoted to the development of third approach based on adsorption processes where different materials with novel properties are developed for effective removal of fluoride through straightforward and cost-effective experimental setups. In this regard, materials including alumina, zeolites, alum, lime, spent catalyst, clay minerals, coconut shell carbon, charcoal and ash have been extensively researched with appropriate contact bed that aids in the elimination of fluoride ions from the passing water as they get adsorbed by the materials used. Especially in developing countries, the focus has been toward the use of low-cost adsorbents which are being explored with environmentally benign materials such as biomass (of chitosan polymer) Eichhornia crassipes, aligned carbon nanotubes (CNTs), amorphous alumina supported on CNTs or by a hybrid process combining Donnan dialysis & adsorption in a few approaches (Sathish et al. 2008, 2006; Ramamurthy et al. 2011).

The use of low-cost adsorbents which present high adsorption efficiency for metal ion impregnation with a goal to accomplish higher fluoride ion removal has been gaining increased awareness in developing countries. In this regard, generally discarded biomaterials that include coconut shell and fiber carbon are being increasingly explored. Initial efforts in this direction have been with the use of lanthanum-modified chitosan and metal ion loaded zeolites. Furthermore, because of the high affinity of zirconium toward the fluoride ions, these materials and their appropriate complexes are being used in different techniques where fluoride ions are detected and adsorbed. The high affinity is because of the high charge density of fluoride ion which results in the generation of fluoro-complexes (which are chemically stable) with the tetravalent zirconium (IV). While these materials are impregnated with zirconium, the fluoride removal efficiency increases by 3–5 times as compared to the generally used activated carbon. Activated carbon impregnated with zirconium is a suitable material that is extensively used for removal of phosphate, arsenic, selenium, mercury and chromium from water samples. Further, different metal ions such as $ZrOCl_2$, CaO, Alum, $CaCl_2$ and Borax have been explored to enhance the fluoride removal efficiency with the use of different low-cost sources for activated charcoal (including groundnut shell, coconut shell and coconut fiber) (Sathish et al. 2007; Sathish et al. 2008; Ramamurthy et al. 2011). The use of such materials makes avenue for new routes to add value to generally discarded materials, for instance the coconut shell (Cocos nucifera L.), is widely accessible in rural India from the agronomical segment and the outlay hence involved in using them as the substrate material is insignificant.

It has been observed that among the several metal ions explored, zirconium ion-impregnated coconut shell carbon (ZICSC) and zirconium ion-impregnated coconut fiber carbon (ZICFC) have ascertained to be the very efficient defluoridating representative for remedying discharges as compared to the groundnut shell charcoals. Various parameters such as pH, agitation time and adsorbent dosage have been optimized to accomplish the best results and the details are presented in earlier works (Sathish et al. 2007; Sathish et al. 2008; Ramamurthy et al. 2011). Through these observations, we emphasize that while detection of hazardous analytes in various water samples is relevant in developing countries (as discussed in previous sections), the usage of advanced nano-engineering techniques for effective removal of such water contaminants needs to go hand in hand in order to render clean water supply in urban as well as rural settings, especially in developing countries.

9.10 Concluding Remarks

The issues pertaining to water pollution and related challenges have been gaining great attention from industrialists and academicians at national and international portfolios. The scenario in the under-developed and developing countries has been improved in the past few years with the advancement of newer technologies for detecting water contaminants as well as with the innovation driven toward development of novel water purification techniques. In both the approaches, nanomaterials have been gaining great interest where advanced interfaces are engineered to ameliorate detection and purification technologies. In light of these observations, this chapter discusses a few recently developed methodologies based on photoplasmonics for the detection of water pollutants. We present an elaborate discussion by considering case studies where new technologies such as SPCE and PCCE are being developed for detection of hazardous chemical moieties in drinking water samples. Different approaches using nano-engineering over these platforms to enable detection of rhodamine, allura red, tannic acid, Cu^{2+}, Zn^{2+}, Hg^{2+}, Al^{3+} and I^- have been discussed. The utility of cost-effective, user-friendly and facile sensing methodologies is developed by integrating the photoplasmonic platforms with the smartphone-based detection methods, amenable for resource-limited settings. Further, we briefly discuss the advancements in water purification techniques by considering a few examples where the nanomaterials play a crucial role in enhancing the performance of the devices used in water decontamination. We believe that the details presented in this chapter would encourage researchers to plan, execute and develop technologies with high performance (in terms of sensitivity) that are suitable for low-and-middle-income countries.

9.11 Consent for Publicaton

Authors provide consent for the publication.

Acknowledgements Authors acknowledge support from Tata Education and Development Trust [TEDT/MUM/HEA/SSSIHL/2017- 2018/0069-RM-db], Prasanthi Trust, Inc., U.S. (22-06-2018), DST-Technology Development Program (IDP/MED/ 19/2016), Life Sciences Research Board (LSRB), and DST-Inspire Fellowship (IF180392), Govt. of India. S.B. thanks the IGB postdoctoral fellowship support from Carl R. Woese Institute for Genomic Biology (IGB). We especially acknowledge SSSIHL-CRIF for extending the usage of the required instrumentation facility. Guidance from Bhagawan Sri Sathya Sai Baba is gratefully acknowledged.

Conflict of Interest Authors have no conflict of interest.

References

Abe I, Iwasaki S, Tokimoto T, Kawasaki N, Nakamura T, Tanada S (2004) Adsorption of fluoride ions onto carbonaceous materials. J Colloid Interface Sci 275:35–39. https://doi.org/10.1016/j.jcis.2003.12.031

Abel PD (2014) Water pollution biology, 2nd ed. CRC Press, London. ISBN 978–0–429–17866–5

Ayoob S, Gupta AK (2006) Fluoride in drinking water: a review on the status and stress effects. Crit Rev Environ Sci Technol 36:433–487. https://doi.org/10.1080/10643380600678112

Badugu R, Descrovi E, Lakowicz JR (2014) Radiative decay engineering 7: Tamm State-coupled emission using a hybrid plasmonic-photonic structure. Anal Biochem 445:1–13. https://doi.org/10.1016/j.ab.2013.10.009

Bagheri H, Afkhami A, Saber-Tehrani M, Khoshsafar H (2012) Preparation and characterization of magnetic nanocomposite of schiff base/silica/magnetite as a preconcentration phase for the trace determination of heavy metal ions in water, food and biological samples using atomic absorption spectrometry. Talanta 97:87–95. https://doi.org/10.1016/j.talanta.2012.03.066

Bauch M, Toma K, Toma M, Zhang Q, Dostalek J (2014) Plasmon-enhanced fluorescence biosensors: a review. Plasmonics 9:781–799. https://doi.org/10.1007/s11468-013-9660-5

Bhaskar S, Ramamurthy SS (2019) Mobile phone-based picomolar detection of tannic acid on Nd2O3 nanorod-metal thin-film interfaces. ACS Appl Nano Mater 2:4613–4625. https://doi.org/10.1021/acsanm.9b00987

Bhaskar S, Ramamurthy SS (2021a) Synergistic coupling of titanium carbonitride nanocubes and graphene oxide for 800-fold fluorescence enhancements on smartphone based surface plasmon-coupled emission platform. Mater Lett 298:130008. https://doi.org/10.1016/j.matlet.2021.130008

Bhaskar S, Kowshik NCSS, Chandran SP, Ramamurthy SS (2020a) Femtomolar detection of spermidine using Au decorated SiO2 nanohybrid on plasmon-coupled extended cavity nanointerface: a smartphone-based fluorescence dequenching approach. Langmuir 36:2865–2876. https://doi.org/10.1021/acs.langmuir.9b03869

Bhaskar S, Moronshing M, Srinivasan V, Badiya PK, Subramaniam C, Ramamurthy SS (2020b) Silver Soret nanoparticles for femtomolar sensing of glutathione in a surface plasmon-coupled emission platform. ACS Appl Nano Mater 3:4329–4341. https://doi.org/10.1021/acsanm.0c00470

Bhaskar S, Das P, Moronshing M, Rai A, Subramaniam C, Bhaktha SBN, Ramamurthy SS (2021a) Photoplasmonic assembly of dielectric-metal, Nd2O3-gold soret nanointerfaces for dequenching the luminophore emission. Nanophotonics 10:3417–3431. https://doi.org/10.1515/nanoph-2021-0124

Bhaskar S, Jha P, Subramaniam C, Ramamurthy SS (2021b) Multifunctional hybrid soret nanoarchitectures for mobile phone-based picomolar Cu2+ Ion sensing and dye degradation applications. Phys E Low-Dimens Syst Nanostruct 132:114764. https://doi.org/10.1016/j.physe.2021.114764

Bhaskar S, Rai A, Mohan GK, Ramamurthy SS (2022a) Mobile phone camera-based detection of surface plasmon-coupled fluorescence from streptavidin magnetic nanoparticles and graphene oxide hybrid nanointerface. ECS Trans 107:3223. https://doi.org/10.1149/10701.3223ecst

Bhaskar S, Rai A, Ganesh KM, Reddy R, Reddy N, Ramamurthy SS (2022b) Sericin-based bio-inspired nano-engineering of heterometallic AgAu Nanocubes for Attomolar mefenamic acid sensing in the mobile phone-based surface plasmon-coupled interface. Langmuir 38:12035–12049. https://doi.org/10.1021/acs.langmuir.2c01894

Bhaskar S, Das P, Srinivasan V, Bhaktha SBN, Ramamurthy SS (2022c) Plasmonic-silver Sorets and dielectric-Nd2O3 nanorods for ultrasensitive photonic crystal-coupled emission. Mater Res Bull 145:111558. https://doi.org/10.1016/j.materresbull.2021.111558

Bhaskar S, Thacharakkal D, Ramamurthy SS, Subramaniam C (2023a) Metal-dielectric interfacial engineering with mesoporous nano-carbon florets for 1000-fold fluorescence enhancements: smartphone-enabled visual detection of perindopril erbumine at a single-molecular level. ACS Sustain Chem Eng 11:78–91. https://doi.org/10.1021/acssuschemeng.2c04064

Bhaskar S, Srinivasan V, Ramamurthy SS (2023b) Nd2O3-Ag nanostructures for plasmonic biosensing, antimicrobial, and anticancer applications. ACS Appl Nano Mater 6:1129–1145. https://doi.org/10.1021/acsanm.2c04643

Bhaskar S, Ramamurthy SS (2021) High refractive index dielectric TiO2 and graphene oxide as salient spacers for > 300-fold enhancements. In: Proceedings of the 2021 IEEE international conference on nanoelectronics, nanophotonics, nanomaterials, nanobioscience & nanotechnology (5NANO), pp 1–6

Bhaskar S, Singh AK, Das P, Jana P, Kanvah S, Bhaktha BNS, Ramamurthy SS (2020) Superior resonant nanocavities engineering on the photonic crystal-coupled emission platform for the detection of femtomolar iodide and zeptomolar cortisol. ACS Appl Mater Interfaces 12:34323–34336. https://doi.org/10.1021/acsami.0c07515

Bhaskar S, Lis SSM, Kanvah S, Bhaktha BNS, Ramamurthy SS (2023) Single-molecule cholesterol sensing by integrating silver nanowire propagating plasmons and graphene oxide π-plasmons on a photonic crystal-coupled emission platform. ACS Appl Opt Mater 1:159–172. https://doi.org/10.1021/acsaom.2c00026

Bhaskar S, Visweswar Kambhampati NS, Ganesh KM, P MS, Srinivasan V, Ramamurthy SS (2021) Metal-free, graphene oxide-based tunable soliton and plasmon engineering for biosensing applications. ACS Appl Mater Interfaces 13:17046–17061. https://doi.org/10.1021/acsami.1c01024

Bhaskar S, Das P, Srinivasan V, Bhaktha BNS, Ramamurthy SS (2020) Bloch surface waves and internal optical modes-driven photonic crystal-coupled emission platform for femtomolar detection of aluminum ions. J Phys Chem C 124:7341–7352. https://doi.org/10.1021/acs.jpcc.9b11092

Cao S-H, Cai W-P, Liu Q, Li Y-Q (2012) Surface plasmon-coupled emission: what can directional fluorescence bring to the analytical sciences? Annu Rev Anal Chem 5:317–336. https://doi.org/10.1146/annurev-anchem-062011-143208

Cao S-H, Cai W-P, Liu Q, Xie K-X, Weng Y-H, Huo S-X, Tian Z-Q, Li Y-Q (2014) Label-free aptasensor based on ultrathin-linker-mediated hot-spot assembly to induce strong directional fluorescence. J Am Chem Soc 136:6802–6805. https://doi.org/10.1021/ja500976a

Cao S-H, Weng Y-H, Xie K-X, Wang Z-C, Pan X-H, Chen M, Zhai Y-Y, Xu L-T, Li Y-Q (2019) Surface plasmon coupled fluorescence-enhanced interfacial "Molecular Beacon" to probe biorecognition switching: an efficient, versatile, and facile signaling biochip. ACS Appl Bio Mater 2:625–629. https://doi.org/10.1021/acsabm.8b00751

Chawla I, Karthikeyan L, Mishra AK (2020) A review of remote sensing applications for water security: quantity, quality, and extremes. J Hydrol 585:124826. https://doi.org/10.1016/j.jhydrol.2020.124826

Chen M, Cao S-H, Li Y-Q (2020) Surface plasmon-coupled emission imaging for biological applications. Anal Bioanal Chem 412:6085–6100. https://doi.org/10.1007/s00216-020-02635-3

Chouler J, Di Lorenzo M (2015) Water quality monitoring in developing countries; can microbial fuel cells be the answer? Biosensors 5:450–470. https://doi.org/10.3390/bios5030450

Chowdhury MH, Ray K, Geddes CD, Lakowicz JR (2008) Use of silver nanoparticles to enhance surface plasmon-coupled emission (SPCE). Chem Phys Lett 452:162–167. https://doi.org/10.1016/j.cplett.2007.12.047

Chowdhury S, Mazumder MAJ, Al-Attas O, Husain T (2016) Heavy metals in drinking water: occurrences, implications, and future needs in developing countries. Sci Total Environ 569–570:476–488. https://doi.org/10.1016/j.scitotenv.2016.06.166

Chu W, Zhang Y, Li D, Barrow CJ, Wang H, Yang W (2015) A biomimetic sensor for the detection of lead in water. Biosens Bioelectron 67:621–624. https://doi.org/10.1016/j.bios.2014.09.077

Cooper RN, Houghton JT, McCarthy JJ, Metz B (2002) Climate change 2001: the scientific basis. Foreign Aff 81:208. https://doi.org/10.2307/20033020

Development U.K.D. for I.; Development, E.C.C.D.-G.; UNDP; IBRD Linking Poverty Reduction and Environmental Management :: Policy Challenges and Opportunities : A Contribution to the World Summit on Sustainable Development Process : Consultation Draft (2002)

Developments in Hydrometric Network Design (2009) A Review-Mishra-2009-Reviews of Geophysics-Wiley Online Library. https://agupubs.onlinelibrary.wiley.com/doi/full/10.1029/2007RG000243. Accessed 10 Feb 2023

Diallo MS, Duncan JS, Savage N, Street A, Sustich R (2009) Chapter 38-nanotechnology solutions for improving water quality. In: Savage N, Diallo M, Duncan J, Street A, Sustich R eds Nanotechnology applications for clean water. Micro and nano technologies. William Andrew Publishing, Boston, pp 585–587. ISBN 978–0–8155–1578–4

Dosi C, Tomasi T eds (1994) Nonpoint source pollution regulation: issues and analysis. Springer Netherlands, Dordrecht. ISBN 978–90–481–4468–6

Dutta Choudhury S, Badugu R, Nowaczyk K, Ray K, Lakowicz JR (2013) Tuning fluorescence direction with plasmonic metal–dielectric–metal substrates. J Phys Chem Lett 4:227–232. https://doi.org/10.1021/jz301867b

Dutta Choudhury S, Badugu R, Lakowicz JR (2015) Directing fluorescence with plasmonic and photonic structures. Acc Chem Res 48:2171–2180. https://doi.org/10.1021/acs.accounts.5b00100

Fluorides and Oral Health (1994) Report of a WHO expert committee on oral health status and fluoride use. World Health Organ Tech Rep Ser 846:1–37

Goel PK (2006) Water pollution: causes, effects and control. New Age International. ISBN 978–81–224–1839–2

Gonzalez-Serrano E, Cordero T, Rodriguez-Mirasol J, Cotoruelo L, Rodriguez JJ (2004) Removal of water pollutants with activated carbons prepared from H3PO4 activation of lignin from kraft black liquors. Water Res 38:3043–3050. https://doi.org/10.1016/j.watres.2004.04.048

Gray N (2017) Water technology, 3rd ed. CRC Press, London. ISBN 978–1–315–27610–6

Gryczynski I, Malicka J, Gryczynski Z, Lakowicz JR (2004) Radiative decay engineering 4. Experimental studies of surface plasmon-coupled directional emission. Anal Biochem 324:170–182. https://doi.org/10.1016/j.ab.2003.09.036

Gupte T, Pradeep T (2022) Chapter 3-nanosensors for water quality monitoring. In: Ahuja S (ed) Separation science and technology. Separations of water pollutants with nanotechnology. Academic Press, vol 15, pp 37–53

Howard G, Bartram J (2003) Domestic water quantity, service level and health

Hussain CM, Mishra AK (2018) Nanotechnology for sustainable water resources. John Wiley & Sons. ISBN 978–1–119–32383–9

Jackson RB, Carpenter SR, Dahm CN, McKnight DM, Naiman RJ, Postel SL, Running SW (2001) Water in a changing world. Ecol Appl 11:1027–1045. https://doi.org/10.1890/1051-0761(2001)011[1027:WIACW]2.0.CO;2

Janardhana C, Rao G, Ramamurthy S, Lakshman V (2006) Study on defluoridation of drinking water by impregnation of metal ions in activated charcoal. Indian J Chem Technol 13:414–416

Kamble SP, Jagtap S, Labhsetwar NK, Thakare D, Godfrey S, Devotta S, Rayalu SS (2007) Defluoridation of drinking water using chitin, chitosan and lanthanum-modified chitosan. Chem Eng J 129:173–180. https://doi.org/10.1016/j.cej.2006.10.032

Lakowicz JR (2004) Radiative decay engineering 3 surface plasmon-coupled directional emission. Anal Biochem 324:153–169. https://doi.org/10.1016/j.ab.2003.09.039

Lakowicz JR, Ray K, Chowdhury M, Szmacinski H, Fu Y, Zhang J, Nowaczyk K (2008) Plasmon-controlled fluorescence: a new paradigm in fluorescence spectroscopy. Analyst 133:1308–1346. https://doi.org/10.1039/B802918K

Li J-F, Li C-Y, Aroca RF (2017) Plasmon-enhanced fluorescence spectroscopy. Chem Soc Rev 46:3962–3979. https://doi.org/10.1039/C7CS00169J

Magar HS, Ghica ME, Abbas MN, Brett CMA (2017) Highly sensitive choline oxidase enzyme inhibition biosensor for lead ions based on multiwalled carbon nanotube modified glassy carbon electrodes. Electroanalysis 29:1741–1748. https://doi.org/10.1002/elan.201700111

Martins R, Quintal C, Cruz L, Barata E (2016) Water affordability issues in developed countries–the relevance of micro approaches. Util Policy 43:117–123. https://doi.org/10.1016/j.jup.2016.04.012

Moronshing M, Subramaniam C (2018) Room temperature, multiphasic detection of explosives, and volatile organic compounds using thermodiffusion driven Soret colloids. ACS Sustain Chem Eng 6:9470–9479. https://doi.org/10.1021/acssuschemeng.8b02050

Moss B (2007) Water pollution by agriculture. Philos Trans R Soc B Biol Sci 363:659–666. https://doi.org/10.1098/rstb.2007.2176

Mulpur P, Podila R, Lingam K, Vemula SK, Ramamurthy SS, Kamisetti V, Rao AM (2013) Amplification of surface plasmon coupled emission from graphene–Ag hybrid films. J Phys Chem C 117:17205–17210. https://doi.org/10.1021/jp406122s

Mustafa G, Shahzeb Khan M, Asif MI, Ullah A, Khan I, Ullah I (2021) Chapter 6-water pollutants and nanosensors. In: Abd-Elsalam KA, Zahid M (eds) Aquananotechnology. Micro and Nano Technologies. Elsevier, pp 105–133. ISBN 978–0–12–821141–0

Nagar A, Pradeep T (2020) Clean water through nanotechnology: needs, gaps, and fulfillment. ACS Nano 14:6420–6435. https://doi.org/10.1021/acsnano.9b01730

Nanomaterials for Sensing and Destroying Pesticides (2023) Chemical reviews. https://pubs.acs.org/doi/full/https://doi.org/10.1021/cr300020c. Accessed 10 Feb 2023

Hu A, Apblett A eds (2014) Nanotechnology for water treatment and purification. In: Lecture notes in nanoscale science and technology. Springer International Publishing, Cham, vol 22. ISBN 978–3–319–06577–9

Nguyen H, Shih W-C (2019) Multiplex sensing of lead and mercury in drinking water using smartphone nano-colorimetry. In: Proceedings of the biophotonics congress: optics in the life sciences congress 2019 (BODA, BRAIN, NTM, OMA, OMP). paper DW1B.6; Optica Publishing Group, April 14 2019; p DW1B.6

Noronha VT, Aquino YMLO, Maia MT, Freire RM (2019) Chapter 3-sensing of water contaminants: from traditional to modern strategies based on nanotechnology. In: Nascimento do RF, Ferreira OP, De Paula AJ, Sousa Neto de VO (eds) Nanomaterials applications for environmental matrices. Advanced Nanomaterials, Elsevier, pp 109–150. ISBN 978–0–12–814829–7

Owa FD (2013) Water pollution: sources, effects, control and management. Mediterr J Soc Sci 4:65

Pan X-H, Cao S-H, Chen M, Zhai Y-Y, Xu Z-Q, Ren B, Li Y-Q (2020) In situ and sensitive monitoring of configuration-switching involved dynamic adsorption by surface plasmon-coupled directional enhanced Raman scattering. Phys Chem Chem Phys 22:12624–12629. https://doi.org/10.1039/D0CP01567A

Peirce JJ, Vesilind PA, Weiner R (1998) Environmental Pollution and Control. Butterworth-Heinemann. ISBN 978–0–08–053111–3

Pooja D, Kumar P, Singh P, Patil S (eds) (2020) Sensors in water pollutants monitoring: role of material. In: Advanced functional materials and sensors. Springer, Singapore. ISBN 9789811506703

Qu X, Brame J, Li Q, Alvarez PJJ (2013) Nanotechnology for a safe and sustainable water supply: enabling integrated water treatment and reuse. Acc Chem Res 46:834–843. https://doi.org/10.1021/ar300029v

Rai A, Bhaskar S, Ramamurthy SS (2021a) Plasmon-coupled directional emission from soluplus-mediated AgAu nanoparticles for attomolar sensing using a smartphone. ACS Appl Nano Mater 4:5940–5953. https://doi.org/10.1021/acsanm.1c00841

Rai A, Bhaskar S, Reddy N, Ramamurthy SS (2021b) Cellphone-aided attomolar zinc ion detection using silkworm protein-based nanointerface engineering in a plasmon-coupled dequenched emission platform. ACS Sustain Chem Eng 9:14959–14974. https://doi.org/10.1021/acssuschemeng.1c05437

Rai A, Bhaskar S, Ganesh KM, Ramamurthy SS (2021c) Engineering of coherent plasmon resonances from silver soret colloids, graphene oxide and Nd2O3 nanohybrid architectures studied in mobile phone-based surface plasmon-coupled emission platform. Mater Lett 304:130632. https://doi.org/10.1016/j.matlet.2021.130632

Rai A, Bhaskar S, Ganesh KM, Ramamurthy SS (2022a) Gelucire®-mediated heterometallic AgAu nanohybrid engineering for femtomolar cysteine detection using smartphone-based plasmonics technology. Mater Chem Phys 279:125747. https://doi.org/10.1016/j.matchemphys.2022.125747

Rai A, Bhaskar S, Ganesh KM, Ramamurthy SS (2022b) Cellphone-based Attomolar tyrosine sensing based on Kollidon-mediated bimetallic nanorod in plasmon-coupled directional and polarized emission architecture. Mater Chem Phys 285:126129. https://doi.org/10.1016/j.matchemphys.2022.126129

Rai A, Bhaskar S, Mohan GK, Ramamurthy SS (2022c) Biocompatible gellucire® inspired bimetallic nanohybrids for augmented fluorescence emission based on graphene oxide interfacial plasmonic architectures. ECS Trans 107:4527. https://doi.org/10.1149/10701.4527ecst

Rai A, Bhaskar S, Ganesh KM, Ramamurthy SS (2022) Hottest hotspots from the coldest cold: welcome to nano 4.0. ACS Appl. Nano Mater 5:12245–12264. https://doi.org/10.1021/acsanm.2c02556

Ramamurthy SS, Chen Y, Kalyan MK, Rao GN, Chelli J, Mitra S (2011) Carbon nanotube-zirconium dioxide hybrid for defluoridation of water. J Nanosci Nanotechnol 11:3552–3559. https://doi.org/10.1166/jnn.2011.3806

Rathnakumar S, Bhaskar S, Rai A, Saikumar DVV, Kambhampati NSV, Sivaramakrishnan V, Ramamurthy SS (2021) Plasmon-coupled silver nanoparticles for mobile phone-based attomolar sensing of mercury ions. ACS Appl Nano Mater 4:8066–8080. https://doi.org/10.1021/acsanm.1c01347

Rawtani D, Khatri N, Tyagi S, Pandey G (2018) Nanotechnology-based recent approaches for sensing and remediation of pesticides. J Environ Manage 206:749–762. https://doi.org/10.1016/j.jenvman.2017.11.037

Richardson SD, Ternes TA (2018) Water analysis: emerging contaminants and current issues. Anal Chem 90:398–428. https://doi.org/10.1021/acs.analchem.7b04577

Roy BN (2003) The costs of water pollution in india. Mark Dev Water Amp Waste Technol

Sarkar AK, Kalita D, Baruah S (2021) Chapter 4-nanomaterials for aquatic contamination sensing and remediation. In: Abd-Elsalam KA, Zahid M (eds) Aquananotechnology. Micro and Nano Technologies. Elsevier, pp 67–89. ISBN 978–0–12–821141–0

Sathish RS, Sujith U, Rao GN, Janardhana C (2006) Fluoride ion detection by 8-Hydroxyquinoline–Zr(IV)–EDTA complex. Spectrochim Acta A Mol Biomol Spectrosc 65:565–570. https://doi.org/10.1016/j.saa.2005.12.011

Sathish RS, Sairam S, Raja VG, Rao GN, Janardhana C (2008) Defluoridation of water using zirconium impregnated coconut fiber carbon. Sep Sci Technol 43:3676–3694. https://doi.org/10.1080/01496390802222541

Sathish RS, Raju NSR, Raju GS, Nageswara Rao G, Kumar KA, Janardhana C (2007) Equilibrium and kinetic studies for fluoride adsorption from water on zirconium impregnated coconut shell carbon. Sep Sci Technol 42:769–788. https://doi.org/10.1080/01496390601070067

sitecontrol Progress on Drinking Water and Sanitation (2014) Update https://data.unicef.org/resources/progress-on-drinking-water-and-sanitation-2014-report/. Accessed 9 Feb 2023

Soni R, Soni M, Shukla DP (2020) Emerging techniques and materials for water pollutants detection. In: Pooja D, Kumar P, Singh P, Patil S (eds) Sensors in water pollutants monitoring: role of material. Advanced Functional Materials and Sensors, Springer, Singapore, pp 277–297. ISBN 9789811506710

Stillinger FH (1980) Water revisited. Science 209:451–457. https://doi.org/10.1126/science.209.4455.451

Street A, Sustich R, Duncan J, Savage N (2014) Nanotechnology applications for clean water: solutions for improving water quality. William Andrew. ISBN 978–1–4557–3185–5

Teodoro KBR, Sanfelice RC, Migliorini FL, Pavinatto A, Facure MHM, Correa DS (2021) A review on the role and performance of cellulose nanomaterials in sensors. ACS Sens 6:2473–2496. https://doi.org/10.1021/acssensors.1c00473

Thakkar S, Dumée LF, Gupta M, Singh BR, Yang W (2021) Nano-enabled sensors for detection of arsenic in water. Water Res 188:116538. https://doi.org/10.1016/j.watres.2020.116538

Treacy J (2019) Drinking water treatment and challenges in developing countries. IntechOpen. ISBN 978–1–78985–804–4

Verma R, Dwivedi P (2013) Heavy metal water pollution-a case study. Recent Res Sci Technol 5:98–99

Walekar L, Dutta T, Kumar P, Ok YS, Pawar S, Deep A, Kim K-H (2017) Functionalized fluorescent nanomaterials for sensing pollutants in the environment: a critical review. TrAC Trends Anal Chem 97:458–467. https://doi.org/10.1016/j.trac.2017.10.012

Wang H, Zhang B, Zhao Y, Chen X, Zhang Z, Song H (2019) Integrated effects of near-field enhancement-induced excitation and surface plasmon-coupled emission of elongated gold nanocrystals on fluorescence enhancement and the applications in PLEDs. ACS Appl Electron Mater 1:2116–2123. https://doi.org/10.1021/acsaelm.9b00489

Wanger TC (2019) Biodiversity and climate change adaptation in tropical islands Edited by Chandrakasan Sivaperuman, Ayyam Velmurugan, Awnindra Kumar Singh, and Iyyappan Jaisankar. Q Rev Biol 94:108–108. https://doi.org/10.1086/702390

Warren CE (1971) Biology and water pollution control W

Weng Y-H, Xu L-T, Chen M, Zhai Y-Y, Zhao Y, Ghorai SK, Pan X-H, Cao S-H, Li Y-Q (2019) In Situ monitoring of fluorescent polymer brushes by angle-scanning based surface plasmon coupled emission. ACS Macro Lett 8:223–227. https://doi.org/10.1021/acsmacrolett.8b00882

World Bank World Development Report (2008) Agriculture for development. World Bank, Washington, DC, p 2007

Xie K-X, Jia S-S, Zhang J-H, Wang H, Wang Q (2019a) Amplified fluorescence by carbon nanotube (CNT)-assisted surface plasmon coupled emission (SPCE) and its biosensing application. New J Chem 43:14220–14223. https://doi.org/10.1039/C9NJ03458G

Xie K-X, Xu L-T, Zhai Y-Y, Wang Z-C, Chen M, Pan X-H, Cao S-H, Li Y-Q (2019b) The synergistic enhancement of silver nanocubes and graphene oxide on surface plasmon-coupled emission. Talanta 195:752–756. https://doi.org/10.1016/j.talanta.2018.11.112

Xie K-X, Liu Q, Song X-L, Huo R-P, Shi X-H, Liu Q-L (2021) Amplified fluorescence by hollow-porous plasmonic assembly: a new observation and its application in multiwavelength simultaneous detection. Anal Chem 93:3671–3676. https://doi.org/10.1021/acs.analchem.0c05219

Xie Z, Ye C, Li C, Shi X, Shao Y, Qi W (2022) The global progress on the non-point source pollution research from 2012 to 2021: a bibliometric analysis. Environ Sci Eur 34:121. https://doi.org/10.1186/s12302-022-00699-9

Xu L-T, Chen M, Weng Y-H, Xie K-X, Wang J, Cao S-H, Li Y-Q (2022) Label-free fluorescent nanofilm sensor based on surface plasmon coupled emission: in situ monitoring the growth of metal–organic frameworks. Anal Chem 94:6430–6435. https://doi.org/10.1021/acs.analchem.1c05349

Yunus IS, Harwin KA, Adityawarman D, Indarto A (2012) Nanotechnologies in water and air pollution treatment. Environ Technol Rev 1:136–148. https://doi.org/10.1080/21622515.2012.733966

Zhang X, Yan S, Chen J, Tyagi RD, Li J (2020) 3-physical, chemical, and biological impact (Hazard) of hospital wastewater on environment: presence of pharmaceuticals, pathogens, and antibiotic-resistance genes. In: Tyagi RD, Sellamuthu B, Tiwari B, Yan S, Drogui P, Zhang X, Pandey A (eds) Current developments in biotechnology and bioengineering. Elsevier, pp 79–102. ISBN 978–0–12–819722–6

Zhao J, Shao Q, Ge S, Zhang J, Lin J, Cao D, Wu S, Dong M, Guo Z (2020) Advances in template prepared nano-oxides and their applications: polluted water treatment, energy, sensing and biomedical drug delivery. Chem Rec 20:710–729. https://doi.org/10.1002/tcr.201900093

Seemesh Bhaskar obtained his Ph.D. from STAR Lab, CRIF, SSSIHL, India. Further he completed his Post-Doctoral higher studies at IIT Bombay and is currently serving as Post-Doctoral Research Associate at Carl R. Woese Institute for Genomic Biology (IGB), Holonyak Micro and Nanotechnology Laboratory, University of Illinois at Urbana-Champaign (UIUC). He has pursued research internships and projects at SASTRA University and IISc & CIIRC Bangalore. His collaboration with research groups at various universities such as IIT-Bombay, IIT-Kharagpur, IIT-Gandhinagar have resulted in 40+ publications in reputed national and international peer-reviewed journals, and 10+ conference proceedings. He is the recipient of the prestigious DST-Inspire scholarship & fellowship, AWSAR, K. V. Rao Scientific Society (KVRSS), Young Achiever, M.Sc. Chemistry Gold medal and All-rounder Gold medal awards. His research work focuses on building effective sustainable nano-engineering protocols for photo-plasmonic biosensing and is the pioneer in the development of photonic crystal-coupled emission (PCCE) platform, Cryosoret and FerroPlasmon-on-Mirror (FPoM) technologie. Link: https://nano.ece.illinois.edu/seemesh-bhaskar/

Sai Sathish Ramamurthy has completed his Ph.D. from SSSIHL, India and postdoctoral studies from University of Maryland Baltimore County, USA. His collaboration with research groups at various universities in USA, Australia, Japan and India have resulted in 120 publications in reputed journals, conferences, patents, books and book chapters. He created the Surface Plasmon based Translational and Advanced Research (STAR) laboratory that has developed multiple biomedical technologies targeting the whitespaces in the medical diagnostics capable for usage in resource-limited settings, especially for the bottom of the pyramid. Link: https://www.starlab-sssihl.com/drrss

Chapter 10
Sustainable Treatment of Contaminants of Emerging Concerns in Water and Wastewater

Sneha Yadav and Shivaraju Harikaranahalli Puttaiah

Abstract Contaminants of emergent concerns (CECs) are contaminants found in trace amounts in the environment that have not before been recognized and pose significant environmental and health hazards. Pharmaceutical and personal care items, endocrine-disrupting chemicals, disinfection by-products, and other CECs are among them. The current chapter describes the many contamination sources of CECs, encompassing point sources, as well as the kind of contamination, from the perspective of environmental health and safety issues. Traditional water and wastewater treatment procedures, as well as innovative remediation approaches, are thoroughly examined, together with their benefits and drawbacks. From the viewpoint of resilience and implantation, prospective solutions in the treatment of CECs such as nano-based adsorption, and membrane technologies are fully revealed. Sustainable management and treatment options, such as citizen science, regulatory, networking, and technical strategies that have been established for real-time application, are well-clarified for CECs in various environments.

Keywords Contaminants of emerging concerns (CECs) · Nanotechnology · Nano-adsorption · Nanomembranes · Public health

S. Yadav · S. H. Puttaiah (✉)
Department of Environmental Sciences, JSS Academy of Higher Education and Research, Mysuru, Karnataka 570015, India
e-mail: shivarajuenvi@gmail.com

S. Yadav
e-mail: snehayadav@jssuni.edu.in

S. H. Puttaiah
Center for Water, Food and Energy, GREENS Trust, Harikaranahalli Village, Dombaranahalli Post, Turuvekere Taluka, Tumkur District, Karnataka 572215, India

S. Suriyanarayanan et al. (eds.), *Water Management in Developing Countries and Sustainable Development*, Water Resources Development and Management,
https://doi.org/10.1007/978-981-99-8639-2_10

Abbreviations

CEC	Contaminants of emergent concern
AOP	Advanced oxidation process
PPCP	Pharmaceuticals and personal care product
EDC	Endocrine-disrupting chemicals
ASW	Artificial sweeteners
ROS	Reactive oxygen species

10.1 Introduction

The problem targets a major effect on water inadequacy, public health, and economic affluence. Moreover, the increase in population and urbanization is making more complications in imparting safe and sufficient drinking water (Domínguez et al. 2021). The majority of Earth's freshwater is consumed by humans, be it for industrial sectors or the domestic sector. The industrial sector considerably consumes a lot percentage of freshwater while generating tons of waste, which when untreated is discharged into the aquatic bodies creating severe pollution which affects all the biotic species (Weerasooriya et al. 2021). Over the years conventional treatments have been assigned to treat water pollution, but the condition becomes more complicated due to the discharge of organic contaminants into the aquatic system. Organic pollutants can be endocrine-disrupting chemicals (EDCs), artificial sweeteners (ASW), pharmaceuticals and personal care products (PPCPs), agricultural products, etc. These diverse groups are classified as contaminants of emerging concerns (CECs). The term 'emerging' is added because they are often identified as newly manufactured or spontaneously detected in nature in various environments at inconsequential levels. This is the reason why most of the CECs do not have any regulatory standards and lack toxicological statistics (Shivaraju et al. 2022a).

The level of CECs detected in water mainly depends upon utility patterns, treatment methods, nature of persistence in the environment, and catchment physiognomies (Salimi et al. 2017). Though the concentrations of CECs in water may be low, small concentrations with persistent exposure will cause harmful effects on the environment. A study done by James and their team studied the presence and concentration of CECs inside an estuary ecosystem. The occurrence of 29 CEC analytes was detected in the water system, signifying a high potential for bioaccumulation among the biotic species growing in the water system. The study also stated the presence of CECs in lower concentrations, which in prolonged duration can be detrimental to the aquatic species (Meador et al. 2016a). From this perspective, various methodologies like advanced oxidation processes, biological treatments, membrane separation processes, etc., are used to augment the removal of CECs from the water system. Despite many treatment processes, the two major constraints are high energy consumption and high cost.

Because of unknown toxicity, there have only been a few studies carried out regarding the comparison between CECs toxicology data and concentration in the environment. Also, studies on CECs have been limited to restricted geographical areas or classes (Yadav et al. 2021). Though there is a much number of ongoing studies related to the removal or degradation of CECs from the water system. Furthermore, the formation of the intermediate by-products after the degradation done by advanced processes should be given more consideration because of the complexity of their structure and the nature of toxicity (Shah et al. 2020).

The major concern of untreated water getting dumped into freshwater channels remains. Moreover, the removal or degradation of organic pollutants from the water system needs more studies along with toxicological studies. Despite all challenges, the freshwater source is achieved via new advanced technologies, which ultimately aim to improve the quality of the environment (Kim et al. 2018). Nanotechnology is a rapidly growing interdisciplinary field that endeavors novel techniques to produce nanoscale prospects to use science and engineering to upkeep biotic health. Nanotechnology deals with the capacity to produce nanoscale materials (less than 100 nm) in various applications as a demonstration for a new and advanced sustainable environment (Alvarez et al. 2018).

This chapter focuses on understanding the capability of using new and emerging sustainable Nanotechnological approaches for the treatment of water and wastewater necessary for a future sustainable environment. The various applications of nano-based materials developed for the treatment that deeds and the challenges faced in the treatment process have also been reviewed. The sources and occurrence of CECs in water and wastewater treatment plants along with their environmental impacts have been focused on.

10.2 Source and Fate of CEC

Because of the huge number of compounds and their repeated presence in various environmental matrices, the presence of CECs in the environment has become a major topic of concern. Pharmaceutical and personal care products, hormones, pesticides, and industrial chemicals are the main growing sections, some of which are naturally occurring, artificial, or manufactured, and pose risks to the biotic world. In most cases, these CECs are created to meet the demands of everyday civilization before being released into the environment. Their prevalence has been consistently documented in low quantities. Bioaccumulation of CECs in aquatic animals has been reported by many researchers (Jasinska et al. 2015).

Contaminants of emerging concern pose a significant threat to aquatic life, jeopardizing the delicate balance of ecosystems. These pollutants, often arising from diverse sources, can have detrimental effects on both the physical and biological components of water bodies. Emerging pollutants encompass a broad range of chemicals, including pharmaceuticals, personal care products, pesticides, industrial chemicals, and microplastics (Hou et al. 2020). These contaminants can enter aquatic environments through direct sources, such as industrial discharges, wastewater treatment plants, and agricultural runoff. Indirect sources include atmospheric deposition and surface runoff from contaminated soils. Leachate from landfills and sewage sludge can also introduce emerging pollutants into water systems. The complex classification of these contaminants highlights the urgent need for comprehensive monitoring and mitigation strategies to safeguard aquatic life and ensure the long-term health of our water resources (Arora et al. 2018).

Wastewater generated from the urban environment is one of the major sources of CECs deposition in the environment. Because conventional techniques employed in urban wastewater treatment plants (WWTPs) are ineffective at entirely removing CECs. Also, household wastewater released into the aquatic environment is another source of the presence of CECs in the environment (Diamond et al. 2011). Furthermore, wastewater recycling results in the widespread presence of CECs in the agricultural sectors and thus their subsequent diffusion into the food chain. CECs can have many negative consequences such as endocrine disruption, lethal and sub-lethal toxicity, reproduction among aquatic animals, superbugs in the form of antibiotic-resistant bacteria, and antibiotic-resistant genes. Regular monitoring for the detection of harmful substances in the environment becomes impossible (Kasonga et al. 2021). In contrast, supplying data on the occurrence of CECs is vital and provides a basis for decision-makers to design mitigation strategies for a more sustainable water policy. The major CECs are narrowed down to 5 divisions namely, pharmaceuticals, personal care products, pesticides, industrial chemicals, and hormones. Studies report that the highest concentration CECs is detected in wastewater, followed by surface water and groundwater. The main source of the pollutants released into the ecosystem is the wastewater treatment plants. The untreated CECs run downstream and contaminate the surface water and further contaminate the groundwater. Studies confirm the presence of pesticides mostly in surface and groundwater (Joseph et al. 2019).

Direct sources of emerging pollutants involve the deliberate release or unintentional leakage of chemicals from industrial processes, manufacturing facilities, and commercial activities. These direct sources contribute to the contamination of water bodies through effluents and discharges. Additionally, wastewater treatment plants can inadvertently become sources of emerging pollutants when they are unable to effectively remove these chemicals from the water before it is released back into the environment. Indirect sources of emerging pollutants include atmospheric deposition and surface runoff. Atmospheric deposition occurs when pollutants are transported through the air and deposited into water bodies through precipitation or dry deposition. This can include contaminants such as heavy metals, persistent organic pollutants, and microplastics. Surface runoff, on the other hand, involves the movement of pollutants from land surfaces into nearby waterways during rainfall or snowmelt events. Agricultural runoff, urban stormwater runoff, and erosion from construction sites are common contributors to surface runoff, leading to the introduction of emerging pollutants into aquatic ecosystems. Understanding the diverse sources and pathways of emerging pollutants is crucial in implementing effective pollution prevention measures and mitigating the impact on aquatic life (Romeyn et al. 2016; Shivaraju et al. 2022b).

Leachate is another significant source of emerging pollutants. It refers to the liquid that percolates through waste materials and carries dissolved or suspended contaminants with it. Landfills, for example, produce leachate as rainwater infiltrates through the waste layers, extracting various pollutants from the decomposing materials. If not properly managed, leachate can seep into groundwater or nearby surface water, leading to the contamination of aquatic ecosystems (Lindamulla et al. 2022).

Runoff plays a crucial role in the transport of emerging pollutants from the land to water bodies. It occurs when rainwater or snowmelt flows over the ground, picking up pollutants along the way and carrying them into streams, rivers, and other waterways. Agricultural runoff is a major contributor, carrying pesticides, fertilizers, and soil particles into nearby water bodies. Urban stormwater runoff, which includes pollutants from roads, rooftops, and other urban surfaces, can also introduce a range of contaminants into aquatic environments. Construction sites and industrial areas are additional sources of runoff that can carry sediment, chemicals, and other pollutants into water systems (Santos et al. 2022).

Understanding the different pathways and sources of emerging pollutants is essential for developing effective strategies to mitigate their impacts on aquatic life. Implementing proper wastewater treatment, improving industrial practices, managing landfills effectively, and adopting sustainable agricultural and urban stormwater management approaches are critical steps toward safeguarding aquatic ecosystems from the threats posed by emerging contaminants (Tondera et al. 2018).

10.3 Environmental Impacts of Contaminants of Emerging Concern

An increase in CEC concentration in the environment can become a harmful source which can affect the food chain at multiple trophic levels, thereafter, causing biomagnification. Also, in some circumstances, consuming contaminated water by animals or humans can result in direct exposure. Long-term exposure has negative consequences on aquatic ecology as well and, in certain situations, can also change how animals and humans functionalize their hormones and metabolism (Vidal-Dorsch et al. 2012).

10.3.1 Impact on Human Health

Even at lower doses, long-term exposure to CECs causes various aberrations in human health. The Conservancy of human health and environmental problems are intimately associated with the case of pharmaceuticals (Starling et al. 2019). For instance, antibiotics led to the formation of antibiotic resistance genes (ARGs) and superbugs as a result of the presence of antimicrobial drugs decreasing the ability of antibiotics to effectively treat animal and human pathogens. Through bioaccumulation, estrogenic substances like 17-estradiol may make their way to people and have an impact on their health via aquatic animals. The development of human embryonic stem cells may be hampered by the presence of statins (HMG-CoA reductase inhibitors). Artificial sweeteners like saccharin and sucralose may also interact with digestive enzymes and microbiomes, resulting in inflammatory bowel illness (Falconer et al. 2006). A study done by Vimalkumar and team studied triclocarban affecting aquatic and human health. Triclocarban is one of the major contaminants that cause CECs. It is a synthetic chemical with a large manufacturing volume that is widely utilized in several pharmaceuticals and personal care products. The study revealed that triclocarban can act as an endocrine disruptor by affecting the immune system as well as reproduction. Higher concentrations of triclocarban have been reported to be found in the aquatic system, which in turn affect and cause bioaccumulation among humans (Vimalkumar et al. 2019). The majority of these published reports are, however, limited, making it very challenging to conclude the toxicological concerns on human health. Consequently, further study is needed to develop appropriate regulatory standards for CECs in aquatic environments.

10.3.2 Toxicity in Aquatic Environment

In an aquatic environment, the base of the food chain is largely composed of planktonic-like algae and other photosynthetic organisms (Pena-Pereira et al. 2021). Thus, any changes to the photo-autotrophic bacteria may have a major impact on the organisms existing at different trophic levels. A study done by Meador and the team assessed the occurrence of CECs in an estuary ecosystem. A total of 25 CEC analytes were detected in the estuary and 42 compounds in the fish tissues. CECs compounds like triclosan, sertraline, fluoxetine, estrone, nonylphenol, and metformin were found in the tissue matrix at crossed concentrations which can cause adversities among fishes. The knowledge gap concluded in the study should be filled in the future by discovering and using mechanism-based biomarkers connected to physiological consequences brought on by exposure to CECs, which would improve prophecies of harmful ecological effects (Meador et al. 2016b). Another study done by Yan and team reported bioaccumulation among aquatic animals in the presence of polyhalogenated carbazoles at Great Lake from the year 2004–2016. Successive median concentrations of carbazoles were found in the lake with an adverse study report for the aquatic flora and fauna. The effect of bioaccumulation was thoroughly studied and reported to find 11 target polyhalogenated carbazoles ranging from 33.4 to 376 ng/g lw. The study also reported the presence of carbazoles, regardless of species and also their presence in the aquatic food chain (Wu et al. 2018).

10.4 Wastewater Treatment with Advanced Technologies

The release of wastewater into the environment by individuals and industries as well as liquid, solid, and gaseous agricultural wastes most one of the solemn risks to the continuation of human civilization. This uncomplimentary behaviour leads to air, land, and water pollution, which thus causes global warming (Bairagi and Ali 2020). Maintaining public health and reducing unendurable levels of environmental harm in the face of growing populations and economies, wastewater treatment and disposal is fundamental. The main aim of sustainability is to maintain both public and economic safety (Zhou and Smith 2011). Additionally, sustainability aims to safeguard the environment, use resources prudently, and advance social justice while taking into account the needs of all people, communities, and the ecosystem (Baker et al. 2020).

In the framework of wastewater treatment, the new challenge is to change the standard from harmful filthy water to a usable resource along with advanced technologies. Conventional wastewater treatment methods are unable to treat organic pollutants. The partial removal of pollutants is not enough, along with the recent issues related to superbugs. The use of advanced treatment processes is the need of the hour for treating CECs as well as superbugs (Upadhye and Joshi 2012).

10.4.1 Nanotechnology in Water Treatment

One of the major focuses of nanotechnology research and development has been obtaining clean and safe water. Nanotechnologies are part of a new and advanced industrial stage that has vast potential to outperform present technology-intensive water treatment methods (Iravani 2020).

These technologies comprehend a wide range of techniques and instruments with their applications in the usage of particles with estimated diameters ranging from a few nanometres to a few hundred nanometers. Nanoparticles have remarkable physiochemical and surface qualities, letting them be used in a variety of applications and contribute to answering the global water crisis. Water treatment, wastewater reuse, and desalination are certain examples of nanotechnology-based solutions to a water crisis (Patanjali et al. 2019). The inherent potential of nanotechnology in wastewater treatment is classified into three categories: sensing and detection, pollution control, and water treatment. Nanotechnology advanced in providing the potential to improve water quality in the following ways: water treatment with ease, and water resource sustainability through the use of improved filter materials that enable more water reuse and recycling (Qu et al. 2013). In terms of sensing and detecting harmful compounds, there is a vast area in the development of new and improved sensors for detecting biological and chemical pollutants at extremely low concentrations. When compared to traditional methods, nanomaterial adsorbents have demonstrated much-improved efficacy and a high rate of removal of pollutants from wastewater

(Goyal et al. 2011). Also, studies on specific types of nanoparticles like Zn, Ag, and Mg along with classes of chemicals like metal chalcogenides, layered-double-hydroxide, etc. are being thoroughly studied in the treatment of wastewater treatment. A varied range of efficient and cost-effective as well as environmentally friendly nanomaterials showcasing various functionalities have been made for use in the degradation of industrial effluents, surface/groundwater, as well as drinking water. Water treatment works with pollutant adsorption and/or photocatalysis, as well as contaminant removal via nanoparticles and bioremediation (Raval and Gohil 2010). Diverse groups of nanomaterials are synthesized for their application in multidisciplinary fields like biomedicine, water/wastewater treatment, etc. The nanoparticles are usually categorized into 1D, 2D, and 3D based on their structural and electronic configurations, each having its unique characteristic feature (Salem et al. 2023).

1D materials refer to structures having only one dimension. These materials are essentially elongated in shape, with their length far greater than their width and height. The properties of 1D materials are unique and can differ significantly from their bulk counterparts. For example, carbon nanotubes, nanowires, molecular chains, etc. Carbon nanotubes are cylindrical structures made of carbon atoms arranged in a hexagonal lattice. They exhibit exceptional mechanical strength and electrical conductivity, making them suitable for varied applications such as composite materials, energy storage, and electronics (Shi et al. 2019). Nanowires are typical ultra-thin wires with diameters on the nanoscale. They can be made from a variety of materials, including metals, semiconductors, and oxides. Nanowires possess distinct electrical, optical, and mechanical properties, and they find applications in nano-electronics, energy devices, and sensors (Garnett et al. 2019) 0.2D materials on the other hand are essentially thin layers of material that exhibit unique properties due to their confinement within two dimensions. These materials have gained significant attention due to their remarkable properties and potential applications (Miró et al. 2014). For example, graphene is a single layer of carbon atoms arranged in a hexagonal lattice. It is known for its extraordinary strength, high electrical conductivity, and excellent thermal properties having applications in electronics, energy storage, and composites (Xu et al. 2013). Another group of materials is transition metal dichalcogenides, such as MoS_2 and WSe_2, which are 2D materials composed of layers of transition metals sandwiched between chalcogen atoms. They tend to exhibit interesting electrical, optical, and catalytic properties, making them suitable for use in electronics, optoelectronics, and energy applications (Chowdhury et al. 2020). Nevertheless, 3D materials are the conventional materials we encounter in everyday life, where all three dimensions (length, width, and height) are substantial. These materials can be naturally occurring or artificially synthesized, and their properties depend on their composition and structure. For example, polymers, are large molecules composed of repeating subunits. They can be found in numerous forms, including plastics, rubber, and fibres. Polymers offer a wide range of properties depending on their composition and structure, making them versatile materials used in packaging, textiles, automotive parts, and more (Arefin et al. 2021). Another group which can be categorized into 3D can be the metals. Common metals like aluminium, copper, and iron are widely used in various industries due to their high strength, ductility, and thermal conductivity having applications under construction, transportation, electrical wiring, and many other fields (Ackermann 2020).

10.4.1.1 Nano-Adsorption

Adsorption technology is a rapid, low-cost, and ubiquitous wastewater treatment solution. Adsorption is defined as a surface phenomenon that is commonly utilized as a polishing step in wastewater treatment to remove inorganic and organic pollutants (Al-Dabbas 2013). When an adsorbable solution comes into contact with a solid with a highly porous surface, intermolecular forces in the liquid–solid surface allow part of the solution's concentrated solute molecules to be deposited on the adsorbent's solid surface. Adsorbate is the solute maintained in adsorption, and an adsorbent is the solid on which the adsorbate is held. Adsorption is thus the buildup of adsorbate on the surface of an adsorbent. Adsorbents that have been engineered are usually very porous materials with large surface areas (Rajendran et al. 2022).

Nano-adsorbents can be created by combining atoms of chemically active substances with strong adsorption capacity on the surface of a nanomaterial. Various kinds of nanomaterial can be used as nano-adsorbents like metallic nanoparticles, metal oxides, zeolites, and nanocatalysts (Khan et al. 2019). These adsorbents offer a variety of outstanding structural and morphological features, making them suitable for usage as excellent adsorbent materials in addressing the issue of water scarcity. Heavy metal adsorption nanoparticles include carbon nanotubes, activated carbon, manganese oxide, zinc oxide, titanium dioxide, ferric oxide, and magnesium oxide. Nano-adsorbents are classified as composites since they are composed of two or more components with considerably differing chemical or physical characteristics. When these features are blended, a material with properties distinct from the separate components is formed (Jiang et al. 2014).

Rafiq and team studied zinc and magnesium oxide-based nano-adsorbent for wastewater treatment. Maximum adsorption for ZnO was found to be 226 mg/g and for MgO was 593 gm/g at a constant temperature of 70 °C. Both ZnO and MgO were found to be effective in the removal of copper from the wastewater (92–98%), but MgO was way much more effective when in comparison to ZnO. Both nano-adsorbents were cost-effective, efficient, and environment-friendly (Rafiq et al. 2014). Another work done by Tatarchuk and the team showcased cobalt-zinc ferrite nanoparticles for the removal of methylene blue and acid orange 7 dye via adsorption technique. The nano-adsorbent was prepared via the co-precipitation method and was further used to study grain conductivities. The results interpreted that the removal efficiency of methylene blue increased with the increase in cobalt dopant. Both the dye was efficiently removed with the help of the nano-adsorbent $ZnFe_2O_4$ reaching 31 mg/g of adsorbent to remove acid orange 7 dye and 3.4 mg/g for $Co_{0.9}Zn_{0.1}Fe_2O_4$ (Tatarchuk et al. 2021).

There are many such nano-adsorbent available which are used to treat water and wastewater. Much focus should be given to understanding the toxic nature of the nanoparticles and their leaching properties.

10.4.1.2 Nanomembranes

Nanomembranes are filter structures with thicknesses ranging from 1 to 100 nm that segregate liquids and gases at the molecular level. Organic polymers are mixed with a nanoparticle mesh to create such membranes. The small size of the pores permits or prevents the different-sized molecules. Nanomembranes display a range of properties like the formation of membrane scaffolds, surface shaping, and surface modification used in a variety of applications like energy sensing, optics, wastewater treatment, plasmonic, etc. Various kinds of nanomembranes have been reported like carbon nanotube membranes, nanoparticle-based membranes, ceramic membranes, nanofibers membranes, etc. (Cigané et al. 2021).

Metal oxide nanoparticle membranes consist of thin films or coatings of metal oxide nanoparticles, such as TiO_2 or ZnO, on a substrate or membrane material. These nanoparticles possess unique properties that can be utilized for various applications, including water purification and pollutant degradation. When exposed to ultraviolet (UV) light or other light sources, metal oxide nanoparticles generate reactive oxygen species (ROS), such as hydroxyl radicals ($OH^{\cdot}$) and superoxide ions (O_2^-), through a process known as photocatalysis. These ROS have strong oxidizing properties, which enable them to break down organic pollutants into harmless byproducts. Metal oxide nanoparticle membranes can be integrated into filtration systems or used as coatings on surfaces to provide effective pollutant removal and degradation (Ng et al. 2013). Ceramic membrane nanocomposites combine the advantages of ceramic membranes with nanomaterials, such as metal oxides or carbon-based nanoparticles, to enhance their performance in pollutant degradation. The nanomaterials can be incorporated into the ceramic membrane structure or applied as coatings on the membrane surface. Ceramic membranes are known for their excellent chemical resistance, high-temperature stability, and durability. By incorporating nanoparticles, such as titanium dioxide or iron oxide, the membrane's photocatalytic properties can be enhanced. The presence of these nanoparticles facilitates the generation of ROS upon exposure to light, enabling efficient degradation of organic pollutants (Li et al. 2020).

Reactive oxygen species (ROS), $OH^{\cdot}$ and O_2^-, are highly reactive and possess strong oxidizing properties. When these ROS come into contact with organic pollutants, they initiate oxidation reactions, leading to the degradation of the pollutants into simpler and less harmful compounds. In the context of water treatment, organic pollutants, such as dyes, pesticides, and pharmaceutical compounds, can be effectively degraded using ROS generated by metal oxide nanoparticles or ceramic membrane nanocomposites. The ROS attack the chemical bonds of the pollutants, breaking them down into smaller molecules or even carbon dioxide and water. This process is known as advanced oxidation or advanced oxidation processes (AOPs). The combination of metal oxide nanoparticle membranes, ceramic membrane nanocomposites, and the generation of ROS through photocatalysis offers a promising approach for the efficient degradation of organic pollutants in water and wastewater treatment. These technologies provide environmentally friendly solutions and have the potential to address pollution challenges in various industries (Li et al. 2022; Mahlambi et al. 2015).

The expected osmotic flow and electric field variations within nanopores are critical for understanding the impacts of pore shape and size of the nanomembrane or changed pore surface. The interaction of these factors governs macromolecule transit via nanopores (Rogers et al. 2011). Nanomaterials can be engineered to create novel nanomembrane processes that rely on passive diffusion processes or active transport systems. New materials may potentially be combined with membranes to generate new nanomembrane reactors. Recent advances in nanotechnology alongside membrane modules have been identified as a practical and efficient technique for improving membrane performance, with synergistic implications in treatment processes. Thus, advancements in nanomembrane technology have provided incalculable facilities and prospects for water purification, even at the ionic level (Turchanin and Gölzhäuser 2016).

A study done by Shawky and the team showcased the synthesis of a multi-walled carbon nanotube infused with an aromatic polyamide. A polymer grafting process was chosen to synthesize the nanotube. The nanotube could efficiently remove the targeted humic acid from water with an initial concentration of 20 ppm (Shawky et al. 2011). Another study done by Kim and their team developed nanosilver-based multi-walled carbon nanotubes for antibacterial application in water systems. The synthesized material could effectively remove bacteria from the taken water sample and further understand the membrane efficiency after and before incubation over selected bacteria SEM analysis was done (Kim et al. 2012).

10.4.1.3 Nanoscale Photocatalysis

Photocatalysis has evolved as a green and sustainable approach for completely mineralizing dangerous organic compounds to CO_2, water, and mineral acids. A photocatalytic reaction is a light-infused reaction that is aided by the introduction of a catalyst. When light strikes the surface of a catalyst, electrons absorb it and get excited. A good photocatalyst should be cost-effective, non-toxic, stable, and efficiently photoactive. TiO_2, CdS, ZnO, ZnS, WO_3, etc., are among the most reported semiconductor nanoparticles, and their composites are ascribed to photocatalytic degradation of organic pollutants in water and wastewater (He et al. 2018).

Table 10.1 provides a list of photoactive materials developed in recent times for the degradation of organic pollutants.

Table 10.1 List of semiconductive materials used for removal of organic pollutants

Sl. no	Target pollutants	Photocatalyst	Removal efficiency (%)	References
1	$MgTiO_3$	Lomefloxacin	83	(Sneha et al. 2022)
2	ZnV_2O_6/GCN	Tetracycline	87.2	(Ravindra et al. 2020)
3	La_2CuO_4	Doxycycline	85	(Prabagar et al. 2022)
4	ZrRGOCuFe-LDH	Methylene blue	95.2	(Kumar et al. 2021)
5	Ag_3PO4	Rhodamine B	~100	(Ge et al. 2012)
6	Fe(VI)/Na_2SO_3	Methylene blue	80	(Sun et al. 2019)

10.5 Conclusion

The need for clean and safe drinking water demand is fast increasing throughout the world as people become more concerned about their health and the environment. The present concern under pollutants is majorly focused on CECs. The CECs remain persistent in the environment and are further not degraded by conventional methods. The need for advanced treatment technologies is an important wing to address because of the growing pollution. Nanotechnology is one aspect which can resolve the technical and economic capacity constraints to supplying clean water to present and future generations for attaining sustainability.

Nevertheless, nanotechnology offers promising opportunities for water treatment, but it also faces several challenges that need to be addressed for its successful implementation. Some of the major challenges include:

1. Cost: The cost of nanomaterials, fabrication processes, and equipment associated with nanotechnology can be relatively high. The development of cost-effective manufacturing methods and large-scale production is crucial for the widespread adoption of nanotechnology in water treatment.
2. Scalability: While many nanomaterials and nano-devices have shown excellent performance at the laboratory scale, scaling up these technologies to meet the demands of commercial-scale operations remains a challenge. Ensuring the reproducibility and scalability of nanotechnology-based water treatment systems is essential for their practical implementation.
3. Safety and Environmental Concerns: Nanomaterials used in water treatment should be thoroughly evaluated for their potential environmental impacts and human health risks. Proper assessment of the toxicity and fate of nanomaterials is crucial to ensure their safe application and prevent unintended consequences.
4. Stability and Durability: Nanomaterials used in water treatment should possess sufficient stability and durability to maintain their performance over time. Challenges exist in maintaining the integrity and functionality of nanomaterials under harsh operating conditions, such as high temperatures, pH variations, and fouling.
5. Recyclability and Reusability: Efficient methods for the recovery, regeneration, and reuse of nanomaterials are essential to minimize waste and reduce

the overall cost of water treatment processes. Developing techniques for the efficient separation and recycling of nanomaterials from treated water is a significant challenge.

6. Fouling and Membrane Scaling: Fouling, deposition of contaminants, or scaling of minerals on nanomaterial surfaces and membranes can significantly impact their performance and lifespan. Strategies to mitigate fouling and scaling effects, such as surface modification or anti-fouling coatings, need to be developed to ensure long-term operation.
7. Regulatory Framework: The regulatory framework for the use of nanomaterials in water treatment is still evolving. Standards, guidelines, and regulations need to be established to ensure the safe and responsible use of nanotechnology and its products in water treatment applications.

Addressing these challenges requires multidisciplinary collaborations among researchers, industry, regulatory bodies, and policymakers. Continued research and development efforts are essential to overcome these hurdles and unlock the full potential of nanotechnology in water treatment while ensuring its sustainability and affordability. More advanced methods for water treatment will be accessible in the future as science and technology improve. Green engineering, which uses natural resources to create nanomaterials, will overcome the environmental and economic issues associated with nanomaterial production. The toxicological study of nanomaterials is one topic which needs attention in research. The future of using nano-based material in clearing out issues related to water and wastewater treatment is very promising and very important for attaining a sustainable future.

Acknowledgements The authors would like to extend their gratitude to JSS Academy of Higher Education and Research, Mysuru and NAM S&T Centre, New Delhi for providing an opportunity to publish this Chapter.

References

Ackermann L (2020) Metalla-electrocatalyzed C-H activation by earth-abundant 3d metals and beyond. Acc Chem Res 53(1):84–104. https://doi.org/10.1021/acs.accounts.9b00510

Al-Dabbas MA (2013) The performance of NANO adsorption solar cooling generator unit. 2013 1st international conference and exhibition on the applications of information technology to renewable energy processes and systems. IT-DREPS 2013:55–59. https://doi.org/10.1109/IT-DREPS.2013.6588150

Alvarez PJJ, Chan CK, Elimelech M, Halas NJ, Villagrán D (2018) Emerging opportunities for nanotechnology to enhance water security. Nat Nanotechnol 13(8):634–641. https://doi.org/10.1038/s41565-018-0203-2

Arefin AME, Khatri NR, Kulkarni N, Egan PF (2021) Polymer 3D printing review: materials, process, and design strategies for medical applications. Polymers 13(9). https://doi.org/10.3390/polym13091499

Arora NK, Fatima T, Mishra I, Verma M, Mishra J, Mishra V (2018) Environmental sustainability: challenges and viable solutions. Environ Sustain 1(4):309–340. https://doi.org/10.1007/s42398-018-00038-w

Bairagi S, Ali SW (2020) Conventional and advanced technologies for wastewater treatment. Environ Nanotechnol Water Purif 33–56.https://doi.org/10.1002/9781119641353.CH2

Baker BR, Mohamed R, Al-Gheethi A, Aziz HA (2020) Advanced technologies for poultry slaughterhouse wastewater treatment: a systematic review, vol 42, no 6, pp 880–899. https://doi.org/10.1080/01932691.2020.1721007. https://doi.org/10.1080/01932691.2020.1721007

Chowdhury T, Sadler EC, Kempa TJ (2020) Progress and prospects in transition-metal dichalcogenide research beyond 2D. In Chemical reviews, vol 120, Issue 22, pp 12563–12591. American Chemical Society. https://doi.org/10.1021/acs.chemrev.0c00505

Cigané U, Palevičius A, Janušas G (2021) Review of nanomembranes: materials, fabrications and applications in tissue engineering (bone and skin) and drug delivery systems. J Mater Sci 56(24):13479–13498. https://doi.org/10.1007/S10853-021-06164-X

Diamond JM, Latimer HA, Munkittrick KR, Thornton KW, Bartell SM, Kidd KA (2011) Prioritizing contaminants of emerging concern for ecological screening assessments. Environ Toxicol Chem SETAC 30(11):2385–2394. https://doi.org/10.1002/etc.667

Domínguez DC, Chacón LM, Wallace D (2021) Anthropogenic activities and the problem of antibiotic resistance in Latin America: a water issue. Water 13(19):2693. https://doi.org/10.3390/W13192693

Falconer IR, Chapman HF, Moore MR, Ranmuthugala G (2006) Endocrine-disrupting compounds: a review of their challenge to sustainable and safe water supply and water reuse. Environ Toxicol 21(2):181–191. https://doi.org/10.1002/TOX.20172

Garnett E, Mai L, Yang P (2019) Introduction: 1D nanomaterials/nanowires. In: Chemical reviews, vol 119, Issue 15, pp 8955–8957. American Chemical Society. https://doi.org/10.1021/acs.chemrev.9b00423

Ge M, Zhu N, Zhao Y, Li J, Liu L (2012) Sunlight-assisted degradation of dye pollutants in Ag 3PO 4 suspension. Ind Eng Chem Res 51(14):5167–5173. https://doi.org/10.1021/IE202864N/SUPPL_FILE/IE202864N_SI_001.PDF

Goyal AK, Johal ES, Rath G (2011) Nanotechnology for water treatment. Curr Nanosci 7(4):640–654.https://doi.org/10.2174/157341311796196772

He R, Xu D, Cheng B, Yu J, Ho W (2018) Review on nanoscale Bi-based photocatalysts. Nanoscale Horiz 3(5):464–504. https://doi.org/10.1039/C8NH00062J

Hou D, David C, Igalavithana A, Daniel SAi, Luo J, Daniel CW, Sparks YY, Rinklebe J, Yong SO (2020) Metal contamination and bioremediation of agricultural soils for food safety and sustainability. Nature.Com

Iravani S (2020) Nanomaterials and nanotechnology for water treatment: recent advances, vol 51, no 12, pp 1615–1645. https://doi.org/10.1080/24701556.2020.1852253

Jasinska EJ, Goss GG, Gillis PL, Van Der Kraak GJ, Matsumoto J, de Souza Machado AA, Giacomin M, Moon TW, Massarsky A, Gagné F, Servos MR, Wilson J, Sultana T, Metcalfe CD (2015) Assessment of biomarkers for contaminants of emerging concern on aquatic organisms downstream of a municipal wastewater discharge. Sci Total Environ 530–531:140–153. https://doi.org/10.1016/J.SCITOTENV.2015.05.080

Jiang Y, Hua M, Wu B, Ma H, Pan B, Zhang Q (2014) Enhanced removal of arsenic from a highly laden industrial effluent using a combined coprecipitation/nano-adsorption process. Environ Sci Pollut Res 21(10):6729–6735. https://doi.org/10.1007/S11356-014-2590-8

Joseph L, Jun BM, Jang M, Park CM, Muñoz-Senmache JC, Hernández-Maldonado AJ, Heyden A, Yu M, Yoon Y (2019) Removal of contaminants of emerging concern by metal-organic framework nanoadsorbents: a review. Chem Eng J 369:928–946. https://doi.org/10.1016/J.CEJ.2019.03.173

Kasonga TK, Coetzee MAA, Kamika I, Ngole-Jeme VM, Benteke Momba MN (2021) Endocrine-disruptive chemicals as contaminants of emerging concern in wastewater and surface water: a review. J Environ Manage 277:111485. https://doi.org/10.1016/J.JENVMAN.2020.111485

Khan NA, Khan SU, Ahmed S, Farooqi IH, Dhingra A, Hussain A, Changani F (2019) Applications of nanotechnology in water and wastewater treatment: a review. Asian J Water Environ Pollut 16(4):81–86. https://doi.org/10.3233/AJW190051

Kim ES, Hwang G, Gamal El-Din M, Liu Y (2012) Development of nanosilver and multi-walled carbon nanotubes thin-film nanocomposite membrane for enhanced water treatment. J Membr Sci 394–395:37–48. https://doi.org/10.1016/J.MEMSCI.2011.11.041

Kim S, Chu KH, Al-Hamadani YAJ, Park CM, Jang M, Kim DH, Yu M, Heo J, Yoon Y (2018) Removal of contaminants of emerging concern by membranes in water and wastewater: a review. Chem Eng J 335:896–914. https://doi.org/10.1016/J.CEJ.2017.11.044

Kumar OP, Ashiq MN, Shah SSA, Akhtar S, Mudhar MA, Mujtaba IM, Rehman Aur (2021) Nanoscale ZrRGOCuFe layered double hydroxide composites for enhanced photocatalytic degradation of dye contaminant. Mater Sci Semicond Process 128:105748. https://doi.org/10.1016/J.MSSP.2021.105748

Li L, Han Q, Wang L, Liu B, Wang K, Wang Z (2022) Dual roles of MoS_2 nanosheets in advanced oxidation processes: activating permonosulfate and quenching radicals. Chem Eng J 440https://doi.org/10.1016/j.cej.2022.135866

Li C, Sun W, Lu Z, Ao X, Li S (2020) Ceramic nanocomposite membranes and membrane fouling: a review. In: Water research, vol 175. https://doi.org/10.1016/j.watres.2020.115674

Lindamulla L, Nanayakkara N, Othman M, Jinadasa S, Herath G, Jegatheesan V (2022) Municipal solid waste landfill leachate characteristics and their treatment options in tropical countries. In Current pollution reports, vol 8, Issue 3, pp 273–287. Springer Science and Business Media Deutschland GmbH. https://doi.org/10.1007/s40726-022-00222-x

Mahlambi MM, Ngila CJ, Mamba BB (2015) Recent developments in environmental photocatalytic degradation of organic pollutants: the case of titanium dioxide nanoparticles-a review. In Journal of Nanomaterials, vol 2015. Hindawi Publishing Corporation. https://doi.org/10.1155/2015/790173

Meador JP, Yeh A, Young G, Gallagher EP (2016a) Contaminants of emerging concern in a large temperate estuary. Environ Pollut 213:254–267. https://doi.org/10.1016/j.envpol.2016.01.088

Meador JP, Yeh A, Young G, Gallagher EP (2016b) Contaminants of emerging concern in a large temperate estuary. Environ Pollut 213:254–267. https://doi.org/10.1016/J.ENVPOL.2016.01.088

Miró P, Audiffreda M, Heine T (2014) An atlas of two-dimensional materials. Soc Rev 43:6537–6554. Pubs.Rsc.Org. https://doi.org/10.1039/C4CS00102HChem

Ng LY, Mohammad AW, Leo CP, Hilal N (2013) Polymeric membranes incorporated with metal/metal oxide nanoparticles: a comprehensive review. In Desalination, vol 308, pp 15–33. https://doi.org/10.1016/j.desal.2010.11.033

Patanjali P, Singh R, Kumar A, Chaudhary P (2019) Nanotechnology for water treatment: a green approach. Green Synth Charact Appl Nanoparticles 485–512.https://doi.org/10.1016/B978-0-08-102579-6.00021-6

Pena-Pereira F, Bendicho C, Pavlović DM, Martín-Esteban A, Díaz-Álvarez M, Pan Y, Cooper J, Yang Z, Safarik I, Pospiskova K, Segundo MA, Psillakis E (2021) Miniaturized analytical methods for determination of environmental contaminants of emerging concern–a review. Anal Chim Acta 1158:238108. https://doi.org/10.1016/J.ACA.2020.11.040

Prabagar JS, Yashas SR, Gurupadayya B, Wantala K, Diganta DB, Shivaraju HP (2022) Degradation of doxycycline antibiotics using lanthanum copper oxide microspheres under simulated sunlight. Environ Sci Pollut Res 2022:1–11. https://doi.org/10.1007/S11356-022-19842-3

Qu X, Brame J, Li Q, Alvarez PJJ (2013) Nanotechnology for a safe and sustainable water supply: enabling integrated water treatment and reuse. Acc Chem Res 46(3):834–843. https://doi.org/10.1021/AR300029V/ASSET/IMAGES/MEDIUM/AR-2012-00029V_0003.GIF

Rafiq Z, Nazir R, Durr-E-Shahwar SMR, Ali S (2014) Utilization of magnesium and zinc oxide nano-adsorbents as potential materials for treatment of copper electroplating industry wastewater. J Environ Chem Eng 2(1):642–651. https://doi.org/10.1016/J.JECE.2013.11.004

Rajendran S, Priya AK, Senthil Kumar P, Hoang TKA, Sekar K, Chong KY, Khoo KS, Ng HS, Show PL (2022) A critical and recent developments on adsorption technique for removal of heavy metals from wastewater-a review. Chemosphere 303:135146. https://doi.org/10.1016/J.CHEMOSPHERE.2022.135146

Raval HD, Gohil JM (2010) Nanotechnology in water treatment: an emerging trend. Int J Nucl Desalin 4(2):184–188. https://doi.org/10.1504/IJND.2010.035176
Ravindra YS, Puttaiah SH, Yadav S, Prabagar JS (2020) Evaluation of polymeric g-C3N4 contained nonhierarchical ZnV2O6 composite for energy-efficient LED assisted photocatalytic mineralization of organic pollutant. J Mater Sci Mater Electron 31(19):16806–16818. https://doi.org/10.1007/S10854-020-04235-4
Rogers JA, Lagally MG, Nuzzo RG (2011) Synthesis, assembly and applications of semiconductor nanomembranes. Nature 477(7362):45–53. https://doi.org/10.1038/nature10381
Romeyn TR, Harijanto W, Sandoval S, Delagah S, Sharbatmaleki M (2016) Contaminants of emerging concern in reverse osmosis brine concentrate from indirect/direct water reuse applications. Water Sci Technol 73(2):236–250. https://doi.org/10.2166/wst.2015.480
Salem SS, Hammad EN, Mohamed AA, El-Dougdoug W (2023) A comprehensive review of nanomaterials: types, synthesis, characterization, and applications. Biointerface Res Appl Chem 13(1):41. https://doi.org/10.33263/BRIAC131.041
Salimi M, Esrafili A, Gholami M, Jonidi Jafari A, Rezaei Kalantary R, Farzadkia M, Kermani M, Sobhi HR (2017) Contaminants of emerging concern: a review of new approach in AOP technologies. Environ Monitor Assess 189(8):1–22. https://doi.org/10.1007/S10661-017-6097-X
Santos VS, Anjos JSX, de Medeiros JF, Montagner CC (2022) Impact of agricultural runoff and domestic sewage discharge on the spatial–temporal occurrence of emerging contaminants in an urban stream in São Paulo, Brazil. Environ Monitor Assess 194(9). https://doi.org/10.1007/s10661-022-10288-1
Shah AI, Din Dar MU, Bhat RA, Singh JP, Singh K, Bhat SA (2020) Prospectives and challenges of wastewater treatment technologies to combat contaminants of emerging concerns. Ecol Eng 152:105882. https://doi.org/10.1016/J.ECOLENG.2020.105882
Shawky HA, Chae SR, Lin S, Wiesner MR (2011) Synthesis and characterization of a carbon nanotube/polymer nanocomposite membrane for water treatment. Desalination 272(1–3):46–50. https://doi.org/10.1016/J.DESAL.2010.12.051
Shi C, Owusu KA, Xu X, Zhu T, Zhang G, Yang W, Mai L (2019) 1D carbon-based nanocomposites for electrochemical energy storage. In Small, vol 15, Issue 48. Wiley-VCH Verlag. https://doi.org/10.1002/smll.201902348
Shivaraju HP, Sneha Y, Yashas SR, Das Diganta B (2022a) Source, treatment, and management options of contaminants of emerging concerns (CECs) in water, pp 3–19. https://doi.org/10.1007/978-3-030-95443-7_1
Shivaraju HP, Sneha Y, Yashas SR, Das Diganta B (2022b) Source, treatment, and management options of contaminants of emerging concerns (CECs) in water, pp 3–19. https://doi.org/10.1007/978-3-030-95443-7_1
Sneha Y, Yashas SR, Thinley T, Prabagar Jijoe S, Puttaiah Shivaraju H (2022) Photocatalytic degradation of lomefloxacin antibiotics using hydrothermally synthesized magnesium titanate under visible light-driven energy sources. Environ Sci Pollut Res 1–12.https://doi.org/10.1007/S11356-022-20540-3
Starling MCVM, Amorim CC, Leão MMD (2019) Occurrence, control and fate of contaminants of emerging concern in environmental compartments in Brazil. J Hazard Mater 372:17–36. https://doi.org/10.1016/J.JHAZMAT.2018.04.043
Sun M, Huang W, Cheng H, Ma J, Kong Y, Komarneni S (2019) Degradation of dye in wastewater by Homogeneous Fe(VI)/NaHSO3 system. Chemosphere 228:595–601. https://doi.org/10.1016/J.CHEMOSPHERE.2019.04.182
Tatarchuk T, Shyichuk A, Sojka Z, Gryboś J, Naushad M, Kotsyubynsky V, Kowalska M, Kwiatkowska-Marks S, Danyliuk N (2021) Green synthesis, structure, cations distribution and bonding characteristics of superparamagnetic cobalt-zinc ferrites nanoparticles for Pb(II) adsorption and magnetic hyperthermia applications. J Mol Liq 328:115375. https://doi.org/10.1016/J.MOLLIQ.2021.115375

Tondera K, Blecken G-T, Tournebize J, Viklander M, Österlund H, Andersson Wikström A, Tanner CC (2018) Emerging contaminants: occurrence, treatment efficiency and accumulation under varying flows, pp 93–109. https://doi.org/10.1007/978-3-319-70013-7_6

Turchanin A, Gölzhäuser A (2016) Carbon nanomembranes. Adv Mater 28(29):6075–6103. https://doi.org/10.1002/ADMA.201506058

Upadhye VB, Joshi SS (2012) Advances in wastewater treatment-a review. In International journal of chemical sciences and applications, vol 3, Issue 2. http://www.bipublication.com

Vidal-Dorsch DE, Bay SM, Maruya K, Snyder SA, Trenholm RA, Vanderford BJ (2012) Contaminants of emerging concern in municipal wastewater effluents and marine receiving water. Environ Toxicol Chem 31(12):2674–2682. https://doi.org/10.1002/ETC.2004

Vimalkumar K, Seethappan S, Pugazhendhi A (2019) Fate of Triclocarban (TCC) in aquatic and terrestrial systems and human exposure. Chemosphere 230:201–209. https://doi.org/10.1016/J.CHEMOSPHERE.2019.04.145

Weerasooriya RR, Liyanage LPK, Rathnappriya RHK, Bandara WBMAC, Perera TANT, Gunarathna MHJP, Jayasinghe GY (2021) Industrial water conservation by water footprint and sustainable development goals: a review. Environ Dev Sustain 23(9):12661–12709. https://doi.org/10.1007/S10668-020-01184-0

Wu Y, Tan H, Zhou C, Crimmins BS, Holsen TM, Chen D (2018) Bioaccumulation and spatiotemporal trends of polyhalogenated carbazoles in great lakes fish from 2004 to 2016. Environ Sci Technol 52(8):4536–4545. https://doi.org/10.1021/ACS.EST.8B00427/SUPPL_FILE/ES8B00427_SI_001.PDF

Xu M, Liang T, Shi M, Chen H (2013) Graphene-like two-dimensional materials. In Chemical reviews, vol 113, Issue 5, pp 3766–3798. https://doi.org/10.1021/cr300263a

Yadav D, Rangabhashiyam S, Verma P, Singh P, Devi P, Kumar P, Hussain CM, Gaurav GK, Kumar KS (2021) Environmental and health impacts of contaminants of emerging concerns: recent treatment challenges and approaches. Chemosphere 272:129492. https://doi.org/10.1016/J.CHEMOSPHERE.2020.129492

Zhou H, Smith DW (2011) Advanced technologies in water and wastewater treatment, vol 28, no 1, pp 49–66. https://doi.org/10.1139/L00-091. https://doi.org/10.1139/L00-091

Chapter 11
An Evaluation of the Potential Use of Environmental Coagulants to Decontaminate Leachates

Adwaitha Sunilkumar, Gajendran Chellaiah, P. P. Sivan, U. Surendran, Mahendran Chellaiah, C. Meiaraj, Athulya Thankachan, and Dhayanand Jayaseelan

Abstract The increasing urbanization and growth in population has led to a rise in environmental pollution that harms communities by allowing municipal waste from homes, businesses, and industries to seep into groundwater and contaminate water quality. Leachate's composition can vary depending on location, making the choice of treatment method crucial. This work aimed to rate the ability of various

A. Sunilkumar
Gujarat Environmental Institute of Management, Gandhinagar, Gujarat, India

G. Chellaiah (✉)
Department of Civil Engineering, SNS College of Technology, Coimbatore, Tamil Nadu, India
e-mail: cgajendran@gmail.com

P. P. Sivan
Department of Civil Engineering, Government Engineering College, Thrissur, Kerala, India

U. Surendran
Centre for Water Resources Development and Management (CWRDM), Kunnamangalam, Kerala, India
e-mail: suren@cwrdm.org

M. Chellaiah
Department of Electronics and Communication Engineering, Alagappa Chettiar Government College of Engineering and Technology, Karaikudi, Tamil Nadu, India

C. Meiaraj
Department of Civil Engineering, Government College of Technology, Coimbatore, Tamil Nadu, India
e-mail: cmeiaraj@gct.ac.in

A. Thankachan
Department of Agricultural Engineering, Sri Shakthi Institute of Engineering and Technology, Coimbatore, Tamil Nadu, India
e-mail: athulyathankachan@karunya.edu.in

D. Jayaseelan
Tamil Nadu, 3Rd Street Vadavalli, Coimbatore, India

S. Suriyanarayanan et al. (eds.), *Water Management in Developing Countries and Sustainable Development*, Water Resources Development and Management,
https://doi.org/10.1007/978-981-99-8639-2_11

extraction methods and the purifying abilities of plant extracts on the chemical and physiological parameters of leachate through the use of flocculation and Soxhlet methods in a dose-dependent manner. The results of this study showed that plant extract coagulants such as sago trunk, oil palm trunk starch, and jackfruit seeds are effective in treating leachate, including TSS, turbidity, COD, heavy metals, NH3-N, TDS, and color. It is essential to prioritize and continue to develop the use of these efficient natural coagulants in order to provide cost-effective solutions and stabilize leachate. By effectively treating leachate, the risk of groundwater contamination can be greatly reduced.

Keywords Natural coagulant · Landfill leachate · Soxhlet extraction · Jar test · Efficacy analysis

11.1 Introduction

Municipal solid trash creation continues to rise as the human population grows and economic development accelerates. As per (Agbor et al. 2020) by 2025 municipal solid waste generation is predicted to expand further, it would be generated by 4.3 billion city residents at a rate of 2.2 billion tonnes yearly or per day 1.42 kg per person. (Ahmed et al. 2020) states that Municipal solid waste disposal methods include dumping into hog feeding, sewers, landfilling, milling, incineration, and with other methods which include grinding, composting, open dumps, etc. Due to its simple operational techniques and minimal cost, this paper outlines an effective option among various municipal waste disposal options, according to a study of solid waste management in 90 nations (Al-Hamadani et al. 2011). One of the most important issues with landfilling is landfill leachate.

11.1.1 Classification of Leachates

Leachates are liquids that result from the percolation of water through waste materials, such as landfills or contaminated soils. These leachates can contain various pollutants, including organic and inorganic compounds, heavy metals, and pathogens. Based on their origin and composition, leachates can be classified into different types:

1. Municipal Solid Waste (MSW) Leachate: Generated from the decomposition of organic waste in municipal landfills. It typically contains organic matter, nutrients, heavy metals, and various dissolved substances.
2. Industrial Waste Leachate: Arises from industrial activities and can vary widely in composition depending on the specific industry. Industrial leachates may contain organic chemicals, heavy metals, toxic substances, and high concentrations of salts.

3. Hazardous Waste Leachate: Produced from the percolation of water through hazardous waste disposal sites, such as hazardous waste landfills or storage tanks. These leachates are highly contaminated and may contain toxic chemicals, heavy metals, and other hazardous substances.

11.1.2 Factors Associated with Leachate Composition

The composition of leachates is influenced by several factors, including:

1. Waste Characteristics: The type and nature of the waste material greatly affect leachate composition. Different wastes have varying organic and inorganic content, chemical composition, and degradation rates, leading to variations in leachate properties.
2. Environmental Conditions: Factors such as temperature, rainfall, humidity, and soil characteristics can impact leachate generation and composition. Environmental conditions influence the infiltration of water into waste materials, the degradation of organic matter, and the release of pollutants.
3. Age of the Waste: The age of the waste material in a landfill or disposal site affects leachate composition. Fresh waste tends to produce leachate with high organic content, while older waste generates leachate with a higher concentration of degradation by-products.
4. Waste Management Practices: The design and operation of waste management facilities, such as landfill liners, cover systems, and leachate collection systems, can influence leachate composition. Proper waste management practices can minimize the potential for leachate contamination.

11.1.3 Rationale for Studying Leachates

Research on leachates is essential for several reasons:

1. Environmental Impact: Leachates can pose significant risks to ecosystems and human health if they contaminate surface water, groundwater, or soil. Studying leachates helps understand the potential environmental impacts and develop mitigation strategies.
2. Waste Management Practices: Effective waste management relies on understanding leachate generation and composition. Research helps optimize waste management practices, such as landfill design, leachate collection, and treatment systems.
3. Regulation and Policy: Studies on leachates provide scientific data for the development of regulations and policies related to waste disposal, pollution control, and water quality standards.

4. Treatment and Remediation: Understanding the composition and characteristics of leachates aids in the development of effective treatment technologies and remediation strategies for contaminated sites.
5. Risk Assessment: Assessing the potential risks associated with leachate contamination requires knowledge of their composition, fate, and transport. Research helps quantify the risks and implement appropriate risk management measures. The name given to a liquid that is produced when water permeates through waste is called Leachate; it typically comprises suspended or floating particles as well as dissolved materials from the trashed objects. Biological oxygen demand (BOD), Total Kjeldahl nitrogen (TKN), pH, Total suspended solids (TSS), Ammonium, Chemical oxygen demand (COD), and Dissolved oxygen (DO) are the main characteristics of leachate. Nitrogen (NH3-N), heavy metals, and other factors should be taken into account. The leachate's composition and the quality of standard effluent vary depending on the landfill site and legislation. The making up of a lot of factors influence leachate, including trash content, the local meteorological circumstances, the physicochemical parameters of the dump, and the age of the landfill. The most essential factors determining treatment technique selection are the structure and properties of the leachate. Many types of leachate have been shown to be effectively treatable through the use of biological therapy.

The effectiveness of this technique is in doubt, nevertheless, because of aged leachate with ammonium concentrations and high COD, adequate removal results were not observed despite the presence of several heavy metals and a low biodegradability characterized by a high COD to BOD ratio. Alternately, it was found that older leachate contains and that, when compared to juvenile leachate, older leachate can be removed more successfully using chemical and physical methods. In general, it is simple to expect long-term leachate, but short-term leachate is more challenging to estimate because of how precipitation affects it (Awang and Aziz 2012).

In general, leachate is having a high amount of non-biodegradable chemicals at the early phases of waste decomposition like volatile fatty acids, biodegradable, and BOD_5. The leachate from old landfills, on the other hand, usually comprises a high number of non-biodegradable chemicals like fulvic and humic-like substances chemical substances. In addition, the bacteria produce inorganic substances such as NH3-N. Nitrogen-containing fractions of biodegradable waste substrates are hydrolyzed and fermented in stabilized leachate from older landfills. When leachate is collected and brought to a facility for treatment after the waste has been stabilized, washout has an impact on the NH3 concentration that has been building up. Applying physicochemical to the leachate, such as polish or pre-treat, is a common procedure. Flocculation-coagulation is a widely recognized chemical technique for pre-treating leachate (Yabroudi et al. 2013).

Normally, this process is used to remove difficult-to-remove chemicals and suspended particles. Only a few examples include heavy metals, humic acids, absorbable chemical compounds and polychlorinated biphenyls (PCBs) and leachate-derived halides (AOX). The pre-treatment approach mechanism is basically both cationic hydrolysis and charge neutralization between negatively charged colloids are involved. The next step is impurity amalgamation by flocculation. This method is crucial in eliminating total suspended solids (TSS) and particulate matter, which are the most important elements.

In the flocculation-coagulation process, a wide variety of chemicals are utilized as flocculants and coagulants, with inorganic metal salts being the most common. On the other hand, the environment can suffer from the over use of these inorganic metal salts. They endanger human health and degrade the environment, therefore minimizing and eliminating them The creation and usage of natural flocculants/coagulants in the clean-up of landfill leachate have significant effects on living things and the environment. Even though various studies have been carried out, and numerous natural coagulants have been developed or derived from different sources such as microbes, plants, animals, and more, it is widely believed that they do not pose any threat to human health (Aziz et al. 2022).

11.2 Commonly Utilized Techniques for Treating Landfill Leachate

Treating landfill leachate is more challenging than treating sewage or industrial wastewater, The complex fluctuations in the composition of solid waste from municipalities, the physical and chemical characteristics of the landfill, the surrounding ecosystem, and the considerable amount of leachate produced, are the reasons for this. However, Fig. 11.1 provides a brief overview of common methods for stabilizing leachate treatment.

11.3 Coagulation-Flocculation Treatment of Semi-aerobic Landfill Leachate

Even though the wide range of composition makes it challenging to build up an effective leachate treatment plan, the efficiency of treating semi-aerobic landfill leachate has been demonstrated to be higher using coagulation-flocculation technology (Badrus 2018). Flocculation process involves the destabilization of colloidal components and neutralization of negatively charged colloids using coagulants, which can either be naturally occurring or artificially synthesized. This mechanism results in the formation of flocs. The effectiveness of coagulation may be influenced by a variety of factors, including changing the mixing speed, coagulant dosage, retention time, pH, and others. However, as the properties of the coagulants utilized have the ability to either boost or decrease therapeutic efficacy, these factors are most important in determining coagulation performance (Camacho et al. 2016). Coagulants must be less expensive, create less sludge, be non-toxic, and not affect people negatively physiologically. Due to the large volume of hazardous sludge and

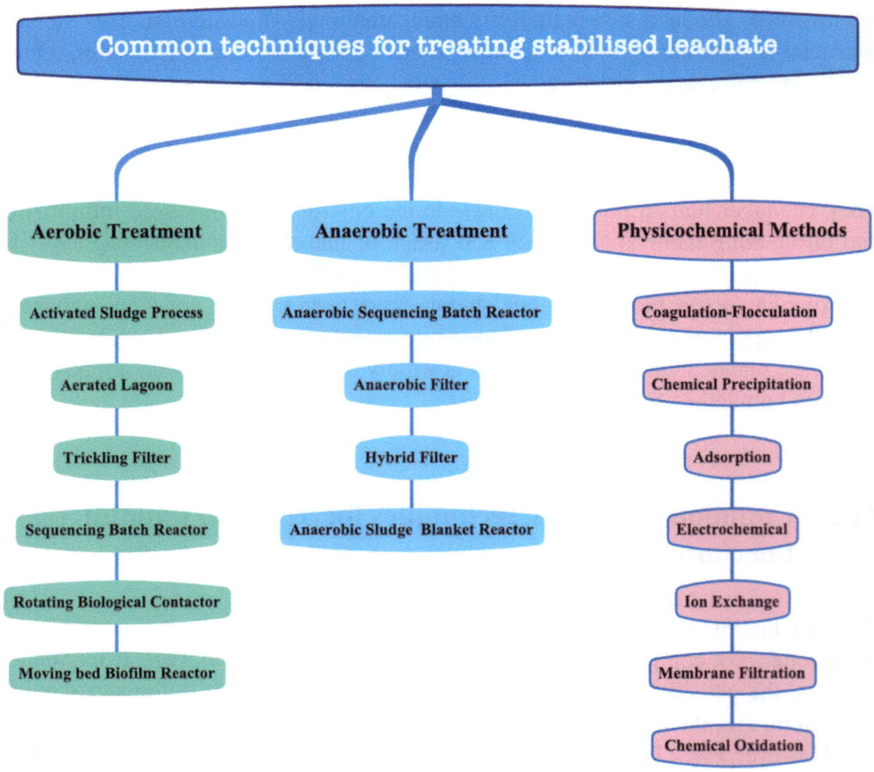

Fig. 11.1 Common techniques for treating stabilized leachate. *Source* Liu et al. (2015)

increasing concerns over chemical coagulants, researchers are facing pressure to create new, environmentally friendly coagulants that are likely to be derived from natural sources (Cheng et al. 2020).

11.4 Unfavorable Consequences of Chemical Coagulants

For a very long time, wastewater has been treated using traditional chemical coagulants, such as ferric chloride, aluminum sulphate, lime, and others. Nevertheless, the side effects of these coagulants, particularly the buildup of residual aluminum in the brain which has been associated with neurological diseases like Alzheimer's, have caused concern (Saravanan et al. 2017). Common inorganic coagulants also create an excessive amount of sludge, which makes disposal more complicated, expensive, and difficult to handle (Gao et al. 2015). Even after being disposed of, the produced sludge seeps into the groundwater and increases metal toxicity (Unda-calvo et al. 2017). The application of synthetic organic polymers, which are commonly used in the coagulation procedure, could potentially endanger human health because of their non-biodegradable nature. Some of these polymers can degrade into monomers, such as acrylamide, which are recognized as neurotoxic and carcinogenic (Badrus 2018). However, the operational costs of the treatment are increased by the nearly unreasonably high cost of chemical coagulants (Simate and Ndlovu 2014). In order to promote clean development methods and reduce environmental and human health issues, the development of environmentally acceptable coagulating agents is a critical task for long-term leachate and wastewater treatment (Impact 2019).

11.5 Landfill Leachate Treatment with Natural Coagulants

Despite the harm that chemical coagulants cause to living things, switching to natural polymeric materials seems like a good idea (Journal et al. 2010). Natural coagulants, in particular plant extracts, are constantly accessible and widely acknowledged as a non-toxic substitute (Mohd-Asharuddin et al. 2017). Due to their renewable nature, studies have extensively explored the utilization of natural organic coagulants such as cassava peel, Chitosan, oil palm starch, and Hibiscus Rosa-Sinensis (Camacho et al. 2016). Although there is insufficient data on the application of natural coagulants in eliminating refractory substances like humic acids from stabilized leachate, it is deemed safe for both terrestrial and aquatic environments, even in secluded regions. This section focuses on the latest developments in utilizing natural coagulants, either independently or combined, to treat stabilized leachate, categorized based on the specific process (Ahmed et al. 2020).

11.5.1 Hibiscus-Rosa Sinensis

The Chinese hibiscus, commonly referred to as *Hibiscus Rosa-Sinensis*, shoe black plant, or China rose, is a flowering plant deeply rooted in Chinese culture. It is an evergreen bush, small tree, or bushy shrub that is native to East Asia. It can reach heights of 1.5–3 m (5–10 ft) and 2.5–5 m (8–16 ft) in width. The plant features glossy leaves and large, solitary red blooms that appear during the summer and fall. The flowers, which measure 10 cm (4 m) in diameter, have five petals and are notable for their orange-tipped, scarlet anthers (Agbor et al. 2020).

An investigation was carried out to assess the efficiency of using Hibiscus rosa-Sinensis leaf extract as a coagulant supplement in treating semi-aerobic landfill leachate. The extract is a well-known biodegradable compound derived from nature (Awang and Aziz 2012), the ability of the extract to remove suspended particles, decrease turbidity, fade color, reduce ammonia nitrogen (NH3-N) levels, and remove iron (Fe^{3+}) from landfill leachate was evaluated. The research findings revealed that a blend of 500 mg/L Hibiscus rosa-Sinensis leaf extract and 4,000 mg/L alum led to a complete removal of Fe^{3+}, which was higher than the rate obtained when only

4,000 mg/L alum was used, where only 60% of Fe^{3+} was eliminated. The study also revealed that 500 mg/L of HBaqs was the ideal concentration to serve as a coagulant aid in leachate treatment and that it effectively removed 61% of the color, 28% of NH3-N, and a percentage of total suspended solids (TSS) from the water. Furthermore, HBaqs discovered that when alum was used as the only coagulant at a concentration of 4,000 mg/L, it eliminated roughly 23% of the leachate at pH 10, thus making HBaqs a suitable coagulant at high pH. The improved flocculation efficacy of HBaqs is credited to the functional groups along the HBaqs chain, such as hydroxyl (O–H) and carboxyl (–COOH) groups that form connections (Ahmed et al. 2020).

11.5.2 Cycas Revoluta Starch

The treatment of semi-aerobic landfill leachate was explored using a combination of *Cycas revoluta*, also known as Commercial Sago Starch (CSS) and Native Sago Trunk Starch (NSTS), as the primary coagulant and coagulant aid, and Polyaluminum Chloride (PAC) to eliminate various pollutants such as suspended solids, color, turbidity, NH3-N, organic UV254, Cd, Chemical Oxygen Demand and Ni. (Aziz and Sobri 2015). 400 mL of water was treated with the addition of a 0.5% w/v aqueous solution of sodium metabisulfite prior to processing small cubes of sago trunks. In order to purify sago trunks, a 0.5% w/v aqueous solution of sodium metabisulfite was added to 400 mL of water. After filtering and sieving the mixture through a 212-μm filter and letting it sit for 60 min, the supernatant was removed and the remaining mixture was mixed with 1 L of a 0.5% w/v aqueous solution of sodium metabisulfite. The supernatant was removed once again, and the mixture was mixed with 100 L of distilled water. The starch extract was purified by centrifuging the mixture twice for 5 min at 3000 g, filtering it using a GF/C filter under vacuum, and rinsing it twice with 50 mL of acetone to remove any residual water. The purified starch was dried overnight in an oven at 45 °C, resulting in a fine powder. The study found that when using 2000 mg/L PAC with 6000 mg/L native sago trunk starch (NSTS) and 2000 mg/L PAC with 5000 mg/L commercial sago starch (CSS), the removal rates for color, suspended solids (SS), and turbidity were 94.7%, 99.2%, and 98.9%, respectively. The same results were achieved with 3100 mg/L PAC alone. The use of NSTS and CSS in combination with PAC resulted in removal rates of 36%, 69.5%, and 53.8% for chemical oxygen demand (COD), organic UV, and Ni, respectively. However, NSTS and CSS were not effective in removing ammonia nitrogen (NH3-N), with a removal rate of only 8%. Despite this, the results indicate that the use of NSTS can reduce the amount of PAC needed by 33%, which could provide a long-term solution for leachate treatment (Ahmed et al. 2020).

11.5.3 Isabgol

A *psyllium* husk solution was prepared by mixing three grams of dry *psyllium* husk powder with one litre of distilled water, and then left to sit for ten minutes. The solution was intended to function as both a coagulant and coagulant aid. The study showed that when PAC was utilized at a concentration of 7.2 g/L at a pH of 7.5, it effectively removed 91% of total suspended solids (TSS), 55% of chemical oxygen demand (COD), and 80% of color. Alum was also tested as a coagulant, with an optimal dose of 11 g/L at a pH of 6.5, resulting in the removal of 78% of TSS, 58% of COD, and 79% of color. When *psyllium* husk was added as a coagulant aid to PAC, the removal of TSS, COD, and color improved to 96%, 64%, and 90%, respectively, surpassing the performance of alum as a coagulant aid. However, the organic nature of *psyllium* husk and its tendency to release carbohydrates resulted in it being less effective in removing biological oxygen demand (BOD) (Ahmed et al. 2020).

11.5.4 Jackfruit Seeds Starch

The efficacy of jackfruit seed starch (JSS) as a natural coagulant was studied in response to increasing concerns about the use of synthetic coagulants for leachate remediation (Yusoff and Mohamad Zuki 2015). A solution was created by spinning Jackfruit Seed Starch (JSS) in a 0.05 M NaOH solution for 6 h at 4 °C, then removing the supernatant. The remaining sediment was combined with distilled water and filtered through a screen with a mesh size of 0.15 mm to eliminate fibers. The solution was then neutralized to a pH of 7.0 with 0.1 M HCl and centrifuged for 20 min at 4 °C. The residual starch cake was cleaned with distilled water, dried, and the supernatant was drained off. The jar test method was used to evaluate the removal efficiency. This study evaluated the use of Poly Aluminum Chloride (PAC) and JSS as coagulants to reduce leachate contamination. The coagulants were dosed at pH 5, with PAC ranging from 400 to 1100 mg/L and JSS from 500 to 3000 mg/L. The effectiveness of PAC and JSS was also assessed in collaboration. The results showed that using JSS as a coagulant aid reduced the PAC dosage from 900 to 600 mg/L, and improved the removal of Chemical Oxygen Demand (COD), color, Ammonical Nitrogen (NH3-N), turbidity, and Suspended Solids (SS) to 33.6%, 93.6%, 13.1%, 94%, and 92%, respectively. Furthermore, the results indicate that JSS was effective in reducing the amount of PAC used in the treatment process of landfill leachate by 33.3%, which can benefit the environment by reducing the amount of sludge produced (Ahmed et al. 2020).

11.5.5 Starch from Cross-Linked Oil Palm Trunks (C-OPTS) and from Palm Oil Trunks

The capability of coagulation of starch derived from Palm Oil Trunks (OPTS) and Cross-linked Oil Palm Trunks (C-OPTS) was assessed through the use of Poly Aluminum Chloride (PAC) as a coagulating agent. The process involved slowly incorporating 300 g of dried OPTS meal into 400 mL of a 0.5% (w/v) sodium bisulfide solution in water to break down the protein-starch matrix and prevent microbial growth for a duration of 60 s. The filtrate obtained was subjected to pressing, and the residual starch was gathered. Impurities were filtered out using a 212-m filter, and the filtrate was allowed to settle for 60 min. To the starch suspensions, 1 L of a 0.5% aqueous sodium bisulfide solution was added, and the mixture was left to settle for 30 min before discarding the supernatant. After discarding the supernatant, the remaining material was diluted with 100 mL of distilled water. To purify the extracted starch, the mixture was centrifuged twice at 3,000 RPM for 5 min and then filtered under vacuum using a fiberglass filter with a pore size of 1.6 m. The purified starch sediment was washed twice with 50 mL portions of acetone to eliminate water before being dried in an oven overnight at 450 °C. In a stirring glass kettle, 20 g of OPTS were combined with 0.3 g of sodium chloride (NaCl), 2 mL of epichlorohydrin, and 150 mL of deionized water. The mixture was continuously stirred at 25 °C (ECH). After 15 min, the uniform slurry was combined with a solution of potassium hydroxide (1.6 g KOH in 8 mL water) and continuously stirred for 18 h at 25 °C. The final concentration of ECH in the mixture was 1.3% (w/w). The pH was adjusted through vacuum filtration using a fiberglass filter with a pore size of 1.6 m and distilled water and 95% ethanol. The C-OPTS was dried overnight in an oven at 45 °C before being ground and sieved for use as a coagulant. The physical characteristics of both OPTS and C-OPTS were also investigated.

Coagulation and flocculation experiments were performed using a standard jar-test setup. Both OPTS and C-OPTS were tested at various doses ranging from 100 to 2000 mg/L, at pH levels between 5 and 9. The removal efficiencies for suspended solids, COD, turbidity, NH3-N, color, Mn, Fe, Zn, Cu, and PO4 were 45.1%, 38.0%, 12.9%, 24.6%, 95.6%, 82.9%, 91.4%, 100%, and 72% respectively. However, the removal efficiencies for turbidity and suspended solids were not satisfactory, with 65.1% and 25% respectively, and a removal efficiency of only 39.1% and 5% was achieved. The results indicate that both OPTS and C-OPTS are effective in treating stabilized leachate, but further investigation is required to fully understand their capabilities (Ahmed et al. 2020).

11.5.6 Nicotiana Tabacum

The coagulating properties of *Nicotiana Tabacum*, commonly referred to as Tobacco leaf, were investigated to determine its effectiveness in removing dust and insoluble contaminants from stabilized leachate. The dried tobacco leaves were first cleaned with distilled water, then blended with 550 mL of distilled water in a household blender at a high speed of 120 g. The mixture was filtered to remove particles, and the leaf suspension was collected in a beaker while the remaining solids were dried in an oven at 105 °C for 48 h. The dried residue was weighed again, and the filtrate was mixed for 30 min at room temperature using a magnetic stirrer to create the final tobacco leaf extract. The structure of the tobacco leaves was analyzed using scanning electron microscopy (SEM). The ability of Nicotiana Tabacum, commonly known as Tobacco leaf, to remove dust and insoluble contaminants from stabilized leachate was evaluated by studying its coagulation properties. The tobacco leaf extract was made by blending dried leaves with distilled water and then filtering the mixture to obtain the leaf suspension. This was then mixed with the leachate at different concentrations of 10,000, 5000, 2000, 1000, 750 and 500 mg/L along with the addition of Poly aluminum chloride (PAC) as a primary coagulant and coagulant aid. The results showed that tobacco leaf performed better in terms of COD removal (64%) compared to PAC (62%). Moreover, when mixed with 1000 mg/L PAC, 1000 mg/L COD, and 1000 mg/L NH3-N, respectively, tobacco leaf demonstrated 86, 91, and 54% reduction in color. However, its removal efficiency for SS (48%) and turbidity (21%) was found to be less effective. Overall, the findings suggest that tobacco leaf can act as a coagulant aid and enhance the pollutant removal from stabilized leachate, as well as its physical features, through a synergistic effect. The morphology of the tobacco leaves was analyzed through scanning electron microscopy (SEM) (Ahmed et al. 2020).

11.5.7 Dragon Fruit White-Fleshed (DFPW) and Dragon Fruit Purple Pink-Fleshed (DFPP)

This experiment employed two varieties of dragon fruit peel: *Hyloceruspolyrhizus*, which has white flesh, and Hylocerusundatus (DFPW), which has purple-pink flesh. The peels were chopped into small pieces, washed with tap water to facilitate drying, and then ground through a 250 mesh sieve. The powder was dried through roasting, filtered using a Whattman 42 filter paper after being suspended in 100 mL of distilled water, and the resulting extract was collected to be used as a coagulant. The coagulation properties were examined using a jar floc test apparatus, with the pH being adjusted to neutral in every experiment, which is the ideal pH for river water (Idris et al. 2013). The performance of the dragon fruit peel coagulant was evaluated by conducting jar flocculation tests at a range of concentrations (10, 30, 50, 70, and 90 mg/L). The mixture was agitated for 4 min at a high speed of 100 rpm, followed by 25 min of slower mixing at 40 rpm, and then allowed to settle for 30 min. The clarity of the resulting water was measured using various instruments, including a Hach Turbid meter, a handheld TDS meter (model HACH), and a TSS meter with a vacuum filtration system (Muyibi et al. 2002; Ismail et al. 2018).

11.5.8 M.oleifera

Moringa oleifera is a highly versatile plant species that thrives in low-lying tropical areas, including desert regions. It is able to flourish in soils of moderate fertility and under low humidity conditions. The tree is native to the Northern Indian Himalayan foothills, and is commonly cultivated in tropical and subtropical regions for its various medicinal properties, as well as for its edible leaves and immature seed pods. In addition, *Moringa oleifera* leaves can be used to cleanse water and as a hand wash. The extract in this study was obtained using the n-hexane solvent and the soxhlet extraction method (using a 1:10 ratio of plant leaves to hexane solvent) (Pandey et al. 2020).

The *genus Moringa*, which is known as drumstick or Golden shower, is part of the *Moringaceae* plant family. One of its species, *Moringa oleifera*, is native to North India and is known for its fast-growing nature and ability to reach a height of 12 m. The tree has a thick grey bark, which peels in some areas similar to cork. Moringa is rich in minerals and vitamins, and can be used for both human and animal nutrition. It is also known for its medicinal properties and its ability to purify dirty water. Moreover, *Moringa Oleifera* is frequently employed as a coagulant, often in combination with aluminum sulfate (Agbor et al. 2020).

To make a coagulant from *Moringa oleifera* seeds, the mature trees were allowed to dry and the seeds were harvested. The seeds were sun-dried, their outer shells were removed, and the remaining dried and shelled seeds were ground into a fine powder using a blender. The powder was measured, diluted with distilled water to make a 50 g/L solution, and stirred with a magnetic stirrer for 20 min. The mixture was then filtered through a Whatman filter No. 40 to produce a "crude extract" or "stock solution" that could be used to treat water without any additional processing (Agbor et al. 2020).

In this study, the impact of coagulant dosage on the removal of turbidity in synthetic turbid water was investigated using a combination of reduced amounts of *Moringa oleifera* and varying doses of alum. The synthetic turbid water had a low turbidity range of 50 NTU, a medium range of 100 NTU, and a high range of 500 NTU. The results showed that a combination of 7.5 mg/L alum and 5 mg/L *M. oleifera* was effective in reducing turbidity, even at the lowest range of 50 NTU. The highest level of turbidity removal was achieved when 7.5 mg/L alum was paired with 5 mg/L *M. oleifera* as a coagulant aid for low (50 NTU), medium (100 NTU), and high turbid water. This research highlights the significance of coagulant dosage in removing turbidity in different levels of turbid water (Pandey et al. 2020).

11.5.9 Okra

Okra or Lady's finger, scientifically known as *Abelmoschus esculentus L.* (Moench), is a widely grown vegetable in tropical and subtropical regions around the world.

This crop can be grown both in large commercial farms and in home gardens. Okra gum has the ability to dissolve in cold water, making it an effective ingredient for stabilizing and maintaining the frothy consistency of products in the food industry. The clarity of the solution improves with an increase in pH, despite the presence of gumbo mucilage. This is demonstrated by the noticeable reduction in turbidity that results from a higher use of mucilage. The stickiness of mucin is due to its polymer molecule composition. The flocculation can be initiated by various factors, such as chemical reactions or intricate structures. Fresh okra pods from the market can be used to extract okra gum by aqueous extraction from the seedpod of the okra plant. To prepare the mucilage, 5 g of fresh inner okra gum was mixed with 100 mL of water. To eliminate floating particles, the thick liquid (mucilage) was spun in a centrifuge for 20 min at 600 revolutions per minute (Yabroudi et al. 2013).

11.5.10 C.Procera

Calotropis procera, commonly known as *Calotrope*, is a tropical shrub that is often used as an ornamental plant. It belongs to the *Aclepiadaceae* family and the *C.Procera* species. According to Ogundiwin and Oke, it contains Calotropin, a non-toxic proteolytic enzyme that is highly effective in curdling milk protein and is commonly used in Nigerian traditional cheesemaking. The use of *C. procera* latex for enzyme purification is recommended as a budget-friendly option due to financial difficulties and a decline in revenue.

This study aimed to assess the efficacy of *C. Procera* as a coagulant. The optimal *C. Procera* dosage of 7.5 mg/lit was combined with various alum dosages in batch coagulation tests. The results showed that additional turbidities were detected after 30 min of jar testing, and it demonstrated how different dosages impact turbidity reduction. The study found that the highest amount of turbidity removal was attained when using 10 mg/l of alum and 5 mg/l of okra gum as a coagulant aid, resulting in maximum turbidity reduction rates of 98, 97, and 97.2% for low, medium, and high synthetic turbid water samples respectively. As per the study's findings, the coagulant aid should be added a minute after the alum (Sou KantabaiBhavarlalji et al. 2013).

11.5.11 Plantago Ovata

Plantago ovata, also known as ispaghul, blond *psyllium*, desert Indian wheat, and blond plantain, is a species of medicinal plant that is native to Western and Southern Asia. *Plantago ovata* typically contains *psyllium* and dietary fiber, and its indigestible seed husks can serve as a source of soluble fiber. When tested as coagulants for decreasing water turbidity, *Plantago ovata* extract was found to be more effective in a range of pH values and turbidity levels compared to starch extract. The study utilized coagulant agents such as distilled water, normal water, sodium chloride, and ammonium acetate to reduce turbidity from *Plantago ovata* seed extracts. Distilled NaCl was determined to have the strongest coagulating ability at low turbidity, and the reduction in efficiency was observed for all starting turbidity values (500, 150, and 50 NTU). The study found that the distilled NaCl coagulant could remove 98.2, 94.9, and 80.2% of the starting water's turbidity, respectively (Badrus 2018).

11.5.12 Tamarindus Indica

Tamarindus indica, commonly known as Tamarind, is a tropical African leguminous tree that bears edible pulp and is often used in cooking. It belongs to the Fabaceae family and bears pod-like fruits. The tree is known for its multiple uses, including in traditional medicine, metal polishing, and furniture making. Tamarind seeds can serve as a natural coagulant, and oil can be extracted from them. The tamarind tree is widely grown in tropical and subtropical regions for its various applications.

Turbidity is removed between 64 and 99 percent by crude pulp extract (CPE) from *Tamarindus indica* fruit. (Sa'id et al. 2016). The powder made from mature-dried *Tamarindus* indica seeds, when combined with 0.5 percent PAM, was found to be the most effective at lowering the turbidity in contaminated river water, resulting in a 91.16% reduction in turbidity (Polyacrylamide). However, in terms of treating the primary pollutants of concern, such as total dissolved solids (TDS), pH, total solids (TS), turbidity and total suspended solids (TSS), *Moringa Oleifera* was found to be more effective than *Tamarindus* Indica, achieving a turbidity reduction efficiency of around 78% (Prasad and Rao 2016). *Tamarindus* Indica seed powder has a larger impact on the removal of heavy metals while *Moringa Oleifera* seed powder has a stronger impact on physicochemical characteristics (Badrus 2018).

11.5.13 Durio

The popular term for durian is durio. In Southeast Asia, durio is often referred to as the King of Fruits and has a strong aroma. It can reach a length of more than 30 cm and a diameter of 15 cm, weighing between 1 and 3 kilos. Depending on the species, durio husks range in color from green to brown, and the meat of the seeds can be cooked and consumed.

An analysis of durian seed flour (whole) revealed a composition of 6.5% water, 3.1% ash, 6.0% protein, 52.9% total dietary fiber, and 10.1% crude fiber. Upon dehulling the durian seed, the components were found to be 7.6% protein, 4.8% crude fiber, 3.8% ash, 76.8% total dietary fiber, 6.6% water, and 0.4% fat, respectively.

The research determined that Durio seeds were efficient in managing landfill leachate and that the optimal treatment conditions were at a pH of 6 and a concentration of 4000 mg/L. This led to a 34% decrease in true color and a 36.9% decrease in turbidity (Yusoff et al. 2018).

Recent studies aim to uncover the behavior and processes behind cross-linked *Durio zibethinus* seed starch flocculants for the management of landfill leachate. Through cross-linking modification, the effectiveness of starch flocculants made from *Durio zibethinus* seed waste significantly increased. The flocculants also improved the elimination of turbidity, suspended solids, chemical oxygen demand (COD), and color (Zamri et al. 2018; Badrus 2018).

11.5.14 Nephelium Lappaceum *Seeds*

The *Nephelium lappaceum* fruits were processed by removing their seeds, rinsing them with tap water, and drying them in an oven at 105 °C for 10–15 min. This procedure helps to make the seed coat of the *Nephelium lappaceum* seeds easily removable, preserving the seeds' natural composition from any potential impact of heightened moisture content. The seeds were then pulverized into powder using a home blender and kept in airtight containers until needed. To prepare a 50 g/L stock solution, 50 g of *Nephelium lappaceum* seed powder was mixed with one litre of distilled water and stirred for two minutes. The mixture was filtered through muslin fabric to produce the stock solution, which was then analyzed for molecular weight, pH, particle size, functional groups, zeta potential, and surface morphology using various instruments including Perkin-Elmer System 2000 FTIR spectrometer, ZEISS SUPRA 35VP Field Emission Scanning Electron Microscopy, and Malvern Zetasizer Nano ZS (Aziz et al. 2022).

11.5.15 Longan Seed

The extraction process was modified somewhat from (Katayon et al. 2006) by distilled water will be heated to 100 °C using a hot plate and stirrer for a duration of 30 min. To start the process, 500 g of freshly peeled Dimocarpus longan seeds had their arils separated. They were cleaned using tap water and allowed to air dry for a period of 48 h. The covering of the seeds was removed to uncover the seeds themselves. To ensure that the seeds were completely dry, they were left to air dry for an additional 48 h before being pounded into powder using a ring mill for 15 s. The resulting fine powder was stored in a dry condition for future research use (Aziz et al. 2018).

11.5.16 Tapioca Starch Flour

This experiment involves the use of tapioca starch flour and pre-hydrolyzed iron (PHI) solution, both of which were made from the highest quality analytical reagents. To ensure freshness, a new 0.1 mol/L FeCl3 solution was made. Using an auto-titrator with a high stirring speed at room temperature, PHI was produced by slowly adding 0.5 mol/L NaOH to the 0.1 mol/L FeCl3 solution at a flow rate of 0.0038 mL/s, based on the desired basicity (basicity = 0.1). The PHI mixture was then left to mature for 24 h at room temperature before being combined with the tapioca starch flour (TF).

To avoid the creation of an insoluble product, the PHITF was formed by adding a heated TF solution at a rate of 0.3 ml/min into a 2.85 g/L Fe stock solution of PHI, using magnetic stirring (700–800 rpm) and a temperature of 50 °C (in a water bath). The TF/Fe ratio used to calculate the amount of TF to be added was 0.7. The necessary amount of TF was dissolved in ultrapure water and heated to reach the gelatinization temperature. The resulting composite coagulant was then left to mature for 24 h at room temperature before being stored in a refrigerator. This process of producing the composite coagulant is an essential step (Moussas and Zouboulis 2009).

In this experiment, the effectiveness of coagulation was evaluated using a jar test apparatus (SW6 Stuart, manufactured by Bibby Scientific Limited in the UK). The jar test was the sole method utilized to analyze the pH and dosage of PHITF. The PHITF was added at a specified dosage (g/L Fe) and the pH of the leachate was adjusted as needed using NaOH and HCL. 500 mL of leachate was placed in a 1 L beaker, followed by 500 mL of PHITF, and then mixed rapidly while being stirred gently. The leachate was then allowed to settle and 50 mL of the supernatant was collected from the leachate at a depth of 3 cm for analysis of suspended solids (SS), color, chemical oxygen demand (COD), and ammonia elimination (Zin et al. 2014).

The calculation for determining the percentage removed was performed using the following formula:

$$\%\,\text{removal} = \frac{(Ci - Cf)}{Ci} \times 100 \tag{11.1}$$

where "Ci" represents the initial concentration of the removal parameter and "Cf" represents the final concentration of the removal parameter.

11.5.17 Guar Gum Dose

The experiment on coagulation-flocculation was performed using the Lovibond Model Floc Tester ET750 apparatus. The leachate sample was transferred to cylindrical beakers of a one-liter capacity for mixing purposes. A fresh stock solution of the guar gum coagulant was prepared daily before the jar test to avoid mold growth. The mixing was carried out in two stages, starting with rapid mixing for five minutes at a speed of 150 RPM, followed by a slower mixing for 15 min at 60 RPM. After mixing, the flocs were allowed to settle for 30 min undisturbed. To avoid foam generation, a plastic syringe was used to collect a supernatant liquid sample from approximately 2 cm below the sample surface for further analysis. To obtain an average reading, three samples of the treated water were taken and equation was used to determine the removal efficiencies for suspended particles, turbidity, and chemical oxygen demand (COD) (Zin et al. 2014).

$$\% \text{ removal} = \frac{(Ci - Cf)}{Ci} \times 100 \quad (11.2)$$

where "Ci" represents the initial concentration and "Cf" represents the final concentration of the removal parameter (Tables 11.1, 11.2 and 11.3).

11.6 Conclusion

In summary, the proper treatment of landfill leachate is critical for the protection of the environment. Knowing the characteristics of leachate, such as its COD levels, BOD5/COD ratios, and landfill age, is key in choosing the right treatment methods. Neglecting the proper treatment of leachate can lead to groundwater contamination and pose significant risks to the environment. Chemical coagulants are effective in improving the quality of leachate by reducing its levels of COD, heavy metals, color, TSS, and turbidity. Plant extracts, on the other hand, can remove harmful and pathogenic microbes from leachate, while also enhancing its quality parameters with minimal negative impacts. By prioritizing research into natural coagulants like Palm Oil trunk starch, jackfruit seeds, Cycas revoluta, and Nicotiana Tabacum extract, cost-effective treatment options can be developed and implemented to stabilize leachate, helping to secure a more sustainable future by effectively managing its potentially harmful effects and reducing groundwater contamination. In conclusion, the treatment of landfill leachate requires careful consideration of the pros and cons associated with physical, chemical, and biological treatment methods. Here is a summary of the advantages and disadvantages of each method:

Physical Treatment:

Pros:

- Simple and cost-effective
- Removes suspended solids and improves clarity
- Requires minimal chemical usage.

Cons:

- Limited removal of dissolved contaminants
- Minimal impact on nutrient content
- Potential sludge generation.

Table 11.1 Summary of research on the use of natural compounds for leachate treatment

Sr No	Substances used for coagulation	Optimal pH value	Optimal dosage	Pollutant reduction efficiency (in %)	References
1	*Hibiscus-Rosasinensis*	6	Alum—4000 mg/L + 500 mg/L	Fe^{3+}—100%, SS—72%, turbidity—60%, NH_3-N—28%. Color—61%	Camacho et al. (2016)
2	Dragon fruit purple pink-fleshed (DFPP)	–	10–90 mg/l	Turbidity—67%, TDS—69%, At an optimal dosage of 90 mg/L, 36% of Total Suspended Solids (TSS) was removed	Ismail et al. (2018)
3	*Cycas revoluta*	6	A combination of PAC at a concentration of 2000 mg/L and NSTS at a concentration of 6000 mg/L was used	The removal rates for color, total suspended solids (TSS), ammonium nitrogen (NH3-N), turbidity, chemical oxygen demand (COD), organic ultraviolet (UV), and cadmium (Cd) were 94.7%, 99.2%, 2.4%, 98.9%, 35.5%, 69.5%, and 53.8% respectively	Awang and Aziz (2012)
4	Dragon fruit white-fleshed (DFPW)	–	10–90 mg/l	Turbidity—60% TSS—23%.The ideal amount of coagulant for 50 mg/L and 65% TDS was found to be 90 mg/L	Ismail et al. (2018)
5	*Isabgol*	7.5	PAC—7.2 g/L + 0.4 g/L of Isabgol (psyllium husk)	COD—64%, Color—90%, TSS—96%	Aziz and Sobri (2015)

(continued)

Table 11.1 (continued)

Sr No	Substances used for coagulation	Optimal pH value	Optimal dosage	Pollutant reduction efficiency (in %)	References
6	*M.Oleifera*	–	7.5 mg/l	Turbidity—76%, optimum dosage-7.5, Min. Residual Turbidity(50 NTU)-12, Min. Residual Turbidity(100 NTU)-13, Min. Residual Turbidity(250 NTU)-20	Sou Kantabai Bhavarlalji et al. (2013)
7	Jackfruit seeds starch	5	A concentration of 500 mg/L JSS and 600 mg/L PAC was used	Removal of COD was 33.5%, with a decrease in color by 93.6%, SS by 94.5%, turbidity by 92.3%, and NH3-N by 14.1%	Al-Hamadani et al. (2011)
8	*Okra*	–	10 mg/l	54% of turbidity, optimum dosage-10 Min. Residual Turbidity(50 NTU)-23, Min. Residual Turbidity(100 NTU)-32, Min. Residual Turbidity(250 NTU)-65	Sou Kantabai Bhavarlalji et al. (2013)
9	Palm oil trunk starch	7	500 mg/L	The experiment resulted in a 33.5% removal of COD, along with a decrease in color by 93.6%, SS by 94.5%, turbidity by 92.3%, and NH3-N by 14.1%	Yusoff and Mohamad Zuki (2015)

(continued)

Table 11.1 (continued)

Sr No	Substances used for coagulation	Optimal pH value	Optimal dosage	Pollutant reduction efficiency (in %)	References
10	*C.Procera*	–	15 mg/l	64% of turbidity, optimum dosage-15 Min. Residual Turbidity(50 NTU)-18, Min. Residual Turbidity(100 NTU)-27, Min. Residual Turbidity(250 NTU)-33	Sou Kantabai Bhavarlalji et al. (2013)
11	Cross-linked Oil Palm Trunk Starch (COPTS)	8.3	1000 mg/L	The reduction in turbidity was 43.2%, with a decrease in SS by 25%, COD by 65.1%, color by 39%, NH3-N by 5%, Fe by 0%, an increase in Mn by 100%, a decrease in Cu by 49%, a decrease in Zn by 79.5%, and an increase in PO4 by 100%	Yusoff and Mohamad Zuki (2015)
12	*Plantago ovata*	6–8	–	Turbidity >90%, turbidity concentrations (500, 150, and 50 NTU) 98.2, 94.9, and 80.2%	Badrus (2018)
13	*Nicotiana tabacum*	6	1000 mg/L Nicotiana tabacum +1000 mg/L PAC	Color—86%, COD—91% turbidity—21%, SS—48% and NH_3-N—54%	Ahmed et al. (2020)
14	Nephelium lappaceum seeds	–	2.00 g/l	Colour-19.48, COD-35.62	Aziz et al. (2022)

(continued)

Table 11.1 (continued)

Sr No	Substances used for coagulation	Optimal pH value	Optimal dosage	Pollutant reduction efficiency (in %)	References
15	*Tamarindus indica*	–	–	Using 0.5% PAM (Polyacrylamide), colour 41.90, and COD 5.90, turbidity was reduced with a 91.16% efficiency	Badrus, (2018)
16	Longan seed	–	2.00 g/l	SS-29.5, Colour-15.10, COD-28.00	Aziz et al. (2022)
17	Guar gum dose	8.500	45 mg/l	COD removal-21.793	Cheng et al. (2020)
18	*Citrullus Lanatus* seeds	7	250 mg/L	Turbidity-81%	Singh and Saxena (2020)
19	Tapioca starch flour	5	0.2 g/L	COD-60%	Zin et al. (2014)
20	Durio	6	4000 mg/L	The rates of turbidity removal and true colour removal were 34 and 36.9%, respectively	Badrus (2018)

Chemical Treatment:

Pros:

- Enhanced removal of a wide range of contaminants
- pH adjustment capability
- Potential for sludge valorization.

Cons:

- Higher operational costs due to chemical usage
- Chemical handling and disposal requirements
- Generation of chemical by-products or residuals.

Biological Treatment:

Pros:

- Cost-effective and sustainable
- High removal efficiency for organic pollutants
- Potential for resource recovery.

Table 11.2 Effectiveness comparison of natural coagulants in removing severe pollution factors

Coagulant names	Turbidity	COD	Colour	TSS	NH3-N	Heavy metals	TDS
1. Hibiscus-rosasinensis	√	–	√	√	*	√	–
2. Dragon fruit purple pink-fleshed (DFPP)	√	–	–	Q	–	–	√
3. Cycas revoluta	√	*	√	√	Q	√	–
4. Dragon fruit white-fleshed (DFPW)	√	–	–	Q	–	–	√
5. Isabgol	–	√	√	√	–	–	–
6. *M.Oleifera*	√	–	–	–	–	–	–
7. Jackfruit seeds starch	√	*	√	√	Q	–	–
8. Okra	√	–	–	–	–	–	–
9. Palm oil trunk starch	*	√	*	√	Q	√	–
10. *C.Procera*	√	–	–	–	–	–	–
11. Cross-linked oil palm trunk starch (COPTS)	√	√	√	*	Q	–	–
12. Plantago ovata	√	–	–	–	–	–	–
13. Nicotiana tabacum	*	√	√	√	√	–	–
14. Nephelium lappaceum seeds	*	–	Q	–	–	–	–
15. Tamarindus indica	√	Q	√	–	–	–	–
16. Longan seed	–	*	Q	–	–	–	–
17. Guar gum dose	–	√	–	–	–	–	–
18. *Citrullus Lanatus* seeds	√	–	–	–	–	–	–
19. Tapioca starch flour	–	–	–	–	–	–	–
20. Durio	*	–	*	–	–	–	–
High	(√)		Medium		(*)		Low (Q)

Table 11.3 An overview of the comparison of pollutant removal between natural and chemical coagulants

Leachate parameters	Hibiscus rosasinensis	Dragon fruit purple pink-fleshed (DFPP)	Cycas revoluta	Dragon fruit white-fleshed (DFPW)	Isabgol	*M.Oleifera*	Jackfruit seeds starch	Okra	Palm oil trunk starch	*C.Procera*
COD mg/l	–	–	36	–	64	–	34	–	45	–
Color	61	–	–	–	90	–	94	–	25	–
Turbidity NTU	60	67	98	–	–	76	92	54	38	64
TSS mg/l	72	36	99	60	96	–	94	–	72	–
Heavy metals mg/l	Fe-100	–	Cd-54	23	–	–	–	–	Mn-100, Zn-55	–
NH3-N mg/l	28	–	–	–	–	–	14	–	13	–
TDS mg/l	–	69	–	–	–	–	–	–	–	–
Leachate parameters	Cross-linked oil palm trunk starch (COPTS)	Plantago ovata	Nicotiana tabacum	Nephelium lappaceum seeds	Tamarindus indica	Longan seed	Guar gum dose	Citrullus Lanatus seeds	Tapioca starch flour	Durio
COD mg/l	65.1	–	91	35.62	5.90	28	21.793	–	60	–
Color	39	–	86	19.48	41.90	15.10	–	–	–	34
Turbidity NTU	43.2	90	21	–	91.16	–	–	81	–	36.9
TSS mg/l	25	–	48	–	–	29.5	–	–	–	–

(continued)

Table 11.3 (continued)

Leachate parameters	Cross-linked oil palm trunk starch (COPTS)	Plantago ovata	Nicotiana tabacum	Nephelium lappaceum seeds	Tamarindus indica	Longan seed	Guar gum dose	Citrullus Lanatus seeds	Tapioca starch flour	Durio
Heavy Metals mg/l	100% Removal of Mn, 49% Removal of Cu, 79.5% Removal of Zn, 100% Removal of PO4	–	–	–	–	–	–	–	–	–
NH3-N mg/l	5	–	–	–	–	–	–	–	–	–
TDS mg/l	–	–	91	–	–	–	–	–	–	–

Cons:

- Longer treatment duration compared to other methods
- Sensitivity to operational conditions
- Limited removal of inorganic compounds.

In conclusion, each treatment method has its advantages and limitations. Physical treatment methods are simple and cost-effective but have limitations in removing dissolved contaminants. Chemical treatment offers enhanced contaminant removal but comes with increased operational costs and potential environmental concerns. Biological treatment methods are cost-effective, efficient for organic pollutant removal, and offer resource recovery possibilities, but they may require longer treatment durations and are less effective for inorganic contaminants.

To choose the most suitable treatment method for landfill leachate, factors such as the composition of leachate, desired removal efficiencies, operational considerations, and environmental impacts must be carefully evaluated. Implementing a combination of treatment methods or integrating multiple stages may be necessary to achieve the desired treatment goals effectively.

References

Agbor RB, Edu NE, Odok EN (2020) Decontamination of leachates using moringa oleifera and hibiscus rosa-sinensis as coagulants aids 39(4):917–924

Ahmed Z, Yusoff MS, Mokhtar KNH (2020) Application of natural coagulants for sustainable treatment of semi- aerobic landfill leachate. AIP conference proceedings, vol 2267(September). https://doi.org/10.1063/5.0017406

Al-Hamadani YAJ, Yusoff MS, Umar M, Bashir MJK, Adlan MN (2011) Application of psyllium husk as coagulant and coagulant aid in semi-aerobic landfill leachate treatment. J Hazard Mater 190(1–3):582–587. https://doi.org/10.1016/j.jhazmat.2011.03.087

Awang NA, Aziz HA (2012) Hibiscus rosa-sinensis leaf extract as coagulant aid in leachate treatment. Appl Water Sci 2(4):293–298. https://doi.org/10.1007/s13201-012-0049-y

Aziz HA, Rahim NA, Ramli SF, Alazaiza MYD, Omar FM, Hung YT (2018) Potential use of Dimocarpus longan seeds as a flocculant in landfill leachate treatment. Water (switzerland) 10(11):1–15. https://doi.org/10.3390/w10111672

Aziz HA, Rahmat NS, Alazaiza MYD (2022) The potential use of nepheliumlappaceum seed as coagulant–coagulant aid in the treatment of semi-aerobic landfill leachate. Int J Environ Res Public Health 19(1). https://doi.org/10.3390/ijerph19010420

Aziz HA, Sobri NIM (2015) Extraction and application of starch-based coagulants from sago trunk for semi-aerobic landfill leachate treatment. Environ Sci Pollut Res 22(21):16943–16950. https://doi.org/10.1007/s11356-015-4895-7

Badrus Z (2018) Potential of natural flocculant in coagulation-flocculation wastewater treatment process. E3S Web Conf 73:2–6. https://doi.org/10.1051/e3sconf/20187305006

Camacho FP, Sousa VS, Bergamasco R, Teixeira MR (2016) The use of Moringa oleifera as a natural coagulant in surface water treatment. Chem Eng J. https://doi.org/10.1016/j.cej.2016.12.031

Cheng SY, Show PL, Juan JC, Ling TC, Lau BF, Lai SH, Ng EP (2020) Sustainable landfill leachate treatment: Optimize use of guar gum as natural coagulant and floc characterization. Environ Res 188(June). https://doi.org/10.1016/j.envres.2020.109737

Gao J, Oloibiri V, Chys M, Audenaert W, Decostere B, He Y, Van Langenhove H, Demeestere K, Van Hulle SWH (2015) The present status of landfill leachate treatment and its development trend from a technological point of view. Rev Environ Sci Biotechnol 14(1):93–122. https://doi.org/10.1007/s11157-014-9349-z

Idris J, Som AM, Musa M, Ku Hamid KH, Husen R, Muhd Rodhi MN (2013) Dragon fruit foliage plant-based coagulant for treatment of concentrated latex effluent: comparison of treatment with ferric sulfate. J Chem. https://doi.org/10.1155/2013/230860

Impact ITS (2019) A review of wastewater treatment using natural material and its potential as aid and composite coagulant. 48(1):155–164

Ismail SNS, Mahiddin NAK, Praveena SM (2018) The used of dragon fruit peels as eco-friendly wastewater coagulants. Futur Food: J Food, Agric Soc 6(Specialissue):112–117

Journal I, Araujo H, Pat H (2010) The Leachate Treatment by using natural coagulants (pine bark and chitosan related papers the leachate treatment by using natural

Katayon S, Noor MJMM, Asma M, Ghani LAA, Thamer AM, Azni I, Ahmad J, Khor BC, Suleyman AM (2006) Effects of storage conditions of Moringa oleifera seeds on its performance in coagulation. Biores Technol 97(13):1455–1460. https://doi.org/10.1016/j.biortech.2005.07.031

Liu ZP, Wu WH, Shi P, Guo JS, Cheng J (2015) Characterization of dissolved organic matter in landfill leachate during the combined treatment process of air stripping, Fenton, SBR and coagulation. Waste Manage 41:111–118. https://doi.org/10.1016/j.wasman.2015.03.044

Mohd-Asharuddin S, Othman N, Mohd Zin NS, Tajarudin HA (2017) A chemical and morphological study of cassava peel: a potential waste as coagulant aid. MATEC Web Conf 103:1–8. https://doi.org/10.1051/matecconf/201710306012

Moussas PA, Zouboulis AI (2009) A new inorganic-organic composite coagulant, consisting of Polyferric Sulphate (PFS) and Polyacrylamide (PAA). Water Res 43(14):3511–3524. https://doi.org/10.1016/j.watres.2009.05.015

Muyibi SA, Megat Johari MJM, Leong TK, Loon LH (2002) Effects of oil extraction from Moringa oleifera seeds on coagulation of turbid water. Int J Environ Stud 59(2):243–254. https://doi.org/10.1080/00207230210924

Pandey P, Khan F, Ahmad V, Singh A, Shamshad T, Mishra R (2020) Combined efficacy of Azadirachta indica and Moringa oleifera leaves extract as a potential coagulant in ground water treatment. SN Appl Sci 2(7):1–8. https://doi.org/10.1007/s42452-020-3124-2

Prasad SVM, Rao BS (2016) Influence of plant-based coagulants in waste water treatment. Ijltemas V(iii):45–48

Saravanan J, Priyadharshini D, Soundammal A, Sudha G, Suriyakala K (2017) Wastewater treatment using natural coagulants. Int J Civ Eng 4(3):40–42. https://doi.org/10.14445/23488352/ijce-v4i3p109

Sa'id S, Mohammed K, Adie DB, Okuofu CA (2016) Turbidity removal from surface water using *Tamarindus indica* crude pulp extract. Bayero J Pure Appl Sci 9(1):236. https://doi.org/10.4314/bajopas.v9i1.37

Simate GS, Ndlovu S (2014) Acid mine drainage: challenges and opportunities. J Ind Eng Chem. https://doi.org/10.1016/j.jiec.2014.03.031

Singh P, Saxena I (2020) Yamuna river water treatment using a natural coagulant of citrullus lanatus seeds 149(September):52–63

Sou KantabaiBhavarlalji L, Renuka Binayke MA, Mvjadhav P (2013) Application of natural coagulants in water purification. Int J Adv Technol Civ Eng 2(1):15

Unda-calvo J, Martínez-santos M, Ruiz-romera E (2017) Chemical and physiological metal bioaccessibility assessment in surface PBET, TCLP and BCR sequential extraction methods. Ecotoxicol Environ Saf 138(December 2016):260–270. https://doi.org/10.1016/j.ecoenv.2016.12.029

Yabroudi SC, Morita DM, Alem P (2013) Landfill leachate treatment over nitritation/denitritation in an activated sludge sequencing batch reactor. APCBEE Procedia 5(L):163–168. https://doi.org/10.1016/j.apcbee.2013.05.029

Yusoff MS, Aziz HA, Zamri MFMA, Suja' F, Abdullah AZ, Basri NEA (2018) Floc behavior and removal mechanisms of cross-linked Durio zibethinus seed starch as a natural flocculant for landfill leachate coagulation-flocculation treatment. Waste Manag 74:362–372. https://doi.org/10.1016/j.wasman.2018.01.016

Yusoff MS, Mohamad Zuki NA (2015) Optimum of treatment condition for *Artocarpus heterophyllus* seeds starch as natural coagulant aid in landfill leachate treatment by RSM. Appl Mech Mater 802:484–489. https://doi.org/10.4028/www.scientific.net/amm.802.484

Zamri MFMA, Suja F, Yusoff MS, Aziz HA, Bahru R (2018) The comparison of Durio Zibethinus seed starch extraction for landfill leachate treatment. Mater Res Express 5(7). https://doi.org/10.1088/2053-1591/aacfcd

Zin NSM, Aziz HA, Adlan MN, Ariffin A, Yusoff MS, Dahlan I (2014) Treatability study of partially stabilized leachate by composite coagulant (prehydrolyzed iron and tapioca flour). Int J Sci Res Knowl 313–319. https://doi.org/10.12983/ijsrk-2014-p0313-0319

Chapter 12
Significance of Renewable Energy in Water Management and Irrigation

Jyoti Bhattacharjee and Subhasis Roy

Abstract New, sustainable, eco-friendly, and renewable energy-based resources are employed to generate water supply and irrigation services in developing nations. Solar energy, wind turbines, hydroelectric power plants, biomass energy, and geothermal energy provide a reliable alternative to fossil fuels, efficiently use water, and significantly reduce farming expenses. This chapter focuses on innovative and eco-friendly irrigation water delivery methods. In solar water pump systems, photovoltaic power is generated by hydrogel-based solar panels. Photovoltaic cells capture photons from the sun and transform them into electrical energy, which powers a motor that pumps out clean water. Water-pumping windmills like the ones shown here can be installed in isolated, windy places where other standard water-pumping methods are impractical. Mechanical wind water pumps use wind turbine generators to supply water to cattle, land drainage, fish farming, and agricultural irrigation. Most geothermal fields are found in rural developing areas with active agriculture, making them a suitable match to cut the cost of irrigation water transportation. Most geothermal sites are found in rural developing areas with active agriculture, making them an ideal match to reduce the cost of irrigation water transportation. After desalinating ionic water, geothermal energy can be directly used for drying crops and aquaculture farms. Slow pyrolysis of lignin-based bioenergy crops such as sorghum, sugarcane, reed canary grass, and others produces charcoal, water, and heat, hence improving soil quality. Geothermal energy can be directly used for drying crops and aquaculture farms after desalinating ionic water. Slow pyrolysis of lignin-based bioenergy crops such as sorghum, sugarcane, reed canary grass, etc., produces biochar, water, and heat, enhancing soil quality. Despite the hefty price, green hydrogen helps decarbonize the economic system and enhances the availability of pure water in arid agricultural lands. The new advancements and challenges of using renewable energy-based water supply for irrigation are covered in this chapter.

J. Bhattacharjee · S. Roy (✉)
Department of Chemical Engineering, University of Calcutta, 92 A. P. C. Road, Kolkata 700009, India
e-mail: srchemengg@caluniv.ac.in

S. Suriyanarayanan et al. (eds.), *Water Management in Developing Countries and Sustainable Development*, Water Resources Development and Management,
https://doi.org/10.1007/978-981-99-8639-2_12

Keywords Agricultural irrigation · Bioenergy crops · Decarbonize · Desalination · Hydrogel-based solar panels · Wind-solar-hydrogen hybrid power

12.1 Introduction

The agricultural sector, like other businesses, is growing more reliant on energy sources and consumes a significant quantity of energy in many nations worldwide (Lokesh and Srivastava 2022). Irrigation is a crucial economic driver in India due to its ability to give food to the inhabitants of the ecosystem. Modernizing numerous agricultural production methods, including irrigation and others, contributes to this dependence. Around 62 terawatt hours of energy are used by electric irrigation pumps each year globally (Ding et al. 2022; Winter et al. 2022). When rainfall is insufficient to meet the evapotranspiration demand of the atmosphere, the irrigation serves the primary purpose of providing crops with controlled water, which can then be extracted from surface reservoirs or aquifers. However, the lack of water in arid and semiarid regions has made it more challenging to irrigate crops. Tea leaves include biomolecules such as polyphenols and chlorophyll that can be isolated and potentially used in solar cell construction. These biomolecules have intriguing optical and electrical features that could improve the efficiency of solar cells. Due to their outstanding optoelectronic capabilities, perovskite materials have received a lot of attention in the field of photovoltaics. Hybrid perovskites are a family of materials that blend organic and inorganic components, providing a high degree of tunability and the promise of low-cost manufacturing (Dey et al. 2016). Fuel-based systems are not the best option in remote locations where the cost of delivery can be prohibitive (Qin et al. 2021). Recent study examines the use of solar energy in agricultural systems, specifically solar-powered irrigation technology, and their impact on water use efficiency (Hou et al. 2023). Renewable energy sources have received a lot of attention as a replacement for fossil fuels or as a supplement in hybrid systems. They are useful in isolated regions where maintaining a consistent fuel supply is difficult and competent maintenance workers are difficult to obtain (Bose et al. 2022). As a result, efficient energy use is critical to sustaining agricultural productivity since it results in cost savings, the preservation of fossil resources, and a reduction in air pollution. This review will concentrate on pumping systems to analyze renewable energy systems for both small- and large-scale irrigation in semi-arid environments. This review presents and summarizes numerous studies that used renewable energy sources. The main problem with current irrigation systems is the inequitable water distribution to crops, which leads to excess electricity consumption. To get around these drawbacks and restrictions, the review chapter discusses irrigation systems powered by renewable energy (Zhou et al. 2021).

12.1.1 Solar-Powered Advanced Irrigation System

Depending on the amount of sunlight, solar pumps, and automatic irrigation systems pump water from a bore well into a ground-level storage tank. Because the water is pumped into an underground tank, our system only requires one stage of energy consumption. The water that flows into the field is then managed by a simple valve system. In contrast, traditional approaches include employing two pumps to pump water from a bore well into a well and then from that well onto a field. This can save a significant amount of electricity while successfully utilizing renewable energy sources. Smart software manages a valve that controls the amount of water that enters the field (Huang et al. 2021).

Solar irrigation is not only good for the environment nevertheless, it can also help farmers adjust their farming practices to account for shifting weather directions. A power revolution might be sparked by further price cuts on photovoltaic panels, which have already experienced considerable and ongoing price reductions. To support the sustainable development of Indian farmers, this chapter will present a thorough analysis of the development of an irrigation system and its design using solar energy (Heo et al. 2021). There have been numerous studies and attempts to use photovoltaic (PV) solar energy systems for irrigation. An irrigation system maximizes water utilization by reducing waste and requiring farmers to work less. A solar water pumping system is appropriate for rural locations without grid electricity (Li et al. 2022a).

The amount of solar energy received from the sun at a specific location determines the power of a solar module. The amount of water pumped and the intensity of the sun's radiation determine the array's (or solar modules') size. The entire solar PV pumping system is controlled by a microcontroller (PIC 16F8768P). The control system charges the battery, balances the voltage and current from the solar panel, and prevents overcharging and undercharging scenarios. It also controls the operation of pumps according to needs. Thanks to the microcontroller, the motor will start when it is dry and stop when it is wet. The LCD panel that shows the status of the soil is also connected to the microcontroller (Alam et al. 2022). Integration of GSM technology might be improved. Implementation regarding solar-powered irrigation systems (SPIS), the federal government and individual farmers must address two crucial, antagonistic objectives. Farmers want to increase output and income while using fewer workers to achieve economic prosperity and meet the expanding needs of their families. The government wants to reduce subsidies and the environmental impact of its operations.

The adoption of micro irrigation systems and the introduction of solar power have produced encouraging results that are briefly discussed in Fig. 12.1.

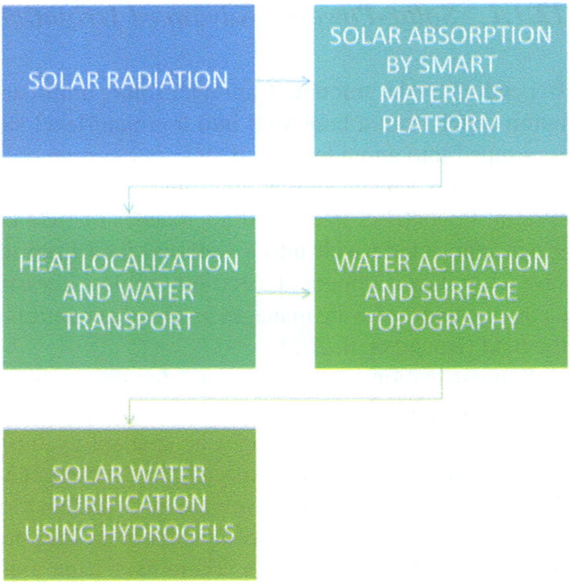

Fig. 12.1 Schematic flowsheet of a new smart materials platform for solar water purification using hydrogels (Lu et al. 2022)

12.1.2 Advantages of Solar Irrigation

The idea of a solar-powered irrigation system has drawn significant interest worldwide. It is a tactic to encourage using a renewable, reusable energy source.

Solar power protects the system more than traditional electricity. There are typically fewer blackouts when power plants are spread out across networks. Solar energy is adequate to safeguard the grid from unplanned natural disasters (Jamesh et al. 2021). For industrial needs, the Genus solar inverter might be the best choice. Solar energy is less expensive than fossil fuels and electricity.

Additionally, fuel prices are constantly changing. Therefore, solar energy is more affordable than traditional energy sources, particularly for small-scale farmers. In addition, solar energy is the best and most cost-effective irrigation option that also lasts a very long time in rural areas where electricity is still not accessible (Guo and Yu 2022). It lessens the farmers' reliance on expensive energy sources to run their water distribution system. Therefore, countries without oil reserve systems may find this solar irrigation helpful. The field's crops are consistently irrigated thanks to the solar energy irrigation system.

Solar power, unlike other energy sources such as coal and biogas, does not emit greenhouse gases, reducing the negative impacts on the ozone layer and the environment. Solar irrigation, as opposed to power generators, is the most environmentally benign energy source. Aside from that, this effective technique consumes relatively little water (Aleid et al. 2022). As a result, more water may be saved and used for farming and crop growth. Modern solar-powered systems, with electronic control devices that can provide real-time inputs about storage tank levels, pump speed, and borehole water levels, provide essential instruments for enhancing water governance. These inputs may potentially lead to legislative judgments prohibiting excessive remote use. This strategy discussed here is being tested in Egypt, India, and Latin America (Díaz-Marín et al. 2022a). A multitude of services for solar energy in irrigation systems is possible. Cold storage facilities, mills, water filtration plants, and other equipment can all be powered by the excess energy it has stored.

Additionally, farmers can sell the extra electricity to the utility by tying it to the electrical grid. Modern advancements have enabled the solar-powered irrigation system to self-regulate for the land based on crop water requirements, ambient conditions, and water availability (Jin et al. 2021). However, if one wants a steady and efficient energy source right away, a solar inverter home can be set up. Solar PV-powered water pumping systems don't produce noise or pollutants. It lasts a lot longer, good quality solar panels are warranted for 25 years, and investors typically 6–8 years.

However, some of the challenges associated with adopting and using solar-powered irrigation systems are also examined in this research (SPIS) (Li et al. 2021a). It concludes that many countries still have issues with access to financing, particularly for small farmers, as well as the availability of high-quality goods and services. Additional capacity-building efforts are required to guarantee that users can handle daily operations and maintenance and have a fundamental understanding of the configuration and operations of the system (Wang et al. 2021a). Solar irrigation has a high initial material and installation cost. A sizable battery bank is necessary because the efficiency is not yet 100%, and no solar power is available at night. Solar-powered cars are slower and less powerful than conventional gas-powered cars; panel theft is possible because they are occasionally valued as desirable products. And production may be less in the winter (Deng et al. 2022). Despite the challenges, the solar irrigation system is the most complex advancement in innovative agriculture techniques that innovation has ever made possible. Due to these smart technologies, landowners and farmers can do more with fewer resources and expenses (Maitra et al. 2021). Irrigated agriculture will be crucial in supplying food for the world's rising population in the future. Due to the demand for more crops per drop, pressurized pipe networks have replaced conventional open-channel irrigation water delivery systems. These networks increase the efficiency of irrigation systems on farms and in irrigation districts. However, they are very energy-intensive (Zhao et al. 2022). The high cost of conventional energy sources has a substantial negative impact on the sustainability of farms that use pressurized irrigation. These energy sources not only cost a lot of money but also significantly impact global warming, which is currently the most urgent environmental issue in the world. The review chapter covers a wide range of subjects, including the design of hybrid, solar, and wind pumps, intelligent irrigation management systems employing renewable energy sources, and hydraulic energy recovery in irrigation networks using pumps and microturbines as turbines (Shi et al. 2022). Several publications on the life cycle of these types of facilities, the environmental impact of crops irrigated with renewable energies, cost–benefit assessments, and strategies to encourage the adoption of renewable energies for irrigation would

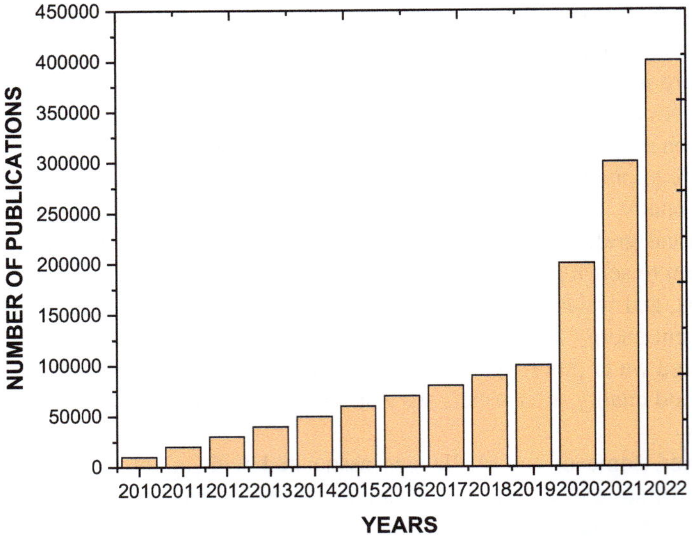

Fig. 12.2 Schematic flowsheet of the number of publications vs years from 2010–2022, depicting purifying water and other items by employing solar water evaporation (Li et al. 2021b)

complement the strictly technological contributions well, which can be depicted in Fig. 12.2.

12.2 Wind-Powered Irrigation System

The agricultural microgrid is a low-cost method of supplying energy to rural areas. Since natural gas is so expensive, farmers are exploring alternative irrigation energy sources. The economic potential of electric and hybrid (electric/wind) irrigation systems as replacements for natural gas irrigation is examined. This review chapter looks at an integrated wind-pumped storage agricultural microgrid that may supply both water and energy. When compared to an off-grid microgrid, the establishment of a microgrid connected to the grid can improve the elasticity and dependability of critical infrastructure systems (Díaz-Marín et al. 2022b). However, one must be aware of the wind power fluctuation characteristics and forecasting uncertainty in the microgrid connected to the network. This is because these characteristics impose constraints on the extensive use of wind energy in microgrids and impact the power system's stable and dependable operation. Experiments show that the annual wind speed data for coastal areas like Andhra Pradesh can be stochastically described more accurately using the Weibull probability distribution (Zhang et al. 2022). To accommodate various daily irrigation demands at various risk levels, the size and scope of the entire irrigation system were established using a mass curve analysis. An agricultural microgrid's operation also involves irrigation and power systems. A difficult optimization problem, the power system considers turbine flow rate, irrigation time and volume, wind power, pump power, electricity demand, water demand, and wind power. The irrigation system also includes an electrical source (Li et al. 2022b). Power system operators have had to deal with various complex problems due to the unpredictable nature of wind energy. For the last 10 years, one of the most common solutions to this problem has been the joint operation of wind power generation and pumped storage units. Pumped-storage systems can help generation businesses boost their earnings in a competitive market. Several researchers have worked at predicting the joint operation of wind and hydropower plants in this regard (Ai et al. 2019). The graphic depicts a model of how wind farms and hydroelectric stations can coexist by modeling, simulation, and evaluation of turbines (Fig. 12.3).

The microcontroller requires the proper voltage, and wind energy is needed to lift water. Due to the use of sensors for water level and soil conductivity, these projects

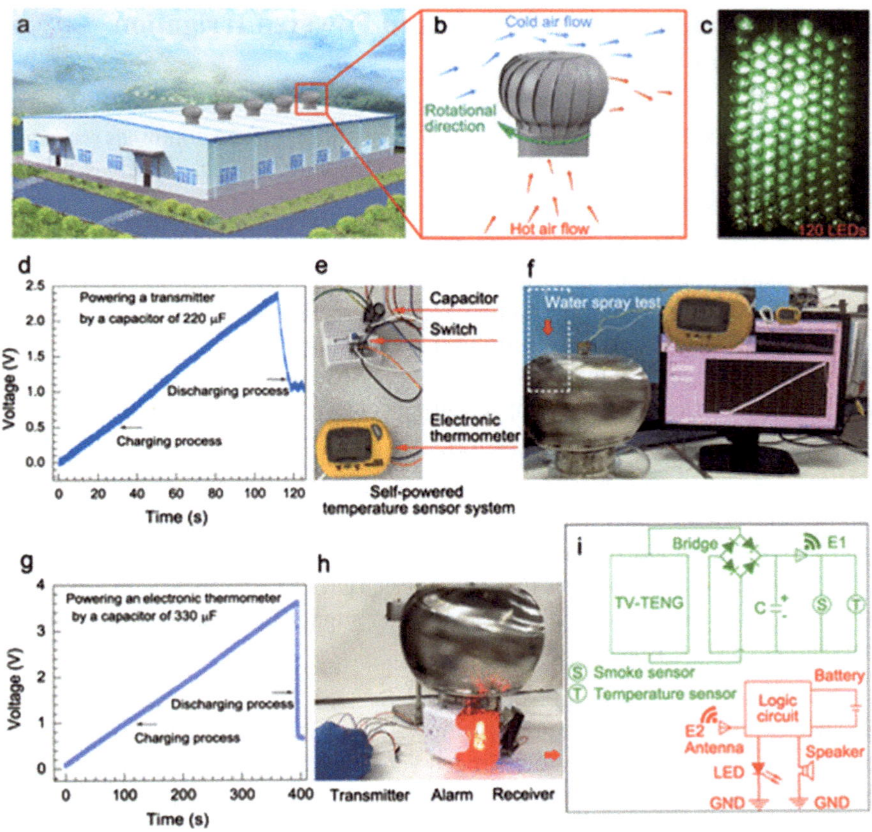

Fig. 12.3 Wind energy harvesting system using a triboelectric nano-generator in an on-site industrial monitoring system (Zhang et al. 2021). (Reproduced with permission)

are less expensive than conventional systems. This review addresses the problem of excessive irrigation, which stems from ineffective water management or distribution and may lead to water contamination. This paper's main contribution is coordinating the irrigation system for a farm microgrid to operate as a pumped-storage generating facility. According to the planned plan, the irrigation system's reservoirs should be at varied levels and serve as pumped-storage reservoirs. The microgrid must be connected to the grid at the grid supply point (GSP), where it exchanges power with the network via a substation transformer. A wind turbine on the farm supplies the base load of the microgrid. The joint operation of pumped storage and wind units in the DA market is intended to reduce the cost of microgrid operation while also delivering agricultural water. To enhance the match between wind and solar energy sources and irrigation power needs, it is desirable to produce a combination of a winter crop (such as winter wheat) and a summer crop (such as maize) (Jamesh et al. 2021).

12.3 Hybrid Solar-Wind-Hydrogen Powered Irrigation System

A hybrid novel solar-wind-hydrogen-powered irrigation system combines renewable energy sources to power irrigation systems efficiently. A few, examples of such systems are discussed here:

12.3.1 Solar-Wind-Hydrogen Pumping System: This system utilizes solar panels and wind turbines to generate electricity. The electricity is used to power a pump that draws water from a well or a water source. Excess electricity is stored in batteries and is also used to power an electrolyzer that produces hydrogen gas from water. The stored hydrogen can then be used during periods of low solar and wind activity to generate electricity for the pump, ensuring continuous irrigation.

12.3.2 Solar-Wind-Hydrogen Drip Irrigation System: In this setup, solar panels and wind turbines generate electricity to power a pump that supplies water to a central storage tank. The excess electricity is used for water electrolysis, producing hydrogen gas. The hydrogen gas is stored in tanks and can be used in a fuel cell during the night or low solar/wind periods to generate electricity for powering a drip irrigation system. This system ensures a consistent water supply for efficient and controlled irrigation (Yu et al. 2022).

12.3.3 Solar-Wind-Hydrogen Pivot Irrigation System: This system uses solar panels and wind turbines to create electricity to operate a pivot irrigation system. The excess electricity is utilized to generate hydrogen via water electrolysis. During periods of low renewable energy generation, the created hydrogen is stored and used in a fuel cell, allowing the pivot canal system to run continuously (Lei et al. 2021; Loo et al. 2021).

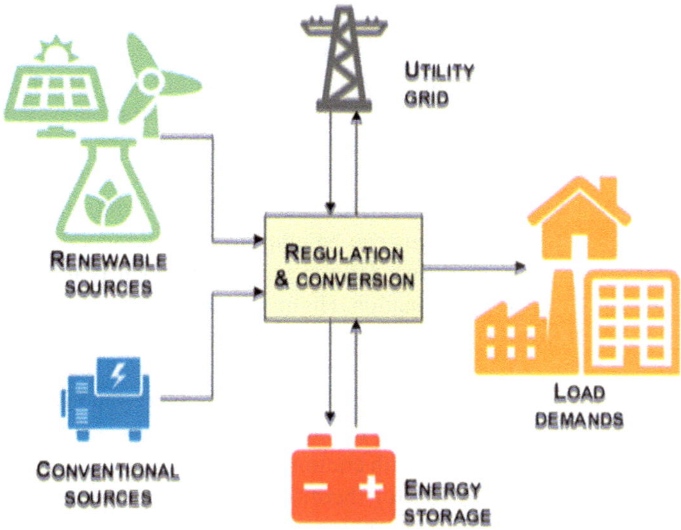

Fig. 12.4 Design hybrid power systems for maximum efficiency (Lee et al. 2018)

12.3.4 Solar-Wind-Hydrogen Rain Gun Irrigation System: In this design, solar panels and wind turbines generate electricity, which is employed to operate a pump that distributes water to a big storage tank. Excess electricity is used to produce hydrogen. During periods of low solar and wind activity, the stored hydrogen is used in a fuel cell to create energy for a rain gun irrigation system. The contributors examine the greenhouse gas energy consumption and evaluate the efficacy of the system in terms of water pumping capacity and crop production (Wang et al. 2021b). As needed, hydrogen fuel cells transform the stored hydrogen gas into power. The fuel cells can provide a continuous and dependable supply of power for the irrigation system, especially during periods of low solar or wind energy generation. As illustrated in Fig. 12.4, AI is increasingly being used to build novel hybrid systems.

12.3.1 Geothermal-Powered Irrigation System

Geothermal water is a source of energy that can supply both heat and electricity. Because of its high ionic content, the residual water can be used for irrigation following treatment. Most geothermal fields are located in rural agricultural areas. This would be a perfect match for lowering irrigation water delivery costs. Desalination for agriculture requires more energy, resulting in more post-treatment procedures. To fulfill energy demands, fossil fuel prices are growing, and greenhouse gas emissions are damaging to the environment (Peeters et al. 2021). Alternative water sources, such as geothermal water, may be harmful because they include toxic elements such as boron and arsenic. Both times, the direct heat and electricity generation technologies should be regarded as developed and suitable for desalination processes. The main barriers that could limit the use of geothermal water are temperature and the efficiency of the energy generation system. Various desalination processes are utilized depending on the temperature of the geothermal resource. If the temperature of the geothermal sources is between 40 and 70 °C, low-temperature desalination methods like membrane distillation, multi-effect distillation units, and simple evaporation basins can be employed. At temperatures exceeding 70 °C, geothermal resources can be used to power processes such as multi-effect evaporation and multi-stage flash desalination. Higher geothermal resource temperatures (between 120 and 200 °C) may be considered in cogeneration schemes (Wang et al. 2021c). Because geothermal energy is their major energy source, these systems have a reduced environmental impact and require less maintenance than RO membranes. Geothermal resources do not require thermal energy storage since they provide constant temperature and flow rate with continuous flow. Geothermal sources, therefore, seemed to be the best option for thermal desalination processes. It has been discovered that the Qinshui Basin's geothermal distribution characteristics influence the reservoir temperature of coalbed methane, affecting the gas's in-situ adsorption, desorption, and diffusion behaviors (Wang et al. 2021d).

Despite numerous research in recent years, the major reasons for variances in groundwater storage (GWS) across India remain unknown. In this context, it is employed the Water Gap model shows that GWW has a stronger domination than GWR on GWS alteration in India. The majority of electric generating plants require water to operate. In the US, 40% of all freshwater withdrawals are used for thermoelectric purposes. Water is typically used in steam turbines in power plants to remove the facility's waste heat. It is also used as a source of hydrogen in gasification facilities to produce syngas, which is then applied in gas turbines. Water is necessary for hydroelectric plants to run, and this water can come from either a reservoir or a run-of-the-river. Water is used to construct fields, power plants, and occasionally cooling in geothermal plants. Geothermal fluid, mined from the earth, is used in all geothermal plants. This is frequently salt water and occasionally steam (Chamberlayne and Zare 2022; Lee et al. 2022). There are benefits to using geothermal energy directly in terms of money, the environment, and energy efficiency. These benefits include cheaper heating costs by lowering the use of electricity, oil, or gas; fewer CO_2 and sulfur; and fewer nitrogen oxide emissions by reducing the use of fossil fuels. Better resource management is achieved by reducing reliance on high-grade fuels (such as natural gas) for low-grade heating and through minimal ongoing expenses after installation (Benjamin et al. 2022).

12.3.2 Benefits of Using Geothermal Energy for Irrigation Water Management

(12.3.2.1) It is a renewable source of energy.

(12.3.2.2) Geothermal energy is a very environmentally beneficial form of renewable energy.

(12.3.2.3) No waste or by-products are created.

(12.3.2.4) Geothermal energy can be used directly. In the past, people used this energy source for cooking, home heating, and other purposes.

(12.3.2.5) When compared to fossil-fired power-producing facilities of equal capacity, geothermal power plants require far less maintenance (Zou et al. 2023).

(12.3.2.6) Geothermal plants help protect the environment because they don't require much room.

(12.3.2.7) Geothermal energy systems can be set up to function in various settings. They can be used to heat, cool, or particularly power homes, communities, or commercial structures and processes.

(12.3.2.8) Reliance on fossil fuels decreases as geothermal energy use rises. Due to the skyrocketing oil prices, many nations are putting pressure on businesses to use these clean energy sources. Global warming is brought on by greenhouse gases released during the combustion of fossil fuels.

(12.3.2.9) Unlike other renewable energy sources, geothermal energy systems are not weather dependent. They are not affected by changing conditions such as wind or sunlight and can operate in both summer and winter. Geothermal power plants provide heat or electricity 24 h a day, seven days a week (Chen et al. 2023).

12.4 Bioenergy for Irrigation System

BECCS, which stands for "bioenergy with carbon capture and storage," is one of the most significant detrimental emissions (NEs) technologies. By considering the extensive usage of irrigated biomass plantations, global warming by the end of the twenty-first century might be reduced to 1.5 °C, as compared to the 3 °C climate change scenario (Meng et al. 2022a). BECCS is based on the development of rapidly growing plant species, which are then harvested for their biomass and converted into biofuels (replacing liquid fossil fuels) or burned to produce energy (compensating for coal or gas power plants) while the released CO_2 is (at least partially) captured. The procedure would reduce anthropogenic greenhouse gas emissions by extracting CO_2 from the atmosphere. However, at the scale required, biomass production will probably be more detrimental to the local ecosystem and the environment as a whole.

For plants like these to achieve expected sequestration levels, irrigation may be required due to increased competition for limited freshwater resources. It is crucial to comprehend how increasing water usage for bioenergy is from a sustainability perspective. Bioenergy (BP) plantations can either be completely rainfed or partially irrigated. The first sort of plantation would only utilize "green" precipitation water stored in the soil. Nonetheless, the second sort of plant would need "blue" water from lakes, rivers, reservoirs, and aquifers to varying degrees. The world's bioenergy electricity generation capacity is expected to grow from over 50 GW in 2009 to 560 GW in 2050, with 50 GW fitted with carbon capture and storage (CCS) technology. The quantity of electricity produced by bioenergy will increase from 290 TWh in 2009 to 3 100 TWh in 2050, with approximately 300 TWh coming from facilities including carbon capture storage (Meng et al. 2022b).

A combined heat and power system was implemented, also known as cogeneration, in which the heat generated during the electricity generation process is captured and used to heat water or meet other thermal needs within the irrigation system. This improves the overall energy efficiency of the bioenergy system. Integrate the bioenergy system with the irrigation system, including automation and control systems to ensure smooth operation. This includes monitoring devices, sensors, and controllers to optimize energy generation and distribution based on irrigation demands. A bioenergy system reduces reliance on fossil fuels, assists in greenhouse gas reduction by capturing and utilizing methane emissions, and provides a self-sufficient energy solution for irrigation, which requires geospatial maps of crop residue biomass hosted in a geoportal with other thematic layers, as shown in Fig. 12.5.

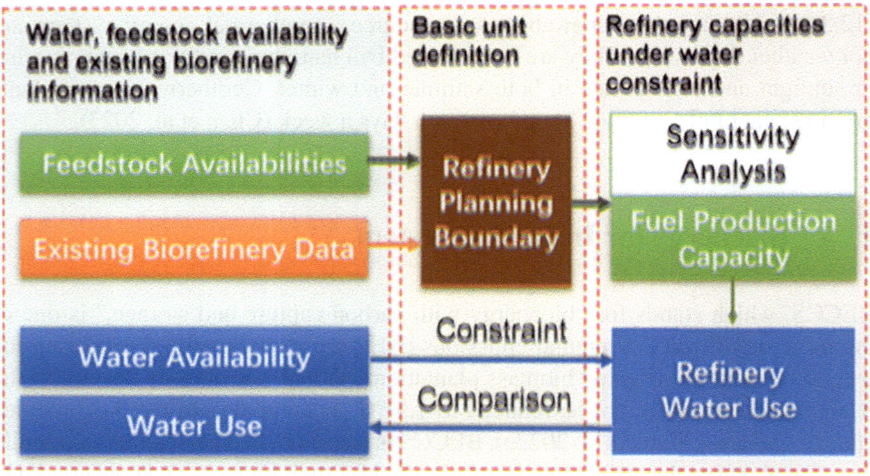

Fig. 12.5 The effects on India's distribution of bioenergy production (Chamberlayne and Zare 2022). (Reproduced with permission)

12.5 Hydrogen Energy

Hydrogen energy is included in this chapter as a widely used, non-toxic, clean fuel energy in water management and irrigation. As blue and green hydrogen is quickly being explored, are predicted to grow, and begin to replace more conventional energy sources, water use, demand, and management will receive increasing attention (Kalia et al. 2023). An essential component of the energy shift required to decarbonize our economy and protect the environment is the growth of a sustainable hydrogen sector. The management and use of water are crucial to understanding how climate change will affect communities and the environment, as well as how our economies and systems will function to support human civilization. It is essential to comprehend and adequately manage the interdependence of our water, food, and energy systems. Compared to pumped storage systems, the installation of the hydrogen production unit requires much less room, and no specific topographic parameters are needed. Despite having no emissions, hydrogen is not an environmentally friendly fuel because it is made from conventional fuels, which generate a lot of CO_2 emissions. Due to the intense demand placed on traditional water sources, communities in developing nations utilize rainwater, stormwater, brackish, and coastal. A new drive for addressing water supply issues might be generated by implementing green hydrogen production. Oversizing the water purification system at an electrolyzer facility will help flow clean drinking water. In general, supplying green hydrogen and water could aid global efforts to reduce carbon emissions and achieve public health objectives in many world regions (Cho et al. 2022). India's energy system is changing, and green hydrogen will be crucial to achieving energy independence and self-sufficiency for the nation.

A hydrogen-based energy molecule, which is totally "made in India," has the potential to increase the country's long-term economic competitiveness and energy security. India has a wonderful potential to capitalize on this ground-breaking technology and establish itself as a global leader in the production and use of ecologically benign hydrogen.

As India's energy system evolves, green hydrogen will be critical to the potential of the country to achieve energy independence and self-sufficiency. Indeed "Made in India," the energy molecule hydrogen has the potential to boost the country's long-term economic competitiveness and energy security. India can benefit from this new technology and establish itself as a world leader (Zheng et al. 2022). Green hydrogen production is far more expensive than hydrogen produced from conventional fuels. Even though there is still a lot of work to be done, green hydrogen will someday be economically viable due to declining costs for renewable energy sources and economies of scale. Hydrogen's sourcing and supply chain are more intricate when compared to oil and gas because it may be manufactured using various methods and is utilized in a wide range of industries. In addition, as was already indicated, moving and storing hydrogen still poses substantial challenges and will call for significant infrastructure upgrades.

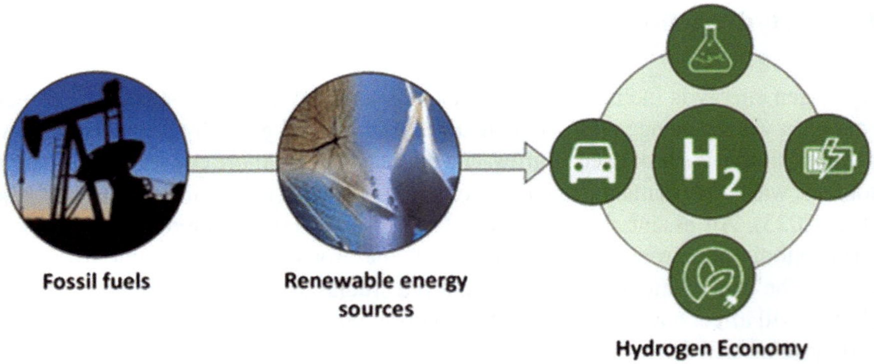

Fig. 12.6 Modern two-dimensional materials for green hydrogen generation for water management and crop cultivation (Wang et al. 2022). (Reproduced with permission)

Hydrogen production through water electrolysis can also support widespread renewable electricity generation and act as a form of energy storage. Additionally, the cost of producing hydrogen will decrease as renewable energy prices do, making hydrogen more competitive. Lastly, hydrogen can help the domestic labor market and reduce the nation's reliance on foreign oil (Dharamalingam et al. 2022). Additionally, it provides the opportunity to benefit economically from the ensuing global energy transition, as shown in Fig. 12.6.

12.6 Conclusion

It is now crucial that fuel and electricity be created in tandem since energy can be utilized in two different ways (as electrons or molecules). The amount of water needed for electrolysis would be equivalent to 1.8% of the world's water consumption if all existing fossil fuel uses were switched to green hydrogen. Saving water, avoiding the need to produce fuel from petroleum or biomass, and using fewer traditional thermal power plants would help meet this new demand. Additionally, the same water used during electrolysis is released back into the environment when combustion engines and fuel cells oxidize green hydrogen. For the promotion and display of SPIS technology, the Indian government and foreign donors are providing grants and subsidies. 2015 saw the installation of 50,000 solar-powered pumping systems. The government combines credit, technical assistance, and contributions to promote PV irrigation. A crucial finding is that farmers should ideally receive technical and agronomic assistance from a single source (one institution) to facilitate the adoption of PV-powered drip irrigation systems and better irrigation practices. All major components are now produced locally, creating jobs in a new industry, as the private sector of India has started offering SPIS components in recent years. Few manufacturers also provide turnkey solutions to farmers, which unquestionably increases the technology's overall performance and system effectiveness. Water and energy are related because both are needed for the extraction, filtration, distribution, and collection of water and its use afterward. Water is used in the production, transportation, and utilization of all forms of energy to some extent. The effectiveness of photovoltaic, wind, and biomass use, as well as solar thermal pumping systems, were the main subjects of the study. A review of this literature reveals the importance of renewable energy for small-scale irrigation, particularly in arid and semi-arid countries. Solar photovoltaic systems have been employed more frequently for small-scale irrigation than wind systems. Both biomass and solar thermal system pumping are still in their infancy. India can employ solar, biomass, geothermal, green hydrogen, and wind energy, thus it should focus more on irrigation using both technologies, particularly in dry and semi-arid regions, to enhance agricultural output. The most efficient way to use water is with an autonomous irrigation system, which reduces waste and farmer input. Green sources have major advantages in water management and irrigation. Its clean and sustainable nature, cost savings, adaptability for off-grid locations, aptitude for water pumping and treatment, climate change mitigation potential, and resilience make it an essential component of modern water management systems. Embracing renewable energy solutions, it can be from this chapter the efficient, sustainable, and ecologically friendly use of water resources in agricultural practices in the circular economy.

References

Ai J, Jin R, Liu Z, Jiang J, Yuan S, Huang G, Li N, Li X (2019) Three-dimensionally ordered macroporous FeP self-supported structure for high-efficiency hydrogen evolution reaction. Int J Hydrog Energy 44(12):5854–5862. Elsevier BV. https://doi.org/10.1016/j.ijhydene.2018.12.202

Alam N, Majumder S, Ray SJ, Sarma D (2022) A wide bandgap semiconducting magnesium hydrogel: moisture harvest, iodine sequestration, and resistive switching. Langmuir 38(34):10601–10610. American Chemical Society (ACS). https://doi.org/10.1021/acs.langmuir.2c01464

Aleid S, Wu M, Li R, Wang W, Zhang C, Zhang L, Wang P (2022) Salting-in effect of zwitterionic polymer hydrogel facilitates atmospheric water harvesting. ACS Mater Lett 4(3):511–520. American Chemical Society (ACS). https://doi.org/10.1021/acsmaterialslett.1c00723

Benjamin SE, LaVerne JA, Sigmon GE, Burns PC (2022) Ozone-facilitated formation of uranyl peroxide in humid conditions. Inorg Chem 61(51):20977–20985. American Chemical Society (ACS). https://doi.org/10.1021/acs.inorgchem.2c03454

Bose A, Lazouski N, Gala ML, Manthiram K, Mallapragada DS (2022) Spatial variation in cost of electricity-driven continuous ammonia production in the United States. ACS Sustain Chem Eng 10(24):7862–7872. American Chemical Society (ACS). https://doi.org/10.1021/acssuschemeng.1c08032

Chamberlayne CF, Zare RN (2022) Simple estimate of the potential drop across an amphiprotic liquid–liquid interface. J Phys Chem B 126(40):8112–8118. American Chemical Society (ACS). https://doi.org/10.1021/acs.jpcb.2c05696

Chen H, Wang R, Xu J, Yuan X, Zhang D, Zhu Z, Marshall M, Bowen K, Zhang X (2023) Spontaneous reduction by one-electron on water microdroplets facilitates direct carboxylation with

CO_2. J Am Chem Soc 145(4):2647–2652. American Chemical Society (ACS). https://doi.org/10.1021/jacs.2c12731

Cho E-C, Chang-Jian C-W, Huang J-H, Huang T-Y, Wu N-J, Li M-T, Chen Y-L, Hsu S-C, Weng HC, Lee K-C (2022) Preparation of Ni(OH)2/CuO heterostructures for improved photocatalytic degradation of organic pollutants and microorganisms. Chemosphere 300:134484. Elsevier BV. https://doi.org/10.1016/j.chemosphere.2022.134484

Deng F, Chen Z, Wang C, Xiang C, Poredoš P, Wang R (2022) Hygroscopic porous polymer for sorption-based atmospheric water harvesting. Adv Sci 9(33):2204724. Wiley. https://doi.org/10.1002/advs.202204724

Dey A, Abdul Moyez Sk, Kumar Mandal M, Roy S (2016) Fabrication of solarcellsusing extracted biomolecules from tea leaves and hybrid perovskites. Mater Today Proc 3(10):3498–3504. Elsevier BV. https://doi.org/10.1016/j.matpr.2016.10.033

Dharamalingam K, Ramasundaram S, Ponnusamy VK, Bhuvaneswari K, Ramalingam G, Balasankar A, Jeyaram S, Pazhanivel T, Florence SS, Thangavel E, Oh TH (2022) Facile synthesis of polymer-based magnesium hydroxide nanocomposites for photocatalytic degradation for methylene blue dye and antibacterial application. In: Biomass conversion and biorefinery. Springer Science and Business Media LLC. https://doi.org/10.1007/s13399-022-02770-0

Díaz-Marín CD, Zhang L, Lu Z, Alshrah M, Grossman JC, Wang EN (2022) Kinetics of sorption in hygroscopic hydrogels. Nano Lett 22(3):1100–1107. American Chemical Society (ACS). https://doi.org/10.1021/acs.nanolett.1c04216

Díaz-Marín CD, Zhang L, Fil BE, Lu Z, Alshrah M, Grossman JC, Wang EN (2022) Heat and mass transfer in hygroscopic hydrogels. Int J Heat Mass Transf 195:123103. Elsevier BV. https://doi.org/10.1016/j.ijheatmasstransfer.2022.123103

Ding Y, Wang C, Zheng R, Maitra S, Zhang G, Barakat T, Roy S, Su B-L, Chen L-H (2022) Three-dimensionally ordered macroporous materials for photo/electrocatalytic sustainable energy conversion, solar cells, and energy storage. EnergyChem 4(4):100081). Elsevier BV. https://doi.org/10.1016/j.enchem.2022.100081

Guo Y, Yu G (2022) Materials innovation for global water sustainability. ACS Materials Letters (Vol. 4, Issue 4, pp. 713–714). American Chemical Society (ACS). https://doi.org/10.1021/acsmaterialslett.2c00187

Heo SJ, Sanders M, O'Hayre R, Zakutayev A (2021) Double-site substitution of Ce into (Ba, Sr)MnO3 perovskites for solar thermochemical hydrogen production. ACS Energy Lett 6(9):3037–3043. American Chemical Society (ACS). https://doi.org/10.1021/acsenergylett.1c01214

Hou X, Yang H, Cao J, Feng W, Zhang Y (2023) A review of advances in groundwater evapotranspiration research. Water 15(5):969. MDPI AG. https://doi.org/10.3390/w15050969

Huang X, Tian F, Chen G, Wang F, Weng R, Xi B (2021) Preparation and characterization of regenerated cellulose membrane blended with ZrO2 nanoparticles. Membranes 12(1):42. MDPI AG. https://doi.org/10.3390/membranes12010042

Jamesh M-I, Moni P, Prakash AS, Harb M (2021) ORR/OER activity and zinc-air battery performance of various kinds of graphene-based air catalysts. Mater Sci Energy Technol 4:1–22. Elsevier BV. https://doi.org/10.1016/j.mset.2020.12.001

Jin H, Ma C, Wang W, Cai Y, Qi J, Wu T, Liao P, Li H, Zeng Q, Xie M, Huang J, Yan Y (2021) Using molecules with superior water-plasticity to build solid-phase molecular self-assembly: room-temperature engineering mendable and recyclable functional supramolecular plastics. ACS Mater Lett 4(1):145–152. American Chemical Society (ACS). https://doi.org/10.1021/acsmaterialslett.1c00724

Kalia R, Pirzada BM, Kunchala RK, Naidu BS (2023) Noble metal-free efficient photocatalytic hydrogen generation by TaON/CdS semiconductor nanocomposites under natural sunlight. Int J Hydrog Energy. Elsevier BV. https://doi.org/10.1016/j.ijhydene.2023.01.126

Lee J-Y, Aviso KB, Tan RR (2018) Optimal sizing and design of hybrid power systems. ACS Sustain Chem Eng 6(2):2482–2490). American Chemical Society (ACS). https://doi.org/10.1021/acssuschemeng.7b03928

Lee K, Lee H-R, Kim YH, Park J, Cho S, Li S, Seo M, Choi SQ (2022) Microdroplet-mediated radical polymerization. ACS Centl Sci 8(9):1265–1271). American Chemical Society (ACS). https://doi.org/10.1021/acscentsci.2c00694

Lei C, Guo Y, Guan W, Yu G (2021) Polymeric materials for solar water purification. J Polym Sci 59(24):3084–3099. Wiley. https://doi.org/10.1002/pol.20210688

Li H, Yang H, Lv W, Liu X, Bai S, Li L, Zhao S, Wang Y, Guo X (2021) Hygroscopic hydrogels for removal of trace water from liquid fuels. Ind Eng Chem Res 60(47):17065–17071. American Chemical Society (ACS). https://doi.org/10.1021/acs.iecr.1c03438

Li Z, Xu X, Sheng X, Lin P, Tang J, Pan L, Kaneti YV, Yang T, Yamauchi Y (2021) Solar-powered sustainable water production: state-of-the-art technologies for sunlight–energy–water nexus. ACS Nano 15(8):12535–12566. American Chemical Society (ACS). https://doi.org/10.1021/acsnano.1c01590

Li A, Xiong J, Liu Y, Wang L, Qin X, Yu J (2022) A rapid-Ab/desorption and portable photothermal MIL-101(Cr) nanofibrous composite membrane fabricated by spray-electrospinning for atmosphere water harvesting. Energy Environ Mater 6(1). Wiley. https://doi.org/10.1002/eem2.12254

Li R, Wu M, Aleid S, Zhang C, Wang W, Wang P (2022) An integrated solar-driven system produces electricity with fresh water and crops in arid regions. Cell Rep Phys Sci 3(3):100781. Elsevier BV. https://doi.org/10.1016/j.xcrp.2022.100781

Lokesh S, Srivastava R (2022) Advanced two-dimensional materials for green hydrogen generation: strategies toward corrosion resistance seawater electrolysis–review and future perspectives. Energy Fuels 36(22):13417–13450. American Chemical Society (ACS). https://doi.org/10.1021/acs.energyfuels.2c02013

Loo S, Vásquez L, Athanassiou A, Fragouli D (2021) Polymeric hydrogels—a promising platform in enhancing water security for a sustainable future. Adv Mater Interfaces 8(24):2100580. Wiley. https://doi.org/10.1002/admi.202100580

Lu K, Liu C, Liu J, He Y, Tian X, Liu Z, Cao Y, Shen Y, Huang W, Zhang K (2022) Hierarchical natural pollen cell-derived composite sorbents for efficient atmospheric water harvesting. ACS Appl Mater Interfaces 14(29):33032–33040). American Chemical Society (ACS). https://doi.org/10.1021/acsami.2c04845

Maitra S, Sarkar A, Maitra T, Halder S, Kargupta K, Roy S (2021) Solvothermal phase change induced morphology transformation in $CdS/CoFe_2O_4@Fe_2O_3$ hierarchical nanosphere arrays as ternary heterojunction photoanodes for solar water splitting. New J Chem 45(28):12721–12737. Royal Society of Chemistry (RSC). https://doi.org/10.1039/d1nj00864a

Meng Y, Gnanamani E, Zare RN (2022) Catalyst-free decarboxylative amination of carboxylic acids in water microdroplets. J Am Chem Soc 145(1):32–36. American Chemical Society (ACS). https://doi.org/10.1021/jacs.2c12236

Meng Y, Gnanamani E, Zare RN (2022) Direct C(sp3)–N bond formation between toluene and amine in water microdroplets. J Am Chem Soc 144(43):19709–19713. American Chemical Society (ACS). https://doi.org/10.1021/jacs.2c10032

Peeters R, Vanderschaeghe H, Rongé J, Martens JA (2021) Fresh water production from atmospheric air: technology and innovation outlook. iScience 24(11):103266. Elsevier BV. https://doi.org/10.1016/j.isci.2021.103266

Qin Y, Kang G, Cao Y (2021) Finely tuned polyamide structure with green plasticizers to construct ultrafast water channels for effective desalination. Sci Total Environ 784:147089. Elsevier BV. https://doi.org/10.1016/j.scitotenv.2021.147089

Shi W, Guan W, Lei C, Yu G (2022) Sorbents for atmospheric water harvesting: from design principles to applications. Angew Chem Int Ed 61(43). Wiley. https://doi.org/10.1002/anie.202211267

Wang W, Wu C, Zhu K, Chen F, Zhou J, Shi Y, Zhang C, Li R, Wu M, Zhuo S, Zhang H, Wang P (2021) Real-time personal fever alert monitoring by wearable detector based on thermoresponsive hydrogel. ACS Appl Polym Mater 3(4):1747–1755). American Chemical Society (ACS). https://doi.org/10.1021/acsapm.0c01275

Wang B, Zhou X, Guo Z, Liu W (2021) Recent advances in atmosphere water harvesting: Design principle, materials, devices, and applications. Nano Today 40:101283. Elsevier BV. https://doi.org/10.1016/j.nantod.2021.101283

Wang W, Xie S, Pan Q, Dai Y, Wang R, Ge T (2021) Air-cooled adsorption-based device for harvesting water from island air. Renew Sustain Energy Rev 141:110802. Elsevier BV. https://doi.org/10.1016/j.rser.2021.110802

Wang M, Sun T, Wan D, Dai M, Ling S, Wang J, Liu Y, Fang Y, Xu S, Yeo J, Yu H, Liu S, Wang Q, Li J, Yang Y, Fan Z, Chen W (2021) Solar-powered nanostructured biopolymer hygroscopic aerogels for atmospheric water harvesting. Nano Energy 80:105569. Elsevier BV. https://doi.org/10.1016/j.nanoen.2020.105569

Wang H, Shao B, Chi Y, Lv S, Wang C, Li B, Li H, Li Y, Yang X (2022) Engineering of Ni(OH)2 modified two-dimensional ZnIn2S4 heterostructure for boosting hydrogen evolution under visible light illumination. Nanomaterials 12(6):946. MDPI AG. https://doi.org/10.3390/nano12060946

Winter LR, Cooper NJ, Lee B, Patel SK, Wang L, Elimelech M (2022) Mining nontraditional water sources for a distributed hydrogen economy. Environ Sci Technol 56(15):10577–10585. American Chemical Society (ACS). https://doi.org/10.1021/acs.est.2c02439

Yu Z, Li S, Liu M, Zhu R, Yu M, Dong X, Sun Y, Fu S (2022) A dual-biomimetic knitted fabric with a manipulable structure and wettability for highly efficient fog harvesting. J Mater Chem A 10(1):304–312. Royal Society of Chemistry (RSC). https://doi.org/10.1039/d1ta08295g

Zhang J, Sun Y, Yang J, Jiang T, Tang W, Chen B, Wang ZL (2021) Irregular wind energy harvesting by a turbine vent triboelectric nanogenerator and its application in a self-powered on-site industrial monitoring system. ACS Appl Mater Interfaces 13(46):55136–55144. American Chemical Society (ACS). https://doi.org/10.1021/acsami.1c16680

Zhang M, Liu R, Li Y (2022) Diversifying water sources with atmospheric water harvesting to enhance water supply resilience. Sustainability 14(13):7783. MDPI AG. https://doi.org/10.3390/su14137783

Zhao X, Wang T, Wang R, Lan Y, Jiang Y, Pan J (2022) Super wetting photothermal membranes enabled by polyphenol-mediated nanostructured coating with raspberry-like architectures for solar-driven interfacial evaporation. Desalination 542:116046. Elsevier BV. https://doi.org/10.1016/j.desal.2022.116046

Zheng D, Chen M, Chen Y, Gao W (2022) In-situ preparation of hollow CdCoS2 heterojunction with enhanced photocurrent response for highly photoelectrochemical sensing of organophosphorus pesticides. Analytica Chimica Acta 1212:339913. Elsevier BV. https://doi.org/10.1016/j.aca.2022.339913

Zhou X, Zhao F, Zhang P, Yu G (2021) Solar water evaporation toward water purification and beyond. ACS Mater Lett 3(8):1112–1129. American Chemical Society (ACS). https://doi.org/10.1021/acsmaterialslett.1c00304

Zou H, Shi H, Hao S, Hao Y, Yang J, Tian X, Yang H (2023) Boosting catalytic selectivity through a precise spatial control of catalysts at pickering droplet interfaces. J Am Chem Soc 145(4):2511–2522. American Chemical Society (ACS). https://doi.org/10.1021/jacs.2c12120

Chapter 13
Securing and Preserving the Support of the Community for the Implementation of the Continuous Water Supply System in India

Dhayanand Jayaseelan, Gajendran Chellaiah, Amit Neogi, Nicolas Bockhoff, P. P. Sivan, Adwaitha Sunilkumar, and Athulya Thankachan

Abstract The Indian water supply system refers to the infrastructure and systems in place for the distribution and management of water in India. This includes dams, canals, water treatment plants, and distribution networks that supply water to households, businesses, and agricultural regions. In India, the water supply system comprises not only the physical infrastructure but also the governmental policies and regulations that oversee the utilization and conservation of water resources. The governing body or organization responsible for managing the water supply system

D. Jayaseelan
Department of Civil Engineering, KITS, Coimbatore, Tamil Nadu, India

G. Chellaiah (✉)
Department of Civil Engineering, SNS College of Technology, Coimbatore, Tamil Nadu, India
e-mail: cgajendran@gmail.com

A. Neogi · N. Bockhoff
Suez India Pvt, Ltd, Coimbatore, Tamil Nadu, India
e-mail: Amit.neogi@suez.com

N. Bockhoff
e-mail: nicolas.bockhoff@suez.com

P. P. Sivan
Department of Civil Engineering, Government Engineering College, Thrissur, Kerala, India

A. Sunilkumar
Environment Division, Gujarat Environmental Institute of Management, Gujarat, India
e-mail: adwaithasunilkumar@karunya.edu.in

A. Thankachan
Department of Agricultural Engineering, Sri Shakthi Institute of Engineering and Technology, Coimbatore, India
e-mail: athulyathankachan@karunya.edu.in

S. Suriyanarayanan et al. (eds.), *Water Management in Developing Countries and Sustainable Development*, Water Resources Development and Management,
https://doi.org/10.1007/978-981-99-8639-2_13

is striving to earn and uphold the community's backing for their endeavors to establish and sustain a dependable and uninterrupted water supply system in the country. This includes making sure that the community is aware of the changes and benefits that the system will bring, and addressing any concerns or issues that may arise. This paper examines the water supply system adopted in India with its features in terms of technical and socioeconomic. Intermittent water supply can be challenging for households and communities as it can make it difficult to plan for and use water effectively, while continuous water supply is considered more desirable as it provides more flexibility and convenience. The primary objective of transitioning to a continuous water supply system is to enhance the availability of safe and consistent drinking water to every individual residing within the region.

Keywords Water supply system · Intermittent water supply · Continuous water supply · Water management · Community support

13.1 Introduction

A water supply system is a network of infrastructure and processes designed to provide safe and reliable access to clean water for human consumption, sanitation, and other purposes. The system includes the collection, treatment, storage, and distribution of water, as well as the management of wastewater and sewage. In general, a water supply system can be divided into two main parts: the source and the distribution network. The source can be a natural water body such as a river or lake, or it can be a man-made source such as a dam or reservoir. Once the water has been collected, it is treated to remove contaminants and impurities, such as bacteria, viruses, and chemicals. After treatment, the water is stored in tanks or reservoirs before being distributed through a network of pipes to homes, businesses, and other facilities. The distribution network includes pumps, valves, and other infrastructure that ensure the water is delivered at a suitable pressure and volume. In addition to the physical infrastructure, a water supply system also requires effective management and governance to ensure its sustainability and resilience. This includes monitoring water quality, managing demand, and maintaining the infrastructure to prevent leaks and other losses. A water distribution system is responsible for bringing treated water from the treatment plant to the consumers, and also provides storage for the same. It also helps in maintaining appropriate flow and pressure levels for firefighting purposes. The distribution system needs to meet fire suppression standards, which leads to increased infrastructure, like a bigger pipe network, pump station, and storage reservoir capacity. The water supply system is a critical infrastructure for any community or area, as it provides a vital resource for drinking, cooking, cleaning, and many other essential activities. It is also important for fire protection and industrial processes. The system must be designed and maintained to ensure that it can withstand natural disasters, such as floods and earthquakes, and that it can meet the demands of the community or area. In most places, the water needed to fight fires is higher compared to regular home and commercial usage. Components in the water supply system include the source of water which is either from the treatment plant or the reservoir, transportation, storage, and distribution to the consumer. Water distribution refers to the process of delivering water from a source, such as a treatment plant or reservoir, to the point of use, such as homes and businesses. The distribution system is a network of pipes, valves, pumps, and storage tanks that transport and distribute water (Ray et al. 2018). The system must be designed to ensure that water is delivered with adequate pressure and flow, and that it meets quality standards. In urban areas, water distribution systems are typically managed by a public utility or private company, while in rural areas, distribution may be managed by local governments or community organizations (Kalbar and Gokhale 2019). The distribution systems are also monitored for leaks and breaks and repaired as necessary to minimize loss. The water supply system is typically managed and operated by a local government agency or private company. They are in control of ensuring safe water to drink, and that it is available to the community or area at all times. They also work to improve the system by upgrading and expanding it as needed, as well as implementing conservation measures to reduce waste and improve efficiency. The water distribution system is a major component of the overall water supply system, as it ensures that water is delivered to the desired location and in the desired quantity. Some water supply systems use the energy of falling water to generate electricity, which can be used to power pumps and other equipment (Andey and Kelkar 2009). The water supply system may also include facilities to treat and discharge wastewater, which can help to protect the environment and prevent water pollution. These structures can be used to store water and regulate its flow (Ghorpade et al. 2021). Aqueducts or canals are used to transport water over long distances. Many water supply systems include metering equipment to measure the amount of water used by each customer, as well as billing systems to charge customers for the water they use (Tabesh et al. 2008). Water supply systems often include emergency response plans and equipment to quickly address any problems that may arise, such as leaks or contamination. Overall, a water supply system is a complex network that requires careful planning, design, and management to ensure that it provides a reliable and safe source of water to all of its users.

13.2 Evolution of Indigenous Water Management

With the emergence of human settlements around 6000 years ago, people have been facing the dual challenge of managing water (Anil and Narain 1997). One part of this struggle involved protecting against floods while the other involved ensuring a reliable and safe water supply for domestic use and irrigation. As a result, hydraulic structures, such as dams, canals, and aqueducts, became some of the first technological innovations created by humans.

The practice of storing water in cisterns predates the construction of the first cities. The ancient cisterns discovered are from Greece and Palestine, which were used to collect precipitation from rooftops, paved areas, and underground water sources. The ancient cisterns were carved into rock, but later on, masonry cisterns were developed (Anon 2006). Around 2000 BC, mortar was used to seal cisterns. In the mid-first century AD, cisterns with large storage capacities of up to 75,000 cubic meters (cum) were built, and all the key principles of dam construction were already known and applied.

The oldest known dams built to hold huge quantities of water were built in Wadi Garawi, Egypt around 2600 BC, and Jawa, Jordan around 3000 BC. However, it is not well understood where the science of dam construction originated, and there is little evidence to suggest that it was developed in one specific place and then spread to other regions. Many ancient civilizations and states around the world built dams, each contributing to the collective human knowledge and experience in this field (Anon 2006). According to Gunther Gfarbrecht, a water management specialist from Braunschweig Technical University in Germany, there is no evidence to suggest that ancient dams were constructed based on mathematical and geometrical calculations. Instead, it is likely that they were built through a combination of practical experience, technical knowledge, and an intuitive comprehension of hydrodynamic forces.

The four oldest civilizations, **Mesopotamia, Egypt, the Indus Valley, and China,** were able to thrive due to their sophisticated water management systems (Anil and Narain 1997). The Indus Valley Civilization (30,001,500 BC), which was discovered through the excavations of Mohenjo-daro and Harappa in the early 1920s, covered a vast area. The Indus–Mesopotamia civilizations had close trade relations and greatly influenced each other. Both of these ancient cultures were known for their advanced water supply and sewage systems (Anon 2009). Even 2,000 years later, their hydraulic structures were not surpassed by the Romans, whose water supply systems are magnificent. Similarly to Mesopotamia, protection against annual floods from the Indus River, a good irrigation system to increase crop yield, and to drain large alluvials were essential to the kingdoms in the Indus Valley for their survival.

The earliest texts of Hinduism, written around 800–600 BC, tells about the people's relationship with hydrological concern (Anon 2004). The Vedic hymns, that of those in the Rig Veda, contain information about irrigation agriculture, river courses, dykes, water reservoirs, wells, and water lifting structures. The connection between the environment and water systems was understood even in ancient times (Anon 2004). For example, the Greek philosopher Plato (427–347 BC) wrote a detailed description of erosion he observed in Attica, Greece, demonstrating an understanding of the relationship between the environment and water systems.

13.3 Indian Water Knowledge System

Over the centuries, Indians have developed a wide range of techniques to harvest every possible form of water, depending on the resources available to them. This includes rainwater, groundwater, stream and river water, and floodwater (Anon 2006). In areas with streams, particularly in hilly and mountainous regions, individuals would employ straightforward engineering methods to redirect water from the main watercourse for agricultural use. They would construct artificial channels to convey the water directly to the fields. As streams grew into rivers, more extensive and advanced diversion systems were devised with the assistance of engineering techniques. A notable illustration of such a structure is The Grand Anicut on the Cauvery River, showcasing intricate engineering methods.

In floodplains, people developed clever techniques to not only use the floodwater to irrigate their fields but also to fertilize them (Iyegar 2007). They also used this floodwater to control diseases such as malaria by making use of fish present in the water to consume mosquito larvae. This method of utilizing flooding as a double-edged sword is one of many examples of the resourcefulness and ingenuity of the people in the history of water management, throughout the world.

In coastal areas, where the periodic influx of saltwater from tides would make river water unsuitable for agriculture, people developed innovative solutions. One example is the Khazana lands of Goa, where systems were developed to control the flow of high salinity river water, thus controlling its effect on the rice agroecosystems and long-term soil fertility (Iyegar 2007). These types of solutions are an example of how human ingenuity and technology have been used to overcome the obstacles presented by the environment and to sustain agriculture.

In areas with a good aquifer, Indians used dug wells to harvest rainwater and developed various methods using locally available materials to get water from the ground, which is used for irrigation. Wells were considered as a primary source of irrigation where the regions had plenty of groundwater like the Indo-Gangetic plains, for example. In areas where water was scarce, people also developed other ways to harvest groundwater (Iyegar 2007). In arid areas such as Rajasthan, people constructed various water storage structures such as step wells and wells beneath tanks. This served as a solution for times when the tank water evaporated rapidly during the dry season or when it wasn't sufficient to meet the local water needs. By utilizing the wells and step wells that contained clean groundwater, people were able to access drinking water. In fact, in many locations, the clean groundwater from these wells and step wells was the preferred source of drinking water. In the Eastern Ghats region, people employed a technique called "Qanats," borrowed from the Middle East, to create underground structures or horizontal wells known as "Surangams." These structures were designed to capture water flowing down from the hilly region, serving as a reliable source of drinking water. When no other alternatives were available, people relied on rainwater for their survival. To provide irrigation water, they used to build rain-fed tanks in downstream command areas or cultivate the tank bed itself (Anil and Narain 2001). In the haveli system of Madhya Pradesh, farmers discovered the benefits of storing rainwater in their agricultural fields, thanks to the suitable soil conditions and traditional crops. In certain areas, individuals built embankments to capture monsoon runoff from specific catchment areas, effectively collecting water in a storage structure. The stored water would gradually seep into the soil, providing adequate moisture for successful crop growth during subsequent dry periods. People adapted their practices to the local ecology, utilizing a diverse range of techniques to sustain their agriculture and livelihoods.

But in very dry areas such as deserts, the primary need of the people was to obtain drinking water where there will be no streams and less or no rainfall. In India places like Rajasthan, for example, people used to collect rainwater using rooftops as catchment areas. If the area is larger in size, people were used to develop rainwater harvesting structures called "Kundis" to get drinking water (Anil and Narain 2001). Kundis are man-made wells designed to collect and store rainwater by utilizing a specially constructed catchment area surrounding the well. When precipitation falls on the catchment, it flows into the well and is stored there. In the Thar Desert of Rajasthan, Kundis are the prevailing structures used to retain water, showcasing that rainwater harvesting is feasible in any terrain and serves as an excellent method for obtaining drinking water.

The nomadic Malharis of the inhospitable Kutch region of Gujarat have developed an interesting system for obtaining potable fresh water, even in drought-prone areas where rainwater is scarce and groundwater is salty. They have learned that the density of freshwater is less than that of salt water, and therefore, it is theoretically possible to keep the harvested freshwater stored in a way that it will float on top of the saltwater, providing them with an opportunity to live and survive in such a challenging region. They have, in fact, through their own experience and knowledge, developed a system locally known as "virda" (Anil and Narain 2001). This system is a remarkable example of human ingenuity in the face of scarce resources, and their ability to make use of the properties of water to ensure their survival in harsh environments.

For transporting water across a terrain, the people living in the Northeast region of India are traditionally skilled with the effective usage of bamboo. They have developed some of the most interesting and artistic systems using bamboo sticks. In the eastern Himalayas and the northeastern hill ranges, individuals constructed basic bamboo pipelines to transport water from natural springs to easily accessible points for drinking purposes. The people residing in southern Meghalaya, near the Bangladesh border, have elevated this practice to a more sophisticated level. They have developed intricate, interconnected networks of bamboo pipelines to supply water to betel leaf plantations situated in rocky terrains where conventional channels cannot be established (Anil and Narain 2001). The entire mechanism functions akin to a contemporary drip irrigation system, delivering water directly to the plant roots. Throughout history, people have continually devised innovative methods to harvest and distribute water for irrigation and everyday use, whether it is rainwater, stream water, floodwater, or groundwater.

13.4 Piped Water Supply System in India

In India, the government has been actively working toward addressing water supply challenges and ensuring access to piped water for the entire population. The central government, in addition to state initiatives, provides support through various programs like the Jal Jeevan Mission, Atal Mission for Rejuvenation and Urban

Transformation, and Jal Shakti Abhiyan. These programs aim to provide tap water connections to rural and urban households and promote rainwater conservation.

As of March 2023, approximately 60% of rural households in India, out of a total of 194.26 million, have been connected to tap water. However, there are variations among states and union territories, with some achieving 100% coverage, including Goa, Andaman and Nicobar Islands, Dadra and Nagar Haveli, Daman and Diu, Haryana, Gujarat, Puducherry, Punjab, and Telangana. On the other hand, nine states, such as Madhya Pradesh, Kerala, Meghalaya, Assam, Chhattisgarh, Rajasthan, Uttar Pradesh, Jharkhand, and West Bengal, have less than 50% of rural households connected to tap water.

In urban areas of India, the situation appears relatively better, with over 80% of urban households having access to water facilities within their premises. However, there are significant inequalities in water supply, particularly within cities and among different urban areas. Unplanned developments, including slums and unauthorized colonies, often lack access to piped water (data from the Ministry of Jal Shakti, India).

While efforts are being made to expand piped water networks to cover the remaining rural households, urban areas, especially densely populated cities and mega-urban regions, are facing a water crisis. With rapid urbanization, water supply agencies are struggling to meet the increasing water demands of the growing population.

13.5 Intermittent Water Supply

In an Intermittent water supply system, water is supplied to the consumer's tap at specific intervals, either for a fixed time per day or on alternate days. In this type of system, the entire community is divided into different zones, and water is supplied to each zone for a fixed time per day or on alternate days (Ghorpade et al. 2021). This method can help to ensure that the water supply can meet the demand, especially in communities where the water supply is limited and there are many users. However, it can also have its drawbacks as it can create inconvenience for the consumer and could lead to water wastage as people might store water for future use. Intermittent Water Supply (IWS) is prevalent in most developing countries.

The types of supply patterns of IWS systems can be categorized as:

- Variable time—the supply time is not known, but the volume of water is known
- Fixed time—the supply time and volume of water are known
- Seasonal—during dry seasons
- Unreliable—the supply time and volume of water are not known

13.5.1 *Intermittent Water Supply: An Inconvenience?*

Cities often face water shortages due to mismanagement of available resources (Andey and Kelkar 2007). Leakage loss is a major contributor to this problem, with metropolitan cities typically experiencing a leakage loss of 30–40%, with some cities reporting leakage losses as high as 44% (Ghorpade et al. 2021). This leads to a reduction in per capita availability and limited hours of supply, often restricted to only a few hours per day. As a result, citizens must be aware of and plan for the limited water supply, often fetching water during the early morning or late at night. Leakage loss can be caused by various factors like corrosion, damage to pipes, faulty valves, and illegal connections, and can be addressed through regular maintenance, upgrading of infrastructure, and monitoring.

The quality of water services in Indian cities has declined over time due to population growth and the structural, managerial, and technological shortcomings of the current supply system. A significant amount of water is wasted and lost as a result of "unaccounted for" losses caused by poor maintenance and operation, exacerbating the already problematic situation of home water supply. In almost all of the nation's cities, distribution losses are believed to be in the range of 30–50% (Ray et al. 2018). In many urban areas, there is a prevalent inability to detect and address both physical and financial leaks. The situation is exacerbated by the absence of reliable metered connections in cities, making it challenging to identify leaks accurately. Aside from losses within the system, non-revenue water (NRW) is also attributed to metering errors, unaccounted consumption, unauthorized tapping, and unauthorized usage. In India, the average NRW is estimated to be around 45 percent, which represents a significantly high figure.

Acquiring data on unaccounted-for water (UFW) can be challenging, as it is typically estimated by technical personnel based on their subjective assessment of the situation. In numerous cities, the absence of bulk meters or meters at the end-user's connection further complicates the calculation of UFW. Consequently, the figures relating to UFW should be regarded as the most accurate estimates provided by the technical staff of the water-supply agency (Farmani et al. 2021). Smaller towns that supply water from nearby sources or use groundwater sources have reported very small quantities of UFW, indicating that larger cities have greater quantities of UFW compared to smaller cities (Klingel 2012). This can be due to the fact that smaller towns have better infrastructure and management, thus fewer losses, whereas larger cities have more complex systems and more challenges to tackle like population density, encroachment of water bodies, illegal tapping, etc.

Intermittent water supply systems may encounter different types of losses, such as physical losses caused by leaks in the distribution network, operational losses due to insufficient maintenance or ineffective management, and commercial losses resulting from water theft or unauthorized connections. Furthermore, intermittent water supply can lead to low water pressure and compromised water quality, thereby increasing expenses associated with treatment and distribution. To address these losses, it is important to regularly inspect and maintain the distribution network, implement effective management and billing systems, and enforce regulations to prevent water theft (Nauges and Thomas 2003). Another major loss in intermittent water supply systems is the loss of water through evaporation and seepage. This can occur when water is stored in open reservoirs or tanks and is exposed to the elements. To minimize this type of loss, it is important to use covered storage tanks and to regularly inspect and repair any leaks or cracks in the tanks (Tabesh et al. 2008). Intermittent water supply can also lead to increased costs for customers, as they may need to purchase water from alternative sources during periods of low supply (Babel et al. 2007). Additionally, the constant need to turn the water supply on and off can cause damage to household plumbing and appliances. To minimize losses and improve the overall efficiency of an intermittent water supply system, it is important to conduct regular audits and assessments, implement effective water conservation and demand management programs, and invest in upgrading and expanding the distribution network (Totsuka et al. 2004). It is also important to involve the community in the management and maintenance of the water supply system. Community-based organizations can help to identify and report leaks, unauthorized connections, and other issues, and can also assist in the development and implementation of water conservation programs (Fig. 13.1).

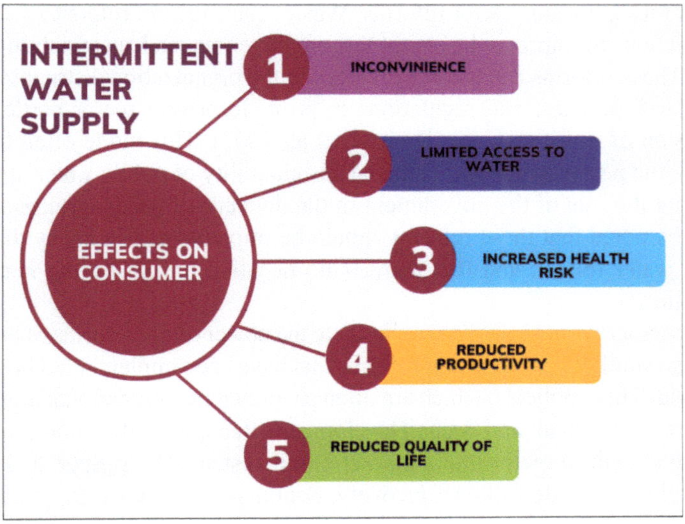

Fig. 13.1 Effects on consumers due to IWS

13.6 Participation of Private Sector

The Indian government has acknowledged the significance of involving the private sector in addressing the water challenges faced by the country. This recognition is evident in the National Water Policy, which was adopted on April 1, 2002, and emphasizes the encouragement of private sector participation "wherever feasible." Various state governments have also incorporated this policy into their own frameworks. By promoting private sector involvement, the government aims to leverage the technical and financial resources necessary for expanding, rehabilitating, operating, and maintaining water supply systems. This approach can introduce new technologies and management practices to enhance the efficiency and sustainability of such systems. Nonetheless, it is crucial to ensure that private sector participation does not compromise the accessibility and affordability of water for vulnerable communities. To prevent any exploitation, adequate regulatory mechanisms should be established.

In India, privatization in the water sector is taking place through two different modes (Dwivedi et al. 2007). The first mode is through outright privatization of water resources, either through management contracts or build-operate-transfer (BOT) projects. This mode is primarily used for industrial water supply and urban water supply projects. The second mode, which is more subtle and has a more far-reaching impact, is through water sector reforms. Water sector reform refers to a set of policies and actions to improve the efficiency, effectiveness, and sustainability of water services. These reforms can include changes to the organizational structure of water utilities, tariff policies, and regulations to promote private sector participation in the provision of water services (Dwivedi et al. 2007). This mode often focuses on improving the performance and financial sustainability of public water utilities, and on reducing the role of the government in the delivery of water services. However, it should be noted that these reforms should be implemented in a way that ensures access to water for all, and that there is no negative impact on the marginalized communities.

Water sector reforms in India are similar to the power sector reforms of the country, and are also similar to water sector reforms that have been implemented in other parts of the world. These policies, which are often promoted by international organizations such as the World Bank and Asian Development Bank, have the underlying goal of converting the entire sector into a market-based system (Dwivedi et al. 2007). The typical policies include full cost recovery, elimination of subsidies, public–private partnerships, and allocation of water to the highest value through market mechanisms. These policies are often justified on the grounds that they are necessary to improve the efficiency of the sector, and that subsidies ultimately hurt the poor. However, it's important to note that these policies should be implemented in a way that balances the needs of the poor and vulnerable communities with the need to improve efficiency and sustainability. It should be done in a way that ensures access to water for all and that the poor are not adversely impacted by these reforms.

Water sector reforms in India are being primarily promoted through World Bank loans to various Water Sector Restructuring (WSR) projects. In India, water is a state subject, and the state governments play a significant role in shaping policy, projects, and programs related to water. Several states have already implemented WSR projects, and more are expected in the future (Dwivedi et al. 2007). Madhya Pradesh is one such example where the project is planned to be sanctioned in December 2003. One of the important features of these loans is the promotion of privatization through so-called "Public–Private Partnerships" (PPPs). The corporate sector is an important player in this, with several major companies such as Suez, Vivendi, Thames Water, and Bechtel present in India. These companies, as well as other private players, are expected to participate in the provision of water services through PPPs, but it is important that the implementation of these policies should be done in a way that is transparent, fair, and protects the rights of the consumers.

13.6.1 Key Characteristics of the Type of Organizational Structure/model

Build-Own-Operate (BOO)—The Visakhapatnam Industrial Water Supply Project is a private sector project that aims to supply 300 million liters per day (MLD) of water to industries located near Visakhapatnam, Andhra Pradesh. The project will utilize the existing 153-km-long Yeleru Left Bank Canal (YLBC) system. The project is implemented by the Vizag Industrial Water Supply Company (VIWSCo), which is a private company. The Visakhapatnam Industrial Water Supply Project and Co. will obtain the required concession rights from the Government of Andhra Pradesh to build, operate, and maintain the project's infrastructure. This includes responsibilities such as managing billing and collecting payments for water supply and wastewater disposal services offered to different customers. The company will finance the project by recuperating its investment and operational expenses through the sale of water to consumers. This approach aims to ensure that the water supply remains sustainable, efficient, and affordable for consumers.

Lease/Management Contract—A management contract is a type of Public–Private Partnership (PPP) in which a private company is contracted by a public entity, such as a civic authority, to manage and operate a public facility, such as a water supply or sewage treatment system. In a management contract, the ownership of the facility remains with the public entity, but the private company is responsible for the day-to-day operations, maintenance, and management of the facility (Dwivedi et al. 2007). They are also responsible for ensuring that the facility is operated efficiently, cost-effectively, and in compliance with the regulations and standards set by the public entity. However, the private company is not responsible for making new investments or expansions of the facility.

An example of this type of PPP is the Delhi water supply and sewerage project, which was managed by the Delhi Jal Board (Dwivedi et al. 2007). The board had invited pre-qualification bids for management contracts for water supply and sanitation in operation zones South II and III on 12 February 2005. The population served in these zones is respectively 8,00,000 and 6,00,000. The Delhi Jal Board had shortlisted four water companies for management contracts—Suez, Saur, Bechtel, and Veolia. Nevertheless, the project encountered widespread demonstrations by individuals and non-governmental organizations (NGOs) who opposed the participation of private corporations in the administration of a public facility. As a result, the Delhi government decided to retract its loan applications from the World Bank Group, effectively terminating the project.

DCCOM—Develop Construct Commission Operate and Maintain—The Gujarat Infrastructure Development Board (GIDB) is inviting proposals for the construction of a large-scale seawater desalination plant. The plant will be capable of producing 100–150 million liters of water per day and will be built to serve the water needs of industries located in the Kutch region (Dwivedi et al. 2007). The project will be executed through the Build-Own-Operate-Transfer (BOOT) model, where a private company will be responsible for the construction, ownership, operation, and maintenance of the plant, and will transfer the ownership back to the government after the contract period ends.

EOI—Expression of Interest—The Department of Industries in the state of Jharkhand, India has plans for a Public–Private Partnership (PPP) to improve the water supply and sanitation systems in Adityapur. The project includes the rehabilitation and augmentation of the existing systems, as well as the operation and maintenance of the water supply and sanitation infrastructure (Dwivedi et al. 2007). The operation of the system, including tasks such as system improvements, minimizing water losses, enhancing metering and billing systems, collecting payments, and managing consumer relations, will be entrusted to a private company. Additionally, the operator may be required to support the implementation of significant capital investments and make their own investments. The project was opened for bidding on July 15, 2006, and the Infrastructure Development Corporation Ltd. has been appointed as the consultant for project development.

Service Contract—This involves brief contractual agreements for particular services such as meter reading and bill preparation. Typically, the private company does not make any investments, eliminating financial risks and avoiding a direct legal association with the user. Example—In Bangalore, the water supply and sewage board has contracted out a systemic leakage reduction program to companies such as Thames Water Asia and Larsen and Toubro. The program involves an investment of 48 crores and aims to improve the system over a period of 18 months, M/s Larsen and Toubro, and M/s Thames Water Asia Pvt Ltd, India executed the project.

B-O-O-T—Build Own Operate and Transfer—In this scenario, a private company constructs certain components of the infrastructure, such as the activated sludge plant or effluent treatment plant, and operates it in exchange for a recurring fee. Typically, these contracts are of long duration and involve a purchase agreement that ensures a minimum level of demand. These agreements often include "take-or-pay" clauses, where the customer is obligated to either consume a specified amount or pay for it regardless. A notable example of this is the desalination plant in Chennai.

D-B-O-O-T—Design Build Own Operate and Transfer—It is a type of PPP project where a private contractor is responsible for the design, construction, and operation of a project on behalf of a public client. The private contractor manages the project for an agreed period of time before transferring ownership and maintenance back to the public client. On 18th November 2004, the Chennai Metropolitan Water Supply and Sewerage Board initiated a bidding process for a seawater desalination project of 100/200 MLD capacity, to be implemented on a BOOT (Build-Own-Operate-Transfer) basis. The project was ultimately awarded to Chennai Water Desalination Ltd (CWDL), a special purpose vehicle (SPV) established by IVRCL Infrastructures & Projects Ltd., a publicly traded Indian company, which holds a 75% stake in the project. The remaining 25% ownership of the project company belongs to the Befessa Construction Tecnologia Ambinetal, S A U (Befesa), an engineering and construction company listed on the Madrid Stock Exchange.

13.7 24× 7 Water Supply

A continuous water supply system is a system in which water is supplied to a community or area on a continuous basis, rather than at specific times or on a schedule. This type of system typically uses a network of pipes, pumps, and storage tanks to ensure that a steady flow of water is available to users at all times. The water is typically treated at a water treatment plant before being distributed to users through the system. This type of system is commonly used in urban areas and is considered to be more efficient and reliable than intermittent water supply systems. CPHEEO (Central Public Health and Environmental Engineering Organisation) provides technical assistance and guidance to State Governments and Union Territories in the areas of public health and environmental engineering. This includes providing standards and guidelines for drinking water, sewage and sanitation, and solid waste management, as well as capacity building for engineers and other professionals working in these fields (CPHEEO 1999).

After three years of implementing reforms, surveys conducted in the 24 × 7 demonstration wards indicated that services were largely uninterrupted, though some minor disruptions occurred. Most households with previously unauthorized connections were provided with meters and were receiving monthly bills. However, the availability of free public water sources underwent a significant decline, with the closure of all 84 handpumps, an unspecified portion of the 110 motorized borewells, and all 115 public standpipes within the 24 × 7 zones. Furthermore, access to Unkal Lake, which was previously accessible for non-potable uses and for domestic buffaloes, is now limited and requires a nominal fee. This "recreational area" is now fenced off. As a result of the 24 × 7 project, there was a noticeable reduction in waterborne illnesses.

Implementing a continuous water supply system involves several key steps, including:

1. Assessing the existing water supply system: The first step is to evaluate the existing water supply system, including the quality and quantity of water that

is available, the condition of the pipes and infrastructure, and the needs of the community.

2. Developing a plan: Based on the assessment, a plan should be developed that outlines the steps needed to implement a continuous water supply system. This plan should include details such as the location and size of new pipes, the type of infrastructure that will be required, and the estimated costs of the project.
3. Upgrading infrastructure: Upgrading the existing water infrastructure is a critical component of implementing a continuous water supply system. This may involve replacing old or damaged pipes, installing new pumps and storage tanks, and upgrading treatment facilities.
4. Installing meters: Meters should be installed in every household or business that is connected to the water supply system, in order to accurately measure water usage and charge customers accordingly.
5. Educating the community: It is important to educate the community about the benefits of a continuous water supply system, as well as how to conserve water and reduce waste.
6. Monitoring and maintenance: Once the system is up and running, it is important to monitor it regularly and perform routine maintenance to ensure that it is functioning properly.

Further, the sustainability and reliability of continuous water supply can be ensured by diversifying water sources, implementing water conservation measures, and investing in water infrastructure. Implementing a continuous water supply system can be a complex and costly undertaking, but it can offer many benefits over traditional intermittent water supply systems. With careful planning, investment in infrastructure, and community engagement, it is possible to provide a reliable and sustainable supply of water to a community. Continuous water supply systems are increasingly being adopted by cities and communities around the world, as they offer many advantages over traditional intermittent water supply systems. Singapore, Japan, South Korea, the United Arab Emirates, and Australia are few examples of countries that have implemented or are planning to implement continuous water supply systems. As the demand for water continues to grow around the world, it is likely that more and more communities will turn to these systems as a way to ensure a reliable and sustainable water supply (Fig. 13.2).

13.7.1 Water Quality

In intermittent water supply systems, the water quality can be adversely impacted by various factors. When water does not flow continuously, it can become stagnant in the pipes, creating an environment conducive to the growth of bacteria and other microorganisms. Moreover, intermittent water supply systems are more vulnerable to contamination from external sources, such as during repairs or maintenance work. In

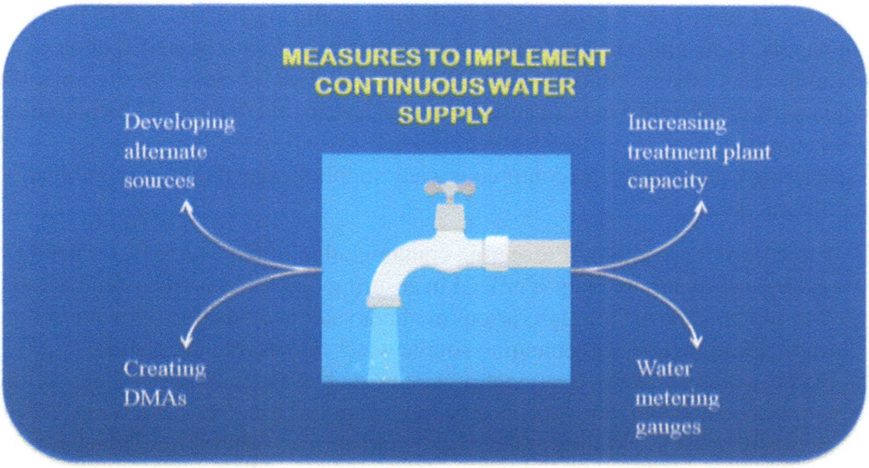

Fig. 13.2 Focus areas for CWS

contrast, continuous water supply systems contribute to maintaining a more consistent and dependable water quality. This is because water keeps flowing throughout the system, minimizing the risk of stagnation and microbial growth. Continuous water supply systems can also incorporate advanced water treatment technologies like ultraviolet disinfection to further enhance water quality. However, it's important to acknowledge that even in continuous water supply systems, water quality can still be influenced by factors such as external contamination, aging infrastructure, and insufficient maintenance. Therefore, implementing robust water quality monitoring and management systems is crucial to ensure the long-term safety and reliability of the water supply. Water samples taken from sources supplying both the continuous and intermittent water supply of the piped network were found to be of similar quality (Ercumen et al. 2015). However, when testing for indicator bacteria, a higher presence and concentration were found in samples taken from taps with intermittent water supply (IWS) compared to those with 24/7 supply. This difference was statistically significant (Kumpel and Nelson 2013). The detection of E. coli was significantly lower in continuous water supply systems, with only 0.7% of tap samples testing positive for the bacteria, compared to a higher percentage of 31.7% in tap samples from intermittent water supply systems. After rainfall events, tap samples from both continuous and intermittent supply systems showed higher concentrations of total coliform. Continuous water supply did not show any association with diarrhea, bloody diarrhea, or in children under 5 years old. In subgroup analyses based on socioeconomic status, children under 5 years old in lower-income households with continuous water supply had a 37% lower prevalence of bloody diarrhea compared to lower income households with intermittent supply. However, for households with higher incomes, the presence of continuous or intermittent water supply did not show a notable correlation with child diarrheal illnesses. Furthermore, areas with continuous water supply experienced a 42% reduction in the number of households reporting at least one instance of typhoid fever compared to areas with intermittent supply. No significant associations were discovered with hepatitis, cholera, or mortality rates among children under 2 years old. Nevertheless, the findings did indicate lower mortality rates in children under 2 years old in areas with continuous water supply.

Ensuring water quality in a continuous water supply system requires a combination of different strategies:

1. Source water protection: Preventing contamination of the source water before it reaches the treatment plant is crucial in ensuring that water is free from bacteria. This can be achieved through various methods such as protecting the water source from pollution, and implementing source water monitoring programs.
2. Water treatment: Water treatment processes such as coagulation, sedimentation, filtration, and disinfection are used to remove or kill bacteria in the water. Disinfection is the most common method used to ensure that water is free from bacteria, and it involves using chlorine, chloramines, or ultraviolet light to kill bacteria.
3. Distribution system management: Once the water is treated, it needs to be distributed to the consumers without being contaminated. This can be achieved by maintaining the water distribution network, and by detecting and repairing leaks and breaks in the pipeline.
4. Consumer education: Educating consumers on how to maintain their own internal plumbing and detect potential sources of contamination can help to ensure that water remains free from bacteria once it reaches their homes.
5. Regular monitoring: Regular monitoring of water quality is essential to ensure that the water is free from bacteria. Water samples are collected and analyzed to detect the presence of bacteria, and to ensure that water treatment processes are working effectively.

13.7.2 Water Management Costs

The research conducted a survey among households that had access to both intermittent water supply (IWS) and continuous water supply (CWS). The aim was to collect data on the expenses related to household water management, which encompassed the time and energy invested in waiting for, collecting, storing, and potentially treating water from different sources like municipal or private suppliers. This included regular expenditures such as utility bills, purchasing bottled or tanker-supplied water, and acquiring consumables for water treatment (Charalambous, Laspidou, IWA 2017). Furthermore, the research study gathered information regarding the investments made by households in infrastructure, including overhead tanks or private borewells, as well as durable equipment such as buckets and barrels. This data was utilized to compare the costs associated with household water management between households that had undergone the 24 $\times$ 7 upgrade and those that had not. With a continuous water supply, individuals no longer need to bear the expenses of pumping water from wells or alternative sources, which can be costly. Moreover, a continuous water supply reduces the likelihood of contracting water-borne diseases, subsequently mitigating the burden of significant medical expenses.

13.7.3 Water Conservation and Loss Reduction

The provision of a continuous water supply system, where water is available 24/7, was found to result in less water consumption than expected. This was due to the fact that people tended to be more conscious of water conservation when they were sure of a reliable supply and were made aware of the value and cost of water. There were fewer leakages and overflowing tanks, people were less likely to leave water taps open, and there was less need to store extra water (Charalambous, Laspidou, IWA 2017). Furthermore, the initial investments made in the distribution network and efficient operations and maintenance have played a crucial role in minimizing water losses. In the demonstration zones, technical losses decreased significantly from an estimated 50–7%. To reduce water loss in a continuous water supply system, a comprehensive approach is required. This includes identifying and resolving the root causes of water loss, implementing robust leak detection and repair programs, and ensuring accurate billing and metering. Conducting regular monitoring and audits of the distribution network and internal plumbing systems can also aid in identifying and addressing potential sources of water loss. It is important to highlight that reducing water loss in a continuous water supply system can yield substantial benefits, such as enhancing the efficiency and sustainability of the water supply system while reducing the expenses associated with water production and treatment.

13.7.4 Recovery of O&M Costs

The primary objective of the 24 $\times$ 7 water supply project was to enhance the financial sustainability of water supply systems by increasing the retrieval of the Operation and Maintenance (O&M) costs. Before the project commenced in 2003, the urban local bodies were able to cover less than half of the operation and maintenance (O&M) costs. The project's objective was to achieve a minimum of 80 percent coverage of O&M costs within the initial two-year maintenance period and reach a billing and collection efficiency of 70 percent by the project's completion. By 2011, at the end of the project, significant progress had been made in terms of the number of consumers making payments, enabling the municipal corporations to recover both O&M costs and outstanding payments. This accomplishment exceeded the average recovery rate in other Indian cities, where, in 2010, only 60 percent of consumers were making bill payments and 67% of O&M costs were being recovered (data from the Ministry of Urban Development's first Service Level Benchmarking Data Book).

13.8 Case Study

Facilitating a Seamless Shift to the Continuous Water Supply System: The Significance of Allowing Consumers Ample Time for Adaptation

The case of the Hubli Demonstration Zone highlights the significance of allowing consumers an adequate adjustment period when implementing a new continuous water supply system. The sudden introduction of volumetric tariffs by the Hubli-Dharwad Municipal Corporation resulted in confusion and complaints from unprepared consumers who received unexpectedly high bills. As a consequence, the implementation of water tariffs in other areas faced delays. In contrast, the Gulbarga City Corporation took a different approach by issuing dummy bills based on the new tariffs for several months before their actual implementation. This strategy facilitated a smoother transition. Approximately 700 consumers raised concerns about their inflated bills, which were attributed to leakages in the old network, consumers' premises, and unauthorized connections. Resolving the issue took several months, and adjustments had to be made to the volumetric tariffs. This setback not only affected the Hubli Demonstration Zone but also caused delays in introducing water tariffs in other regions. The key lesson learned was the necessity of allowing consumers ample time to develop confidence in the reliability of the 24/7 water supply, understand their monthly bills, and take measures to reduce consumption and address leaks. The Government of Karnataka intends to expand the 24/7 water supply system to other areas within the three cities, and there is also interest from other cities across the country to replicate the system as part of the Jawaharlal Nehru National Urban Renewal Mission (JNNURM).

Significance of Separating Outstanding Payments from New Water Supply Charges

Separating arrears from new water supply charges is crucial for efficient financial management and equitable service provision. By delinking outstanding payments from the charges associated with the new water supply, it ensures that customers are not burdened by past debts when accessing essential water services. This approach promotes transparency, as it enables clear and accurate billing for current usage while addressing arrears through separate arrangements. It also encourages a more sustainable revenue collection system and fosters better customer relations, as it allows for a fresh start in terms of payment obligations. Ultimately, delinking arrears from new water supply charges benefits both the service provider and the customers by promoting fairness, financial stability, and uninterrupted access to clean water. Delinking arrears from new charges can increase the transparency and fairness of the new system and can help to build trust between consumers and the water supplier.

13.9 Conclusion

Adapting to a new supply system in water management can be challenging for some people, particularly those who are not familiar with the supply system or who may not have the necessary skills or resources to implement it. It is important for organizations and individuals responsible for water management to provide education and training to help people understand and use the new water supply system. This can include workshops, webinars, and other forms of training to help people understand how this method works and how it can be used to improve water management. Additionally, it's important to consider the cost of new technology, as the cost can be a barrier for some organizations and individuals. To address this, organizations should work to identify funding sources and provide financial assistance to those in need. Communication and community engagement are also important for the successful adoption of a new water supply system. It is important to involve the community in the decision-making process, and to provide clear and transparent information about the system and its benefits. It is necessary to keep in mind that, water is a natural resource and should be accessed by people with diverse backgrounds, cultures, and levels of education. It is important to have a constant and dependable supply of clean water available 24 × 7 in order to enhance the standard of living for city dwellers and boost economic growth. Intermittent water supply often causes water scarcity and water rationing, which can be unfair to the poor as they may have less access to water than more affluent people. Hence, intermittent water supply can exacerbate poverty and inequality, and converting to continuous water supply can help to alleviate these issues. The acceptance of the continuous water supply system by the people can vary depending on a number of factors.

Some of the key factors that can influence the acceptance of Continuous water supply include:

- **Reliability**: People are more likely to accept a continuous water supply system if they perceive it to be reliable and consistent.
- **Affordability**: The cost of the system, including the tariffs and any necessary infrastructure upgrades, will also play a role in determining acceptance. People are more likely to accept the system if they perceive it to be affordable.
- **Education and awareness**: If people are educated about the benefits of a continuous water supply system, and how it will improve their lives, they are more likely to accept it.
- **Perception of a fair and transparent system**: If people perceive the system to be fair, transparent, and without any hidden charges, they are more likely to accept it.
- **Convenience**: A continuous water supply system offers the convenience of having water available at all times, which can be a significant factor in determining acceptance.
- **Quality of water**: People are more likely to accept a continuous water supply system if the water is of good quality and meets their needs.

- **Government's efforts**: Government's efforts to involve the community in the planning and implementation of the system, and its efforts to address any concerns or complaints can greatly affect the acceptance of the system.

Overall, the acceptance of a continuous water supply system by the people is likely to be influenced by a combination of these factors, and it can be improved through effective communication, education, and community engagement.

References

Andey SP, Kelkar PS (2007) Performance of water distribution systems during intermittent versus continuous water supply. J AWWA 99(8):99–106

Andey SP, Kelkar PS (2009) Influence of intermittent and continuous modes of water supply on domestic water consumption. Water Resour Manage 23:2555–2566. https://doi.org/10.1007/s11269-008-9396-8

Anil A, Narain S (1997) Dying wisdom-rise, fall and potential of India's traditional water harvesting systems. CSE, New Delhi

Anil A, Narain S (2001) Making water everybody's business-practice and policy of water harvesting. CSE, New Delhi

Anon (2004) Water harvesting and artificial recharge—Rajiv Gandhi national drinking water mission. Government of India, New Delhi

Anon (2006) Status of water supply, sanitation and solid waste management in urban areas. NIUA, June. Mimeo

Anon (2009) Delhi economic survey 2008–09. Delhi Planning commission, New Delhi

Babel MS, Das Gupta A, Pradhan P (2007) A multivariate econometric approach for domestic water demand modeling: an application to Kathmandu. Nepal. J Water Resour Manage 21:573–589. https://doi.org/10.1007/s11269-006-9030-6

Charalambous B, Laspidou C, IWA (2017) Dealing with the complex interrelation of intermittent supply and water losses

CPHEEO (1999) Manual on water supply and treatment, 3rd edn. Ministry of Urban Development and Poverty Alleviation, Govt. of India, New Delhi.

Dwivedi G et al (2007) Water: private, limited- issues in privatisation, corporation and commercialisation of water sector in India. Manthan Adyanan Kendra, Badwani (MP)

Ercumen A, Arnold BF, Kumpel E, Burt Z, Ray I, Nelson K, Colford Jr JM (2015) Upgrading a piped water supply from intermittent to continuous delivery and association with waterborne illness: a matched cohort study in Urban India. https://doi.org/10.1371/journal.pmed.1001892

Farmani R, Dalton J, Charalambous B, Lawson E, Bunney S (2021) Sarah Cotterill; Intermittent water supply systems and their resilience to COVID-19: IWA IWS SG survey. J Water Supply: Res Technol-Aqua 70(4):507–520. https://doi.org/10.2166/aqua.2021.009

Ghorpade A, Sinha AK, Kalbar PP (2021) Drivers for intermittent water supply in India: critical review and perspectives. Front. Water Sec Water Built Environ 3. https://doi.org/10.3389/frwa.2021.696630

http://www.jubileesouth.org/news/EpZZZlkuAkDMZZOUuf.shtml

https://www.worldbank.org/en/news/feature/2013/01/01/karnataka-three-towns-pilot-water-supply

Indo–US Financial Institutions Reform and Expansion Project—Debt Market Corporation FIRE (D) (2000) Improving the operation of urban water supply system in India. Note no 23, http://www.tcgillc.com/firenotes/23.pdf

Iyegar S (2007) Water nama- A collection of traditional practices for water conservation and management in Karnataka. Communication for Development and Learning, Bangalore

Kalbar P, Gokhale P (2019) Decentralized infrastructure approach for successful water supply systems in India: use of multi-outlet tanks, shafts and manifolds. J Water Supply: Res Technol-Aqua 68(4):295–301. https://doi.org/10.2166/aqua.2019.158

Klingel P (2012) Technical causes and impacts of intermittent water distribution. Water Supply 12(4):504–512. https://doi.org/10.2166/ws.2012.023

Kumpel E, Nelson KL (2013) Comparing microbial water quality in an intermittent and continuous piped water supply. https://doi.org/10.1016/j.watres.2013.05.058

Nauges C, Thomas A (2003) Long-run study of residential water consumption. Environ Resour Econ 26:25–43. https://doi.org/10.1023/A:1025673318692

Ray I, Billava N, Burt Z, Colford JM, Ercümen A, Jayaramu KP, Kumpel E, Nayak N, Nelson K, Woelfle-Erskine C (2018) From intermittent to continuous water supply a household-level evaluation of water system reforms in Hubli–Dharwad

Tabesh M et al (2008) An integrated model to evaluate losses in water distribution systems. J Water Resour Manage. https://doi.org/10.1007/s11269-008-9284-2,Publishedonline:14June2008

Totsuka N, Trifunovic N, Vairavamoorthy K (2004) Intermittent urban water supply under water starving situations. In: 30th WEDC international conference: people-centred approaches to water and environmental sanitation. Vientiane, Lao PD.

Chapter 14
Exploring Ethical Impacts of Nanomaterials for Wastewater Remediation

Ineke Malsch, Panagiotis Isigonis, Evert Bouman, Antreas Afantitis, Georgia Melagraki, Dalila Antunes, and Maria Dusinska

Abstract Nanotechnology and nanomaterials offer potential contributions to the Sustainable Development Goal 6: clean water for all. However, the use of nanomaterials in innovative water treatment solutions can raise ethical concerns as well as benefits. These potential ethical impacts should be assessed before implementing the technology in real life, to ensure that remediation measures can be taken in parallel. Such ethical impact assessment can be supported by guidelines and online tools based on CEN Workshop Agreement part 2 CWA 17,145-2:2017 (E)), applied to nanomaterials and nanoproducts in the RiskGONE project (https://riskgone.eu/). This paper reports on an evaluation of identified ethical impacts of the imagined use of nanomaterials in wastewater remediation in a developing country and formulate recommendations for remedial actions. The case study is based on desk research and the main purpose was to test the online ethical impact assessment tools, not to assess the ethical impacts of a real-life project.

Keywords Nanomaterials · Ethics · Risk governance · Wastewater · Developing countries

---

I. Malsch (✉)
Malsch TechnoValuation, Utrecht, The Netherlands
e-mail: malschtechnovaluation@xs4all.nl

P. Isigonis
UNIVE, Ca'Foscari University of Venice, Mestre-Venezia, Italy

E. Bouman · M. Dusinska
NILU-Norwegian Institute for Air Research, Kjeller, Norway

A. Afantitis · G. Melagraki
Novamechanics, Nicosia, Cyprus

D. Antunes
Factor Social, Lisboa, Portugal

S. Suriyanarayanan et al. (eds.), *Water Management in Developing Countries and Sustainable Development*, Water Resources Development and Management,
https://doi.org/10.1007/978-981-99-8639-2_14

14.1 Introduction

Water is a precious commodity in most developing countries. In some areas in Africa, people have to walk or row for a long time and over long distances to have access to boreholes, many of them with poor water quality and safety standards. In Mozambique, cases have been reported in which people walk 20 km or row for 8 h, to have access to 20 L of water, many times collected in boreholes where water salinity is quite high. World Vision news reported in 2020 that in Sub-Saharan Africa:

> Every day, women and girls spend 200 million hours walking to collect water for their families. That's 8.3 million days. More than 22,800 years... Moms and daughters walk an average 6 Km barefoot or in rubber sandals to collect water from polluted rivers and ponds. More than 3 million children and nearly 14 million women walk more than 30 minutes to collect water. And they often make that trip more than once a day!' (Reid 2020).

In other cases, people directly collect water on rivers for drinking purposes, even in places with poor sanitation systems where people use the riverbeds for sanitation purposes as well. This not only presents direct threats to people's lives due to conflicts with local fauna such as crocodiles and hippos (which in some villages in Africa can represent about 4 attacks per week) but can also lead to the spreading of diseases such as cholera outbreaks.

In this context, innovative wastewater remediation systems incorporating nanomaterials offer promising solutions by cleaning wastewater at the household or village level. But how does the new technological solution compare to alternative solutions? To assess this, increasingly, life cycle assessment (LCA) is used to explore potential ecological, economic, and social impacts of emerging technologies during research and innovation projects. While some such projects raise ethical issues, ethical impacts are not generally considered. For example, in a recent review, Petros Kokkinos explored the use of nanomaterials in (photo) catalytic decontamination of wastewater (Kokkinos et al. 2020). The scope of nanomaterials included academic and commercial carbon and graphene-based materials as well as metal–organic frameworks and metal oxides and composites. Antimicrobial effects as well as removal of chemical and metallic contaminants were discussed. While potential environmental risks and economic aspects were also addressed, ethical issues were not mentioned.

In the EU-funded project RiskGONE, guidelines and tools were developed to support Risk Governance by multiple stakeholders for identifying and assessing the ethical impacts of nanomaterials along with more traditional LCA issues. These guidelines were based on the CEN Workshop Agreement part 2 CWA 17,145-2:2017 (E) on Ethical Impact Assessment (EIA) (CEN 2017). In the present paper, the results are presented of an ethical impact assessment of nanomaterials for photocatalytic decontamination of wastewater in developing countries, using the online EIA tools, to complement the abovementioned review by Kokkinos. This case study refers to the water sector, in itself subject to ethical constraints, considering the commercialization of public goods, which is not accepted in all countries. In this case study, the ethical impacts of the use of nanomaterials in photocatalytic decontamination of wastewater were explicitly addressed. Some issues were related to the properties of nanomaterials, while others were relevant to wastewater remediation in general or the application of wastewater remediation in developing countries. This was specified in the case study.

The analysis started with a self-assessment of the likely ethical impacts of the use of nanomaterials in photocatalytic decontamination of wastewater, using the online Ethical Impact Assessment Screening tool[1] developed in the RiskGONE project. Ethical risks as well as benefits for health, equality, environment, and sustainability were identified (Fig. 14.1). Concerning **health impacts**, the use of nanomaterials in water remediation may introduce health risks if these particles are released and end up in drinking water. The severity of this issue depends on the matrix and design of the product. Conversely, the intended use of nanomaterials in wastewater remediation is to improve access to clean drinking water, thereby improving human health. Concerning **equality**, If the water remediation technology is expensive or primarily offered to wealthy persons, this may negatively impact social justice and equality. This impact is not nano-specific but depends on the societal context. Conversely, if the water remediation technology is primarily offered to poor people in developing countries, this will improve equality. Concerning the **environment,** if nanoparticles are released from the water remediation technology, and end up in surface or groundwater, this may negatively impact the environment. Conversely, the intended use of nanomaterials in wastewater remediation is to purify wastewater before releasing it into the environment, thereby improving the environment. Concerning the **Sustainable Development Goals**[2] negative impacts are expected on SDG 3 and 14. If nanoparticles are released into drinking, surface, or groundwater, this may negatively impact human health and the marine environment. These impacts are already noted under health and environment. Conversely, positive impacts are expected on SDG 6: the use of nanomaterials for wastewater remediation is intended to improve access to clean drinking water.

The identified negative impacts called for further examination in a small-scale Ethical Impact Assessment. Therefore, an EIA plan was prepared, including the selection of sources and methods for identifying and evaluating ethical impacts, drafting remediation measures, and review of the EIA by an independent ethicist. The research questions were: which ethical issues are known to be raised using nanomaterials in wastewater remediation in developing countries? How serious are these issues? Which ethical principles or values are relevant to evaluate the identified ethical issues? Which remedial actions does the outcome of the evaluation suggest? What do stakeholders think about the identified ethical issues and recommendations? Because limited resources were available for this case study, the peer review of this paper is considered to count as the review of our EIA by external ethicists. Complementing our desk research, feedback was solicited from participants in four online events to ensure mandatory engagement with stakeholders in the case study.[3]

---

[1] http://www.enaloscloud.novamechanics.com/riskgone/eia/thresholdanalysis/index.zul.

[2] https://sdgs.un.org/goals

[3] Virtual ANSOLE DAYS & BALEWARE 2021, 4–5 February 2021 http://ansole.com/, online stakeholder workshops on 26–27 January 2021 and 14 April 2021 of the NMBP-13 projects

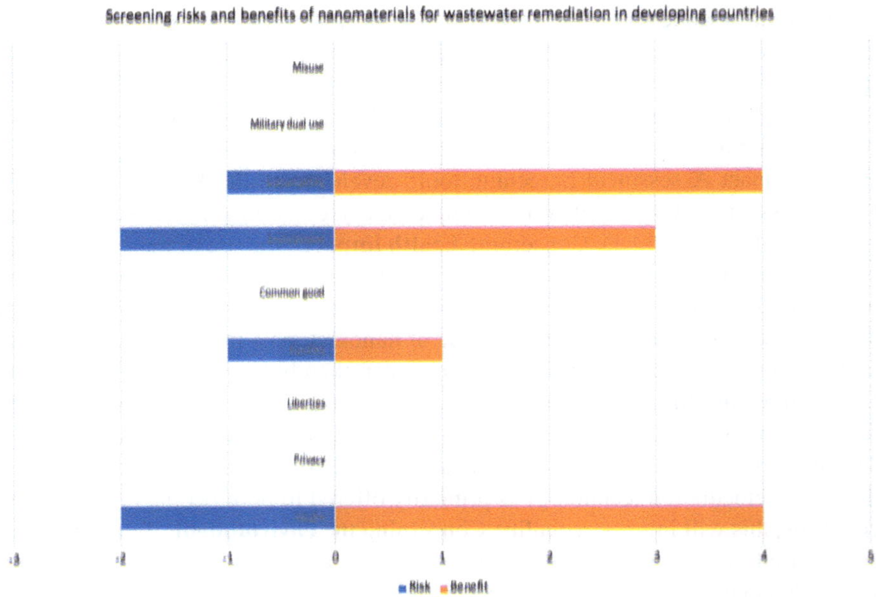

Fig. 14.1 Self-assessment screening of ethical risks and benefits of nanomaterials for wastewater remediation in developing countries

14.2 Identifying Ethical Impacts

Relevant literature was identified through searching a shortlist of journals and internet repositories of grey literature where ethical aspects of nanotechnology for remediation of wastewater were discussed. Paid as well as open-access literature was identified. However, due to budgetary restraints, and assuming the paid literature did not include significant additional issues to the open-access literature, the EIA team only analyzed the collected open-access literature using the checklist in Table 14.1.[4] This open-access literature list was included in the references to this paper.

Three identified issues were deemed to be moderate: the generation of harmful waste and two effects on people in developing countries. These issues are unlikely or expected to have mild potential severity. In this case, the issues were deemed moderate, because the new technology noticeably worsens ethical impacts caused by existing alternatives. Three other issues were deemed to be minor: the impacts on public health or safety, the environment, and the distribution of risks over the population. This means that it is very unlikely that the issue will occur or that the potential severity is very low. In this case, the issues were deemed minor because the

(Gov4Nano, NanoRIGO and RiskGONE), and International Water Climate Summer School, South Africa, 6–10-22, BALEWARE and Water Net, North-West University, Mahikeng Campus, North-West Province, South Africa: https://www.nwu.ac.za/date/international-water-climate-summer-school.

[4] http://www.enaloscloud.novamechanics.com/riskgone/eia/idethimpact/index.zul.

Table 14.1 Analysis of negative ethical impacts of nanotechnology for wastewater remediation found in literature. (1 = minor; 2 = moderate; 3 = medium; 4 = high; 5 = severe)

Issue	Citation	Ethical risk
Public health or safety	Nanomaterials used in wastewater remediation pose health risks if released into drinking water (Balakrishnan and Srivastava 2009)	1
Environment (water, land)	Oil and gas industry wastewater upgraded as irrigation water in arid regions could still be contaminated (Dotson 2019)	1
Harmful waste	Used filters and media containing nanomaterials pose uncertain environmental risks (Balakrishnan and Srivastava 2009)	2
Risk distribution	Nanoparticles may pollute water, air, and soil, raising safety issues and balancing the interests of industry, competing for market shares, against other interests (EGE 2007, p 38)	1
Effect on people in developing countries	Maintenance, training of local technicians and adaptation of local regulatory frameworks may not be properly integrated in the wastewater remediation system (Padma 2009)	2
Effect on people in developing countries	If the wastewater purification system is made of expensive materials, it may be less accessible to poor people (Makoni 2009)	2

new technology does not significantly worsen the ethical impacts caused by existing alternatives.

Stakeholders have been consulted on the draft results of the identification and evaluation of ethical issues and on draft recommendations for remediation during four online events.[5] A nanosafety expert suggested to target the use of TiO_2 nanoparticles, because this material has been demonstrated to be riskier than other nanomaterials. This suggests that a possible recommendation for organizational remediation actions could be to replace TiO_2 with other nanomaterials in wastewater remediation systems.

14.2.1 *Identifying Future Impacts*

The application of nanomaterials in wastewater remediation does not involve frontier research with uncertain long-term ethical impacts. Therefore, no formal foresight exercise was undertaken to identify those future impacts. However, expected impacts

[5] Virtual ANSOLE DAYS & BALEWARE 2021, 4–5 February 2021 http://ansole.com/, online stakeholder workshops on 26–27 January 2021 and 14 April 2021 of the NMBP-13 projects (Gov4Nano, NanoRIGO and RiskGONE), and International Water Climate Summer School, South Africa, 6–10-22, BALEWARE and Water Net, North-West University, Mahikeng Campus, North-West Province, South Africa: https://www.nwu.ac.za/date/international-water-climate-summer-school.

on three SDGs[6] (3, 6, and 14) were identified in the risk–benefit assessment, the affected subgoals were specified. On the risk side, the uncertain health effects of the release of nanomaterials in purified water could negatively impact SDG 3.9: *By 2030, substantially reduce the number of deaths and illnesses from hazardous chemicals and air, water, and soil pollution and contamination.*On life below water, nanomaterials could be released in surface water and end up in oceans, thereby introducing delays in achieving SDG 14.1: *By 2025, prevent and significantly reduce marine pollution of all kinds, in particular from land-based activities, including marine debris and nutrient pollution.* While the health and environmental risks of nanomaterials are uncertain, the amount of nanomaterial likely to be released from wastewater remediation systems is so low, that these impacts are expected to be negligible on a global scale.

On the benefit side, the use of nano-enabled wastewater remediation systems in developing countries could positively contribute to achieving SDG 6.3: *By 2030, improve water quality by reducing pollution, eliminating dumping, and minimizing release of hazardous chemicals and materials, halving the proportion of untreated wastewater, and substantially increasing recycling and safe reuse globally.* The introduction of wastewater remediation systems in regions where this water is currently not cleaned explicitly targets this aim. However, it is unclear how much the use of nanomaterials in these systems contributes to this subgoal. This depends on the future market share of nano-enabled versus other wastewater remediation systems.

14.3 Evaluating Ethical Impacts

The evaluation of the identified ethical impacts started with the selection of a list of relevant ethical principles and values from relevant literature. [7]These principles are at stake where nanomaterials are used in wastewater remediation in developing countries. Some principles are more relevant to the development of the remediation systems, while others are more relevant to the use or waste processing of obsolete systems.

The precautionary principle is used to govern uncertain risks of new technologies (COMEST 2005). In this case, responsive regulations and the precautionary principle can govern uncertain risks of nanomaterials for the environment (Balakrishnan and Srivastava 2009). In addition, the precautionary principle should be applied in risk governance of uncertain health-related impacts of nanomaterials (EGE 2007). To contribute to equal access to the wastewater remediation systems, affordable local materials should be used to produce it (Makoni 2009; Saidi 2016). To ensure participatory technology development, local communities should be engaged in decision-making on implementing nanomaterials in wastewater purification (Grimshaw 2009; Saidi 2016). Overall, the global code of conduct for research in resource-poor settings

[6] https://sustainabledevelopment.un.org/topics/sustainabledevelopmentgoals.

[7] http://www.enaloscloud.novamechanics.com/riskgone/eia/evethimpact/index.zul.

offers guidelines for performing research in developing countries which are relevant to the development of nano-enabled wastewater treatment: fairness, respect, care, and honesty (Global code 2018).

In later stages, the human right to water, adopted by the UN Human Rights Council, applies (see OECD 2008). This stipulates that governments must ensure access to safe drinking water and sanitation under international law. The practical implementation of this human right to water is explained in the water ethics framework, a framework for assessing the ethical impacts of water management, including human dignity and the right to water, equity in availability and accessibility of the right to water, multiple and beneficial use of water, user and polluter pays, and participation (Jie Liu et al. 2011). In addition, the use of affordable local resources remains relevant.

14.3.1 Analysis of the Identified Ethical Issues

In general, three types of academic ethics can be distinguished. Deontological (rule-based) ethics can be interpreted roughly like this: as long as you act in accordance with ethical and legal rules, the outcome of your actions is less important. Examples are research ethics frameworks or codes of conduct to guide the ethical conduct of research. Research ethics includes research integrity, such as promoting transparency of the research process and fighting plagiarism. It also includes discipline-specific research ethics, such as biomedical or engineering ethics.

Virtue ethics implies that an individual should always strive for personal moral values (e.g., prudence, justice, fortitude, temperance). For example, individual researchers can choose to contribute to SDG 6 (clean water and sanitation) or SDG 13 (climate action) in their research.

Consequentialist ethics can be interpreted to suggest that collective actions are ethically sound if the overall consequences are beneficial to individuals, society, and the environment. To improve the likelihood that the technology under development will bring more benefits than detrimental effects, research programs may integrate modules for responsible research and innovation, technology assessment, or stakeholder dialogue, and let the results guide the research strategy. The researchers should be aware that the future remains uncertain and build flexibility to address emerging risks as they become noticeable.

In our ethical impact assessment of nanomaterials in wastewater remediation in developing countries, three sets of principles or values are relevant. Ethical theories and literature were analyzed to clarify which ethical theories were applied and what they meant in this case.

The **precautionary principle** is an example of consequentialism, meaning that people and organizations should take responsibility for potential future consequences of their actions. While different definitions of the precautionary principle are used, the definition proposed by COMEST (2005) is applied here: "*When human activities* *may lead to morally unacceptable harm that is scientifically plausible but uncertain, actions shall be taken to avoid or diminish that harm."* The use of nanomaterials in innovative wastewater remediation technologies, even though preventing some risks related to actual solutions, could introduce uncertain risks to humans and the environment if those nanomaterials are released from the water treatment system into nature or used in agriculture or for human consumption. Precautionary actions can include early risk assessment studies, monitoring the release of nanomaterials during operation of the wastewater systems, and restrictions and delays on the market introduction of the systems.

The **global code** of conduct for research in resource-poor settings (Global code 2018) is an example of virtue ethics. On top of formal legal requirements, researchers should take care that their research project and the resulting technology are beneficial to the local community. It is also an example of procedural ethics since it consists of guidelines for organizing participation and taking steps to ensure respecting local norms and values. The value **fairness** calls for consulting with local stakeholders before starting the project and making sure that the project results are relevant to solving local water issues. Researchers should involve local stakeholders during the whole project and inform them about progress in an understandable way. The project should engage local researchers as much as possible and compensate local research support systems fairly for contributions.

The researchers should explore local values before starting the research and **respect** them. They should fully inform and obtain community assent before starting the research and obtain local ethics approval, respecting local community committees and research ethics committees, if available. It is important to take **care** of individual participants in the research by tailored informed consent and by offering accessible feedback procedures. The researchers should comply with the highest standards and best practices of research ethics worldwide or explain why transgression is justified (e.g., sub-optimal water quality standards may in the short term be an improvement in the local context). The project leader should determine before the start of the research whether the project will occupy local staff which is then no longer available for other local tasks and engage local stakeholders in discussion and monitoring of the issue. Researchers should respect the higher standards of environmental and social protection when local standards are below international standards and manage health and safety risks for researchers. The value of **honesty** calls for agreement on the role and responsibilities of all collaborators and the capacity-building plans before the research starts. The researchers should communicate in plain language understandable to all stakeholders. Corruption and breaches of privacy and EU data protection laws are prohibited.

The **water ethics** framework (Jie Liu et al. 2011) is a framework for assessing the ethical impacts of water management, including human dignity and the human right to water, equity in availability and accessibility of the right to water, multiple and beneficial use of water, user and polluter pays, and participation. It combines three ethical theories: deontology, consequentialism, and virtue ethics. When nanomaterials are used in wastewater remediation in a developing country, the UN Human Rights Council requires governments to ensure access to safe drinking water and sanitation under international law to all people. The project leader must investigate whether the project or product will improve such access or diminish it in the short or long term. Will the project or product allow such access with higher or lower risks than the actual ones, in the short or long term?

Using nanomaterials in wastewater remediation is likely to improve **equity**, because the product is intended to improve the availability and accessibility of potable water to local consumers and other secondary users of the cleaned wastewater.

The impact of **multiple and beneficial use of water** is less clear. While the product is intended to improve the reusability of the cleaned wastewater, if nanomaterials are released in the cleaned water, this may, in the long-term, hamper the beneficial use of this water. It is unclear whether the **users and polluters** of the wastewater will have to **pay** for the cleaning. This depends on national regulations in the country where the product will be used. **Participation** is already expected to be addressed during the project under the code of conduct for research in resource-poor settings.

14.3.2 Ethical Risk–Benefit Assessment

In a semi-quantitative assessment, the strength of the negative and positive impacts was compared of the use of nanomaterials in wastewater remediation in a developing country on the identified values (c.f. Table 14.2). The negative impacts on water ethics principles were deemed minor, because the use of nanomaterials in the wastewater remediation systems does not significantly worsen ethical impacts caused by existing alternatives in the context of a developing country with limited access to clean water. The potential beneficial contributions to water ethics were deemed high because the new system significantly remediates the current violation of water ethics principles in the target countries.

Regarding the likelihood and intensity of violation of the voluntary global code (2018), moderate risks were expected that researchers and companies developing and testing nano-enabled wastewater remediation systems in developing countries will not respect the values of fairness, respect, care, and honesty. On the flip side, a somewhat smaller chance was expected that these values will be respected in practice.

Table 14.2 Assessing the likelihood and intensity of violation of ethical values

Ethics risk	Identified principle or value	Ethics benefit
1	Water ethics: equity, multiple and beneficial use of water, users and polluters pay principle	4
2	Fairness, respect, care, honesty	1
2	Precaution	2

1 = minor; 2 = moderate; 3 = medium; 4 = high; 5 = severe

As for respecting the precautionary principle, both moderate risks and moderate opportunities were expected for respecting this principle in a practical project.

14.3.3 Balancing Trade-Offs

Nanotechnology-enabled wastewater remediation technologies are often developed for contexts or applications in developing countries where adequate solutions are lacking, since they are currently more expensive or less effective than standard water remediation techniques used in developed countries. In those contexts, municipal governments, and industrial or mining companies may not have cost-effective wastewater remediation systems in place and pollute local drinking water resources with microbial, chemical, and metallic contaminants. In those circumstances, trade-offs may occur in the short to medium term between the precautionary principle, which could lead to delays in the development of innovative solutions, and the human right to clean water. The value conflict can be described as follows. In the target countries, many consumers currently do not have access to clean water. The nano-enabled wastewater remediation system is therefore urgently needed. However, the precautionary principle calls for taking risk management measures to address uncertain risks of nanomaterials, which may be released in the cleaned water, which could delay the implementation of the solution. This conflict may be remediated as follows. While the human right to water is more fundamental than the precautionary principle, the project partners do not have the primary responsibility for it. The consortium could engage with the national government or donors to ensure access to clean drinking water from other sources during the project and invest sufficient resources in investigating potential release and health or environmental impacts of the nanomaterials used in the wastewater remediation system.

Other trade-offs are more related to the conflicting interests of different water users, including industry, agriculture, and consumers. Those trade-offs must be tabled in participatory dialogues engaging all stakeholders. No conflicts between fundamental values are at stake in this case study. Therefore, it is not needed to find ways to solve such value conflicts by following the five rules of thumb prescribed in the EIA guidelines.

14.4 Remediating Ethical Issues

In general, three different kinds of approaches can be distinguished for remedying the ethical impacts of research and technology development: technical, legal, and social fixes (Malsch 2018). An example of a technical fix is safe and sustainable by design of nanomaterials (Amodio et al. 2021). Legal fixes can be the adoption by governments of formal (inter)national law, but also non-binding international treaties, or soft law. Stakeholders can agree on voluntary codes of conduct or other forms of self-regulation. Social fixes include public and stakeholder dialogue about the ethical impacts of research.

The first step in drafting remedial actions[8] was to collect recommendations from similar projects. Two projects were identified which had developed comprehensive ethical frameworks for water management and research in resource-poor settings, which are well-balanced frameworks for addressing societal and organizational impacts. A societal recommendation to remediate broad social impacts due to the development of nano-enabled wastewater remediation systems in developing countries was to implement the UNESCO water ethics framework (Jie Liu et al. 2011) in national regulations or policies. An organizational recommendation was to apply the code of conduct for research in resource-poor settings (Global code 2018) to remediate impacts due to the malfunctioning of organizations.

After this, more specific societal recommendations were formulated for the case under study. In some developing countries, the human right to water is currently not fully respected and wastewater is released impurified into rivers. In this value conflict, scientists, industry, environmental NGOs, local communities, and governments are involved. The value conflict is that any project targeting remediation of wastewater will positively contribute to respecting the right to water. However, in the long term, the release of nanomaterials may introduce new uncertain health and environmental risks, calling for a precautionary approach. It was recommended that the consortium should engage actively with the government, local communities, and environmental organizations to offer safe drinking water during the project, until risk assessment of the nanomaterials is performed, and the remediated wastewater is found to be safe.

In addition, organizational recommendations were formulated for different kinds of partners involved in developing and testing nano-enabled wastewater remediation systems in developing countries. These recommendations were mainly limited to traditional risk management. Producers of nano-based wastewater remediation system should perform nano-specific health and environmental impact assessment before full-scale manufacturing of the system. The prices should be affordable for funding organizations while at the same time allowing the manufacturers not to sell below production expenses.

Local industries intending to use the system to clean their wastewater should assess the available alternatives for cleaning wastewater and score them on comparative compliance with water ethics, taking a precautionary approach. They should also demand results of nano-specific health and environmental impact assessment before buying and installing the system. Local waste processing companies should engage their staff in nanosafety training before nano-enabled wastewater remediation systems enter the local market.

---

[8] http://www.enaloscloud.novamechanics.com/riskgone/eia/remedialactions/index.zul.

14.4.1 *Additional recommendations from stakeholder consultations*

In the ANSOLE DAYS & BALEWARE 2021 online conference on 4–5 February, 2021, the participants were mainly academic experts in sustainable energy in Africa and other countries. The response was limited to the question, whether performing the ethical impact assessment of nano-enabled wastewater remediation was relevant in the African context. While the proposed ethical values in the UNESCO water ethics framework and the global code of conduct for research in resource-poor settings are based on experience in non-western contexts, real-life EIAs should more explicitly take the interests and capabilities of the audience as a starting point.

During an online meeting of predominantly nanosafety experts on 14 April, 2021, participants commented on the choice of nanomaterials. The case study was based on a review of different kinds of nanomaterials used in wastewater remediation including carbon and graphene-based materials, metal–organic frameworks, and metal oxides and composites (Kokkinos et al. 2020). According to participants, the case study should be focused on TiO_2 because there is mounting evidence of nanosafety issues of those materials, and not of other nanomaterials used in wastewater remediation.

During the International Water Climate Summer School, South Africa, 6-10-22, social scientists discussed ethical dilemmas in their studies of local communities involved in water and climate research. The focus was on the interpretation of the rules protecting the rights of human participants, avoiding bureaucracy, and how to safeguard scientific quality, while taking responsibility for the impact of policies inspired by the research results on the livelihood of the participants. This social side of research should be explicitly addressed in remediation measures.

14.5 Conclusions

During the pre-screening, ethical issues were identified, related to health and environmental impacts of nanomaterials, equality, and sustainable development. In the EIA plan, the following research questions were formulated to clarify how the proposed introduction of nano-enabled wastewater remediation systems in developing countries could lead to the expected ethical impacts, and to explore their likelihood and magnitude:

- Which ethical issues are known to be raised using nanomaterials in wastewater remediation in developing countries? How serious are these issues?
- Which ethical principles or values are relevant to evaluate the identified ethical issues?
- Which remedial actions does the outcome of the evaluation suggest?
- What do stakeholders think about the identified ethical issues and recommendations?

These research questions were relevant, because they were not restricted to the issues identified in the pre-screening, allowing for identifying issues, which were originally overlooked. The pre-identified issues were however used in the search strategy, and their relevance and seriousness in the case study were clarified during the EIA process. The questions guided us through the identification and evaluation of ethical issues, the formulation of recommendations, and stakeholder consultation.

Since the case study was intended as a test case for developing the RiskGONE Ethical Impact Assessment tools, limited resources were available. In a real-life case, additional budget is needed for organizing stakeholder engagement and for buying paid literature, and more person-months for the EIA team. Since the case study was fictitious, there was no strict time schedule. In a real-life case, the EIA should be performed before starting the demonstration of the innovative system or before the market introduction of a product. This calls for a clear time schedule and monitoring progress during the EIA process.

The case study resulted in the following identified ethical issues. The unintended release of predominantly TiO_2 nanomaterials from the wastewater remediation systems may cause uncertain health and environmental impacts, and issues for waste processing. If the installed wastewater remediation systems release hazardous nanomaterials into the cleaned water, the companies or organizations selling or operating the systems will benefit, while downstream users of the water may unknowingly be exposed to risks. Careless introduction of nano-enabled wastewater remediation systems in developing countries may negatively impact people in developing countries.

Three ethical principles and frameworks were found to be most relevant to evaluating the ethical issues: the precautionary principle, water ethics, and the global code of conduct for research in resource-poor settings. The EIA resulted in the following societal recommendations. Local governments should adapt the local regulatory framework to the technological state of the art of wastewater remediation after consultations with local stakeholders. If innovative nano-enabled wastewater remediation systems are piloted in the field in a developing country, the project consortium should engage actively with the government, local communities, and environmental organizations to provide safe drinking water already during the project.

The following organizational recommendations were formulated. Manufacturers of nano-enabled wastewater remediation systems should replace TiO_2 with other nanomaterials and perform nano-specific health and environmental impact assessment before full-scale manufacturing of the system. They should also select materials and components that are affordable for local customers. Companies selling the systems should offer accompanying maintenance services and training of local technicians. Local industries intending to use the system to clean their wastewater should assess the available alternatives for cleaning wastewater and score them on comparative compliance with water ethics, taking a precautionary approach. They should also demand results of nano-specific health and environmental impact assessment before buying and installing the system. Local waste processing companies should engage their staff in nanosafety training before nano-enabled wastewater remediation systems enter the local market. These results and outcomes are salient because they are specific to the case study and each stakeholder is explicitly addressed in practical recommendations.

Acknowledgements The research reported here was supported by the European Union's Horizon 2020 project RiskGONE, grant agreement no 814425. We gratefully acknowledge comments from participants in the Virtual ANSOLE DAYS & BALEWARE 2021, 4–5 February, 2021 http://ansole.com/and online stakeholder workshops on 26–27 January, 2021 and 14 April, 2021 of the NMBP-13 projects (Gov4Nano, NanoRIGO and RiskGONE), and the International Water Climate Summer School, South Africa, 6-10-2022, Baleware and Waternet, North-West University, Mahikeng Campus, North-West Province, South Africa.

References

Amodio A, Malyska A, Markouli C, Salinas S, Sanfelix J, Van Humbeeck T (2021) Mapping study for the development of sustainable-by-design criteria. European Commission, Brussels. https://doi.org/10.2777/109081

Balakrishnan M, Srivastava N (2009) Nanotech for clean water: New technology, new rules? SciDev. https://www.scidev.net/global/opinions/nanotech-for-clean-water-new-technology-new-rules/

COMEST (2005) Nanotechnologies and Ethics. UNESCO World Commission on the Ethics of Scientific Knowledge and Technology. Paris. https://unesdoc.unesco.org/ark:/48223/pf0000152146 (guiding principles for assessing responsible development of nanotechnology)

Dotson TC (2019) The promise and perils of produced waters: intelligent trial and error as an anticipatory framework for enabling responsible innovation. J Responsible Innov 6(3):305–322. https://doi.org/10.1080/23299460.2019.1603567

EGE (2007) Opinion no. 21 on ethical aspects of nanomedicine. European Group on Ethics in Science and New Technologies. European Commission. Brussels. https://op.europa.eu/en/publication-detail/-/publication/4d7d9c99-2129-42e1-993e-c815b91f256b/language-en/format-PDF/source-77404425

Global code of conduct on research in resource-poor settings (2018). https://www.globalcodeofconduct.org/

Grimshaw D (2009) Nanotechnology for clean water: facts and figures. SciDev. https://www.scidev.net/global/features/nanotechnology-for-clean-water-facts-and-figures/

Kokkinos P, Mantzavinos D, Venieri D (2020) Current trends in the application of nanomaterials for the removal of emerging micropollutants and pathogens from water. Molecules 2020:25. https://doi.org/10.3390/molecules25092016

Liu J, Dorjderem A, Macer D, Fu J, Lei X, Liu H, Qiao Q, Yu L, Zheng Y, Sun A (2011) Water ethics and water resource management. UNESCO Office Bangkok and Regional Bureau for Education in Asia and the Pacific. Bangkok. https://unesdoc.unesco.org/ark:/48223/pf0000192256

Makoni M (2009) Nanosponges: South Africa's high hopes for clean water. SciDev. https://www.scidev.net/global/features/nanosponges-south-africa-s-high-hopes-for-clean-wa/

OECD WPN (2008) Global challenges: nanotechnology and water. working party on nanotechnology. OECD. Paris. http://www.oecd.org/officialdocuments/publicdisplaydocumentpdf/?cote=DSTI/STP/NANO(2008)14&docLanguage=En

Padma TV (2009) Arsenic: when will the clean water start flowing? SciDev. https://www.scidev.net/global/features/arsenic-when-will-the-clean-water-start-flowing-1/

Reid, K (2020). Walk for water. Your 6K versus theirs. World Vision website. https://www.worldvision.org/clean-water-news-stories/walk-water-6k

Saidi T (2016) Travelling nanotechnologies. PhD thesis, Maastricht University

Chapter 15
Using Applied Theatre in Communicating Water Management Challenges and Solutions in African Communities

Emelda Ngufor Samba, Nnenesi Kgabi, Naledi Seheri, and Damian Onwudiwe

Abstract Communicating complex scientific findings to indigenous communities has always been a challenge to researchers worldwide. How to package and disseminate vital information to communities that are essentially illiterate is a concern that is being addressed by Higher Education Institutions as well as non-governmental organisations. This is the case with water-climate action related research where findings on the effects of climate change (drought, global warming, poor water quality and in some cases water shortages) and water policies are constantly being released in book chapters and journal articles. Water management techniques, regional and national water policies and their implementation seem to be some of the central focus in this research area. The challenge most higher eduction and research centres face

E. Ngufor Samba (✉)
Performing Arts and Cinematography, University of Yaounde I, Yaounde, Cameroon
e-mail: emelda.ngufor@univ-yaounde1.cm

N. Kgabi
Unit for Environmental Sciences and Management, North West University, Potchefstroom 2531, South Africa
e-mail: nnenesi.kgabi@nwu.ac.za

N. Seheri · D. Onwudiwe
Materials Science Innovation and Modelling (MaSIM) Research Focus Area, Faculty of Natural and Agricultural Science, North-West University (Mafikeng Campus), Private Bag X2046, Mmabatho, South Africa

Department of Chemistry, Faculty of Natural and Agricultural, Science, North-West University (Mafikeng Campus), Private Bag X2046, Mmabatho 2735, South Africa

N. Seheri
e-mail: Naledi.seheri@nwu.ac.za

D. Onwudiwe
e-mail: damian.onwudiwe@nwu.ac.za

S. Suriyanarayanan et al. (eds.), *Water Management in Developing Countries and Sustainable Development*, Water Resources Development and Management,
https://doi.org/10.1007/978-981-99-8639-2_15

is how to package their findings in ways that are immedeiately accesible, culturally friendly, and easily understood by rural and poor urban communities that are most affected by inadequate water management. Applied theatre, a participatory art form that explores people's communication media in addition to accentuating cultural diversity, seems to be meeting this need in Africa. This chapter is therefore based on an experiment in using Applied theatre in packaging and disseminating water management research findings with participants of the International Water-Climate School in Mafikeng, South Africa in 2022. The chapter asserts that, a blend of indigenous African art forms and applied theatre, which in essence is participatory, can contribute in creating awareness on water management, sanitation and improved drinking water in rural and urban poor communities. The study has shown that, theatre in general and applied theatre in particular, is an effective way of communicating issues of water management challenges and solutions in African communities. Through this tool, community members, who in this case were students and teachers of the summer school and memebers of the Mafikeng community came to the understanding of the health risks involved in consuming polluted water.

Keywords Water management · Applied theatre · Communication tool · Gendered roles

15.1 Introduction

In the last three decades or so, Africa, like other parts of the world, has witnessed climate change and its inadvertent effects on the water cycle; droughts, poor water quality, and in some cases, water shortages. Inadequate environmental management techniques, absence of adequate water policies, and negligence in implementing policies where they exist have been, for the most part, the cause of this demise. In the midst of all these, women seem to be affected the most, for they are the greatest consumers of water. The situation is even more precarious in rural areas and poor urban communities where potable water is often considered a commodity of ostentation rather than a necessity as it should be. This chapter is based on the assumption that participatory applied theatre that integrates indigenous African art forms is a veritable tool to communicate research findings on water management, sanitation, and improved drinking water in rural and poor urban communities.

The Mafikeng environment and water source supplying surrounding communities were selected as a case study for this theatre performance. Water samples were collected from underground water sources in reserved areas (free from anthropological pollution) in a study. The results showed that the concentration of some toxic elements exceeded their permissible limit based on the World Health Organization (WHO) and South African Bureau of Standards (SABS) (Onwudiwe et al. 2023). It was noted that, despite serving as water sources to the surrounding communities, the purity of water is not known to the community members. Leaning on communication theory, this chapter therefore focuses on the use of appropriate communication models that meet the needs of creating awareness on water policies and regulations, sanitation, and improved drinking water, especially among women in rural and poor urban communities.

The challenge of water scarcity in Africa resulting from climate change and poor water management has been reported by researchers (Kgabi 2020; Takem 2005). The World Health Organization (WHO) reported that most African countries face water resource stress, and the need to provide reliable access to water of sufficient quantity and quality in the face of increased urbanization, a changing and unpredictable climate, and economic instability cannot be overemphasized. Decisions on water supply and climate adaptation for local communities continue to be attended on a 'budget availability/allocation' basis instead of on relatively accurate scientific information on the sustainability of available water sources co-produced with local communities.

Co-production of knowledge and sharing of sustainable practices are increasingly promoted as deliberate approaches by fostering the partnership between "producers" and "users" to create a service that is effectively tailored and targeted (Bremer et al. 2019). The history of co-defined challenges and co-produced water and climate management solutions speaks to a range of partners co-producing context-specific information.

Across the globe, several activities from agricultural, industrial, and municipal sectors discharge about 80% of wastewater back into natural water sources (i.e., rivers and oceans) without proper treatment. As a result, the availability of drinking water and freshwater for other uses has been reduced making it imperative for countries to ensure that water demand and its use are adhered to.

The discharge and water usage of industries should therefore not be treated separately but should be considered as connected in order to enhance sustainable development. The UN 2030 Agenda has fostered the worldwide community of developed and developing countries to take proper action, embrace inclusiveness, and work together toward achieving the SDGs among which is the UN Sustainable Development Goal (SDG) 6 on 'Clean water and sanitation for all'. (Angela Renata Cordeiro Ortigara 1, Melvyn Kay 2 and Stefan Uhlenbrook 1.) Applied Theatre Practitioners have taken an interest in SDG 6 among others and are experimenting with this participatory art form to engage communities in greater water awareness.

Thompson (2003), defines Applied Theatre as the theatre for people in a state of bewilderment, a state that renders people not only hopeless but unable to see opportunities in life challenges they are faced with. Hughes and Nicolson (2016 page 3) consider Applied theatre as 'an established term that is widely associated with creative practices that engage with the social, educational and political functions of theatrical processes.' They further assert that:

> "Applied Theatre emerges as a creative force that responds imaginatively to the ways in which the loci of power have become diffused and fragmented in the twenty-first century challenges, setting out fresh agendas by analysing creative forms of compassionate art-making that engage with post-humanist perceptions of a material world." (Ibid page 2.)

Jena (2017) sees it as an alternative approach and medium by which theatre can be of direct service to the marginalized urban and rural masses. Akashoro et al. (2010) also assert that theatre conscientizes and mobilises groups within a society because Africa has a rich and vibrant performing tradition, and it helps to depict social reality and encourages audience participation.

These definitions of applied theatre speak to an experimental theatrical concept that is applied to challenging community concerns (social, political, cultural, natural) to make them more comprehensible and more approachable. Theatre for development, the form of Applied Theatre exploited in this research in addition to creating awareness, focuses on producing mental and attitudinal changes among community members on a specific political or social concern over which they are bewildered.

According to Nda and Ekong (2012: 440):

> Theatre for development is one of the methods that make provision for the communal creation of the story to be performed, the acting of the story by the people themselves, the employment of improvisational creation of the dialogue by the actors and actresses, the performance of the rehearsed play before a live audience, discussions on the problems put forward through the play and a communal agreement to implement the agreed provisions.

Samba (2005: 24) quotes Beyer who posits that:

> Theatre for development makes people both the primary agents and subjects of their own development. As a popular art form, it holds promises to be fulfilled not through a narrow instrumentalism, but through its efficiency in helping them reflect on, and transform their ways of life, their personal and social being, their sameness and differences.

Being an art form that is all inclusive and makes people the determiners of their own development, the theatre for development methodology is flexible and constantly adjusts to the socio-cultural situation of workshop participants as well as their mental dispositions.

In this chapter, we highlight the theatre for development process followed and the lessons learned from the 'Applied theatre workshop' on water-climate action conducted as part of the 'International Water-Climate Summer School' held in Mahikeng, South Africa in October-November 2022. We further analyse the TFD method that was adopted and demonstrate how sensitivity to cultural diversity contributed in creating a play that was African both in content and in form. Additionally, we capture the challenges faced in creating a twenty-five minute play on the most relevant information from a multitude of findings on water management, climate action, sanitation, clean water for all, and the role gender plays in water management. Finally, we articulate the use of participatory theatre to communicate these findings in ways that are gender and culturally friendly and are easily accessible to rural as well as urban poor communities.

15.2 Methods/Study Approach

Theatre for development is a branch of applied theatre that aims at mobilizing community members to create plays on issues that concern them directly. A participatory art form, it brings together mainly members of rural communities and urban poor communities that live in a state of deprivation of some sort to reflect on their problems and challenges, propose solutions to them, and create stories that are eventually dramatized and performed in a public place. Such deprivations could be physical, mental, or material. The main objective of running an applied theatre workshop at the International Water–Climate Summer School was to offer Masters and PhD students specialized in water and climate action an alternative community engagement tool, which they could explore in communicating their research findings to communities suffering the impact of climate change, poor water quality, and water shortage. Our study therefore makes use of an integrated approach whereby the results of the research findings on the quality and quantity of water in the Mafikeng Bio-reserve as well as water management concerns constituted the main idea of the storyline of the play created and performed to an audience at the end of the summer school.

15.2.1 Composition and Activities of the Group

The notion held by Vincent et al. (2021: 42) became evident in the composition of the group and throughout the theatre for development workshop, i.e. 'The benefits of working in a diverse group should be embraced from the beginning, and an ethic of respect for differences should be fostered, taking into account that creativity may be required to enable everyone to feel comfortable to share their perspectives. That said, disagreement and debate should be encouraged (in a safe space) because they are often the starting point for new insights'.

The participants involved in the International Summer School included students from eight African countries (Tanzania, Zimbabwe, Zambia, Namibia, Democratic Republic of Congo, Nigeria, South Africa, and Cameroon) who also constituted the community. The total number of participants was 38 students (5 honours, 22 Masters, eight doctoral, and 3 postdoctoral students), with 20 of them being male and 18 female. These participants came from diverse fields of study, including Physics with Electronics, Water Resources Engineering, Chemistry, GIS, Remote Sensing, and Climate Change, Integrated Water Resource Management, Environmental Engineering, Sanitation and Wastewater Management, Demography, and Population Studies, Water Resources Engineering and Hydrology, and Theatre and Film studies. The participants' age group ranged from 20 to 46 years. It goes without saying that the participants were from different linguistic backgrounds which were reflected in the performance: (i.e. DRC (Lingala and French); South Africa (Setswana &Tshivenda); Nigeria (Yoruba); Cameroon (French and English); Zambia (Bemba); Tanzania (Swahili and Sukuma); Zimbabwe (Shona); and Namibia (Sukuma).

The summer school had three parallel yet integrated activities. First were the academic lectures and presentations on water–climate action, then were the field trips to water points and water management facilities in the Mafikeng neighbourhood, and finally was the applied theatre workshop. The content and TfD methodology adopted for this workshop was greatly influenced by the personal lived experiences of the students and lecturers, the discussions between and among the teachers and the students, and finally the findings during the field trips.

15.2.2 The Study Area

The study was conducted in Groot Marrico Bioreserve in the North-Western province of South Africa. Four points were selected as the study sites in the bioreserve (Fig. 15.1). The study sites were the Groot marrico eye, a stream that flows from the eye, the confluence point where the stream joins another stream that flows into the bioreserve, and a point after the confluence wherein the river flows out from the bioreserve. The bioreserve area lies between 25°29′18"S and 26°23′47"E. It covers an area of 36.17 km^2, which forms about 0.035% of the province's total land size (104 882 km^2). It is bordered on the east by the Marico River, the south by the Molopo River, and on the north and west by the Bechuanaland Border (Hatch 1904).[1] The total population of inhabitants in the region is 3,384, representing about 0.082% of the province's population of 4,122,854 (Statistics South Africa 2011).[2] The natural vegetation of the North-West province of South Africa is noted for its savannah, with grassland in the south-east, an arid climate in the western half of the province, while the eastern part is semi-arid. The North-West Province of South Africa is mainly driven by agriculture and mining, and thus remains a key contributor to the economy of the country. However, the province is regularly affected by droughts which greatly impact farming activities and in turn, have negative socio-economic consequences on the populace.[3]

15.2.3 Sample Collection and Preparation

All the reagents used were of analytical grade. Samples were collected in the summer of 2022; four water samples were collected from the bioreserve using polyethylene bottles. Prior to the sample collection, the bottles were thoroughly washed with

[1] F.H Hatch. The Geology of the Marico District, Transvaal. Transaction of the Geological Society of S. Africa vol. 7, Part 1 (1904) 1 – 9.

[2] www.tourismnorthwest.co.za/platinum-in-the-province/

[3] https://www.suninternational.com/sun-city.
http://www.tourismnorthwest.co.za/dr-ruth-s-mompati-district/#tab=tab-1

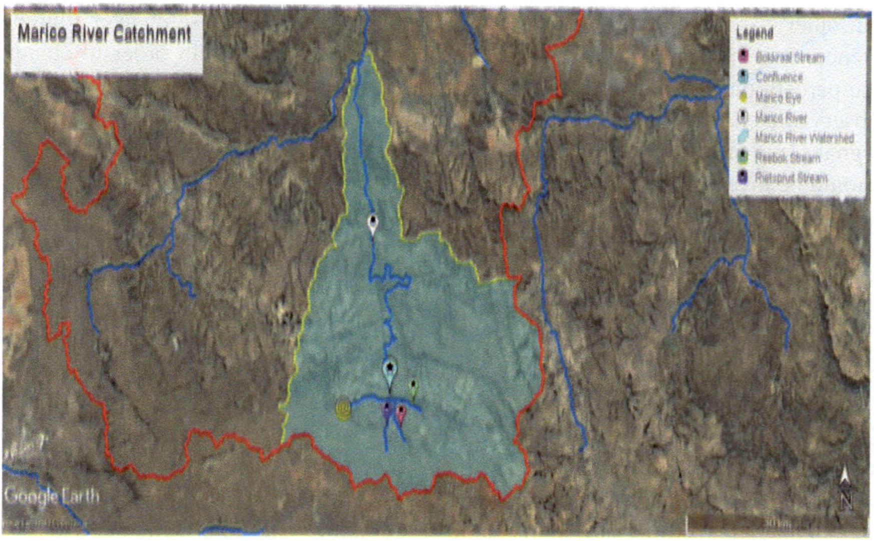

Fig. 15.1 Areal picture of Groot Marico showing the stream array

deionized water. Thereafter, all liquid samples were conveyed to the laboratory in ice bags and stored in a refrigerator (<4 °C).

15.2.4 Analysis

The collected water samples were filtered (Whatman number 4) and analyzed for determination of toxic metals using Inductively Coupled Plasma-Mass Spectrometry (ICP-MS) (Agilent 7500, USA).

Chemical Aspect

The health hazards associated with chemical constituents that are present in drinking water arise primarily from the dissolution of these chemicals in water and their ability to impact negatively on human health after prolonged periods of exposure. While some chemicals can cause serious health problems at low concentrations even from a single exposure, others may require higher concentration.

Humans generally rely mainly on their senses to assess the quality of drinking water. In some cases, water becomes undrinkable due to bad taste, poor appearance, and odour. Factors including chemical and physical constituents of water, and even the microbial components can affect the appearance of water as well as its odour or taste. In most rural communities, the quality and acceptability of the water are evaluated solely based on these criteria. Even though these may not have direct health effects, highly turbid water is characteristically coloured and could have some objectionable taste or odour. Rural communities, in extreme cases, may reject aesthetically poor but otherwise safe water for pleasant-looking but potentially unsafe water sources. Hence, it is important to create awareness of water standards in order to change the perceptions of rural dwellers. Water is odourless, tasteless, and colourless, and observable changes in the taste, odour, or normal appearance of potable water may signal some changes in the water quality or inefficacy in the treatment process and should be investigated.

Assessment of Drinking-Water Quality

Assessment of water quality involves the use of techniques, procedures, or tests to determine the performance of drinking water. This is necessary to evaluate the level of quality and compliance with the stated standards outlined by the health-based organizations and whether there is a need for revalidation.

To evaluate the chemical constituents of the water, samples were collected from the Groot Marico biosphere for analysis. The samples were subjected to both compositional and structural analysis to evaluate the constituent present in the water as well as the type of material contained in the solid samples. Table 15.1 presents the elemental composition of the water samples from different sampling zones and their comparison with the World Health Organization (WHO) and South African National Standard.

The table shows that the water sample directly from the eye has all the elements of interest within the standard limit. However, a water sample from the stream flowing from the eye, a sample collected just before the confluence point where the stream converges with another stream flowing into the bioreserve, and the sample after the confluence point, all present a high concentration of As. Arsenic is toxic, and high concentrations have deadly effects on humans, animals, and microorganisms

Table 15.1 Elemental composition of the water samples and their concentrations

Sample code	GME	CF1	SR	CF2	BH
Analyte					
Cr	0.04019	0.02453	0.03289	Nd	Nd
Ni	0.00283	Nd	Nd	Nd	Nd
As	0.00503	0.02729 #(0.02) *(0.01)	0.03432 #(0.02) *(0.01)	0.018 #(0.02) *(0.01)	0.07349 #(0.02) *(0.01)
Se	0.00029	Nd	Nd	Nd	Nd
Hg	0.00118	0.00118	0.02938 #*(0.006)	0.00047	0.0006
U	0.00072	0.09319 #*(0.03)	0.00153	0.00083	0.00118
Key: #*(both WHO & SANS); #(WHO); *(SANS)					

(Mishra et al. 2019).[4] Mercury was also detected in the sample collected from the stream, and the concentration was beyond the permissible limit. Mercury and arsenic are part of the class of elements referred to as potentially toxic elements (PTEs) due to their potential bioaccumulation and poor biodegradability. Hence, they accumulate in body organs such as the kidneys, brain, liver, and bones even for years thereby causing damage to human health (Mohammadi et al. 2019).[5] Health agencies such as the United States Agency for Toxic Substances and Disease Registry have recognized Hg and As priority contaminants (Kamunda et al. 2016).[6] The results showed that the concentration of these heavy metals in the water was above the WHO as well as the SANS standard, hence unfit for drinking.

15.3 The Workshop Process

Generally considered as 'process theatre', theatre for development entails community members engaging in reflections on their problems or challenges, dialogue on how to resolve or overcome them and decisions on actions toward positive change. Identified problems are analysed, classified and prioritised and the most important ones woven into a short (25–45 min) open-ended play. Participants cast themselves in different roles and rehearsals take place with the help of a facilitator or what Augusto Boal has termed 'a joker'. Insistence is generally not so much on the aesthetic quality of the play but rather on how well the message is communicated using indegenous communication modes; songs, dance, proverbs, music, story-telling techniques, mimes, etc. The process is often considered more important than the end product which is the play. It is during the play creation process that community members engage in reflection and dialogue on specific issues. In the Mafikeng context, the diversity in cultures and research fields of the workshop participants contributed to the rich content and cultural backdrop of the play. It came as no surprise therefore that the play was performed mainly in English and French with interjections of other national languages. The songs carrying messages on water management and environmental protection were mainly in African languages, and to make the play more inclusive, traditional dances from the different African countries involved in the production were integrated in the production. The play that was eventually produced brought across this message since it was a collective production. The play content metamorphosed throughout the summer school period as newly captured ideas introduced at different levels of the school during presentations/lectures and field trips were integrated in the play. Reflections and dialogues pointed to the fact

[4] Mishra S, Bharagava RN, More N et al. (2019) Heavy metal contamination: an alarming threat to environment and human health. In: Environmental biotechnology: for sustainable future. Springer, Singapore, pp 103–125.

[5] Mohammadi AA, Zarei A, Majidi S et al. (2019) Carcinogenic and non-carcinogenic health risk assessment of heavy metals in drinking water of Khorramabad. Iran Methodsx 6:1642–1651.

[6] Kamunda C, Mathuthu M, Madhuku M (2016) Health risk assessment of heavy metals in soils from Witwatersrand Gold Mining Basin, South Africa. Int J Environ Res Public Health 13:663.

that communities in different parts of Africa were undergoing similar challenges with respect to water quality and quantity. So even though th play was on the water points in MAFIKENG, each student had a sense of ownership to the story.

15.3.1 Data Collection, Analysis, and Prioritization

As mentioned earlier, the summer school brought together students from eight African countries who had to participate in 'interactive lectures, engagement of the participants in a theatre performance for outreach to local communities, practical sessions, fieldwork/excursions, a science slam competition, poster session, three-minute thesis competition, and an award ceremony'.[7] The theatre for development workshop was designed and implemented in a way that created room for all the actors to find space to communicate their different reflections of SDG 6 and SDG 13 based on their academic and cultural backgrounds. A blend of Augusto Boal's (1979) Forum theatre and the African theatre for development methodologies was used in the play creation process and the performance.

The first step of the workshop process was to introduce students who had very little or no notion of theatre techniques to script writing, directing, role play, the concept of costumes, etc. Augusto Boal's technique (1970) of transforming people into actors (Stage 1: knowing your body, stage 2: Making your body expressive, and stage 3: Theatre as language) became handy here. The participants were taken through basic actor training techniques and theatre games beginning with exercises for relaxation through exercises of concentration and improvization.

The next stage of the process was data collection. In mixed cultural groups, participants brainstormed on the causes of water–climate change problems and proposed solutions to the identified problems. Participants were placed into four groups making it possible for everyone to contribute to the discussions. While groups 1 and 3 brainstormed on climate action, groups 2 and 4 reflected on water, identifying problems relating to them as seen in Figs. 15.2 and 15.3.

The outcome of the reflections by groups 1 and 3 were combined while that of 2 and 4 were equally combined as seen in Figs. 15.2 and 15.3 above. What came out clearly was the interconnectedness between climate change and water action. In the context of this chapter, we shall focus on water problems, their causes, their impact on the population and suggested solutions. In the applied theatre process, after the identification and analysis of the problem, the next step is prioritization. Collectively, the group prioritized the problems as follows: 1. Accessibility, 2. Water management, 3. Water quality, and 4. Water quantity. General opinion held that in rural areas and poor urban communities in Africa, poor water management is at the centre of the unavailability of water, poor water quality, and inadequate water quantity to meet the needs of the people. What also came across clearly in the discussions was that in the midst of all these, women and children were those who suffer the most for

[7] Extract from the Summer school program, October 2022.

Fig. 15.2 Problems related to climate action and solutions proposed by Groups 1 and 3

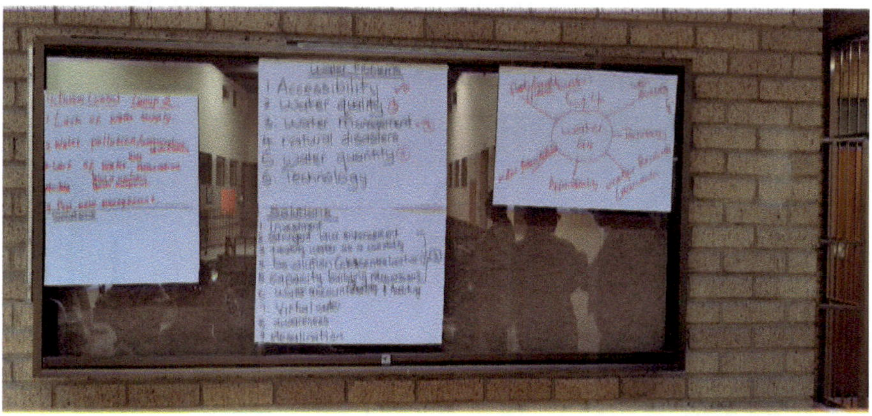

Fig. 15.3 Problems related to water and solutions proposed by Groups 2 and 4

they are those who have to fetch water for household chores and other everyday uses. To resolve these problems, the participants reiterated the need for investment, stringent law enforcement, the treatment of water as a commodity, desalination, capacity building, and awareness creation among others. All these issues made up the subthemes of the play that were collectively created.

Reflections and discussions on the findings during the field visits to some water sources in Groot Marico in the Mafikeng neighbourhood as seen in Figs. 15.4 and 15.5 strengthened the necessity to include problems relating to water accessibility, management, quality, and quantity in the play that was created.

Fig. 15.4 Field visit at the international summer school

Fig. 15.5 Field visit to Groot Marico Eye

15.3.2 The Story Creation Process

The storyline in a theatre for development workshop captures the identified problems and the different levels of analysis. It also captures the solutions proposed and how each participant relates to the different subthemes. This explains why the story is not written by a single playwright as is the case with mainstream theatre but by the entire group. In the Mafikeng situation, groups 1 and 3 worked together while groups 2 and 4 stayed together to create stories related to the discussions and the investigation carried out at the Groot marrico eye. In their groups, their stories were developed into dialogues and participants were cast in various roles. The two plays were later performed to the entire group for collective appraisal and were combined into a single play. The story was beefed up with examples from their different countries on how such problems have impacted the lives of communities and how they were resolved and the final version mirrored the reflections and lived experiences of the different participants. The play creation process took approximately two weeks and the storyline was continuously revisited based on what the students received during the lectures and their findings during the four field trips.

15.3.3 The Storyline

The play opens in a community where the people are in harmony with themselves and their environment. The climate is conducive for animals and crops. Harvest is good and the people are happy. There is good water supply and the women in the village meet at the river to do laundry and fetch water for drinking, cooking, and other household chores. There are trees in abundance that do not only provide food but also shed. The chief of the village goes out from time to time visiting his people on their farms to fellowship with them. Two technicians appear in the village with paper and pen, surveying the land and taking down notes. The villagers are curious but because they are accompanied by the mayor's attendant, they remain civil towards the surveyors. They turn out to be the messengers of two mining investors (male and female) who eventually bribe the mayor (money and a car) in order to install their mine in the community. The investors however promise the mayor that in return for the mining land, they will provide bigger pieces of land for each family and will even build them houses, and provide water and electricity. The mayor gives the money to his pregnant wife and the car to his mistress. Shortly after the investors' visit, the villagers are displaced from their original habitat, the trees are felled and the mining process begins. The heat becomes unbearable, the water changes from colourless to brown, there is sickness in the land and the people now live a life of misery. A man asks for drinking water from his wife and what he gets is some brown liquid. His wife explains that what she has offered is what the entire village now uses as water. The man refuses to drink the water and instead beats his wife for being stubborn. Her cries attract her female neighbours who come to her defence. Together they explain to the man that the water from the source is now dirty and that is what the entire village is now using. The man is still not convinced but fortunately, a friend of his comes to visit and presents the same problem of polluted water.

In the face of this water crisis, the people decide to meet with the chief to complain about the state of the water. His first reaction is that they should offer sacrifices to the gods of the land. But when the people draw his attention to the fact that the problems began with the start of the mining activities, he decides to see the mayor to concert with him. The entire village forcefully follows him to visit the mayor and coincidentally, they meet the two investors and the technicians (members of the community) at the mayor's office. The technicians revealed to the people that they had warned the mayor of the possible adverse effects of the mining activities on the environment and the health of the people. Sensing danger from the mounting expression of anger by the crowd, the investors try to escape but the people bar the way to them. At this point, the 'joker'[8] emerges and asks the audience what should be done to resolve the problem. Audience members propose a number of solutions: imprison the Mayor and his accomplices; kill the mayor and the investors. Another audience member intervenes saying that killing the 'culprits' will not solve the problem. Rather, the entire community should be sensitized to know that managing water sources is the responsibility of the entire community. Everyone should be involved in protecting water sources and ensuring water quality and quantity for 'water is life'. The play comes to an end with participants shouting out the slogan 'water is life' in their different languages. They exit the stage amidst singing and dancing.

15.3.4 The Rehearsal Process

Unlike in mainstream theatre, rehearsals in applied theatre are based on improvization. The storyline is discussed over and over and reshaped as necessary to the point where every participant can tell the story in his/her own way. At this stage, the dialogue can be written down and roles distributed to the various actors. Conversely, community members could cast themselves in roles they feel comfortable in or each actor could be allowed to say the lines as understood by them. In the case of the Mafikeng summer school workshop, the second option was adopted mainly because of the diverse linguistic backgrounds of the participants the play tapped into. Fortunately, the presence of bilingual and multilingual participants facilitated translation mainly from English to French and vice versa.

During rehearsals, the focus was on voice training (volume and articulation), body movement, and the use of stage props. Bearing in mind that this was a training-of-trainers workshop, the facilitator gave turns to participants to either lead warm-up exercises, theatre games, or different stages of the story creation process. It was hoped that with this training, participants will subsequently duplicate the technic in their communities. Additionally, to give everyone a sense of ownership, participants were encouraged to speak in the language they felt most comfortable using. The dialogue and songs were therefore in English, French, Africans, Shona, Lingala, etc. On the last day of rehearsals, there was a run-through with warm-up exercises, voice training, and line jumble. The different scenes were revisited and participants

[8] In Augusto Boal's *Theatre of the Oppressed*, (1979) the joker is the person who interacts with the public to encourage them participate in the play creation process.

Fig. 15.6 Performance at the closing ceremony

brought in costumes and stage properties from their different cultural backgrounds as seen in Fig. 15.6.

15.3.5 The Audience, the Performance, and the Performance Space

The audience in theatre for development is generally the oppressed, the underprivileged, those who are in need of change the most. The improvized play could be used as a reflection and analytical tool for the problems they face in their day to day life. Sometimes it is principally for the sensitization of the workshop participants and at other times it is for a wider audience. Generally, such plays are performed outdoor in open-air spaces (school yards, school fields, marketplaces, etc.) that can easily attract a large number of people. They are also unpaid performances considering that the main objective is the conscientization, social, mental, or attitudinal change. Augusto Boal calls the audiences in applied theatre 'spectator–actors' or 'spec-actors' for they are active participants in the story creation process unlike in mainstream theatre where the fourth wall separates the audience from the actors on stage making them passive recipients of the end product. We would here reiterate that there are no fixed rules with the TfD process. Facilitators would generally adapt to the situation at hand. The Mafikeng production was a hybrid of the mainstream theatre format and the TfD format; although the play was a collective production of the workshop participants, the play was performed indoors to an audience of Mafikeng community members, some Mafikeng officials and lecturers of the summer school.

Interaction with the audience was timid probably because the audience was new to the concept of participatory theatre and also because the 'joker' herself was not confident that if asked to speak or come on stage, the audience members could do that. From the responses by the few who could integrate the performance, it was however evident that the play had effectively communicated the effects of poor water management on communities.

15.4 Results and Discussions

One of the main objectives of integrating applied theatre in the Mafikeng summer school project was to demonstrate the use of this art form in communicating complex scientific concepts to indigenous communities that lack access to reading and writing. Many studies have already been carried out to this effect (Samba 2005, 2014, 2021; Takem 2005) and it has been proven that participatory theatre is a more inclusive approach towards critical consciousness since both the workshop participants and the facilitator whom Augusto Boal has called the 'Joker' resign from their position of objects and take up that of subjects who engage in the process of critical thinking for social transformation. Finn et al. (2017: 467) assert that 'When assisting the participants in the creation of plays or short workshop scenes, the joker does not mandate content, narrate story, or offer solutions, rather she facilitates dialogue providing a process by encouraging the participants to think critically about the nature of the situation illustrated.'

Samba (2005: 24) asserts that, 'Sustainable development stresses the importance of meeting the needs of the present without jeopardising the chances of future generations and the need for ensuring that the poor and the powerless receive their fair share of what is popularly known as "the national cake"'. For any equitable distribution of resources to take place, there is a need for political systems to involve their citizens at all levels of decision-making through participatory methods. One of such methods is theatre, specifically, Theatre for Development (TfD), which engages marginalized and oppressed groups in personal and social construction.

The specificity of the summer school applied theatre workshop created opportunities to experiment with new forms. The multicultural and multilinguistic character of the group composition made room for the inculcation of many different languages in the same play. English was implicitly agreed upon as the main language of communication and participants who were not fluent in English were encouraged to speak in French. Similarly, there were certain emotions that could be communicated most efficiently in languages other than English and French even by participants who were fluent in these languages. Robinson in his book *Language Use in Rural Development, an African Perspective* (1996: 208) asserts that 'The importance of using the people's language lies in its ability to draw people to the [development intervention agent]'. Robinson (248) furthers this argument with Bamgbose's contention that '… the non-use of local languages in development can only reinforce the socio-economic position of "elites". Where development communication ignores the language of the very people it is supposed most to help, local participation and initiation can hardly be expected'. Exploring the different languages of the different participants contributed in producing positive outcomes: a positive group dynamics; it created a sense of ownership of the entire process by the participants; it made the play inclusive; and it gave the production an African colour.

The summer school organizers had made an effort to create a male–female balance in recruiting participants at a ratio of 20–18. The group dynamics took advantage of this and were able to dissect male–female power relationships in their different communities. In the problem identification and analysis phases of the TfD process, it emerged that apart from the linguistic diversity, the different participants had different value systems, customs, and traditions with respect to gender issues. It was generally agreed that, men are often at the centre of policy making while women and the poor are expected to execute the decisions taken. This ideology was challenged with the argument that when placed in positions of power, women equally exhibit corrupt tendencies and take actions that are detrimental to their communities. To present an objective picture of the situation, two mining investors were in the play; a male and a female. The group was however unanimous that women suffer most in the face of poor water quality, water unavailability, and poor water management since they are responsible for fetching water, doing the cooking, and other house chores while the man lives a more relaxed life. The play equally demonstrates this clearly: it is the wife who fetches water and does the cleaning while the husband is seen reading a newspaper in the reading room. The wife gets beaten for bringing home dirty water and serving her husband with it. Interestingly, though she notices that the water is dirty, she is fatalistic like all the other women who come to protect her against her husband. To them, that is the state of the water in the village and there is nothing they can do about it. It takes the reflection of two men for the village to decide to talk to the chief of the village about the poor water condition.

Another specificity the summer school applied theatre workshop took advantage of was the research carried out at the Groot Marrico Bioreserve in the North-Western Province of South Africa. The water analysis proved that water could be colourless and polluted while coloured water could be good enough for drinking. Another remarkable finding was the presence of pollutants in some stretches of the river and their absence in others. These findings confirmed what had been elaborated as water-related problems in other African countries. The group was challenged to present this information in simple and uncomplicated ways that could be easily understood by rural communities. Schrag (2018: 7) is of the opinion that in exploring art forms for a better life, 'What is at stake is not only the application of knowledge, in the form of end products that meet community goals, but also all alternate methods of knowledge production, one which includes and sets within the community, subverting traditional academic boundaries and power relations'. Some of the questions that had been raised during the problem identification and problem analysis phases featured in the play. For instance: When people are displaced for mining activities to take place, how do they get shelter, water, and electricity? What happens to their shrines and grooves? What becomes of the graves of their ancestors? What about their farms and animals?

In the play, as would generally happen in a real situation, the mayor; who is the authority in place; provides alternatives to the people who in their simplicity accept them with jubilation and dancing. When the crisis begins, these same people also explain their understanding of their predicaments in as simple a way as they experience them: children and animals are dying, there is a shortage of water and the little that is available is polluted. Some of the promises of the mayor and the mining investors are not kept. One of the women even challenges the mayor to drink the dirty water she had brought with her. The mayor's refusal to do so tells of many things: He has a different source of water (bottled water) and he is not even aware that the community had been consuming dirty water.

Ethical concerns on the management of communal property also came across strongly. Most of the participants attested to the fact that government officials are guilty of sacrificing communal well-being for personal gainand stronly recommended that this should be articulated in the play. How this was to be done in a way that was both artistic and informative remained a challenge. The mayor is made to accept money and a car from the investors as bribes to allow them to mine in the community. He displaces the people to less accommodating living areas but does not fulfill all the promises he had made. He gives a blind eye to the poor management of the water sources and the catchment areas. The mere fact of accepting bride is an indication of his awareness that handing over the people's lands to miners is unconstitutional and inhuman. He offers the money to his wife and the car to his mistress. Even though the surveyors warn him of the imminent danger: excessive heat, water shortage, and water pollution, he decides to proceed with the plans that eventually lead to the crisis in his community.

15.5 Conclusion

In this chapter, we have examined how Applied Theatre was used to foster Sustainable Development Goal 6 (SDG6)—'Ensure access to water and sanitation for all', and SDG13—Climate Action during the International water–climate summer school that took place in NWU Mafikeng Campus, South Africa in late 2022. We have demonstrated the complementarity between the arts and the sciences and how through this art form, students from different cultural and academic backgrounds could bring together their experiences and knowledge to create a story that is reflective of their individual as well as collective history as Africans grappling with the effects of climate change and poor water management. It became clear that because water is a primary commodity, it is the responsibility of every community member to work towards its sustainability: good water quality and quantity in rural as well as poor urban communities. The paper has also presented the applied theatre methodology as an alternative development communication tool (a combination of improvization, dance, song, and story-telling) students and community agents could use in community water management with Mafikeng used as a sample African community facing water challenges. The question still remains: what amount of scientific knowledge with regard to water quality should be made available to a rural community and to what end? How could we use the research findings of the Groot Marrico eye to address practical issues of inefficient water management Mafikeng itself and in other parts of Africa?

References

Akashoro OG, Kayode J, Husseini S (2010) Theatre and development: opportunities and challenges in a developing world. J Commun 1(2):107–112

Boal A (1979) Theatre of the oppressed. Pluto Press, London

Bremer S, Wardekker A, Dessai S, Sobolowski S, Slaattelid R, Van der Sluijs J (2019) Toward a multi-faceted conception of co-production of cli- mate services. Clim Serv 13:42–50

Finn JL, Capana J, Jacobson M (2017) Participatory research, popular education and popular theatre contributions to group work. In: Garvin CD, Gutierrez LM, Galinsky MJ (eds) Handbook of social work with groups, 2nd edn. The Guilford Press, London, NewYork

Hatch FH (1904) The geology of the Marico district, transvaal. Trans Geol Soc S Afr 7(1):1–9

Hughes J, Nicholson H (EDTS) (2016) Applied theatre: ecology of practices. In: Critical perspectives on applied theatre. Cambridge University Press, Cambridge, United Kingdom, pp 1–12

Jena A (2017) Theatre as a tool for social development. Theatr Str J 1(1):27, 149–155

Kamunda C, Mathuthu M, Madhuku M (2016) Health risk assessment of heavy metals in soils from Witwatersrand Gold Mining Basin, South Africa. Int J Environ Res Public Health 13:663

Kgabi NA (ed) (2020) 'Preface'. In: Kgabi NA (ed) Integrated transboundary water–climate management tools (Water security and climate adaptation in Southern Africa Volume 1), pp xxxix–xlvi, AOSIS, Cape Town. https://doi.org/10.4102/aosis.2020. BK205.00

Mishra S, Bharagava RN, More N et al (2019) Heavy metal contamination: an alarming threat to environment and human health. Environmental biotechnology: for sustainable future. Springer, Singapore, pp 103–125

Mohammadi AA, Zarei A, Majidi S et al (2019) Carcinogenic and non-carcinogenic health risk assessment of heavy metals in drinking water of Khorramabad. Iran Methodsx 6:1642–1651

Nda US, Ekong FU (2012) Communicating climate change in Africa through the theatre for development process. Ethiop J Environ Stud Manag EJESM 5(4) (Suppl.1):437–443. https://doi.org/10.4314/ejesm.v5i4.S1

Robinson CDW (1996) Language use in rural development, an African perspective. In: Fishman JA (ed). Mouton de Gruyter, Berlin, New York

Samba EN (2005) Women in theatre for development in cameroon: participation, contributions, and limitations. band 74.Baureth, Bayreuth African studies

Samba E (2014) Bole Butake's family saga at the interface of Brechtian poetics and the poetics of the Oppressed. In: Perspectives on Cameroonian Art, Archaeolgy, Archaeology and Culture, Papers in Honour of Professor Bole Butake, Ediction CLE, Yaounde, pp 65–77

Samba EN (2021) Rethinking theatre approaches in changing environmental behaviour. In: RiDE: journal of applied theatre and performance, vol 17, Issue 2. Routledge Taylor and Francis Group

Takem TT (2005) Theatre and environmental education in cameroon. Bayreuth African Studies, vol 76. Bayreuth, Germany

Schrag B, Bure KJV (2018) Make arts for a better life: a guide for working with communities. Oxford University Press, Oxford

Thompson J (2003) Applied theatre: bewilderment and beyond. Peter Lang, Stage and Screen Studies

Vincent K, Steynor A, McClure A, Visman E, Waagsaether KL, Carter S, Mittal N (2021) Co-production: learning from contexts, In: Conway D, Vincent K (eds) Climate Risk in Africa, Chapter 3

httpwww.tourismnorthwest.co.za/platinum-in-the-province/

https://www.suninternational.com/sun-city

http://www.tourismnorthwest.co.za/dr-ruth-s-mompati-district/#tab=tab-1

https://doi.org/10.1007/978-3-030-61160-6_3

http://www.tourismnorthwest.co.za/platinum-in-the-province/

https://www.suninternational.com/sun-city

http://www.tourismnorthwest.co.za/dr-ruth-s-mompati-district/#tab=tab-1

https://doi.org/10.1007/978-3-030-61160-6_3

Chapter 16
Using Desalination to Supplement the Groundwater Supply to Arid and Semi-arid Environments

Nnenesi A. Kgabi and Hilma R. Amwele

Abstract Water insufficiency is the main hindrance to socio-economic development in Southern Africa. Water accessibility changes as the quantity and quality get negatively impacted, thereby reducing the affordability and timely access to the supply services. "Desalination of saline and brackish water is a solution that can help reduce current and future water scarcity in many parts of the world" (Isaka, Water desalination using renewable energy. ESTAP and IRENA, 2013). This chapter is a review of groundwater desalination as an augmentation option for the SADC. The chapter puts more emphasis on the energy requirements and efficiency of the desalination process.

Keywords Groundwater desalination · Desalination technologies · Membrane processes · Groundwater supply · Energy for desalination

16.1 Introduction

A study by Gleick (1996) confirmed the fact, that about 23,400,000 km^3 of groundwater exists on Earth; of which 54% is saline and 46% is freshwater. However, Shahzad et al. (2017) later observed that the quantity of groundwater supplies is decreasing due to the annual increase in abstraction. The 2–3% annual extraction

N. A. Kgabi (✉)
Unit for Environmental Science and Management, North-West University, Potchefstroom, South Africa
e-mail: Nnenesi.Kgabi@nwu.ac.za

Centre for Environmental Management, University of the Free State, Bloemfontein, South Africa

H. R. Amwele
Department of Agriculture and Natural Resources Sciences, Namibia University of Science and Technology, Windhoek, Namibia

S. Suriyanarayanan et al. (eds.), *Water Management in Developing Countries and Sustainable Development*, Water Resources Development and Management,
https://doi.org/10.1007/978-981-99-8639-2_16

rate measured by Shahzad et al. (2017) for the world shows the need for intensified research and innovation initiatives to augment groundwater resources. Several programmes including desalination and artificial recharge have been implemented for the preservation and augmentation of groundwater supplies worldwide. Artificial recharge is normally used to increase the quantity of groundwater, and to dilute the contaminants, thereby improving the quality. Desalination provides 1% of the world's water supply today, and by 2030, it is possible that the amount may double (Patel 2018).

More than half of the African continent is classified as semi-arid to arid with a variable climate (see Fig. 16.1) and limited freshwater supplies. Flanders Investment & Trade (2018) recorded "the water usage of 77% surface water, 9% groundwater and 14% reuse of return flows for the African continent". According to MacDonald et al. (2021), "much of Africa's water supplies are based on groundwater, and as demand for safe water rises, groundwater development is increasing". The map in Fig. 16.1 also shows the impeding challenge of water scarcity, which is linked to the aridity and low recharge for the northern and southern parts of the African continent.

The fact that most southern African countries are either arid or semi-arid suggests that the countries are also drought-prone. Also, Foster et al. (2012) reported that groundwater resources typically exhibit substantial drought resistance, though this varies depending on the amount of storage that can be drained from the aquifer. Thus, the sustainability of groundwater resources is crucial for the southern African region. However, the low recharge values depicted in Fig. 16.1 for southern Africa confirm the challenge of water scarcity in the region. Manju and Sagar (2017) also noted

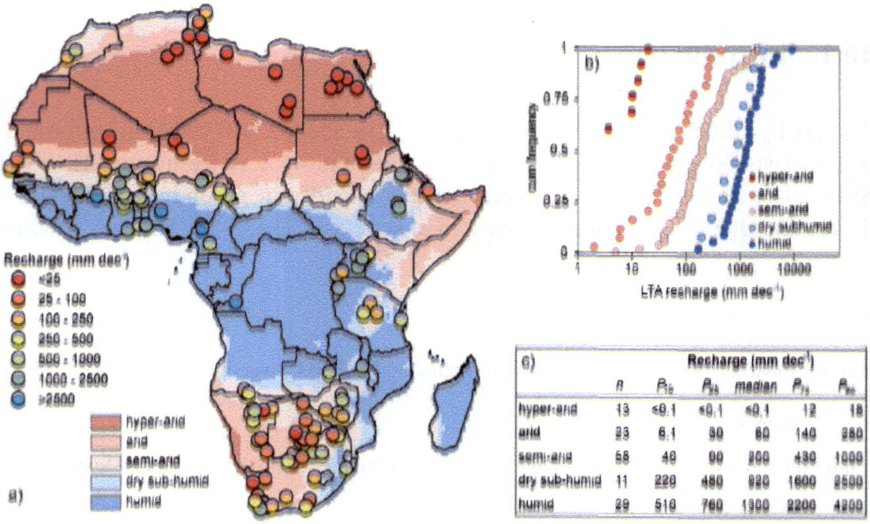

c)	Recharge (mm dec⁻¹)					
	n	P_{10}	P_{25}	median	P_{75}	P_{90}
hyper-arid	13	<0.1	<0.1	<0.1	12	18
arid	23	6.1	30	60	140	280
semi-arid	58	40	90	200	430	1000
dry sub-humid	11	220	480	920	1600	2500
humid	29	510	760	1300	2200	4200

Fig. 16.1 Long-term average (LTA) groundwater recharge for Africa; **a** LTA recharge according to aridity zone; **b** LTA recharge for each aridity zone; and **c** percentiles of recharge estimates. *Source* MacDonald et al. (2021)

that, the over-harvesting of groundwater resources and the increase in environmental contamination exacerbate the scarcity of fresh-water resources across most Southern African countries.

Therefore, there is an urgent need to supplement the groundwater resources in the region.

Desalination

According to Isaka (2013), "desalination of saline and brackish water is a method that can help many regions of the world with their water shortage now and in the future". Hence, the increased demand of brackish water desalination especially in heavy water consuming sectors, such as mining and manufacturing industries. Therefore, this encourages the smaller scale desalination activities to recover the water consumed by industries during processes (Patel 2018).

Desalination of groundwater and seawater depends on the level of salinity in the water. It takes a longer time to remove salts from highly saline water, and the removal takes more energy. This makes the desalination process to be expensive. Gobin et al. (2019) also reported that water supply systems have high energy demand for each step of water treatment (including desalination) level throughout the network. The water quality, desalination, and price were also confirmed by Turner et al. (2015) who reported a direct correlation between the price of producing water, and the water quality. Papapetrou et al. (2010) also observed that the high cost of energy used in desalination plants, especially the ones that employ non-renewable energy, may influence the price of water since water pricing is a factor in energy cost.

The operation process of desalination plants has a negative impact on water prices and, depending on the type of energy utilised, the desalination process may impact the environment negatively through greenhouse gas emissions (Isaka 2012; Papapetrou et al. 2010). In addition, the desalination process of seawater is high priced than bulk water charges, and the costs are usually passed onto the water user (Flanders Investment & Trade 2018). Using renewable energy sources to run desalination processes could alleviate the impacts of water prices and greenhouse gas emissions.

16.2 Study Methods

A review of articles chosen from open-source literature was conducted, focussing on desalination studies published in the past twenty years, i.e. between 2001 and 2021.

16.2.1 Selection of Studies/literature

Desalination studies from South Africa and Namibia were chosen for evaluation using purposeful sampling. The researcher chose the South African and Namibian studies that appeared first in the results of a Google search and were readily available (mainly open access). The type of location (rural or urban) was not considered in the selection because the focus was more on studies that linked desalination processes with the type of energy used, irrespective of the location. There are no implications for rankings or merit in the study selection. There is no intention to rank the study done in the two nations. Additionally, information was accessed from sources outside of organisations, mostly from government databases, published research networks and online sources.

16.2.2 Content Analysis

Comparisons, looking for themes and trends, creating a logical chain of evidence, and creating conceptual theoretical coherence were all part of the content analysis process for generating conclusions. A description of the groundwater desalination plants was deduced from some descriptive information that came from the content analysis.

16.3 Energy Required for Water Desalination

"Desalination uses 75.2 TWh of energy annually, or 0.4% of the world's electricity, hence it is considered the most energy-intensive method of treating water" (Shahzad et al. 2017). The energy needed to pump water also depends on the distance between sources, flow rate and desalination plant efficiency and the storage and/or consumer point. The quality of the water (e.g. salinity, floride levels, etc.) also adds to the energy demand for desalination (Shahzad et al. 2017).

According to Manju and Sagar (2017), "desalination methods utilise a lot of energy, which increases greenhouse gas emissions, thus raising concerns about environmental contamination". Therefore, using renewable energy-powered desalination technologies has a better potential to sustain sufficient fresh water supply to water-scarce communities. Shahzad et al. (2017) also reported that the three most important benefits that the world will gain from utilisation of renewable desalination processes include environmental, future freshwater and energy sustainability.

The availability (Fig. 16.2) and the cost of energy is an important consideration.

According to Avlonitis et al. (2003), "the average energy required to convert saline water to fresh water for a range of salinity and recovery levels = 1 kJ/kg (1.05 kWh/kgal)". It is also crucial to note that water treatment operations require a significant amount of energy (Plappally and Lienhard 2012). The energy utilisation during treatment is usually high when the water is highly contaminated with more impurities (Plappally and Lienhard 2012).

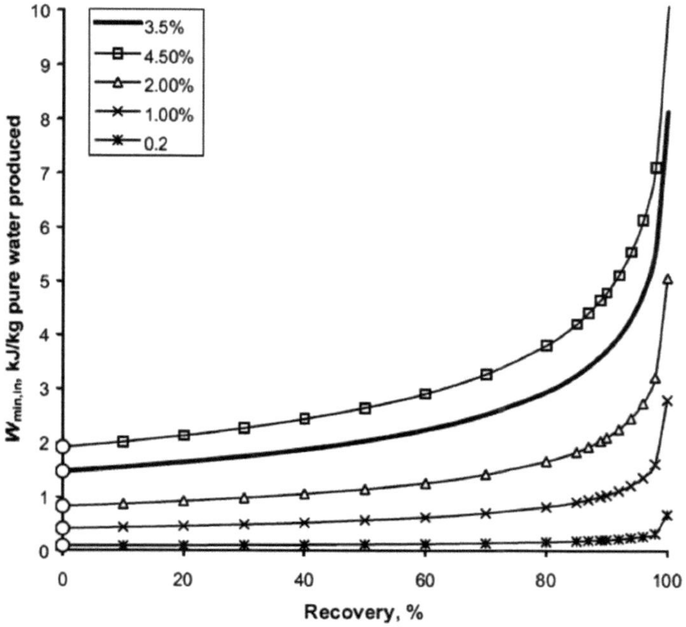

Fig. 16.2 Energy recovery per quantity of water produced during desalination. *Source* Avlonitis et al. (2003)

16.3.1 Renewable Energy Desalination Technologies

It is important to remember that fresh water and energy sustainability of any desalination process also depend on the type of renewable energy and membrane technology used. Thus, it is crucial to consider the two main categories of commercial desalination technologies, namely processes that are thermally driven (multistage flash (MSF), multiple effect distillation (MED), and electrodialysis (ED)) and membrane separation (reverse osmosis (RO)) as outlined by Shahzad et al. (2017). There are 131 renewable-powered desalination plants in the world, and this produces 1% of the desalinated water used worldwide (Shahzad et al. 2017). The status of sustainable power-operated desalination technologies is summarised in Fig. 16.3 in terms of capacity, production cost and technology trends. It is therefore crucial for water suppliers and researchers to learn more and assess the feasibility of using renewable energy as the main power source for desalination processes.

Current use of renewable energy includes "43% solar photovoltaic (PV), 27% solar thermal, 20% wind and 10% hybrid" (Patel 2018). The main challenge with the utilisation of PV is the massive land area required for installation.

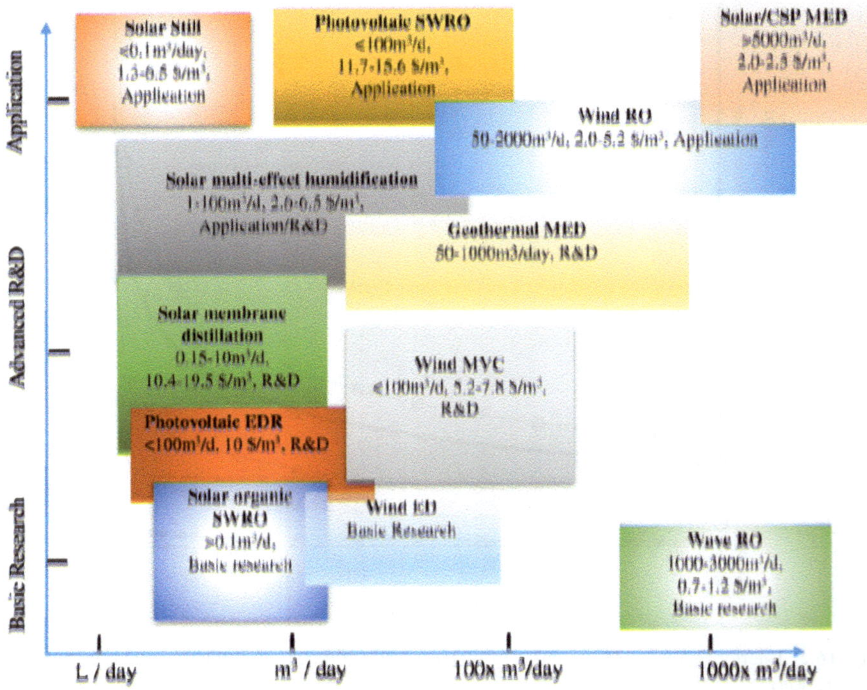

Fig. 16.3 Capacity, production cost and trend of desalination technologies using renewable energy. *Source* Shahzad et al. (2017)

The primary drivers for desalination reported by Patel (2018) include "water supply requirements and socioeconomic development, while the barriers were identified by the authors as; being fit for purpose, cost savings, ease of implementation, environmental sustainability, and energy security".

It is crucial for Southern African countries to note that, though there are challenges of capacity and costs that need to be navigated in relation to desalination plants and technologies, desalination of groundwater remains a feasible solution to the global freshwater deficit (Manju and Sagar 2017). Different renewable energies and types of desalination systems are shown in Fig. 16.4 (Manju and Sagar 2017).

16.3.2 Desalination Membrane Processes

Most countries in Southern Africa choose the type of desalination technologies/processes according to the amount of electrical energy consumed. The siting of desalination plants does not always take into consideration the typical size of the desalination unit (m^3/day) as the space/land is not a challenge. Table 16.1 shows the comparison of the reported energy consumption by various desalination processes.

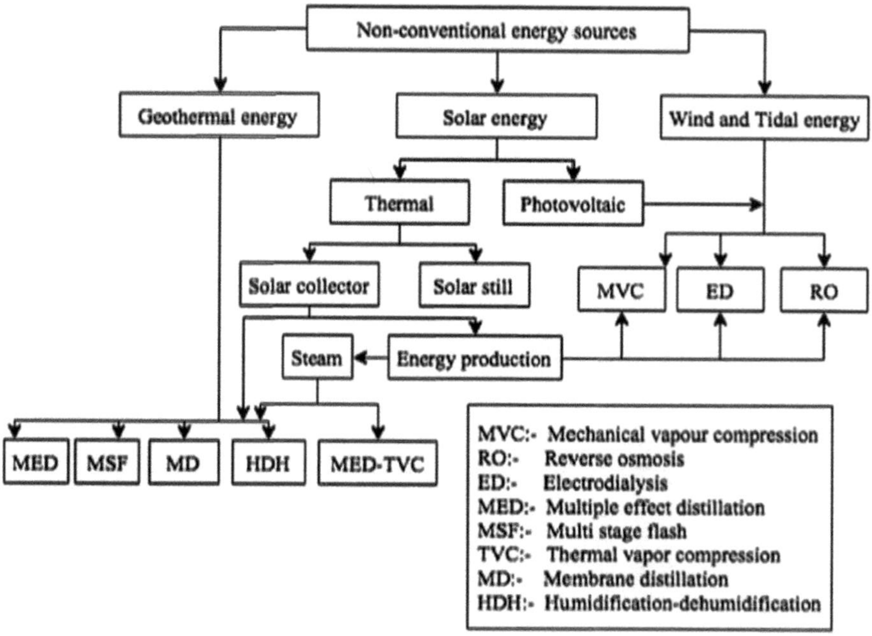

Fig. 16.4 Desalination systems and renewable energies. *Source* Manju and Sagar (2017)

Reverse Osmosis (RO) is the preferred desalination technology in Southern Africa because of the low total equivalent energy consumption of 2–5.50 kWh/m^3 (Table 16.1). Electrodialysis (ED) also uses relatively low energy in the range 2.64–5.50 kWh/m^3 (Table 16.1) however, according to Manju and Sagar (2017), "ED is solely used for brackish water distillation, while RO can be used to distil both brackish water and sea water". The effective removal of salts (water quality improvement) is also a crucial property in the desalination process.

According to Turner et al. (2015), "in Southern Africa, many desalination applications, such as the desalination of groundwater and seawater, favour the reverse osmosis (RO) membrane technique. The RO uses less energy than distillation methods". A study by Dashtpour and Al-Zubaidy (2012) reported that, typically a reverse osmosis plant requires electric energy of 3–10 kWh to produce 1 m^3 of freshwater from seawater (see Fig. 16.5). An increasing trend for seawater desalination was also observed by Shahzad et al. (2017) from "2.0 Mm^3 per day in 2005, to 3.5 Mm^3 per day in 2008" and further increase is also foreseeable in future.

Reverse osmosis processes were also found to be dominant in the brackish water treatment market (Shahzad et al. 2017).

Table 16.1 Energy utilisation in desalination process

	Multi-stage flash (MSF)	Multiple effect distillation (MED)	Thermal vapour compression (TVC)	Mechanical vapour compression (MVC)	Reverse osmosis (RO)	Electrodialysis (ED)
Typical unit size (m^3/day)	50,000–70,000	5,000–15,000	10,000–35,000	100–2,500	24,000	2–145,000
Electrical energy consumption (kWh/m^3)	4–6	1.50–2.50	1.80–1.60	7–12	2–5.50	2.64–5.50
Thermal energy consumption (kJ/kg)	190–390	230–390	187–227	None	None	
Electrical equivalent of thermal energy (kWh/m^3)	9.50–19.50	5–8.50	14.50	None	None	
Total equivalent energy consumption (kWh/m^3)	13.50–25.50	6.50–11	16.26	7–12	2–5.50	2.64–5.50
Water quality (ppm)	≈ 10	≈ 10	≈ 10	≈ 10	200–500	150–500

Source Manju and Sagar (2017)

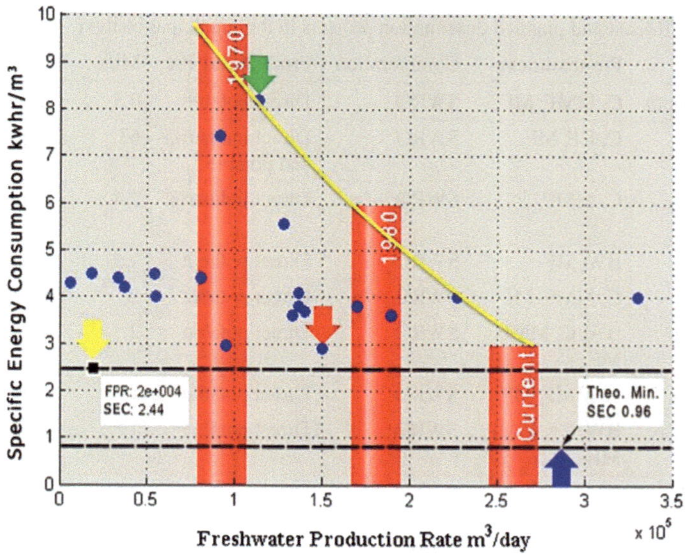

Fig. 16.5 Specific energy consumption during freshwater production. *Source* Dashtpour and Al-Zubaidy (2012)

16.4 Groundwater Desalination in South Africa

South Africa's water supply consists of 77% surface water, with the remaining 22% being groundwater and water reuse (Gobin et al. 2019). The reliance on surface water is not sustainable due to the unfavourable climatic conditions (i.e. rainfall variability, decreasing annual rainfall and high evaporation rates) linked to the aridity of the country (see Fig. 16.1).

Unlike Namibia which does not have access to surface water sources (except for shared watercourses) and depends almost entirely on groundwater, the saline or brackish groundwater in South Africa is said to provide a secondary alternative source of water (Swartz et al. 2006). Flanders Investment & Trade (2018) also observed that utilisation of groundwater desalination technologies has not received sufficient attention in South Africa, and most of the existing desalination plants are industry-owned and service-specific industrial processes, with a limited number of municipal renewable energy-powered desalination plants. For example, the first solar desalination plant in the Western Cape province was launched by the provincial government in the Hessequa Municipality in July 2018 (Flanders Investment & Trade 2018). Other desalination projects (planned and operational) are shown in Table 16.2.

South Africa fairly established approaches to the selection of membrane technologies for desalination. "Since the middle of the 1980s, desalination and membrane technology have been used commercially in South Africa" (Turner et al. 2015). According to Turner et al. (2015), despite historically lagging behind its industrial

Table 16.2 Recent and planned desalination projects in the municipal context

Location	Pre-treatment	Configuration	Product end use	Ml/d	Status
Robben Island	C, MMF, MF	SWRO	Direct potable	0.5	Operational
Trekkopje, Namibia	C, UF, MF	SWRO	Direct industrial and potable	63	Operational
Transnet, Saldanha	C, MMF	SWRO	Direct industrial	2.4	Operational
Sedgefield	BW, MF	SWRO	Direct potable	1.5	Operational
Knysna	C, MMF, MF	SWRO	Direct potable	2	Operational
Plettenberg Bay	BW, C, MMF, MF	SWRO	Direct potable	2	Operational
Mossel Bay	C, MMF, MF	SWRO	Direct potable	15	Operational
Lamberts Bay	BW, C*, MMF, MF	SWRO	Direct potable	1.7	Built
Saldanha Bay	N/A	SWRO	Direct potable	8.5	Feasibility and tender
Port Nolloth	N/A	SWRO	Direct potable	1.5	Feasibility
Umgeni Water	N/A	SWRO	Direct potable	150	Feasibility
City of Cape Town	N/A	SWRO	Direct potable	150	Feasibility
Chintsa, Amathole DM	N/A	SWRO	Direct potable	1.5–4.5	Feasibility and tender

BW = beach wells, C = coagulation (*Lamella settler), UF = ultrafiltration, DMF = dual media filtration, MMF = multi-media filtration, MF = microfiltration, SWRO = seawater reverse osmosis
Source Turner et al. (2015)

counterparts, municipal desalination and reuse installations (see Table 16.2) have seen a major uptake in activity in recent years.

In South Africa, the sectors of water and energy are closely entwined (Gobin et al. 2019). The different desalination procedures and their energy intensity are shown in Table 16.3.

Techno-economic risks associated with desalination include high capital costs, energy expenses, unpredictable future precipitation, and environmental costs (Patel 2018). "Desalination requires a lot of chemicals and electricity; it was discovered that it has the highest cost of producing water. Nearly 51% of the entire running costs for the desalination treatment process are related to the electrical component" (Patel 2018).

Le Guern (2016) described two types of desalination systems operated in South Africa namely, fixed infrastructure plants and portable packaged plants Table 16.4 summarises a few types.

Table 16.5 shows a description of the few desalination projects that were planned in 2017.

Table 16.3 Desalination technologies' energy intensities

Process	Thermal energy (MJ/m^3)	Electrical energy (MJ/m^3)	Total energy (MJ/m^3)
Multi-stage flash (MSF) distillation	9–43.2	9–14.4	18–57.6
Multiple effect distillation (MED)	0.72–25.2	4.32–7.2	5.04–32.4
Seawater reverse osmosis (SWRO)		9–16.6	9–16.6
Brackish water reverse osmosis (BWRO)		1.08–9	1.08–9

Source Gobin et al. (2019)

Table 16.4 Types of existing desalination plants

City/town	Capacity (ML/day)	Desalinated water type	Operating company	Status/comments	Reference/s
Mossel Bay	15		Veolia	Deferred indefinitely	Cloete (2017) WRC (2015a)
Knysna	2	Seawater	Veolia	Incorporates a system for energy recovery to reduce energy expenses	Patel (2018) Veolia (n.d.)
Plettenberg Bay	2	Seawater	Veolia		Veolia (n.d.) WRC (2015a)
Cederberg	1.7	Seawater		Planned to upgrade to 5 ML/day	
Sedgefield			Knysna Municipality		WRC (2015b)
Cannon Rocks and Boknes Communities	750 m^3/day	Groundwater			Veolia (n.d.)
Richard's Bay	10,000 m^3/day			Mobile desalination plant	Patel (2018)

South Africa is host to local manufacturers of desalination technologies whose services are exported to countries/regions such as the Middle East (Patel 2018).

Waste handling: Swartz et al. (2006) also noted the challenge of managing the waste (brine) generated during the desalination process. Disposal of the brine increases production costs for the water service provider. "The quality of feed water, pre-treatment procedures, desalination procedures, water recovery, post-treatment,

Table 16.5 Planned desalination projects in South Africa

Project name	Client	Projected cost	Description
Amatola Water BWROs	Amatola Water	R60 million (US$6.6 million)	"As part of the Ndlambe Bulk Water Supply Scheme, three brackish water reverse osmosis (BWRO) facilities are being built. The existing 750 m^3/day BWRO plant at Cannon Rocks will be expanded and enhanced to a new capacity of 6000–6500 m^3/day. A new 5000 m^3/day BWRO facility is proposed for Port Alfred, while a new 7000 m^3/day BWRO facility is planned for Dabi" (Patel 2018)
Saldanha Bay	West Coast District Municipality	R500 million (US$46 million)	"A 25,500 m^3/day desalination facility was planned to feed 22 West Coast Region communities. Depending on the demand, the plant would be completed in three stages of 8500 m^3/day each, but the accompanying infrastructure, including the intake and outfall, would be developed to handle the full capacity right away" (Patel 2018)
Durban SWRO	Umgeni Water	R5.1 billion (US$382 million)	"Up to two 150,000 m^3/day SWRO plants each to service Pietermaritzburg and Durban. There are two probable locations: one on the south coast at the Lovu River and the other in the city's northern most section next to the Mdloti River. There were signs that this investment will be postponed because of rising electricity prices" (Pillay 2017)
Cape Town SWRO	City of Cape Town	n/d	"Building a desalination plant with a 150,000 m^3/day starting capacity that can be increased to 450,000 m^3/day to serve the Cape Town municipality. The precise capacity and technology were still to be established" (Patel 2018)
Zandkopsdrift rare earths project SWRO	Sedex Desalination	n/d	"A brand-new SWRO facility with a 22,000 m^3/day (up to 8 million m^3/year) capability to provide process and potable water to the Northern Cape's Zandkopsdrift rare earths production. The facility was planned to use an open water intake with coarse and fine screens and dissolved air flotation (DAF)/dual media filtration (DMF) pre-treatment processes" (Patel 2018)

(continued)

Table 16.5 (continued)

Project name	Client	Projected cost	Description
Kouga desalination plant	Kouga Local Municipality	n/d	"The Kouga municipality was looking into the viability of constructing a desalination plant and wants to start by conducting a feasibility study. The viability of a desalination plant would be compared to a number of options, such as the construction of a new wellfield to supplement current groundwater supplies and a recycling facility that would treat sewage effluent to potable levels" (Patel 2018)
Elysium SWRO	Umgeni Water	n/d	"The construction of a 2500 m^3/day SWRO facility to supply potable water for pumping to the Elysium Reservoir south of Durban. In the future, the plant could be expanded to produce 10,000 m^3/day" (Patel 2018)

Source Patel (2018)

and concentrate mixing will all have an impact on the nature of the brine that is produced" (Swartz et al. 2006).

Flanders Investment & Trade (2018) observed that "saline groundwater requires treatment with reverse osmosis, but the challenge of disposal of brine often exceeds municipal wastewater discharge limits for electrical conductivity".

16.5 Groundwater Desalination in Namibia

Namibia is an arid to semi-arid country with scarce and unpredictable rainfall (Von Oertzen and Schultz 2008). Due to the scarcity of rainfall in the country and continual expansion of the population, business and agricultural activity, and the rising "water demand, the balance between water demand and availability has reached a critical level" (Kambanda 2013).

Namibia's total renewable natural water resources are estimated to be 45 km^3/yr, however only 6.2 km^3/yr (4.1 km^3/yr surface water and 2.1 km^3/yr groundwater) are produced domestically (Kunene River Awareness Kit 2021). The maximum potential annual amount of water that is really accessible for the nation at any given time corresponds to the total amount of naturally renewable water resources. Also, in many areas, the quantity and quality of the available water supplies declined due to pollution and overuse of surface and groundwater aquifers (Rahaouia et al. 2017), which further increases the demand for water.

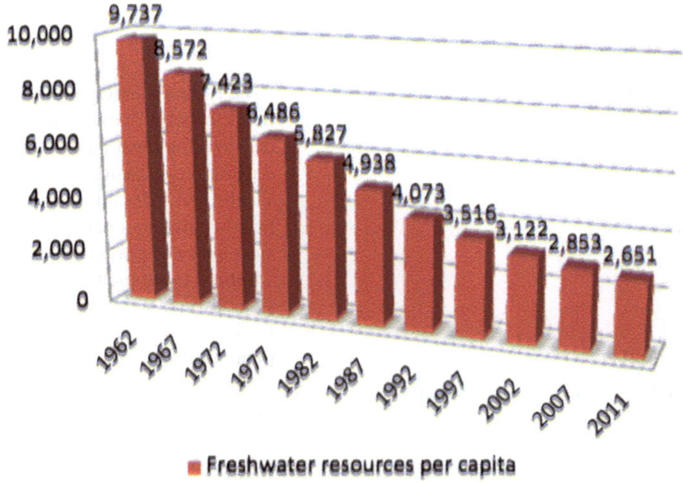

Fig. 16.6 Namibia freshwater resources per capita. *Source* Kgabi and Mashauri (2014)

Figure 16.6 shows that freshwater resources have steadily declined over the previous 49 years, going from 9,737 in 1962 to 2,651 in 2011 based on a report by Kgabi and Mashauri (2014).

In Namibia, different energy sources (e.g. hydroelectricity, coal-fired (electricity), diesel-fired and solar) are used to desalinate both seawater and brackish groundwater. However, the efficiency of desalination facilities powered by renewable energy needs to be increased because of the nation's energy and water shortages.

As part of the CuveWaters project, three cutting-edge small-scale desalination units were set up in the villages of Akutsima (Fig. 16.7) and Amarika in the Omusati region in 2010 with the intention to provide the locals with clean drinking water in a sustainable manner (Liehr et al. 2015). The desalination plants were part of a collaboration between the CuveWaters Project and the Ministry of Agriculture, Water and Forestry (MAWF).

A comparison between the desalination plants is summarised in Table 16.6 putting emphasis on the key technical data including daily raw water demand, daily freshwater production, brine production and specific electric energy demand.

According to Liehr et al. (2015), the communities in the remote rural areas in the Central Northern region of Namibia use small hand-dug wells or tank trucks to carry their water. "These wells fill up during the rainy season, but during the dry season, the chemical and microbiological quality of the water quickly declines, becoming saline and typically unfit for human use" Liehr et al. (2015). The use of humidification-dehumidification (MEH) desalination (see Table 16.6) is therefore well suited for Akutsima village. Zamen et al. (2011) also observed that most distant dry regions that require modest capacity desalination systems can benefit from humidification-dehumidification (MEH) desalination.

Fig. 16.7 Groundwater desalination at Akutsima using: **a** solar energy and **b** geothermal energy. *Source* Liehr et al. (2015)

Table 16.6 Main properties of the CuveWaters desalination plants

Company	Prolaqua	Terrawater	SIJ/IBEU
Technology	Reverse osmosis (RO)	Humidification-dehumidification (MEH)	Multi-stage desalination (MSD)
Location	Amarika	Akutsima	Akutsima
Salinity of raw water	35,000 μS/cm	7,500 μS/cm	7,500 μS/cm
Salinity of product water	980 μS/cm	6 μS/cm	5–10 μS/cm
Daily freshwater production (max. reached in brackets)	3.3 m^3/d (4.7 m^3/d)	1.4 m^3/d (2.1 m^3/d)	0.25 m^3/d (0.6 m^3/d)
Daily raw water demand	14.1 m^3/d	16.7 m^3/d	0.5 m^3/d
Daily brine production	10.8 m^3/d	15.3 m^3/d	0.25 m^3/d
Specific electric energy demand	7.9 kWh/m^3	3.0 kWh/m^3	0 kWh/m^3
Specific thermal energy demand	0 kWh/m^3	436 kWh/m^3	404 kWh/m^3

Source Liehr et al. (2015)

Fresh Water Production

Table 16.6 shows a daily fresh water production of 0.25–0.6 m^3/d with a daily raw water demand of 0.5 m^3/d for MSD; and 1.4–2.1 m^3/d with a daily raw water demand of 16.7 m^3/d for MEH. This suggests that, the MEH can only desalinate a maximum of 12.6% of the raw water on a daily basis, which also implies a relatively low capacity to meet the freshwater demand of the Akutsima community. Zamen et al. (2011) studied the effectiveness and possible use of multi-stage (MSD) desalination process to improve solar humidification-dehumidification (MEH). The study showed that "with a two-stage process, daily production per solar collector area grows by more than 40%, but with a three-stage and four-stage desalination process, it drops to 4% and 1%, respectively". Liehr et al. (2015) observed that desalination expenses are heavily influenced by site factors, such as the level of salt present and the ideal freshwater quality. The MEH and MSD in Table 16.6 show a comparable reduction in salinity, i.e. reduction from salinity of 7500 μS/cm in raw water to 6 μS/cm in product water using MEH and, from 7500 to 5–10 μS/cm using MSD.

Energy Requirements

Desalination by any process involves energy. High salt concentrations demand more energy (Patel et al. 2021) leading to an increase in energy prices as the recovery process is repeated. Thus, as a small amount of water is taken out and cleaned, the remaining water has a larger salt content, requiring more energy to separate the following unit of water (Patel et al. 2021).

Namibia's solar regime is among the best in the world, with an average daily sun radiation of 6 kWh/m^2, and an enormous development potential (Meteonorm 2009). Additionally, there are many on-land wind energy facilities with installed capacities surpassing 100 MW, albeit some system upgrading would need to be done if considerable wind capacity were to be added to the mix of the country's electrical supply (von Oertzen 2009).

Geothermal potential is thought to be "greater than 100 MW, and practical tidal and wave energy potentials are more than 200 MW" (von Oertzen 2009). The real resource base and potential yields however remain largely unquantified.

Groundwater desalination using renewable energy sources seems to be the best option for the augmentation of water supply to the rural communities in Namibia. However, the lack of technical capacity and the difficulty with handling waste (brine) cannot be ignored.

Sustainability of Desalination Plants

Several factors including communication, technical capacity, availability of spare parts and waste handling are crucial to ensure sustainability of desalination plants.

The main challenge for desalination plants in remote areas is the technical capacity to ensure consistency in operation and maintenance. The desalination plants in Amarika and Akutsima are operated, repaired and maintained by a local caretaker on a daily basis, and a professional service provider (based more than 500 km away, in Windhoek) on a monthly basis. In most cases, the spare parts need to be imported from Europe. The limited technical capacity and communication challenges affect the production and supply of desalinated water to rural communities.

An ongoing groundwater desalination initiative by the Namibia Water Corporation Ltd. (also known as NAMWATER) is worth noting by the Southern African region. The pilot desalination plant in Grünau is completed and the one in Bethanie is still in progress. A technical description of the two desalination plants is summarised in Table 16.7.

NAMWATER used the lessons learned from the existing desalination plants in Namibia to inform a motivation for implementing technological approaches with

Table 16.7 Description of desalination plants

	Bethanie	Grünau
Membrane	Reverse osmosis (RO)	MF/RO
Permeate production	23 m^3/h	6 m^3/h
Feed rate		9 m^3/h
Brine/concentrate production rate	7.5 m^3/h	2.5 m^3/h
Evaporation ponds design capacity	54,750 m^3/annum	5,475 m^3/annum
Evaporation ponds description	Three ponds with surface area of 5,645 m^2 and depth of 1.5 m per pond	Three ponds with surface area of 560 m^2 and depth of 1.5 m per pond

a positive effect on socio-health. NAMWATER uses the two (pilot) renewable energy (solar and wind energy) operated desalination (see Table 16.8) to address resources and sustainability challenges in the two towns. The project is funded by the Adaptation Fund (AF).

The energy/power supply considerations and assumptions made in the design of the plants are shown in Table 16.8.

The two pilot plants would also be used to generate: "(1) understanding of how to use hybrid power supply to optimise the design, building, operation, and maintenance of small RO desalination plants, (2) information on the technical and financial feasibility and viability of the two small-scale RO desalination plants, powered by hybrid renewable energy (sun/wind) systems, including the cost of desalinated water from

Table 16.8 Power supply considerations and assumed parameters

Consideration	Assumed parameter	
Hybrid power Hybrid battery voltage Power supply quality	Wind, solar and battery storage combination 48 V DC 400 V AC, 3-phase, 50 Hz	
Power needs for RO plant Ancillary power needs	100% +10%	
Energy demand/m^3 of water Water demand—Bethanie Water demand—Grünau Solar availability Wind availability	4 kW/m^3 460 m^3/d 50 m^3/d Worst month irradiation—site specific Worst month low speed—site specific	
Energy demand	Bethanie	Grünau
Total demand (kWh)	1840	200
Demand/hour (kWh)	77	8

the project plants, and (3) awareness among students of increasing climate change resilience through desalination and renewable energy".

16.6 Discussion

Small-scale desalination plants are used in Namibia and South Africa to address the lack of surface water. The Central north part of Namibia also depends on highly saline groundwater for domestic and economic activity.

South Africa uses groundwater desalination to supplement the existing surface water supply. However, due to water shortage, municipalities have started to increase the uptake of desalination technologies.

"Namibia ranks low on the human development index, and the government encourages increased commercial activity and investment to address the low development status". However, the scarcity of water and energy affects economic activity negatively. Hence, most desalination projects for water-intensive industries such as mining are designed for processing seawater.

Small-scale desalination activities benefit water-intensive industries like mining, e.g. the Trekkopje desalination plant, a seawater reverse osmosis (SWRO) plant described in Table 16.2, producing desalinated water for the Trekkopje Uranium mine in Namibia. Trekkopje desalination plant was built to produce water for mining processes, as well as potable usage for mine workers.

16.7 Conclusion

This study showed that desalination of groundwater can help to improve water quality and availability in the SADC countries provided that the renewable energy demand, socio-economic situation, climatic conditions, communication and technical capacity are taken into consideration. One of the lessons from the chapter is for researchers to give more attention to comprehensive analysis of water quantity and quality requirements for specific desalination plants, and catchment areas.

The study was designed to provide a basis for sustainable water augmentation solutions to water scarcity challenge of the Region. The authors showed that desalination of groundwater can help to improve water quality and availability in the SADC countries provided that the renewable energy demand, socio-economic situation, climatic conditions, communication and technical capacity are taken into consideration.

Water augmentation challenges of the arid environment include continual increase in water demand. The high evaporation leads to increased salinity, which increases the energy demand for desalination, and finally, the production costs. Environmental impacts (e.g. greenhouse gas (GHG) and precursor emissions per energy source) associated with the energy consumption also add to the already high cost of water.

"A thorough examination of the various plants' needs for water, including their demands for water quality" (Turner et al. 2015) is crucial. Also, "the key water-intensive industries, such as mining, should be encouraged to engage in smaller-scale desalination activities in order to recover spent water that has been treated in industrial processes through the desalination of brackish industrial water" (Patel 2018).

Sustainability considerations for groundwater desalination in arid environments should include constant production cost, free energy source, clean energy, renewable energy and alternative uses.

Acknowledgements The authors acknowledge in-kind support by the North-West University and the UNESCO Chair on Sustainable Water Research for Climate Adaptation in Arid Environment at the Namibia University of Science and Technology.

References

Avlonitis SA, Kouroumbas K, Vlachakis N (2003) Energy consumption and membrane replacement cost for seawater RO desalination plants

Cloete K (2017) Experts unpack their experience in desalination solutions. Creamer Media's Engineering News. http://www.engineeringnews.co.za/article/experts-unpack-their-experience-in-desalination-solutions-2017-08-31. Accessed 1 Aug 2018

Dashtpour R, Al-Zubaidy NS (2012) Energy efficient reverse osmosis desalination process. Int J Environ Sci Dev 3(4)

Flanders Investment & Trade (2018) Water research report 2018—South Africa. Flanders State of the Art—Johannesburg, September. www.flandersinvestmentandtrade.com

Foster S, Tuinhof A, van Steenbergen F (2012) Managed groundwater development for water-supply security in Sub-Saharan Africa: investment priorities. Water SA 38(3):359–366. In: International conference on groundwater, special edition 2012. https://doi.org/10.4314/wsa.v38i3.1

Gleick PH (1996) Basic water requirements for human activities: meeting basic needs. Water Int 21:83–92. https://doi.org/10.1080/02508069608686494

Gobin A, Sparks D, Okedi J, Armitage N, Ahjum F (2019) Assessing the energy and carbon footprints of exploiting and treating brackish groundwater in Cape Town. Water SA 45(1):63–74. https://doi.org/10.4314/wsa.v45i1.08

Isaka M (2012) Water desalination using renewable energy. IEA-ETSAP and IRENA, p 28

Isaka M (2013) Water desalination using renewable energy. ESTAP and IRENA, Masdar City

Kambanda K (2013) Evaluation of desalination techniques for treating the brackish water of Olushandja sub-basin. Trier, Datum der wissenschaftlichen Aussprache, Germany

Kgabi N, Mashauri D (2014) Sustainable domestic and industrial water utilisation in Namibia. Eur J Sci Res 127(1):46–57

Kunene River Awareness Kit (2021) http://www.kunene.riverawarenesskit.com/KUNENERAK_COM/MANAGEMENT/WATER_DEMAND/WATERAVAILABILITY.HTM

Le Guern S (2016) KZN, WC to have desalination plants in place within months. Creamer Media's Engineering News. http://www.engineeringnews.co.za/article/kzn-wc-to-have-desalination-plants-in-place-within-months-2016-05-16. Accessed 1 Aug 2018

Liehr S, Papangelou A, Deffner J, von Nidda KA, Urban W (2015) Groundwater desalination in Central-Northern Namibia. Factsheet. CuveWaters, Institute for Social-Ecological Research (ISOE), Germany. www.cuvewaters.net

MacDonald MA, Lark RM, Taylor RG, Abiye T, Fallas HC, Favreau G, Goni IB, Kebede S, Scanlon B, Sorensen JPR, Tijani M, Upton KA, West C (2021) Mapping groundwater recharge in Africa from ground observations and implications for water security. Environ Res Lett 16:034012. https://doi.org/10.1088/1748-9326/abd661

Manju S, Sagar N (2017) Renewable energy integrated desalination: a sustainable solution to overcome future fresh-water scarcity in India. Renew Sustain Energy Rev 73:594–609. https://doi.org/10.1016/j.rser.2017.01.164

Meteonorm (2009) Meteonorm maps of global horizontal radiation and temperature. http://www.meteonorm.com/pages/en/downloads/maps.php

Papapetrou M, Wieghaus M, Biercamp C (2010) Roadmap for the development of desalination powered by renewable energy. PRODES, p 79

Patel M (2018) Desalination in South Africa: panacea or peril for industrial development? Trade & Industrial Policy Strategies (TIPS), July, p 41

Patel SK, Biesheuvel PM, Elimelech M (2021) Energy consumption of brackish water desalination: identifying the sweet spots for electrodialysis and reverse osmosis. ACS ES&T Eng 1(5):851–864. https://doi.org/10.1021/acsestengg.0c00192

Plappally, AK. & Lienhard VJH (2012): Costs for water supply, treatment, end-use and reclamation, Desalination and Water Treatment, DOI:10.1080/19443994.2012.708996

Rahaouia K, Dinga L, Tana L, Mediourib W, Mahmoudia F, Nakoaa K, Akbarzadeha A (2017) Sustainable membrane distillation coupled with solar pond. Energy Procedia 110:414–419

Shahzad WM, Burhan M, Ang L, Choon Ng K (2017) Energy-water-environment nexus underpinning future desalination sustainability. Desalination 413:52–64. https://doi.org/10.1016/j.desal.2017.03.009

Swartz CD, du Plessis JA, Burger AJ, Offringa GA (2006) Desalination guide for South African municipal engineers

Turner KN, Naidoo K, Theron JG, Broodryk J (2015) Investigation into the cost and operation of Southern African desalination and water reuse plants. Volume II: current status of desalination and water reuse in Southern Africa

Veolia (n.d.) Desalination plant. http://www.veoliawatertechnologies.co.za/water-solutions/desalination/desalination-plant/#c379an84s1. Accessed 1 Aug 2018

von Oertzen D, Schultz R (2008) Village-scale and solar desalination technology experience in Namibia. CuveWaters Papers, No. 4, CuveWaters Research Co-operation

von Oertzen D (2009) Green energy in Namibia. Electricity Control Board of Namibia. http://www.ecb.org.na

Water Research Council (WRC) (2015a) Investigation into the cost and operation of Southern African desalination and water reuse plants volume II: current status of desalination and water reuse in Southern Africa

Water Research Council (WRC) (2015b) Energy recovery devices in desalination plants

Zamen M, Soufari SM, Amidpour M (2011) Improvement of solar humidification-dehumidification desalination using multi stage process. Chem Eng Trans 25:1091–1096. https://doi.org/10.3303/CET1125182

Chapter 17
Contextual Engendering of Water and Climate Adaptation in Developing Countries

Nnenesi A. Kgabi, Naledi H. Seheri, and Damian C. Onwudiwe

Abstract The need to find sustainable solutions to the gender-specific burden in developing countries requires context-specific engendering of the water and climate sector. This study aimed to address the inadequate contextual gender mainstreaming of water and climate programmes to promote adaptation for societal and economic development in South Asia (Afghanistan, Bangladesh, India, Pakistan, and Sri Lanka) and Southern Africa (Botswana, Eswatini, Lesotho, Namibia, and South Africa). This is a secondary desktop review study augmented with the collection and analysis of data sourced from publicly accessible databases. The engendering was contextualized in line with three indicators ("the interconnected dots"), namely (1) water–climate linkages, (2) governance and policy systems, and (3) multi-stakeholder partnerships in the water–climate programmes. South Asia and Southern Africa have the highest female proportion of the total economically active population (49–68%, and 55–66%, respectively). However, relatively, high rate of unemployment among females was observed as a hindrance to the engendering process. The Women, Business, and Law (WBL) Index scores for South Asia range from 26 to 72 (below the global average of 76.5), and Southern Africa, 38 to 88. The representation of women in national parliaments varies in terms of the proportion of seats they hold, ranging from 4.9 to 27.7% in South Asia, and, 6.2 to 42.8% in Southern Africa. The proportion of women in ministerial-level positions has fluctuated between 0 and 22% in South Asia and, 12 and 48% in Southern Africa for the past fifteen years. The presence of women in decision-making processes alone is inadequate to ensure meaningful

N. A. Kgabi (✉)
Unit for Environmental Science and Management, North-West University, Potchefstroom, South Africa
e-mail: Nnenesi.Kgabi@nwu.ac.za

N. H. Seheri · D. C. Onwudiwe
Material Science Innovation and Modelling (MaSIM) Research Focus Area, North-West University, Mahikeng Campus, Mahikeng, South Africa

Department of Chemistry, School of Physical and Chemical Sciences, North-West University, Mahikeng Campus, Mahikeng, South Africa

S. Suriyanarayanan et al. (eds.), *Water Management in Developing Countries and Sustainable Development*, Water Resources Development and Management,
https://doi.org/10.1007/978-981-99-8639-2_17

advancements in gender mainstreaming. For both regions, the law guarantees equal access to property yet, in practice, gender inequality in land and assets ownership, and inheritance exists. Analysis of the gender aspects of selected national water policies also showed that in most cases, gender issues are not considered in water management policies.

Keywords Gender mainstreaming · Governance and systems · Multi-stakeholder partnerships · Water–climate linkages

17.1 Introduction

A decade ago, scholarly research on gender and climate change brought attention to the distinct gender-based variations in the understanding of water scarcity and climate change, which include abilities, priorities, perceptions, responses, and preferences within discussions on mitigation and adaptation (Brody et al. 2008; Terry 2009; Agostine and Lizarde 2012). A few years later, studies including Sultana (2018) laid a foundation for the gendered dynamics of climate change, endeavouring to heed the call to action by activists during the International Conferences of the Parties (COP) meetings in Bali, Indonesia December 2007, where the slogan "No climate justice without gender justice" was introduced as a new dimension to the climate dynamics. Sultana (2018) also holds a view that two significant shifts come to the forefront when exploring the relationship between climate change, water, and gender: the transformative impact of climate change on socioecological systems and the concurrent challenges faced by historical patriarchal societies in their gendered power structures. These transformations serve as important perspective to examine the intricate relationship among these three elements.

It is common knowledge that both men and women encounter and adjust or react differently to water insecurity and climate change challenges. However, Sultana (2018) observed that, despite numerous studies conducted on climate change, only a few scholars focussed on gender as an essential component in climate change impacts, adaptation, or mitigation. In contrast, Oluwasanya (2022) noted a distinct observation that several studies have made efforts to quantify the representation of women in the water sector on varying scales. Prior to the studies of Sultana (2018) and Oluwasanya (2022) earlier research had already emphasized the importance of conducting nuanced and context-specific gender analyses within climate change discussions, so as not to reproduce the "women only" narratives that downplay the potential role of women as proficient leaders in the water–climate space (Arora-Jonsson 2011). Dankelman (2010) also emphasized the significance of considering gender in climate change and water discussions and initiatives.

The unease echoed by scholars such as Sultana (2018) is that, most climate change and water academic and policy discourses are predominantly gender-neutral, despite the profound impacts of climate change and water insecurity being felt along gender lines. The scholar purports that the process of adapting to climate change is influenced by gender, and considering that climate change is closely intertwined with water dynamics, any discourse on climate change must thoroughly examine the gender–water relationships that exist and are being enhanced.

According to the Global Gender and Climate Alliance (GGCA nd.), "many countries in Asia are experiencing significant effects associated with climate change, and the capacity of women and men to cope varies widely across settings, depending on local socio-economic and geographic contexts". Furthermore, Sultana (2018) indicated that the presence of both extreme water events (such as floods, storm surges, cyclones, riverbank erosion, and waterlogging), as well as insufficient water occurrences (including pollution, drought, salinity, and desertification), shapes the dynamic relationship between climate change and the societies of South Asia.

For Sub-Saharan Africa, Kgabi (2020a) and SADC (2016) reported that the Region has its own share of challenges of insecurities related to water scarcity and climate change-related impacts. According to Kgabi (2020a), a brief examination of water–climate studies in the SADC region revealed that most of the studies conducted on water and climate are approached independently, without sufficient integration (Brewis et al. 2019) with limited exceptions by Ngoran et al. (2016) and Liehr et al. (2017) that aim to address the intertwined aspects of water and climate simultaneously.

The challenges for Africa and Asia are further intensified by constrained in financial resources amidst the mounting pressures of climate change and the ever-growing demand for water. While the two regions (South Asia and Southern Africa) continue to grapple with country-level socio-economic ills, the uncertainty and frequent occurrence of floods and drought continue to challenge the gender-specific roles and responsibilities within the domestic settings and societal contexts (Cannon 2002; Sultana 2010).

The rationale for this chapter is thus, the need to address the lack or inadequate contextual gender mainstreaming of water and climate programmes to promote water and climate adaptation for societal and economic development in most developing countries. The interconnectedness of the three crucial aspects, i.e. gender, water, and climate cannot be debated as the uneven gender burden of water security in the face of climate change is on the rise. The need to find sustainable solutions to the gender-specific burden requires context-specific engendering of water and climate change. The contextual engendering discussed herein is intended to connect three dots namely (1) Water–climate linkages, (2) Governance and policy systems, and (3) Multi-stakeholder partnerships in the water–climate programmes.

This chapter also affirms the view by Sultana (2018) that despite the fact that climate change is already exacerbating vulnerabilities, marginalization, and sufferings for numerous individuals worldwide, its impacts are unequally experienced among different social groups. Furthermore, factors such as race, gender, and socio-economic conditions play a fundamental role in shaping vulnerability to climate change (Douville et al. 2022). Thus, the intersectionalities of economic and social differences, particularly following gender and class lines, segregate the vulnerabilities and the ways in which water insecurity and climate change impacts are experienced. Bush and Clayton (2022) conducted a study to address the possibility that women express more concern for water–climate change to a larger degree than men within their respective their countries. The study found that the gender gap in climate change is strongly correlated with economic development. In wealthier countries, both men and women express less concern regarding climate change compared to their counterparts in poorer nations, and the decline was found to be sharper amongst men.

The UN ECOSOC (1997) expounded on the idea of gender mainstreaming as the process of evaluating the involvement of both genders (women and men) in planning, policy-making, and legislative actions at all levels of a project with consideration for the values and diversities forming a fundamental element of the planning and management of projects/programmes. Lyle-Gonga (2013) viewed gender mainstreaming simply as a shift from a discourse predominantly shaped by male perspectives to a more balanced dialogue that accurately represents the enhanced status of women in society. The purpose and benefit of gender mainstreaming in this regard carries an inherent socio-economic and political nobleness. This view is further upheld by Caba (2017) who described climate change as an economic, social, and political process with deep consequences for social justice and gender equality.

Gender is integrated into Goal 5 of the Sustainable Development Goals (SDGs), calling for the achievement of gender balance and empowerment of all women, it is essential for the integration and mainstreaming of gender to tackle the economic, social, and cultural norms associated with gender equality, fairness, and empowerment to ensure effective climate change adaptation and water security. Therefore, it is imperative for researchers in the field of water–climate studies to conduct comprehensive and intersectional analyses that are essential for providing valuable insights into academic discussions as well as informing policymaking and implementation processes.

17.2 Materials and Methods

This chapter is the result of a secondary desktop review study design based on an extensive review of literature, collection, and analysis of data sourced from publicly accessible databases.

17.2.1 Selection of Study Sites

The selection of countries and regions for this study was informed by a very wide review of open-access literature and other reports and databases in the public domain. The two regions, South Asia and Southern Africa were chosen as hosts to a list of developing countries, also based on the diversity of climatic regions (see Figs. 17.1 and 17.2).

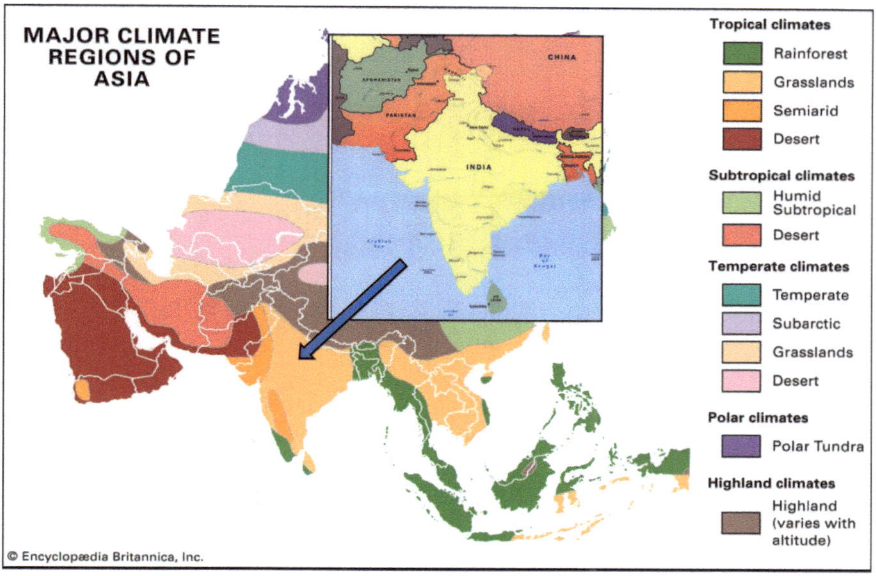

Fig. 17.1 Map of countries and climatic regions of South Asia. Source for the aridity map of Asia—https://www.britannica.com/place/Asia/Climate (nd.); source for map of South Asia—https://www.mapsland.com/maps/asia/south-asia/large-detailed-political-map-of-south-asia-with-major-cities-1998.jpg

Fig. 17.2 Map of countries and climatic regions of Southern Africa. Source for the aridity map of Africa—Koenker et al. (2019); source for map of Southern Africa—https://maps-africa.blogspot.com/2012/05/southern-africa-map-pictures.html

Five countries were chosen from the two regions—South Asia (Afghanistan, Bangladesh, India, Pakistan, and Sri Lanka) and Southern Africa (Botswana, Eswatini, Lesotho, Namibia, and South Africa) based on the population size (Table 17.1) and the female proportion of total population (Figs. 17.3 and 17.4), ranging between 48 and 52% for South Asia and 50% and 53% in Southern Africa for the period 2005–2020.

Table 17.1 Female and male population per country in South Asia and Southern Africa

Female and male population (^million)												
	2005			2010			2015			2020		
	Female	Male	Total pop.	Female	Male	Total pop.	Female	Male	Total pop.	Female	Male	Total population
South Asia												
Afghanistan	12.11	12.30	24.41	13.94	14.24	28.19	16.68	17.07	33.75	19.28	19.69	38.97
Bangladesh	69.57	71.33	140.91	74.17	74.21	148.39	79.29	78.53	157.83	84.35	83.06	167.42
India	557.16	597.47	1154.63	599.04	641.56	1240.61	639.32	683.54	1322.86	675.38	720.99	1396.38
Pakistan	84.43	89.94	174.37	94.46	99.98	194.45	102.98	107.98	210.96	112.38	114.81	227.19
Sri Lanka	9.92	9.62	19.54	10.39	9.86	20.26	10.82	10.14	20.97	11.35	10.56	21.91
Southern Africa												
Botswana	0.96	0.92	1.89	1.06	1.02	2.09	1.17	1.13	2.30	1.28	1.25	2.54
Eswatini	0.55	0.51	1.07	0.56	0.53	1.09	0.57	0.55	1.13	0.59	0.58	1.18
Lesotho	1.01	0.96	1.97	1.03	0.99	2.02	1.07	1.04	2.11	1.14	1.11	2.25
Namibia	1.01	0.95	1.96	1.08	1.01	2.09	1.17	1.10	2.28	1.28	1.20	2.48
South Africa	25.58	23.43	49.01	26.85	24.92	51.78	28.72	27.15	55.87	30.22	28.58	58.80

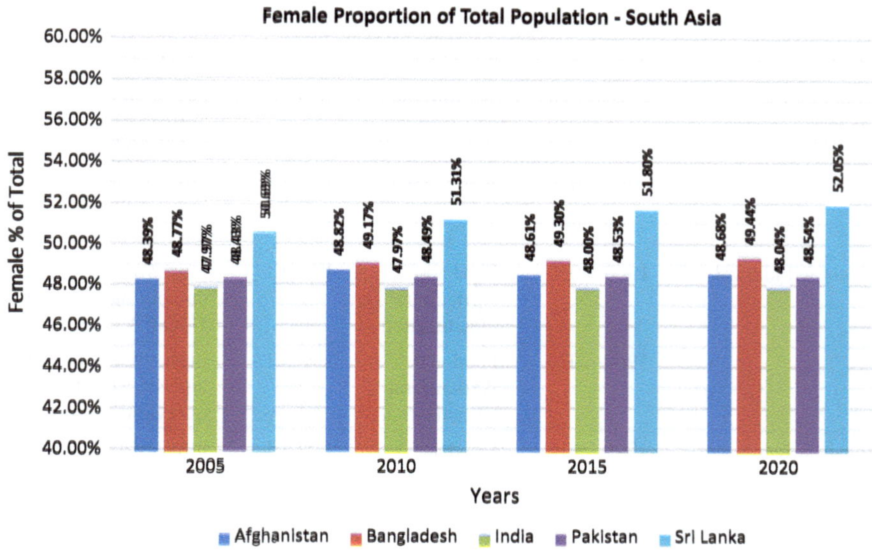

Fig. 17.3 Female proportion of population per country in South Asia

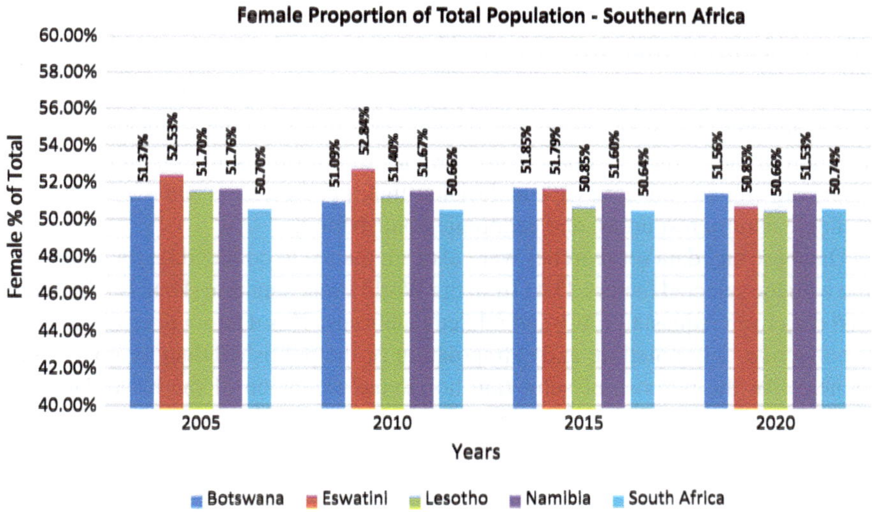

Fig. 17.4 Female proportion of population per country in Southern Africa

Nearly 600 million people are affected by water shortage each year in India (Bhaduri 2021).

Table 17.2 Types and sources of data used for graphs and tables in the study

Focus	Year	Sources	Material	References
Population	1960–2020	World Bank	EXCEL	https://data.worldbank.org/indicator/SP.POP.TOTL
	1960–2020	World Bank	EXCEL	https://www.fao.org/aquastat/statistics/query/index.html
Water withdrawal	1960–2020	Food and Agricultural Organization	EXCEL	https://data.worldbank.org/indicator/ER.H2O.FWTL.K3
	1960–2020	Food and Agricultural Organization	EXCEL	https://www.fao.org/aquastat/statistics/query/index.html
Fresh surface water abstracted	1990–2016	UN data	EXCEL	https://data.un.org/Data.aspx?q=water+&d=ENV&f=variableID%3a24
Population/Water withdrawal	–	Government or statistical bureau	EXCEL	https://doi.org/10.6084/m9.figshare.19387406.v2

17.2.2 Data Collection

A selected number of scientific and open-source literature was reviewed, and the selection of the studies was based on relevance to inform discussion on population, water, and climate, governance and decision-making, the access and management of available natural resources, and economic contribution.

Database and data set searches were also conducted to collect data for analysis in line with the selected factors. Data from the World Bank database (Data Bank), which used data from Aquastat, FAO, and related databases. This data set does the sole role of collection, interpolation, and extrapolation of the published data by government agencies for the purpose of achieving horizontal extension of the basic data (Yan et al. 2022). Table 17.2 shows the input data sets that have been used to generate the population as well as the related socio-economic data, and water withdrawal data.

17.2.3 Data Analysis

The method used to analyse the content and consequently generating conclusions is based on the observation of patterns, trends and themes making comparisons, establishing a logical chain of evidence and drawing conceptual flow of theoretical coherence. Some descriptive population data was yielded by some content analysis water availability, governance, and management of resources, thereby informing discussions of the context and possible engendering of water–climate linkages, governance and systems, and multi-stakeholder partnerships in the water–climate programmes.

Most data points are presented in two decimal places for uniformity and ease of analysis. The data on population and water withdrawal was provided by the World Bank, Food and Agriculture Organization (FAO) and, according to Yan et al. (2022), there is a possibility that various international organizations may diverge from the actual population or water withdrawal data of a country or region due to statistical quality issues. As a result, additional corrections may be required to enhance the accuracy of the time series data.

17.3 Results and Discussion

17.3.1 Water–Climate Linkages

Effective management of the effects of climate on water resources is central to gender mainstreaming in the water and climate sector. Implementation of the framework for managing the impacts of climate on water resources (see Fig. 17.5) developed by Pegasys (2009) provides a context in the developing countries. According to Kgabi (2020b), the adoption of the Framework will ensure the development and adaptation of water–climate management tools within the region. This approach will involve the active participation of all stakeholders, including both men and women, from various sectors. Their contributions will be crucial in addressing, understanding, and contextualizing the water–climate challenges, as well as in the creation and implementation of effective solutions.

Aligning gender mainstreaming initiatives with the framework will require changing of mindsets regarding the decision-making, as well as developing and implementing water–climate solutions.

The gender mainstreaming should take into consideration the relationships among multiple dimensions and modalities of social relationships and economic implications (i.e. intersectionality), particularly in programmes addressing the impacts and vulnerabilities. Sultana (2018) confirmed that, while geographical location provides the contextual backdrop, it is essential to recognize the significant social variations in the manifestation of hazards and vulnerabilities. This distinction becomes particularly pronounced when examining water-related productive and reproductive tasks within agrarian societies, which represent the majority of the developing world.

17.3.1.1 Water Availability and Climate Change Adaptation

Available evidence indicates that climate crisis has tangible impacts on agriculture and the accessibility of water resources (Tantoh et al. 2021).

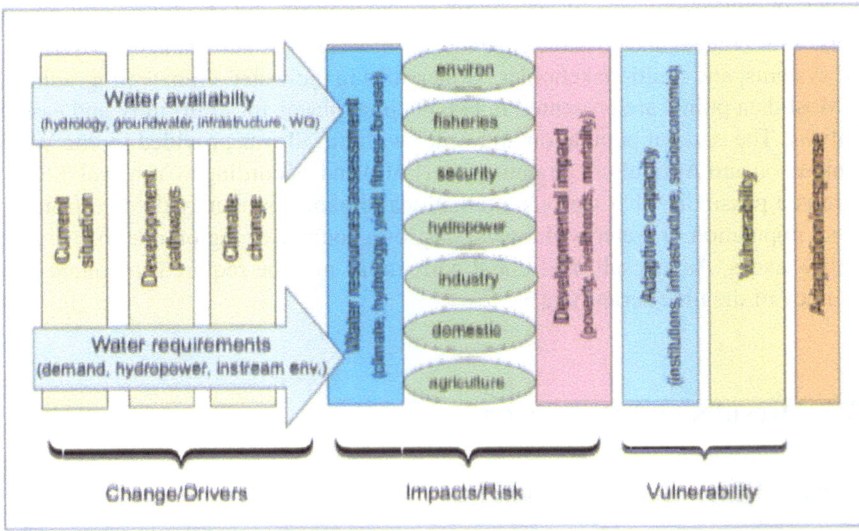

Fig. 17.5 Conceptual framework for managing the impacts of climate on water resources. *Source* Pegasys (2009)

Projections in Fig. 17.6a show water withdrawal as a percentage of total available water by 2025, while Fig. 17.6b shows that all five Southern African countries in this study will be among the highly water-stressed countries in 2040. For South Asia, Afghanistan and Pakistan are projected as extremely water-stressed, while India and Sri Lanka will be highly stressed, and Bangladesh will be in the low to medium water stress category. This suggests that, though the context for gender mainstreaming might be fairly uniform in Southern Africa, it is important to acknowledge the diversity within South Africa when formulating and executing water–climate programmes. Additionally, investing in climate-resilient and enhanced water management initiatives plays a pivotal role in addressing water stress and mitigating the effects of climate change on vulnerable and marginalized communities, including women (Atim 2022).

Values of 20% above water availability represent water stress conditions; 40% above water availability, water scarcity (UN World Water 2012).

Water stress occurs when the demand for water is higher than the available freshwater supply. Figure 17.7 shows the proportions of the global population under water stress per month in 2010.

India and Pakistan appear to be the main contributors affecting the water stress level at a regional level. Aggarwal (2012) confirmed that the projected effects of climatic change, coupled with population growth, industrialization, and urbanization, are expected to decrease per capita water availability in numerous regions of India. Bhaduri (2021) further reported that the demand for water in India is consistently increasing, leading to water stress in most river basins. This rise in demand

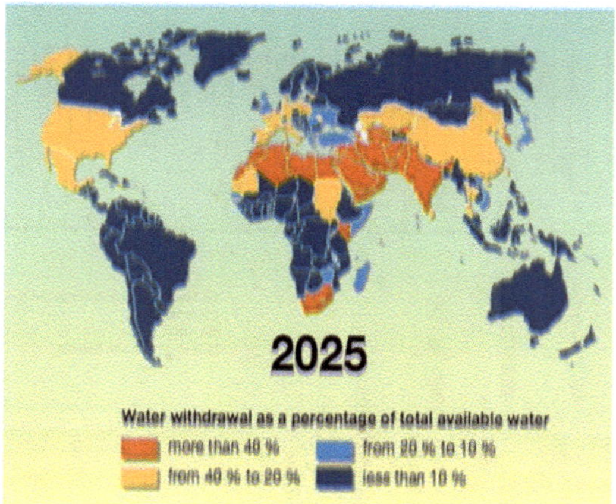

(a)

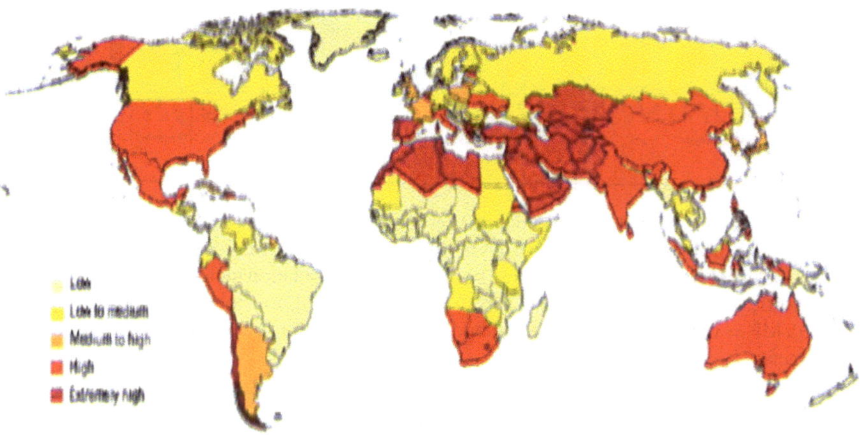

(b)

Fig. 17.6 **a** Water withdrawal as a percentage of total available water. *Source* Urbini (2015). **b** Ranking the world's most water-stressed countries in 2040. *Source* Gassert et al. (2013)

is primarily driven by the growing needs for domestic, industrial, and agricultural purposes.

According to Eckstein (2009), decreased precipitation and water flows have the potential to exacerbate droughts and water scarcity, leading to a decline in agricultural productivity, posing risks to public health, influencing migration and settlement patterns, and significantly burdening livelihoods and social well-being.

Figure 17.8 shows persistent challenge of high levels of water stress (around 78%) for South Asia in 2018 while Sub-Saharan Africa was below 30%. The increase in water stress for South Asia can be attributed to the growing population which is proportional to the growing water demand observed by Aggarwal (2012).

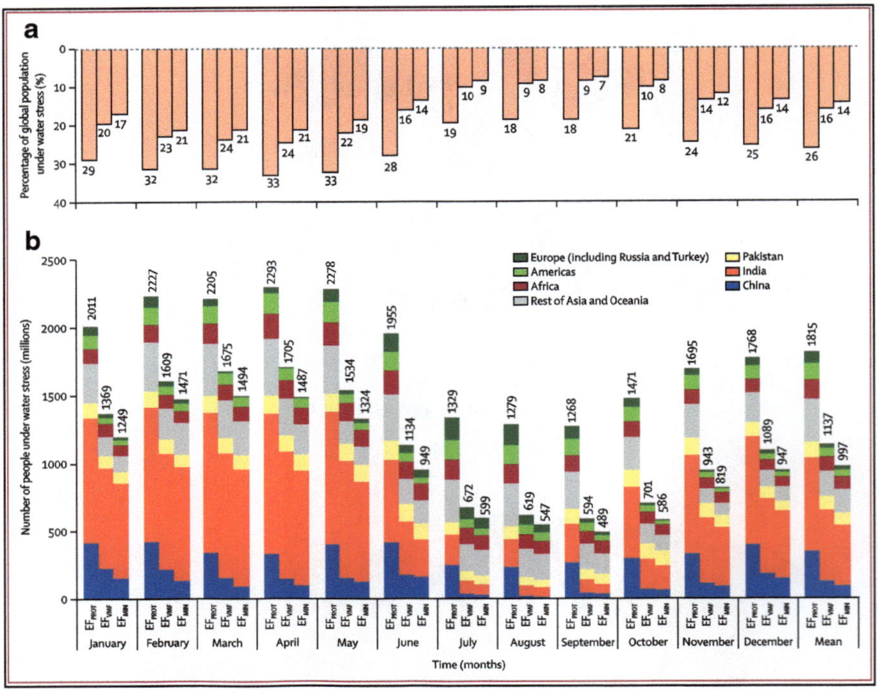

Fig. 17.7 Proportions of the global population under water stress per month in 2010. *Source* Vanham et al. (2021). Abbreviations/acronyms used in the figure are described as follows: percentages (**a**) and absolute numbers in millions (**b**) of the global population under water stress per month for EFPROT, EFVMF, and EFMIN. EFPROT = environmental flow with high ecological protection. EFVMF = variable monthly flow method. EFMIN = environmental flow with minimum flow requirements

Water demand is reported mostly in terms of freshwater annual freshwater withdrawals (see Table 17.3). In terms of water usage, withdrawals for agriculture and industry encompass the total amount of water used for irrigation, livestock production, and direct industrial purposes. On the other hand, withdrawals for domestic purposes encompass drinking water, municipal use or supply, as well as water usage for public services, commercial establishments, and residential households (World Bank 2022).

The data in Figs. 17.6, 17.7, and 17.8, and Table 17.3 show the need to consider the impact of good governance, human development, economic growth, and urbanization on the water stress when engendering the water–climate adaptation programmes at community, national, and regional level. The results also “calls” for the effective management of water, land, and natural resources, in the engendering process.

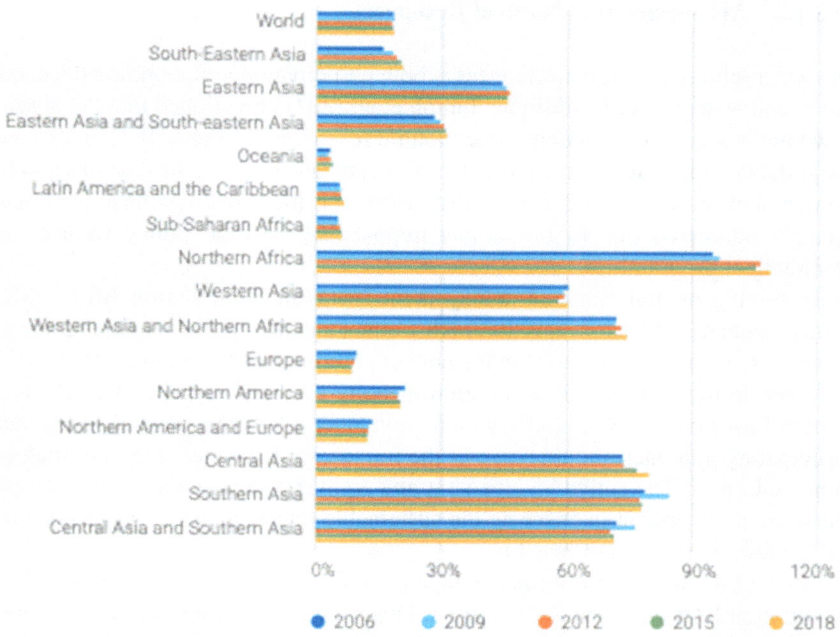

Fig. 17.8 Level of water stress by region and subregion (2018). *Source* FAO (2021)

Table 17.3 Annual freshwater withdrawals per country

Country name	2005	2010	2015
South Asia			
Afghanistan	20.282	20.282	20.282
Bangladesh		35.870	35.870
India	628.950	647.500	647.500
Pakistan	178.939	185.280	196.900
Sri Lanka	12.946	12.946	12.946
Southern Africa			
Botswana	0.195	0.197	0.188
Eswatini	1.068	1.068	1.068
Lesotho	0.044	0.044	0.044
Namibia	0.282	0.282	0.282
South Africa	13.797	14.861	18.660

17.3.1.2 Management of Natural Resources

The water–climate interconnectedness is both participation/collaboration dependent and resource-sensitive. In addition, Tantoh et al. (2021) mentioned that the absence of women's access to and control over natural resources, particularly land, can have a significant implication and thus, it can perpetuate poverty, further marginalize women, and reinforce gender inequality. However, these realities have been inadequately addressed due to limitations imposed by sectoral policy reforms and development programmes.

Examining natural resource management issues in Sub-Saharan Africa (SSA) through a gendered lens reveals that both women and men possess unique roles and interests in the utilization and management of natural resources (Tantoh et al. 2021). However, in the Ministerial Declaration meeting on Gender, Water, and Development in East London, South Africa on 5 November 2014, Ministers affirmed that a participatory approach should serve as the foundation for water resources management (Salo nd.). The active involvement and equitable representation of both men and women are crucial in ensuring the sustainable management of water resources and the fair distribution of benefits.

Table 17.4 shows the high percentage of women living in rural areas with the exception of Botswana and South Africa. This suggests the need to intensify gender mainstreaming in rural areas for the benefit of the majority of vulnerable communities. Land ownership is crucial in the management of natural resources and gender mainstreaming of water and climate change because most countries in South Asia and Southern Africa offer limited access to resources, which are predominantly governed through a top-down management approach, and thus aggravate the vulnerability of impoverished rural women.

According to Caba (2017), it is evident that women in developing countries, particularly, experience a lack of land ownership rights and equitable access to essential public services like water and sanitation. Furthermore, the plight of women and girls who migrate to urban areas to escape environmental degradation often exposes them to even greater challenges than if they had remained in their ancestral homes. These circumstances highlight the need for intervention and support.

Understanding the scope of women's land rights is crucial, given the mounting empirical evidence that securing these rights contributes to reduced poverty and vulnerability, environmental sustainability, and investments in future generations (Doss et al. 2015).

The utilization of land and irrigated farming practices can also play a role in intensifying changes in the water cycle at the regional level. Therefore, it is imperative to exercise cautious management in these areas (Douville et al. 2022). In many South Asian communities, a patrilineal and patrilocal kinship system prevails, and a significant portion of the population follows practices such as benami and naior. Under benami, husbands acquire property under their wives' names while retaining control over the land. Similarly, naior encourages daughters to relinquish their inheritance rights to their brothers (Kieran et al. 2015). Thus, even the figures in Table 17.5 could be much lower in practice. In both regions, legal provisions ensure equal access

Table 17.4 Urban and rural population—South Asia

Gender proportions of urban and rural population—South Asia										
		2005			2010			2015		
		Female (%)	Male (%)	% of total pop.	Female (%)	Male (%)	% of total pop.	Female (%)	Male (%)	% of total pop.
South Asia										
Afghanistan	Rural	37.83	39.28	77.30	37.11	38.20	76.26	36.16	37.14	75.20
	Urban	11.20	11.70	22.70	12.14	12.55	23.74	13.16	13.55	24.80
Bangladesh	Rural	36.11	37.08	73.19	34.78	34.76	69.54	33.17	32.56	65.69
	Urban	12.58	14.23	26.81	14.45	16.01	30.46	16.31	17.97	34.31
India	Rural	34.39	36.37	70.77	33.64	35.43	69.07	32.82	34.44	67.22
	Urban	13.82	15.42	29.24	14.63	16.30	30.93	15.50	17.25	32.78
Pakistan	Rural	31.76	33.51	66.02	30.95	32.45	65.00	29.90	31.34	63.97
	Urban	16.74	17.99	33.98	17.69	18.90	35.00	18.70	20.06	36.03
Sri Lanka	Rural	41.32	40.30	81.70	41.77	39.91	81.77	41.99	39.65	81.74
	Urban	9.08	9.30	18.30	9.15	9.17	18.23	9.24	9.12	18.26
Southern Africa										
Botswana	Rural	22.65	22.28	44.06	21.98	21.78	37.59	21.29	21.24	32.85
	Urban	27.56	27.51	55.94	27.92	28.32	62.41	28.29	29.17	67.16
Eswatini	Rural	40.27	37.65	77.96	40.12	38.36	77.52	39.97	38.72	76.70
	Urban	10.86	11.22	22.04	10.72	10.80	22.48	10.58	10.73	23.30
Lesotho	Rural	39.25	38.47	77.75	37.68	37.58	75.20	36.04	36.65	73.09
	Urban	11.99	10.28	22.25	13.19	11.55	24.80	14.48	12.83	26.91
Namibia	Rural	32.84	30.52	63.37	30.56	27.86	58.38	28.05	25.29	53.10
	Urban	18.34	18.29	36.63	20.93	20.65	41.62	23.33	23.33	46.90
South Africa	Rural	21.30	19.16	40.46	20.01	17.77	37.78	18.65	16.55	35.17
	Urban	29.99	29.55	59.54	31.49	30.72	62.22	32.75	32.05	64.83

to property. However, customary and religious laws prevalent in these areas often contribute to gender disparities in land ownership.

The absence of land ownership or inheritance rights for women in numerous South Asian societies underscores the necessity for discussions on the significance of property rights in land and resources. These rights serve as an engendering mechanism to reduce vulnerability and enhance water, climate, and adaptation efforts (Sultana 2018).

Monitoring of the mainstreaming initiatives can be done through reporting and analysis of the Women, Business, and Law (WBL) index score. The WBL index scores for South Asia and Southern Africa are shown in Figs. 17.9 and 17.10, respectively. Women, Business, and the Law are derived from a series of annual studies, which measure the laws and regulations affecting women's economic potential and

Table 17.5 Percentage of women landholders

	Report period (year)	Sole ownership (%)	Joint ownership (%)
Bangladesh	2005	2.8	
	2008	4.6	
India	2000	10.9	
	2005	11.7	
Pakistan	2012–13	2.1	1.9
Sri Lanka	2002	16.3	

Source Kieran et al. (2015); FAO Gender and Land Rights Database (http://www.fao.org/gender/landrights/home/en/)

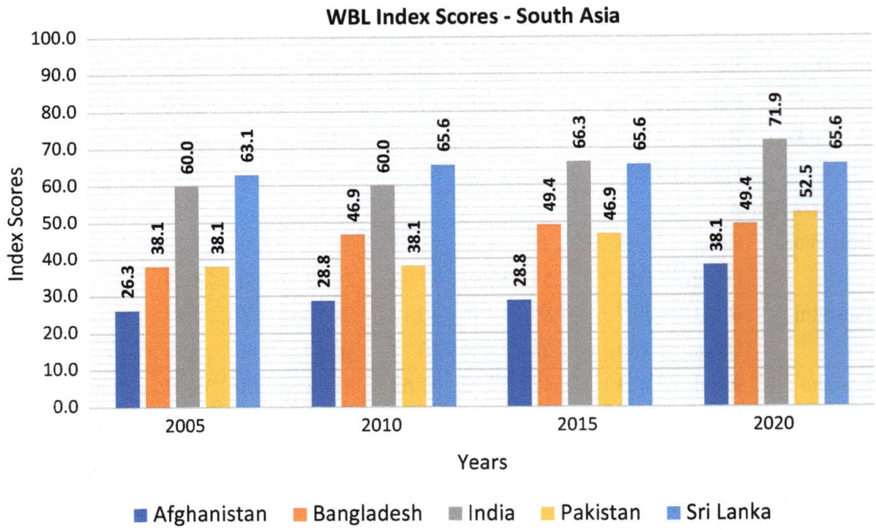

Fig. 17.9 Women, Business, and Law (WBL) index scores for South Asia

growth in different economic sectors. The eight WBL (Women, Business, and Law) indicators, which include mobility, marriage, workplace, parenthood, assets, pay, marriage, entrepreneurship, and pension, are employed to assess women's engagement with the legal system as they advance in their personal and professional journeys (World Bank 2022).

The scores for South Asia range from 26 to 72, and Southern Africa, 38 to 88. Overall, South Asia recorded values are below the global average of 76.5. Despite the implementation of gender mainstreaming and various public policy measures that explicitly recognize the equal rights of women and men, and officially incorporate gender perspectives into research, resource allocation, legislation, and project management/monitoring, this trend persists (Oluwasanya 2022).

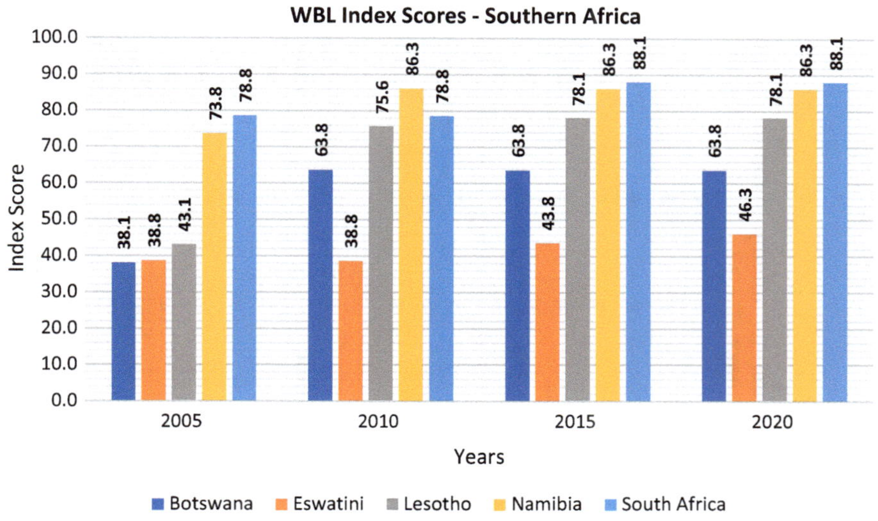

Fig. 17.10 Women, Business, and Law (WBL) index scores for Southern Africa

Effective implementation of reforms to establish a more gender-equitable legislative framework, which ensures equal inheritance rights for both men and women in the legal ability to own a property and the legal treatment of spousal property can increase women's property ownership.

Table 17.6 shows the responses to questions relating to ownership and/or inheritance of assets. More Southern African countries affirmed the presence of laws that provide equal rights for both men and women. All countries except Bangladesh have shown improvements over the years (from 2005 to 2020).

17.3.2 Multi-stakeholder Partnerships in the Water–Climate Programmes

17.3.2.1 Women as Contributing Partners

Developing innovative water–climate solutions involves the creation of partnerships. Effective and sustainable partnerships depend on participation by all parties, which require engagement on equal footing. Seager and Hartmann (2005) also highlighted the crucial point that achieving water security and climate change adaptation will face obstacles without addressing capacity and structural constraints to gender equality. Even if gender mainstreaming discussions have made some progress in developing countries, they are insufficient for attaining these goals.

Observations were made (see Figs. 17.11 and 17.12) that South Asia and Southern Africa have the highest female proportion of the total economically active population

Table 17.6 WBL scores and response to questions on the "assets" indicator

WBL index assets indicator—South Asia

	2005						2010					
	Ownership	Equal inheritance	Spousal inheritance	Authority	Valuation	Indicator score	Ownership	Equal inheritance	Spousal inheritance	Authority	Valuation	Indicator score
AFG	ü	û	û	ü	û	40	ü	û	û	ü	û	40
BGD	ü	û	û	ü	û	40	ü	û	û	ü	û	40
IND	ü	ü	ü	ü	û	80	ü	ü	ü	ü	û	80
PAK	ü	û	û	ü	û	40	ü	û	û	ü	û	40
LKA	ü	ü	ü	ü	û	80	ü	ü	ü	ü	û	80
BWA	û	û	ü	û	û	20	ü	û	ü	ü	û	60
SWZ	û	û	û	û	ü	20	û	û	û	û	ü	20
LSO	û	û	ü	û	ü	40	ü	û	ü	ü	ü	80
NAM	ü	ü	ü	ü	ü	100	ü	ü	ü	ü	ü	100
ZAF	ü	ü	ü	ü	ü	100	ü	ü	ü	ü	ü	100
	2015						2020					
	Ownership	Equal inheritance	Spousal inheritance	Authority	Valuation	Indicator score	Ownership	Equal inheritance	Spousal inheritance	Authority	Valuation	Indicator score
AFG	ü	û	û	ü	û	40	ü	û	û	ü	û	40
BGD	ü	û	û	ü	û	40	ü	û	û	ü	û	40
IND	ü	ü	ü	ü	û	80	ü	ü	ü	ü	û	80
PAK	ü	û	û	ü	û	40	ü	û	û	ü	û	40
LKA	ü	ü	ü	ü	û	80	ü	ü	ü	ü	û	80

(continued)

Table 17.6 (continued)

WBL index assets indicator—South Asia												
BWA	ü	û	ü	ü	û	60	ü	û	ü	ü	û	60
SWZ	ü	û	û	ü	ü	60	ü	û	û	ü	ü	60
LSO	ü	ü	ü	ü	ü	100	ü	ü	ü	ü	ü	100
NAM	ü	ü	ü	ü	ü	100	ü	ü	ü	ü	ü	100
ZAF	ü	ü	ü	ü	ü	100	ü	ü	ü	ü	ü	100

Key:
Do men and women have equal ownership rights to immovable property?
Do sons and daughters have equal rights to inherit assets from their parents?
Do male and female surviving spouses have equal rights to inherit assets?
Does the law grant spouses equal administrative authority over assets during marriage?
Does the law provide for the valuation of nonmonetary contributions?
ü—yes
û—no

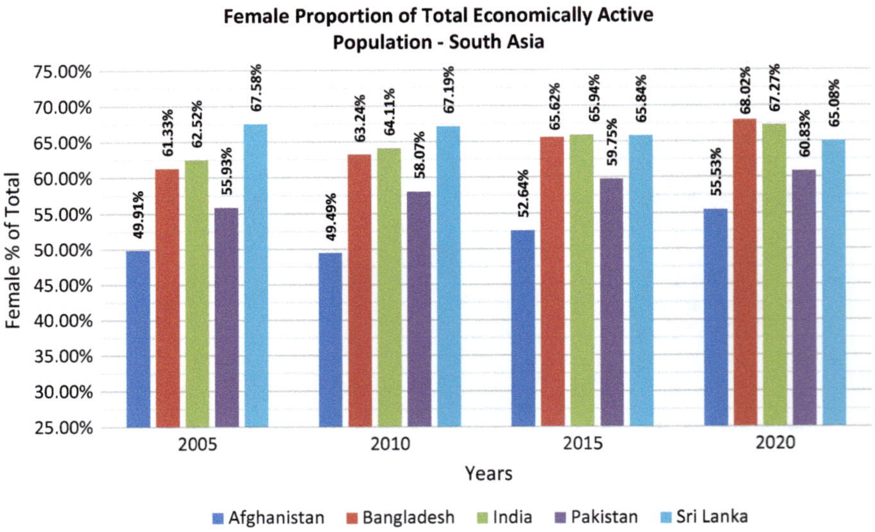

Fig. 17.11 Female proportion of the economically active population in South Asia

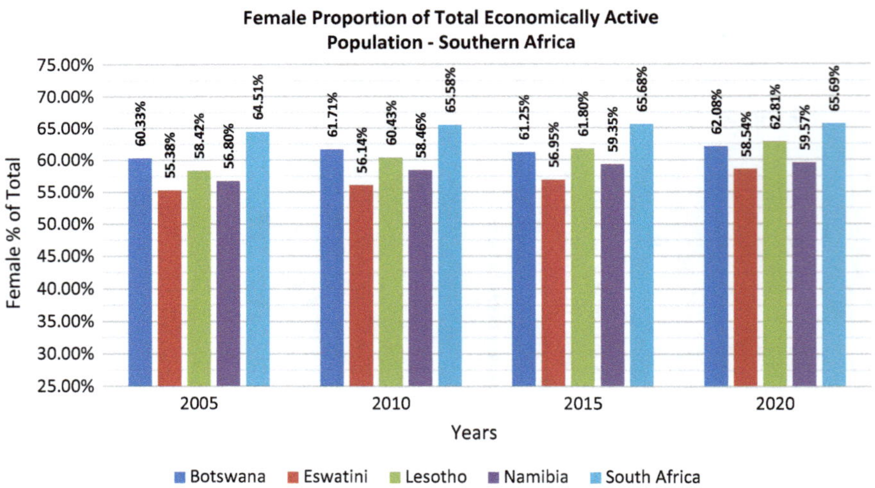

Fig. 17.12 Female proportion of the economically active population in Southern Africa

(49–68%, and 55–66%, respectively). However, one of the hindrances to effective gender mainstreaming in South Asia and Southern Africa is the relatively high rate of unemployment among females as presented in Table 17.7 based on national estimates submitted by government departments (excluding Eswatini and Lesotho) to the World Bank, and Table 17.8 shows estimates modelled by the International Labour Organization, and accessed from the ILOSTAT database.

Table 17.7 National estimates of unemployment (% of the labour force)

	Unemployment (% of labour force)—national estimate											
	Female unemployment (% of female labour force)				Male unemployment (% of male labour force)				Total unemployment (% of total labour force)			
	2005	2010	2015	2020	2005	2010	2015	2020	2005	2010	2015	2020
South Asia												
Afghanistan				16.77				10.45				11.71
Bangladesh	7.03	4.36			3.35	2.95			4.25	3.38		
India	2.92	3.30		4.42	2.23	2.17		4.78	2.42	2.44		4.68
Pakistan	12.06	0.64	6.11		5.96	0.66	2.80		7.05	0.65	3.57	
Sri Lanka	11.89	7.40	7.33	7.50	5.55	3.42	2.94	4.10	7.67	4.78	4.52	5.20
Southern Africa												
Botswana		21.43		25.64		14.59		24.22		17.86		24.93
Eswatini												
Lesotho												
Namibia		23.96				20.34				22.12		
South Africa	36.50	26.98	27.52	31.02	23.56	22.84	23.20	27.74	29.12	24.68	25.15	29.22

Table 17.8 ILO modelled estimates of unemployment (% of labour force)

	Unemployment (% of labour force)—modelled ILO											
	Female unemployment (% of female labour force)				Male unemployment (% of male labour force)				Total unemployment (% of total labour force)			
	2005	2010	2015	2020	2005	2010	2015	2020	2005	2010	2015	2020
South Asia												
Afghanistan	14.30	14.49	14.09	16.86	10.63	10.76	10.42	10.51	11.22	11.35	11.13	11.71
Bangladesh	7.05	4.44	7.42	8.00	3.36	3.00	3.15	4.29	4.25	3.38	4.37	5.41
India	5.94	5.75	5.49	5.51	5.49	5.48	5.42	8.62	5.61	5.55	5.43	8.00
Pakistan	0.33	0.63	6.16	4.99	0.65	0.65	2.82	4.13	0.59	0.65	3.57	4.30
Sri Lanka	11.93	7.47	7.43	8.93	5.57	3.45	2.98	4.34	7.67	4.78	4.52	5.88
Southern Africa												
Botswana	22.26	21.56	22.66	25.66	17.70	14.68	18.58	24.24	19.81	17.86	20.57	24.93
Eswatini	30.07	28.87	25.20	27.38	24.25	23.98	21.57	23.83	27.07	26.41	23.33	25.51
Lesotho	37.03	30.72	25.42	24.40	26.06	23.93	22.54	24.70	31.31	27.08	23.84	24.56
Namibia	25.05	24.00	22.67	20.15	19.68	20.38	19.23	22.70	22.22	22.12	20.91	21.45
South Africa	36.44	27.02	27.55	31.04	23.52	22.87	23.22	27.75	29.12	24.68	25.15	29.22

According to Salo (nd.), initiating gender mainstreaming requires two essential prerequisites: a firm dedication to economic growth coupled with poverty reduction efforts, and a constitutional commitment to gender equality. These factors represent the fundamental minimum requirements for the implementation of gender mainstreaming. The ILO estimates in Table 17.8 were accessed to compensate for the scarcity of data experienced for this unemployment indicator.

When considering the areas measured by Women, Business, and the Law, it is observed that women, on average, possess only three-quarters of the economic rights that men enjoy. Moreover, an alarming statistic reveals that approximately 2.4 billion women of working age reside in countries where they do not have equal economic rights as men. South Africa and India are part of the Pilot Economies in the Women, Business, and Law implementation (World Bank 2022). A set of questions is used to assess one indicator, for example, the score for "Workplace" (Table 17.9) was assigned based on the answers provided by a country on the following;

(1) Is it possible for a woman to secure employment under the same conditions as a man?
(2) Does the law prohibit gender-based discrimination in employment?
(3) Is there specific legislation addressing sexual harassment in the workplace?
(4) Are there criminal or civil consequences for instances of sexual harassment in employment?

Lack of, or slow progress in addressing unemployment of the labour force in Southern African countries (Table 17.8) was observed. However, the findings could inform decisions and plans to address the effects articulated by Sultana (2018)—The ecological and hydrological changes resulting from climate change will further enhance existing development challenges, leading to increased gender disparities in health, income, and education.

Unemployment and lack of capacity contribute to the perpetuation of top-down adaptation, hindering the effective participation of women and vulnerable communities in projects aimed at fostering their development and that of the community (Chant 2016; Tantoh and McKay 2018). Thus, engendering the multi-stakeholder partnerships in the developing countries will require capacity strengthening and empowerment of women so they would not be disempowered victims of quota systems, but rather equal partners contributing and leading the decision-making, planning, and implementation of programmes.

It is important to acknowledge that, alongside the limited access to employment opportunities faced by women in the two regions, the implementation of gender mainstreaming in the water and climate sector faces challenges due to sector-specific silos. These silos are often created when water–climate programmes are managed in a fragmented manner, involving multiple ministries, and often intersecting with agriculture, land, and environment (Salo nd.).

Table 17.9 WBL scores and response to questions on the "workplace" indicator

WBL index workplace indicator—South Asia

	2005					2010				
	Recruitment	Discrimination	Protections	Remedies	Indicator score	Recruitment	Discrimination	Protections	Remedies	Indicator score
South Asia										
Afghanistan	ü	û	û	û	25	ü	û	û	û	25
Bangladesh	ü	û	û	û	25	ü	û	ü	û	50
India	ü	ü	û	û	50	ü	ü	û	û	50
Pakistan	ü	û	û	û	25	ü	û	û	û	25
Sri Lanka	ü	û	ü	ü	75	ü	û	ü	ü	75
Southern Africa										
Botswana	û	û	û	û	0	ü	û	û	û	25
Eswatini	û	ü	û	û	25	û	ü	û	û	25
Lesotho	û	ü	ü	û	50	ü	ü	ü	û	75
Namibia	ü	û	û	û	25	ü	ü	ü	ü	100
South Africa	ü	ü	û	û	50	ü	ü	û	û	50
	2015					2020				
	Recruitment	Discrimination	Protections	Remedies	Indicator score	Recruitment	Discrimination	Protections	Remedies	Indicator score
South Asia										
Afghanistan	ü	û	û	û	25	ü	û	ü	ü	75
Bangladesh	ü	û	ü	û	50	ü	û	ü	û	50
India	ü	ü	ü	ü	100	ü	ü	ü	ü	100

(continued)

Table 17.9 (continued)

WBL index workplace indicator—South Asia										
Pakistan	ü	û	ü	ü	75	ü	ü	ü	ü	100
Sri Lanka	ü	û	ü	ü	75	ü	û	ü	ü	75
Southern Africa										
Botswana	ü	û	û	û	25	ü	û	û	û	25
Eswatini	û	ü	û	û	25	û	ü	û	û	25
Lesotho	ü	ü	ü	û	75	ü	ü	ü	û	75
Namibia	ü	ü	ü	ü	100	ü	ü	ü	ü	100
South Africa	ü	ü	ü	ü	100	ü	ü	ü	ü	100

Key:
Recruitment—Is it possible for a woman to secure employment under the same conditions as a man?
Discrimination—Does the law prohibit gender-based discrimination in employment?
Protection—Is there specific legislation addressing sexual harassment in the workplace?
Remedies—Are there criminal or civil consequences for instances of sexual harassment in employment?
ü—yes, û—no

17.3.2.2 Sectoral Approach

The focus of mitigation and adaptation strategies primarily revolves around sector-specific objectives, resulting in insufficient consideration and coordination among crucial sectors such as water, climate, and land (Tantoh et al. 2021).

Low scores were observed for South Asian (0–25 per question/response) countries (Table 17.10), suggesting a "negative" answer to the following questions

(1) Is there a legal requirement for equal remuneration for work of equal value?
(2) Are women permitted to work during night hours under the same conditions as men?
(3) Can women work in occupations considered hazardous, just like men?
(4) Are women allowed to work in industrial jobs on an equal basis with men? This is an indication that women are not at liberty to choose the sector or organization they want to work for. Southern Africa is doing relatively well with the selected indicators.

Engendered multisectoral cooperation between public and private sectors (Fig. 17.13) addressing the water, agriculture, health, technology, and economy is crucial for climate adaptation and building resilience of water and related systems.

Multi-stakeholder partnerships/collaborations between government and non-governmental organizations/institutions for planning and implementation of engendered, and resilient water–climate systems/organizations. The context for Southern Africa also requires considerations for the transboundary nature of water supply and impact of water and climate related disasters strengthen existing transboundary/cross-border collaborations and programmes, and focus on River Basin commissions/organizations for planning and implementation of national water, sanitation, and hygiene programmes.

Tantoh et al. (2021) also highlighted that while existing research acknowledges the importance of concerted efforts to assist vulnerable communities in developing countries, especially in adapting to climate change, current programmes primarily focus on the national level, neglecting gender-related concerns within vulnerable communities at the local level.

The multi-layered approach is crucial to effective management and good governance because regional-level challenges cannot be addressed at the national and/or provincial level. An example of a significant hurdle in transboundary cooperation and collaborative monitoring is the need to align systems, methodologies, and data formats across aquifer boundaries in order to achieve harmonization (Nijsten et al. 2018), while local challenges might be limited to finances and maintenance of water supply systems and/or wastewater treatment systems.

Table 17.10 WBL scores and response to questions on the "pay" indicator

WBL index pay indicator—South Asia

	2005					2010				
	Remunera-tion	Night shift	Dangerous work	Industrial work	Indicator score	Remunera-tion	Night shift	Dangerous work	Industrial work	Indicator score
South Asia										
Afghanistan	û	û	û	û	0	û	û	û	û	0
Bangladesh	û	û	û	û	0	û	ü	û	û	25
India	û	û	û	û	0	û	û	û	û	0
Pakistan	û	û	ü	û	25	û	û	ü	û	25
Sri Lanka	û	û	ü	û	25	û	û	ü	û	25
Southern Africa										
Botswana	û	ü	ü	ü	75	û	ü	ü	ü	75
Eswatini	û	ü	ü	û	50	û	ü	ü	û	50
Lesotho	ü	ü	ü	û	75	ü	ü	ü	û	75
Namibia	û	ü	ü	ü	75	ü	ü	ü	ü	100
South Africa	û	ü	ü	ü	75	û	ü	ü	ü	75
	2015					2020				
	Remunera-tion	Night shift	Dangerous work	Industrial work	Indicator score	Remunera-tion	Night shift	Dangerous work	Industrial work	Indicator score
South Asia										
Afghanistan	û	û	û	û	0	û	û	û	û	0
Bangladesh	û	ü	û	û	25	û	ü	û	û	25

(continued)

Table 17.10 (continued)

WBL index pay indicator—South Asia									
India	û	Û	û	û	0	û	û	ü	û	25
Pakistan	û	Û	ü	û	25	û	û	ü	û	25
Sri Lanka	û	Û	ü	û	25	û	û	ü	û	25
Southern Africa										
Botswana	û	ü	ü	ü	75	û	ü	ü	ü	75
Eswatini	û	ü	ü	û	50	û	ü	ü	û	50
Lesotho	ü	ü	ü	û	75	ü	ü	ü	û	75
Namibia	ü	ü	ü	ü	100	ü	ü	ü	ü	100
South Africa	ü	ü	ü	ü	100	ü	ü	ü	ü	100

Key: ü—yes, û—no
Is there a legal requirement for equal remuneration for work of equal value?
Are women permitted to work during night hours under the same conditions as men?
Can women work in occupations considered hazardous, just like men?
Are women allowed to work in industrial jobs on an equal basis with men?

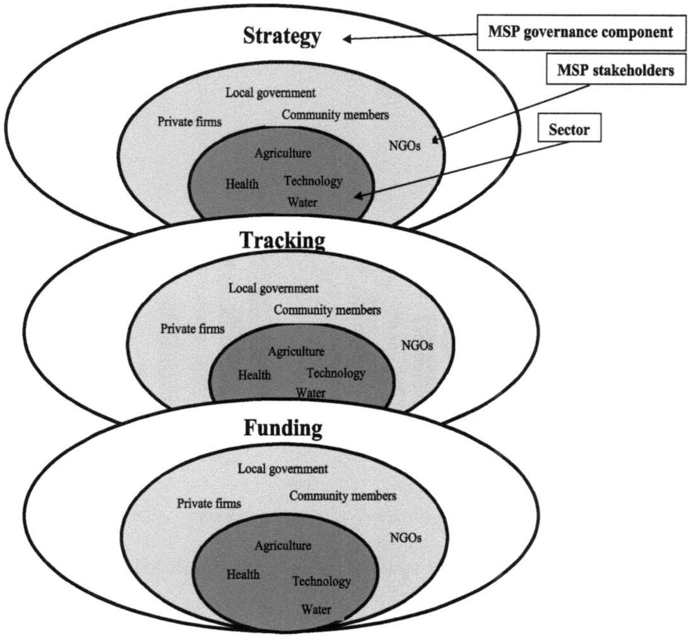

Fig. 17.13 District-level layered approach to building resilience through multi-stakeholder partnerships. *Source* Suresh et al. (2018)

17.3.3 Governance and Policy Systems

Gender mainstreaming is an important aspect of effective governance and strong institutions. In this context, gender mainstreaming is used as a tool/mechanism to ensure water climate linkages to ensure that the needs and interests of women are addressed by policies and programmes, thus enhancing the accountability of organizations/governments for the benefit of all communities.

Establishing a conducive political and institutional environment is recognized as the fundamental foundation for initiating and facilitating policies and advancements in gender mainstreaming. This environment is crucial in strengthening and maintaining gender-sensitive policies within water governance (Salo nd.).

17.3.3.1 Involvement in All Decision-Making Levels

Despite constituting nearly half of the world population, women continue to be inadequately represented in leadership and decision-making positions (Oluwasanya 2022; Atim 2022).

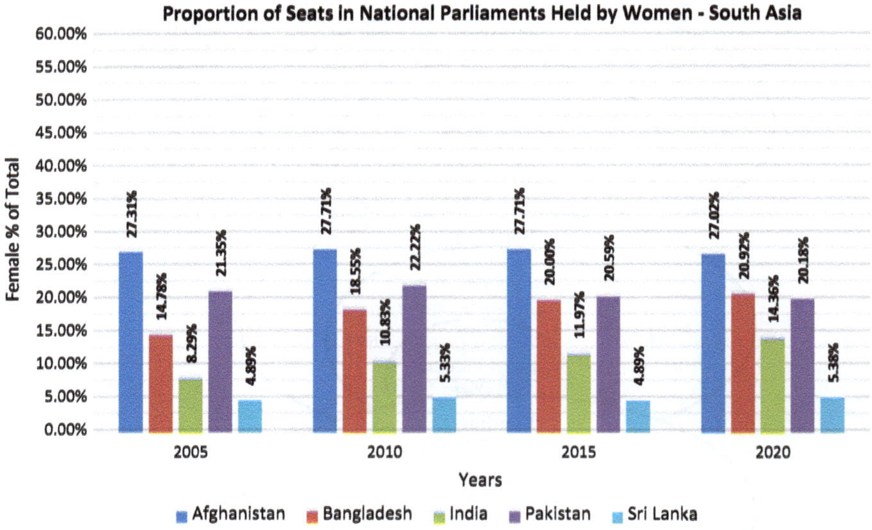

Fig. 17.14 Seats held by women in the national parliament—South Asia

Nevertheless, the focus of existing programmes primarily targets the national level, resulting in limited attention being given to gender issues in vulnerable communities at the local level (Fonjong 2008; UNDP 2010). This disparity is evident in the varying proportions of seats held by women in national parliaments across different regions, ranging from 4.9 to 27.7% in South Asia (Fig. 17.14), and, 6.2 to 42.8% in Southern Africa (Fig. 17.15).

Within the context of Southern Africa, the Ministerial Declaration adopted by Ministers at the Gender, Water, and Development gathering in East London, South Africa on 5 November 2014 (Salo nd.) affirmed the imperative of utilizing participatory, gender-balanced, and gender-sensitive approaches by local governments, utilities, and the public sector. This recognition emphasized the need to foster the development of sustainable, equitable, and affordable water and sanitation systems, as well as water treatment infrastructure. Furthermore, the declaration highlighted the essential role of women as key actors in decision-making, as well as in the developing, implementing, and monitoring of these systems.

The proportion of women in ministerial-level positions has fluctuated between 0 and 22% in South Asia (Fig. 17.16), and, 12 and 48% in Southern Africa (Fig. 17.17) for the past fifteen years. Trivedi (2018) reported that women representatives at the national policy level play a crucial role in designing policies that are more inclusive and addressing previously overlooked issues.

In addition, Salo (nd.) also reported that mere representation in decision-making processes is inadequate to ensure meaningful advancements in gender mainstreaming. To achieve effective progress, it is essential to provide training and capacity-building initiatives, particularly for relevant bureaucrats in the national water–climate sector.

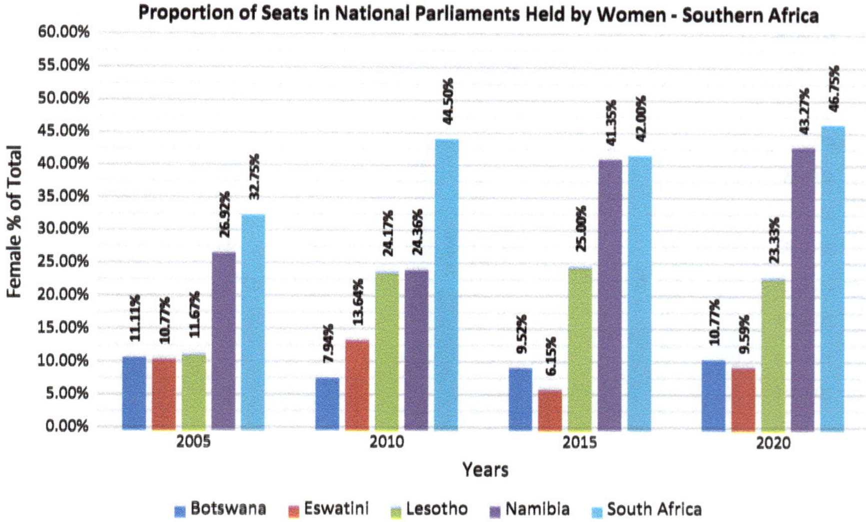

Fig. 17.15 Seats held by women in the national parliament—Southern Africa

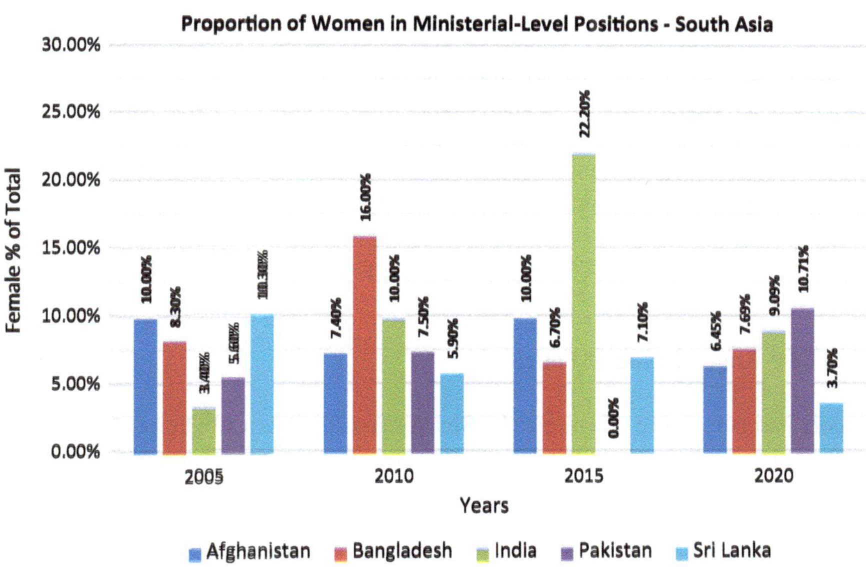

Fig. 17.16 Proportion of women in ministerial level positions—South Asia

Tantoh et al. (2021) observed that local decision-making and resource control predominantly lie in the hands of male members within the society. This arrangement perpetuates inequality and contributes to the suppression of individual rights. Therefore, a multi-layered approach to gender mainstreaming is crucial to apply

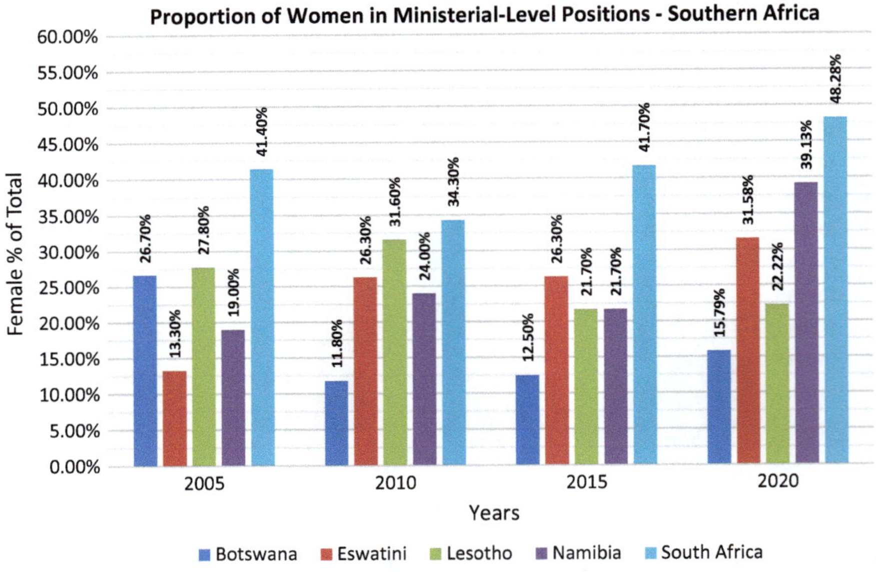

Fig. 17.17 Proportion of women in ministerial level positions—Southern Africa

the multisectoral and multi-stakeholder plans taking into consideration the layers i.e. regional level, country, province/district, city/town/village to manage the monitoring, evaluation, and improvement.

It is crucial for regional ministers' councils, such as the African Ministers Council on Water, to prioritize gender mainstreaming within national water ministries. This ensures that gender remains a prominent focus on the national water agenda and facilitates the initiation of gender mainstreaming efforts.

17.3.3.2 Engendering of Policies

Gender mainstreaming was termed by Sivakumar (2016) as a tool that can potentially be used to radically root out the deep-rooted and often concealed causes of inequality in the climate adaptation and water sector. Sivakumar (2016) found that the inability of progress in improving the status of women has raised concerns among policy-makers and equality advocates regarding the effectiveness of equal opportunities policies.

Furthermore, Salo (nd.) also holds a view that regional gender policies and protocols within the water sector play a crucial role in promoting gender mainstreaming in water and climate governance at the national level. Well-defined policies and strategies, supported by dedicated gender directorates, add an additional layer of accountability to ensure the integration of gender mainstreaming efforts. Table 17.11 is a summary of the gender components/aspects of selected national water policies.

Table 17.11 Summary of the gender components/aspects of selected national water policies

Country (policy)	Policy and supporting	Gender/women priority/leadership/decision-making component in the policy	
Afghanistan	National Rural Water, Sanitation, and Hygiene (WASH) Policy 2010	Principle: To achieve social equity and justice, it is imperative to prioritize gender mainstreaming by actively involving women, especially in Shuras as well as in CDC decision-making processes	
		MIS will incorporate gender reporting to encourage partners to share the gender-related aspects of their programmes. Additionally, it will ensure the availability of gender-segregated data as part of the system	
		Policy strategy: Promoting women's empowerment involves leveraging groups such as women's Shuras to actively participate in the planning and implementation of water and sanitation services at the village level	
Bangladesh	–	–	
India	National Water Policy 2012	None	
Pakistan	National Water Policy 2018	None	
Sri Lanka	National Policy on Protection and Conservation of Water Sources, their Catchments and Reservations in Sri Lanka 2014	None	
	National Drinking Water Policy	None	
Botswana	National Water Policy 2012	Policy principle: Efforts will be made to ensure gender and social equity in accessing water resources, with a particular focus on empowering women to fully participate in matters and decisions related to the sustainable development and management of water resources	"Dublin Principle" Women occupy a central role in the provision, management, and safeguarding of water resources
Eswatini	National Water Policy 2018	4.5.2.1 Gender balance in water resources development and management	The water sector will endeavour to attain gender balance in the development and management of water resources

(continued)

Table 17.11 (continued)

Country (policy)	Policy and supporting	Gender/women priority/leadership/decision-making component in the policy	
		4.5.2.2 Role of women in water resources development and management	The important role of women in water resources management is acknowledged, and it is imperative to ensure their full engagement and equal opportunities at all levels
Lesotho	National Water Resources Management Policy 2007	Objective: To ensure the participation of all gender groupings in the formulation and implementation of all sector programmes	Stakeholder involvement: To guarantee the sustainability of sector programmes, a participatory approach that effectively involves stakeholders at various levels will be ensured in the management and development of water resources
		Objective: To ensure the integration of environmental and other inter-sectoral matters, including HIV/AIDS and gender equity, it is essential to incorporate these aspects into water resources programmes and activities	Water and environment: The protection and conservation of water resources, as well as the mitigation of negative impacts arising from socio-economic development activities, will be prioritized
Namibia	Water Supply and Sanitation Policy 2008	None	
	National Water Policy White Paper 2000	"Dublin Principle" Women occupy a central role in the provision, management, and safeguarding of water resources	
South Africa	National Water Policy	Institutions for water management: 7.3.3 Water and gender Ensuring the representation of women at all levels and in all spheres of water management activities, including political, technical, and managerial positions, holds great significance	

The study found that in most cases, gender issues have not been dealt with in water management policies.

The analysis of national policies revealed that Eswatini has signed gender conventions and aligns with the Dublin Principles, which acknowledge the pivotal role of women in water resource management. Similarly, Lesotho has embraced various frameworks, including the Dublin Principles, Agenda 21, the SADC Declaration, the Helsinki Rules, the Global Water Partnership, the SADC Regional Water Policy, the Southern African Vision for Water and Environment, and the SADC Protocol on Shared Watercourses.

The SADC region was acknowledged by Nijsten et al. (2018) for having a relatively advanced transboundary aquifer management system. However, the success of the transboundary watercourse commissions was hampered by socio-economic factors including unequal access to resources caused by climate change. However, because Southern African countries rely on transboundary shared watercourse commissions and transboundary aquifer systems to ensure the implementation of the Protocol and related policies and strategies, there is an urgent need to rethink the concepts and frameworks of gender mainstreaming (Sivakumar 2016).

17.4 Conclusion

This study contextualized gender mainstreaming in terms of (1) Water–climate linkages, (2) Governance and policy systems, and (3) Multi-stakeholder partnerships in the water–climate programmes. South Asia and Southern Africa have the highest female proportion of the total economically active population (49–68% and 55–66%, respectively). However, a relatively high rate of unemployment among females was observed as hindrance to the engendering process. The Women, Business, and Law (WBL) index scores for South Asia range from 26 to 72 (below the global average of 76.5), and Southern Africa, 38 to 88. Overall, South Asia. The percentage of seats occupied by women in national parliaments varies from 4.9 to 27.7% in South Asia, and, 6.2 to 42.8% in Southern Africa. The proportion of women in ministerial-level positions has fluctuated between 0 and 22% in South Asia and, 12 and 48% in Southern Africa for the past fifteen years. Having representation in decision-making processes alone is inadequate to ensure substantial progress in addressing gender issues. For both regions, the law guarantees equal access to property yet, in practice, gender inequality in land and assets ownership, and inheritance exists. Analysis of the gender aspects of selected national water policies also showed that in most cases, gender issues have not been dealt with in water management policies.

Improve/strengthen existing cross-sectoral and multi-stakeholder collaborations/partnerships.

Develop stronger cross-sectorial collaborations for instance between the water and climate change sectors with a stronger link to gender.

Improve on advocacy, communication, and accountability regarding the role of water and climate in the socio-economic development of the nations/communities.

Build multisectoral adaptation, decision-making, and resources management capacity at all managerial and governance levels including transboundary, national, provincial/district, and local/community.

The study has revealed that while quantifying the presence of women in the water and climate sectors is a significant stride towards enhancing their representation, ensuring intersectionality and inclusive leadership positions for women necessitates a commitment from the existing governance structures. Hence a call for more emphasis on intersectionality when addressing the water–climate and gender issues in developing countries.

The limited (and sometimes absence of) data on crucial indicators such as land and assets inheritance and ownership observed during this study also highlights the inconsistencies and, in some cases, lack of commitment to contextualization of gender mainstreaming at the national and local levels.

It was observed that few countries made provision for gender mainstreaming in the national water policies. These could create transdisciplinary, multisectoral working groups that have an increased involvement of all the relevant stakeholders to improve the governance of water–climate systems, and natural resources management. The working groups would develop policies and guiding instruments to facilitate contextual gender mainstreaming of water and climate programmes, and facilitate implementation of joint programmes and results frameworks that enhance water governance and recognize that good water–climate policy constitutes, good gender aspects.

Partnership between academia and water and climate government ministries/departments in the development of government's supervisory, regulatory, and supporting systems can help with monitoring and implementation of recommended policy system principles (Table 17.12) to guide improvement of engendering initiatives, and provide the scientific data needed for reporting and decision-making.

Table 17.12 Recommended policy system principles to guide improvement of engendering initiatives

Principle	Policy system, water security, and climate adaptation system
Transparency	Policy system: Should allow transparency, openness, participation, and institutional systems
Equality	Policy system: Should consider economic diversity, decentralized decision-making, and social safety nets
Accountability	Policy system: Policy environment and processes, transparency, and participation
Effectiveness and efficiency	Policy system: Emphasize monitoring and evaluation Institutional system: Early warning systems and data/information system Production system: Research and innovation, water–climate–gender linkages, water and climate programmes
Strategic vision	Policy system: Land and natural resource rights, institutional systems (e.g. transboundary water commissions), and joint (transboundary) decision-making
Responsiveness	Policy system: Democracy, transparency, and participation Production system: Should focus on research linkages and capacity

Recommended interventions (based on the findings of the study) for contextual engendering water and climate at the national and local levels in the developing countries are aligned with USAID (nd.) as follows:

- Encourage the active involvement of both men and women in the process of prioritizing and designing climate change adaptation projects, while leveraging their indigenous knowledge.
- Encourage the participation of men and women through interventions in agriculture, industry, and leadership/management.
- Give priority to information, training, and technology transfer for water security, climate change adaptation, and mitigation, making it relevant and accessible to all stakeholders at the local, regional, and national levels.
- Capitalize on women's educational qualifications, training, skills, and knowledge, such as in natural resources management and community-based adaptation, while recognizing the value of their social networks.
- Intergrade the gender issues into national climate change strategies and regulations.
- Ensure that the burdens and opportunities arising from water and climate change adaptation efforts are distributed equitably.

References

Aggarwal S (2012) Climate change and household water insecurity in India: a case of gender inequalities. Climate Change, Agriculture & Food Security, 19 July 2012

Agostine A, Lizarde R (2012) Gender and climate justice. Development 55(1):90–95

Arora-Jonsson S (2011) Virtue and vulnerability: discourses on women, gender and climate change. Glob Environ Chang 21(2):744–751

Atim J (2022) Women and water on the frontline of climate change. Presentation during the webinar on "women and water: on the frontline of climate change", WaterAid, 12 Oct 2022

Bhaduri S (2021) The interlinkages of gender, climate change and water. The Times of India, 25 Aug 2021

Brewis A, Rosinger A, Wutich A, Adams E, Cronk L, Pearson A et al (2019) Water sharing, reciprocity, and need: a comparative study of interhousehold water transfers in sub-Saharan Africa. Econ Anthropol 6(2):208–221. https://doi.org/10.1002/sea2.12143

Brody A, Demetriades J, Esplen E (2008) Gender and climate change: mapping the linkages: a scoping study on knowledge and caps. BRIDGE, Institute of Development Studies, University of Sussex, Brighton

Bush SS, Clayton A (2022) Facing change: gender and climate change attitudes worldwide. Am Polit Sci Rev 2022:1–18. https://doi.org/10.1017/S0003055422000752

Caba MP (2017) The gendered role of climate change. State of the Planet, Columbia Climate School, 13 July 2017. https://news.climate.columbia.edu/2017/07/13/the-gendered-role-of-climate-change/

Cannon T (2002) Gender and climate hazards in Bangladesh. Gend Dev 10(2):45–50

Chant S (2016) Women, girls and world poverty: empowerment, equality or essentialism. Int Dev Plann Rev 38:1–24. https://doi.org/10.3828/idpr.2016.1

Dankelman I (2010) Gender and climate change: an introduction. Earthscan, London

Doss C, Kovarik C, Peterman A, Quisumbing A, van den Bold M (2015) Gender inequalities in ownership and control of land in Africa: myth and reality. Agric Econ 46:1–32

Douville H, Allan RP, Arias PA, Betts RA, Caretta MA, Cherchi A, Mukherji A, Raghavan K, Renwick J (2022) Water remains a blind spot in climate change policies. PLOS Water 1(12):e0000058. https://doi.org/10.1371/journal.pwat.0000058

Eckstein G (2009) Water scarcity, conflict, and security in a climate change world: challenges and opportunities for international law and policy. Wis Int Law J 27:409

FAO and UN Water (2021) Progress on level of water stress. Global status and acceleration needs for SDG indicator 6.4.2. Rome. https://doi.org/10.4060/cb6241en

Fonjong LN (2008) Gender roles and practices in natural resource management in the North West Province of Cameroon. Local Environ 13(5):461–475. https://doi.org/10.1080/13549830701809809

Gassert F, Landis M, Luck M, Reig P, Shiao T (2013) Aqueduct Global Maps 2.0. World Resources Institute, Washington, DC. www.wri.org/publication/aqueduct-metadata-global

GGCA (nd.) Facts from gender and climate change: a closer look at existing evidence, gender and change in Africa. Global Gender and Climate Alliance (GGCA)

Kieran C, Sproule K, Doss C, Quisumbing A, Kim S (2015) Examining gender inequalities in land rights indicators in Asia. Agric Econ 46(supplement):119–138

Kgabi NA (ed) (2020a) A review of selected water–climate studies conducted in Tanzania, Southern Africa. In: Kgabi NA (ed) Integrated transboundary water–climate management tools. Water security and climate adaptation in Southern Africa, vol 1. AOSIS, Cape Town, pp 1–12. https://doi.org/10.4102/aosis.2020.BK205.01

Kgabi NA (ed) (2020b) Postscript. In: Kgabi NA (ed) Integrated transboundary water–climate management tools. Water security and climate adaptation in Southern Africa, vol 1. AOSIS, Cape Town, pp 199–203. https://doi.org/10.4102/aosis.2020.BK205.0p

Koenker H, Taylor C, Burget-Brucker CR, Thwing J, Fish T, Kilian A (2019) Quantifying seasonal variation in insecticide-treated net use among those with access. Am J Trop Med Hyg 101(2):371–382. https://doi.org/10.4269/ajtmh.19-0249

Liehr S, Röhrig J, Mehring M, Kluge T (2017) How the social-ecological systems concept can guide transdisciplinary research and implementation: addressing water challenges in central northern Namibia. Sustainability 9(7):1109. https://doi.org/10.3390/su9071109

Lyle-Gonga M (2013) A critical analysis of gender mainstreaming. Polit Gend 9(2):209–213. https://doi.org/10.1017/S1743923X13000056)

Ngoran SD, Xue XZ, Wesseh PK Jr (2016) Signatures of water resources consumption on sustainable economic growth in Sub-Saharan African countries. Int J Sustain Built Environ. https://doi.org/10.1016/j.ijsbe.2016.04.002

Nijsten G, Villholth KG, Braune E, Gaye CB (2018) Transboundary aquifers of Africa: review of the current state of knowledge and progress towards sustainable development and management. J Hydrol: Reg Stud. https://doi.org/10.1016/j.ejrh.2018.03.004

Oluwasanya G (2022) Quantification of women's roles & representation in the water sector. Presentation during the webinar on "women and water: on the frontline of climate change", WaterAid, 12 Oct 2022

Pegasys (2009) Conceptual framework for managing the impacts of climate on water resources, Pegasys, strategic transboundary water resources assessment. OneWorld Sustainable Investments, Cape Town

Salo E (nd.) Gender and water policies in Africa: synthesis report. Final report on gender and water policies in Africa project

Seager J, Hartmann B (2005) Mainstreaming gender in environmental assessment and early warning. United Nations, New York

Sivakumar I (2016) Gender mainstreaming as a tool for women empowerment. Indian J Women's Stud 09:53–59

Southern African Development Community (SADC) (2016) Regional strategic action plan on integrated water resources development and management phase IV, RSAP IV, Gaborone

Sultana F (2010) Living in hazardous waterscapes: gendered vulnerabilities and experiences of floods and disasters. Environ Hazards 9(1):43–53

Sultana F (2018) Gender and water in a changing climate: challenges and opportunities. In: Fröhlich C et al (eds) Water security across the gender divide. Water security in a new world, chap 2. Springer International Publishing AG, pp 17–33. https://doi.org/10.1007/978-3-319-64046-4_2

Suresh Y, Subhashini N, Bindu Kiranmayi C, Srinivas K, Prasastha Ram V, Chaitanya G, Swathi Vimala B, Srinivasa Rao T (2018) Isolation, molecular characterization and antimicrobial resistance patterns of four different vibrio species isolated from fresh water fishes. Int J Curr Microbiol Appl Sci 7(07)

Tantoh HB, McKay TTJM (2018) Rural self-empowerment: the case of small water supply management in Northwest, Cameroon. GeoJournal 85:159–171

Tantoh HB, McKay TTJM, Donkor FE, Simatele MD (2021) Gender roles, implications for water, land, and food security in a changing climate: a systematic review. Front Sustain Food Syst 5:707835. https://doi.org/10.3389/fsufs.2021.707835

Terry G (2009) No climate justice without gender justice: an overview of the issues. Gend Dev 17(1):5–18

Trivedi A (2018) Women are the secret weapon for better water management. World Resources Institute, 18 Oct 2018. https://www.wri.org/insights/women-are-secret-weapon-better-water-management

Urbini G (2015) Internationalization, education and technological innovation: three key factors to improve the quality of the environment and public health. Rev Ambient Água 10(1). https://doi.org/10.4136/ambi-agua.1574

UNDP (2010) Gender, climate change and community-based adaptation: a guidebook for designing and implementing gender-sensitive community-based adaptation programmes and projects. United Nations Development Programme (UNDP), New York

UN ECOSOC (1997) Coordination of the policies and activities of the specialized agencies and other bodies of the United Nations System. Substantive session of 1997 Geneva. http://www.un.org/esa/documents/ecosoc/docs/1997/e1997-66.htm. Consulted 10 Apr 2021

USAID (nd.) Climate change and gender. Fact Sheet. https://pdf.usaid.gov/pdf_docs/PNADR707.pdf

Vanham D, Alfieri L, Flörke M, Grimaldi S, Lorini V, de Roo A, Feyen L (2021) The number of people exposed to water stress in relation to how much water is reserved for the environment: a global modelling study. Lancet Planet Health 5:e766–e774

World Bank (2022) Women, Business and the Law 2022. World Bank, Washington, DC. https://doi.org/10.1596/978-1-4648-1817-2. License: Creative Commons Attribution CC BY 3.0 IGO

Yan D, Zhang X, Qin T, Li C, Zhang J, Wang H, Weng B, Wang K, Liu S, Li X, Yang Y, Li W, Lv Z, Wang J, Li M, He S, Liu F, Bi W, Xu T, Shi X, Man Z, Sun C, Liu M, Wang M, Huang Y, Long H, Niu Y, Dorjsuren B, Gedefaw M, Li Y, Tian Z, Mu S, Wang W, Zhou X (2022) A data set of distributed global population and water withdrawal from 1960 to 2020. Sci Data 9:640. https://doi.org/10.1038/s41597-022-01760-1

Chapter 18
Water and Sanitation: A Global Priority

J. Usha and S. Nithiya

Abstract This chapter focuses on the domestic and industrial water supply, waste disposal, and the promotion of hygiene and recent technologies in the prevention of diseases. Water sanitation programs offer numerous benefits in the improvement of human health and nutrition, and are considered because they have important implications in the share of the cost that attributes to the health sector. The data revealed the fact that in spite of the existence of traditional wells and bore wells in rural areas, most of the people adopt practices that affect the quality of drinking water. Their social and economic situations lead to the use of alternate forms of water, polluted by solid and liquid wastes, and the consumption of the same leading to waterborne diseases. The main health benefit of water supply, sanitation, and hygiene is a reduction in water-borne diseases, although the effects on other diseases are substantial. On the other hand, the situation of people in the urban areas pose challenge as their livelihood is in cities, where sewerage and waste disposal seem to be difficult and the space for sanitation are all comparably on a higher rate. Inequalities in access are multiplied when sewage is removed and discharged into storm drains, waterways or landfills, rivers, lakes, or the ocean thereby resulting in pollution. The analysis assists in deriving the estimates of the potential effects of water supply and excreta disposal leading to health hazards and diseases, globally and by region, and with cost data to derive cost-effectiveness estimates. As a conclusion, simple human entitlements like potable water and appropriate nourishment must be prioritized by the government, non-governmental, and stakeholders with utmost priority.

Keywords Waste disposal · Water-borne diseases · Human health · Alternate sources · Hygiene and sanitation · Recent technologies

---

J. Usha (✉) · S. Nithiya
Department of Chemistry, Sir Theagaraya College, Chennai, Tamilnadu, India
e-mail: ushaj@stcchennai.edu.in

S. Suriyanarayanan et al. (eds.), *Water Management in Developing Countries and Sustainable Development*, Water Resources Development and Management,
https://doi.org/10.1007/978-981-99-8639-2_18

18.1 Introduction

Waste disposals directly impact the health of humans, paving way to severe consequences when ignored. Most of the developing countries are on the way toward the fastest economies, but when the same comes to sanitation and hygiene indicators, there continues a lag in moving ahead with improvements. With an increased global population, sanitation toward security is a mounting requirement. SLWM is a major key component, with the objective of bringing improvement in a clean environment, hygiene and improvement of the quality of life in both rural and urban areas (Adilov et al. 2021). An introduction to Solid Liquid Waste Management (SLWM) in rural areas is fastened up, particularly for the administrators in assisting the focus on the management along with measurement toward sanitation and hygiene (Government of India (2014) Swachh Bharat Mission, Jalshakti, 2016).

Among many groups, solid and liquid waste management is an important determinant of improved sanitation. Sanitation goals and missions will be considered as a failure when solid and liquid waste management has not been given the needed priority and attention toward improvement of the public health. Especially, this aspect is ignored and neglected due to lack of appropriate infrastructure, lack of education toward sustainable technology, and also due to lack of awareness among common people.

Across the globe, with its great diversity of climatic conditions, terrains, availability of needed sources, livelihood, enormous increase of population, social and economic cultural changes, etc., both the quantity and quality of the wastes generated differ significantly according to places. In general, as indigenous systems are prevailing across various regions, efficiency and steps toward a sustainable environment are needed leading to zero disposal and discharge (Ejalshakti 2015). As per UNO reports, more than eighty percentage of the wastewater is being generated in the developing countries is getting discharged without any treatment into various surface water bodies. On a global scale, nearly two million tons of untreated sewage, industrial as well as agricultural waste is being discharged into almost all forms of waterways. Recent studies report that 245,000 km^2 of aquatic, specifically the marine ecosystems are affected due to fishing, and the food chain gets affected. Around 2 million people lose their life every year from diarrhea, especially children who are under five years in developing countries, which is almost attributable to unsafe water supply and lack of hygiene. An estimated amount of 748 million people do not have accessibility to improved sources of potable water. Nearly, 2.5 billion people lack access to improved sanitation (which is more than 35% of the world's population) (WHO and FAO 2006; UNEP and UN-Habitat 2010; WHO 2014a; CEB 2010; OWG 2014; TEEB 2013; UNEP 2011; UNEP and UN-Habitat 2010; UN-Water 2014; WWAP 2009).

18.2 Wastewater Generation and Characteristics

Generation of wastewater from various households as well as commercial activities can be classified into domestic sewage, industrial sewage, and storm sewage. Domestic sewage carries used water from houses and apartments; it is also called sanitary sewage. Industrial sewage is used water from manufacturing or chemical processes. Storm sewage, or storm water, is runoff from precipitation that is collected in a system of pipes or open channels. The major characteristics of wastewater are classified below:

1. Physical characteristics including color and odor changes, dissolved solids, temperature modifications, etc.
2. Chemical characteristics due to oxygen demands, and other chemicals such as chlorides and sulfates.
3. Biological characteristics due to microbial and bacterial contaminants, wastewaters, and water treatment processes.

18.2.1 Rural Wastewater as a Sustainable Resource

As the majority (more than 80%) of water supplied for domestic purpose are wastewaters, thereby both directly and indirectly leading to health issues in the absence of a proper wastewater management scheme. However, the health and hygiene of rural areas suffer pest- and vector-borne diseases such as dengue, malaria, chikungunya, etc., which are all impossible to eradicate in the absence of a proper wastewater treatment scheme. In recent days, sanitation, hygiene, and cleanliness have received attention and a proper drainage network has become the prerequisite of any rural wastewater management scheme.

18.2.2 Effects of Rural Wastewater Management on Health and Hygiene

Wastewater management and health are well interconnected in developing countries, as nearly more than 75% of illness is directly linked to inaccessibility of consumable cum potable water. In the year 2012, nearly 842,000 deaths in middle- and low-income poor groups occurred due to the contaminated and polluted drinking water, inadequate knowledge of sanitation and hygiene services (WHO 2014b). In addition, it was observed that nearly three lakh deaths among children aged below 5 years could have been prevented by the reduction of risks that are linked to inadequate hygiene, sanitation, and usage of wastewater during the same year (Pruss-Ostun et al. 2014) (Fig. 18.1).

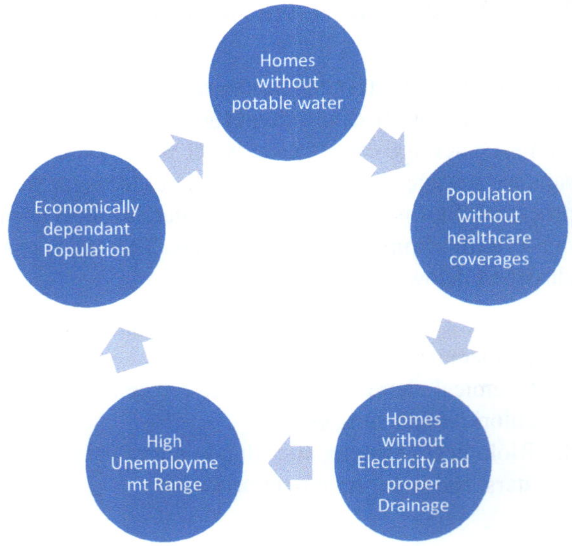

Fig. 18.1 Lack of sanity in rural region

There is much need for low-cost wastewater treatment technologies in developing countries especially in the rural areas. Compared to the prevailing situation of wastewaters and the treatments in urban areas, sustainability toward the wastewater treatment techniques and the proper management in rural areas is considered to be speculative (Raza et al. 2018; Richey et al. 2015). The facts and figures from rural regions at a macro level have rarely been reported or considered even while planning for implementations on the treatment of solid and liquid waste management programs (Machado et al. 2016). Sequential batch reactors and membrane reactors track records of conventional and traditional wastewater treatment technologies such as activated sludge process, etc., thereby making their penetration in resourceless or rural areas.

18.3 Urbanization and Sanitation

Urbanization for water, sanitation, and hygiene management in rural and urban areas face challenges. Urban areas often find difficulty with the cost of the operation and maintenance costs and the decay of the available infrastructure (Shakoor et al. 2015; Szydłowski et al. 2017; Tytła and Kostecki 2019). Urban growth and development in developing countries is facing serious issues in most of the regions to provide adequate services (Cohen 2006). Different applications and drivers are responsible for the transition in various water phases; a few among which include the increase in population and the accelerated consumption that require more water (Mahlknecht et al. 2020; Mateo-Sagasta et al. 2015). In finding the solution to the emerging

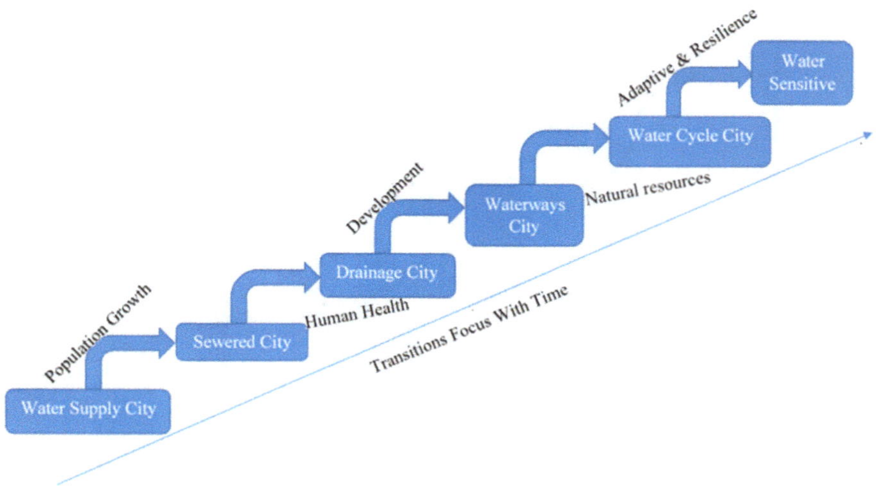

Fig. 18.2 Changing paradigms of water and transition phase in urban areas

problem related to water and security issues, it is vital to modify the responses from the social and political media responsible (Brown et al. 2008) (Fig. 18.2).

18.4 Health Hazards Due to Usage of Wastewaters

It has been concluded by researchers that approximately 80% of all diseases and 20% of the death rates in developing countries occur due to polluted water and inadequate supply of potable water, sanitation, and hygiene. This situation particularly seems to be on a high rate in rural areas where access rates to clean water, sanitation, and hygiene applications and services are typically too low. Among the common water-borne diseases such as diarrhea, cholera, malaria, dysentery, dengue, typhoid, gastroenteritis, etc., diarrhea has been observed to pose a serious risk leading to malnutrition. Consumption of poor drinking water leading to water-borne diseases is enormous, leading to an accelerated development in millions of children each year (Prüss-Ustün et al. 2014).

18.5 Challenges in Water, Hygiene, and Sanitation

People who live in extremely poor conditions are most affected by these economic, health, and social issues especially related to water sanitation. Researchers have observed that nearly 40% of those who are very poor use unimproved drinking water. In undeveloped regions, it is economically difficult to construct huge infrastructure to

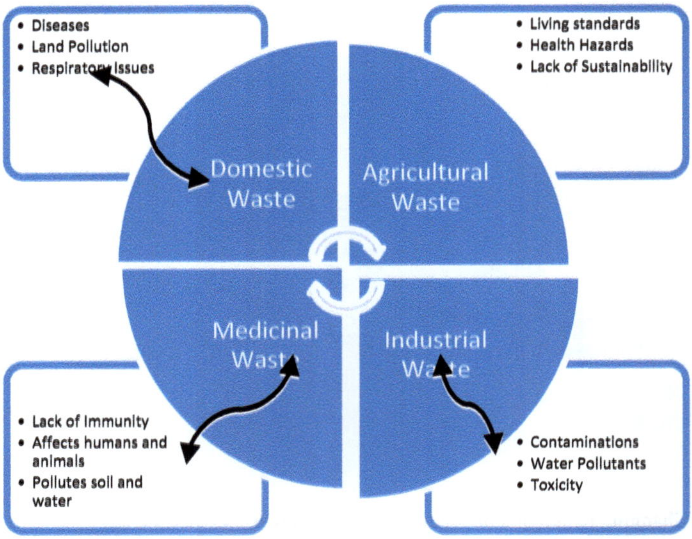

Fig. 18.3 Health issues with using wastewater from domestic, medicinal, irrigation, and industrial applications

replace conventional water treatment and distribution methods. By continuous monitoring, the ground water quality can be checked but is a challenging phenomenon due to the vast spread of villages and inadequate accessibility for the regular monitoring approaches (Carrard et al. 2019) (Fig. 18.3).

It has been calculated by researchers in the year 2016 that nearly 2 million deaths and 123 million disability-adjusted life-years (DALYs) leading to nearly 3.3% of global deaths could be prevented by providing adequate knowledge in hygiene and sanitation methods. Pathogens, rotavirus, cryptosporidium, and Escherichia coli are few of the biological agents responsible for the cases of mild to severe diarrhea in rural areas (Urbaniak et al. 2016). Cholera, a killer disease if not treated on time, is still observed to be endemic in about 69 countries, resulting in an estimated three million cases and 95,000 deaths, worldwide annually.

Reliance on antibiotics due to poor hygienic conditions further drives the force toward antibacterial, antifungal, and antimicrobial resistance.

It has been reported that roughly, 140 million people in around 50 countries are drinking water containing excessive levels of arsenic (World Health Organization 2018).

Long time exposure toward higher levels of arsenic in consumable water and food leads to skin lesions, heart diseases, diabetes, and also cancer (Kasonga et al. 2021). Exposure in early childhood is linked to impaired cognitive development that also leads to deaths in youth (Larsen et al. 2016). Contaminants in pharmaceuticals, endocrine disruptors, and plastics should be stopped before reaching and getting mixed with the drinking water even though the concentrations present in drinking water or the associated sources are unlikely to cause risk to the health of humans (Gworek et al. 2020; Kasonga et al. 2021; World Health Organization 2012, 2019a).

18.6 Monitoring of Drinking Water

Quality water monitoring is now being accorded as a high-priority concern and the institutional mechanisms are being developed at various levels on a global scale. The plans associated with also outline requisite mechanisms in monitoring the quality checks of drinking water frequently, dragging information, institutional and communicational interventions in gathering and disseminating data, and thereby educating people on sanitation, health, and hygiene. Institutionalization and imparting knowledge on hygiene programs in the continual monitoring and the surveillance of sources of drinking water at the base level by Sanitation Committees are being envisaged, followed by frequent checking of the water using Internet of Things (IoT), satellites, and the latest technologies from all zones before reaching consumers (Fig. 18.4).

A major issue in addressing the issues and problems associated with water includes the appropriate means and provisions for distributing water across various regions. As interrelated coordination acts as the strength in dealing with maintaining the standards of water quality, distributers and consumers should also be aware and play a key role in maintaining hygiene near water sources. Even collecting and storage should be properly taken care while operating in both rural and urban areas. While decentralizing the water supply programs, it is very important that all the involved communities and institutions actively participate and get involved in the planning, acting, propagation, implementation, and finally execution of water supply methods (https://www.un.org/waterforlifedecade/waterandsustainabledevelopment2015/images/water_quality_eng.pdf).

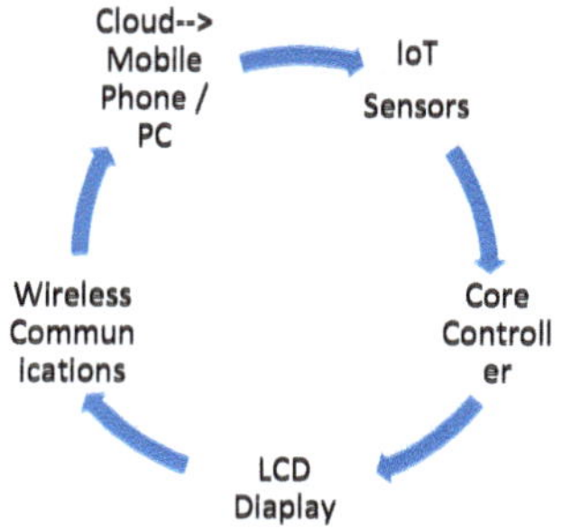

Fig. 18.4 Water quality monitoring using IoT

18.7 Sanitation and Hygiene of Water

Safe water, sanitation, and hygiene (collectively called WASH) are considered to be much crucial for maintaining the health of humans and their well-being. But, even then, people on a global scale are lacking knowledge of adequate WASH services and as a result, suffer from the exposure leading to illnesses that can be prevented. (https://apps.who.int/iris/bitstream/handle/10665/330100/WHO-CED-PHE-WSH-19.149-eng.pdf, Water, sanitation, hygiene, and health: a primer for health professionals. Geneva: World Health Organization (2019b) (WHO/CED/PHE/WSH/19.149). Licence: CC BY-NC-SA 3.0 IGO).

Researchers work with providing access to cleanliness, sanitation, and in promoting basic hygiene practices in rural and urban areas, including the emergency cases and situations. According to them, achievement with WASH results in young children can be achieved by practicing and implementing the following:

While increasing the power of communities and enhancing the focus on sustainability, scientists regulate and promote the communities by providing information on sanitation and hygiene through the utilization of media and also through campaigns likes Global Sanitation Day, which might reach millions of people each year. This approach has associated all the communities in eliminating the hazardous practice of defecation, which is almost open to everyone. The power of communications is directly proportional to health care centers in order to improve the access toward the right usage of water, sanitation, and hygiene facilities, and in establishing the protocols in preventing and controlling of diseases (Villanueva et al. 2014; Zhang et al. 2018; Zhu et al. 2021). Fragile settings assist in preparing for and responding to the humanitarian requirements and emergencies, which include the transportation of water to the needy consumers, thereby working toward the construction of water basis and sanitation facilities that stay according to the needs of the consumers, simultaneously providing the leadership as well as the accountability in the response against humanitarian activities.

Sanitation programs are committed to a sustainable as well as adaptive basis toward the climatic changes. In addition to the sanitation programs, in recent years, thousands of solar-powered water systems in order to address the water scarcity issues that keep increasing have been rehabilitated.

UNICEF works in close collaboration with governmental and non-governmental organizations, private agencies, academic institutions, construction companies, and also needs communities to improvise the systems and the associated practices fulfilling the needs of an individual toward a clean environment with safe water and a great degree of sanitation (https://www.unicef.org/wash).

18.8 Conclusions

Decentralization of managing waters, continuous monitoring of the supply of water and water usage and a modern approach to rural areas were all observed to be important in achieving the sustainability in developing the goals that are objective in providing safe drinking water to people, which acts as a challenge for the researchers and the planning commissions.

Identification of the problems that are sustainably systematic and developing and implementing the policies for water access in the regions play a vital role in sanitation and hygiene; water is available as and how needed and also not contaminated. This kind of cooperation will also ensure sustainability in maintaining public health and bring in the benefits, economy-wise, to rural areas. As a result, analysis of consumer demand for the available water supply systems and the preference of consumers in planning and choosing the source of drinking water contribute to effective plans for water supply, thereby, supporting the sustainability in the development of rural areas.

In spite of these issues, as the need toward potable, nutrient-rich, scalable water distribution models remains high, and demand for clean drinking water will rise more, with intense competition for water from domestic to commercial scale, and increasing scarcity and variability of water resources, continuing financial and technological contributions from the service sectors in the development of sustainability to water solutions stays important. Synergic relationships and communications between the private and public sectors will also be enhanced in providing solutions and to cope up and balance with the upcoming water crisis.

Relationship between the environment and water security is inter-beneficial and acts as a core criterion of providing solutions to secure water in ensuring sufficiency and availability of quality water in supporting the proper functioning of the ecosystem. Even though it is trending, it has to be considered as a stressful component in achieving the overall societal development. It becomes necessary to achieve the water security (Vörösmarty et al. 2018) at a watershed scale level, in order to ensure the conservation of the available ecosystems of our rural and urban regions, as they provide water in terms of quality as well as quantity for maintaining food security and to mitigate with the hydrological hazards.

Attainment of water security and focus on a global scale in a context is either not an easy or "a one step" solution, but can be achieved by expanding the portfolio of the solutions needed from engineered systems to institutional management systems based on the capability to build awareness among communities, wetland managements, and aquifer conservation in incorporating the natural processes for the well-being of humans and other living species. But, it requires well-accepted integration of soft and hard measures in water maintenance, storage, and transportation, and also in protecting the available existing resource itself. Addressing the impacts of human activities over water usage, land, air, and aquatic ecosystems, climatic changes, it can be concluded that water security has multiple interrelated and interdependent dimensions, each involving strong and complex interactions between the society and the environment. Achieving a water-secure world requires stability and balance among social, cultural, political, economic, and environmental components (Mishra et al. 2021).

References

Adilov TT, Sarikulov MK, Artikbaevich RH, Kuchkarova NK (2021) To study the problem of drinking water shortage and public health. Int J Innov Anal Emerging Technol 1(5):192–196. http://openaccessjournals.eu/index.php/ijiaet/article/view/489

Brown R, Keath N, Wong T (2008) Transitioning to water sensitive cities: historical, current and future transition state. In: Proceedings of the 11th international conference on urban drainage, Edinburgh, UK

Carrard N, Foster T, Willetts J (2019) Groundwater as a source of drinking water in Southeast Asia and the Pacific: a multi-country review of current reliance and resource concerns. Water 11(8):1605–1625

CBD (2010) Drinking water, biodiversity and development: a good practice guide. Convention on Biological Diversity, Secretariat of the Convention on Biological Diversity, Montreal. http://www.unwater.org/wwd10/downloads/cbd-good-practice-guide-water-en.pdf

Cohen B (2006) Urbanization in developing countries. Current trends, future projections, and key challenges for sustainability. In: Technology in society, vol 28, pp 63–80

Gworek B, Kijeńska M, Zaborowska M, Wrzosek J, Tokarz L, Chmielewski J (2020) Occurrence of pharmaceuticals in aquatic environment—a review. Desalination Water Treat 184:375–387

https://ejalshakti.gov.in/misc/Docs/ProvenTech_S.pdf (2015)

https://www.un.org/waterforlifedecade/waterandsustainabledevelopment2015/images/water_quality_eng.pdf

Kasonga TK, Coetzee MAA, Kamika I, Ngole-Jeme VM, Benteke Momba MN (2021) Endocrine-disruptive chemicals as contaminants of emerging concern in wastewater and surface water: a review. J Environ Manage 277:111485

Larsen TA, Hoffmann S, Luthi C, Truffer B, Maurer M (2016) Emerging solutions to the water challenges of an urbanizing world. Science 352(6288):928–933

Machado AI, Beretta M, Fragoso R, Duarte E (2016) Overview of the state of the art of constructed wetlands for decentralized wastewater management in Brazil. J Environ Manage 187(1):560–570. https://doi.org/10.1016/j.jenvman.2016.11.015

Mahlknecht J, Gonzalez BR, Loge FJ (2020) Water-energy-food security: a nexus perspective of the current situation in Latin America and the Caribbean. Energy 194(3):1–17. https://doi.org/10.1016/j.energy.2019.116824

Mateo-Sagasta J, Raschid-Sally L, Thebo A (2015) Global wastewater and sludge production, treatment and use. In: Drechsel P (ed) Wastewater: economic asset in an urbanizing world. Amsterdam, pp 3–14. https://doi.org/10.1007/978-94-017-9545-6-2

Ministry of Drinking Water and Sanitation, Government of India (2014) Swachh Bharat Mission (Gramin) guidelines. https://jalshakti-ddws.gov.in/sites/default/files/Primer%20SLWM.pdf

Mishra BK, Kumar P, Saraswat C, Chakraborty S, Gautam A (2021) Water security in a changing environment: concept, challenges and solutions. Water 13:490–511

OWG (2014) Introduction and proposed goals and targets on sustainable development for the post-2015 development agenda. Open Working Group on Sustainable Development Goals, June 2014. http://sustainabledevelopment.un.org/focussdgs.html

Pruss-Ostun A, Bartram J, Clasen T, Colford JM Jr, Cumming O (2014) Burden of disease from inadequate water, sanitation and hygiene in low- and middle-income settings: a retrospective analysis of data from 145 countries. Tropical Med Int Health 19(8):894–905

Prüss-Ustün A, Bartram J, Clasen T, Colford JM, Cumming O, Curtis V, Bonjour S, Dangour AD, France JD, Fewtrell L, Freeman MC, Gordon B, Hunter PB, Johnston RB, Mathers C, Mäusezahl

D, Medlicott K, Neira M, Stocks M, Wolf J, Cairncross S (2014) Burden of disease from inadequate water, sanitation and hygiene in low- and middle-income settings: a retrospective analysis of data from 145 countries. Tropical Med Int Health 19(8):894–905

Raza N, Hashemi B, Kim KH, Lee SH, Deep A (2018) Aromatic hydrocarbons in air, water, and soil: sampling and pretreatment techniques. TrAC Trends Anal Chem 103:56–73

Richey AS, Thomas BF, Lo MH, Famiglietti JS, Swenson S, Rodell M (2015) Uncertainty in global groundwater storage estimates in a total groundwater stress framework. Water Resour Res 51(7):5198–5216

Shakoor MB, Niazi NK, Bibi I, Rahman MM, Naidu R, Dong Z, Shahid M, Arshad M (2015) Unraveling health risk and speciation of arsenic from groundwater in rural areas of Punjab, Pakistan. Int J Environ Res Public Health 12(10):12371–12390

Szydłowski K, Rawicki K, Burczyk P (2017) The content of heavy metals in bottom sediments of the watercourse in agricultural catchment on the example of the river Gowienica. Ecol Eng 18(4):218–224

TEEB (2013) Natural capital accounting and water quality: commitments, benefits, needs and progress. A briefing note. The Economics of Ecosystems and Biodiversity (TEEB). http://www.teebweb.org/wp-content/uploads/2014/01/TEEB_-NaturalCapitalAccountingandwaterQualityBriefingnote_20131.pdf

Tytła M, Kostecki M (2019) Ecological risk assessment of metals and metalloid in bottom sediments of water reservoir located in the key anthropogenic "hot spot" area (Poland). Environ Earth Sci 78(5):179

UNEP (2011) Towards a green economy: pathways to sustainable development and poverty eradication. United Nations Environment Programme (UNEP). www.unep.org/greeneconomy

UNEP and UN-Habitat (2010) Sick water: the central role of wastewater management in sustainable development. United Nations Environment Programme (UNEP) and United Nations Human Settlements Programme (UN-Habitat). http://www.unep.org/pdf/SickWater_screen.pdf

UN-Water (2014) A post-2015 global goal for water: synthesis of key findings and recommendations from UNWater. http://www.zaragoza.es/ciudad/medioambiente/onu/en/detallePer_Onu?id=811

Urbaniak M, Gągała I, Szewczyk M, Bednarek A (2016) Leaching of PCBs and nutrients from soil fertilized with municipal sewage sludge. Bull Environ Contam Toxicol 97(2):249–254

Villanueva CM, Kogevinas M, Cordier S, Templeton MR, Vermeulen R, Nuckols JR, Nieuwenhuijsen MJ, Levallois P (2014) Assessing exposure and health consequences of chemicals in drinking water: current state of knowledge and research needs. Environ Health Perspect 122:213–221

Vörösmarty CJ, Osuna VR, Cak A, Bhaduri A, Bunn SE, Corsi F, Gastelumendi J, Green P, Harrison I, Lawford R (2018) Ecosystem-based water security and the sustainable development goals. Ecohydrol Hydrobiol 18(4):317–333

WHO (2014a) Water. World Health Organization. http://www.who.int/topics/water/en/

WHO (2014b) Preventing diarrhoea through better water, sanitation and hygiene: exposures and impacts in low- and middle-income countries. Work! Health Organization, Geneva

WHO and FAO (2006) Guidelines for the safe use of wastewater, excreta and grey water in agriculture and aquaculture, 3rd edn, vols 1–4. World Health Organization, Geneva. http://www.who.int/water_sanitation_health/wastewater/gsuww/en/

World Health Organization (2012) Pharmaceuticals in drinking-water. World Health Organization, Geneva. https://apps.who.int/iris/handle/10665/44630. Accessed 30 Aug 2019

World Health Organization (2018) Arsenic [fact sheet]. World Health Organization, Geneva. https://www.who.int/news-room/fact-sheets/detail/arsenic. Accessed 30 Aug 2019

World Health Organization (2019a) Microplastics in drinking-water. World Health Organization, Geneva. https://apps.who.int/iris/handle/10665/32649. Accessed 30 Aug 2019

World Health Organization (2019b) Water, sanitation, hygiene and health: a primer for health professionals. World Health Organization, Geneva. https://apps.who.int/iris/bitstream/handle/10665/330100/WHO-CED-PHE-WSH-19.149-eng.pdf (WHO/CED/PHE/WSH/19.149). Licence: CC BY-NC-SA 3.0 IGO

WWAP (2009) Water in a changing world. The United Nations World Water Development Report 3. World Water Assessment Programme, UNESCO, Paris. http://www.unesco.org/new/en/natural-sciences/environment/water/wwap/wwdr/wwdr3-2009/

Zhang H, Huo S, Yeager KM, Xi B, Zhang J, He Z, Wu F (2018) Accumulation of arsenic, mercury and heavy metals in lacustrine sediment in relation to eutrophication: impacts of sources and climate change. Ecol Indic 93:771–780

Zhu T, Chen W, Jafvert CT, Fu D, Cheng H, Chen M, Wang Y (2021) Development of novel experimental and modelled low density polyethylene (LDPE)-water partition coefficients for a range of hydrophobic organic compounds. Environ Pollut 291:118223

Chapter 19
Regional Water Harvesting Structures in Kerala: A Gender Perspective

Vinnakota Yesubabu, P. Aravind, Anu Varughese, and Amit Kumar

Abstract This chapter emphasizes the importance of water conservation and management, focusing on regional and local water harvesting and water reuse as two effective methods for ensuring sustainable water use. Regional water harvesting involves the construction of structures such as dams, reservoirs, and recharge pits to capture and store rainwater, while local water harvesting involves the installation of rainwater harvesting systems on individual buildings. Water reuse refers to the treatment and reuse of wastewater for various purposes. These methods can significantly reduce water demand and pressure on existing water sources and help mitigate the effects of climate change. However, gender issues remain a significant challenge in water management initiatives. Women's limited access to decision-making processes and the burden of water collection often prevent their active participation and exacerbate gender inequalities. The chapter suggests addressing these issues by promoting women's participation in decision-making processes, providing clean water sources that are accessible to women, and involving women in the maintenance and operation of water management systems. The Women's Water Collective in Kerala, India, is presented as an example of a successful women-led organization that promotes gender equality in water management. The chapter concludes by highlighting the importance of addressing gender issues for the long-term sustainability of water management initiatives.

Keywords Regional water harvesting · Water reuse · Madakas · Gender equality

V. Yesubabu (✉) · P. Aravind · A. Varughese
Department of Irrigation and Drainage Engineering, KCAET, Kerala Agricultural University, Malappuram 679573, Kerala, India
e-mail: vinnakota-2020-28-002@student.kau.in

A. Kumar
Department of Farm Machinery and Power Engineering, KCAET, Kerala Agricultural University, Malappuram 679573, Kerala, India

S. Suriyanarayanan et al. (eds.), *Water Management in Developing Countries and Sustainable Development*, Water Resources Development and Management,
https://doi.org/10.1007/978-981-99-8639-2_19

19.1 Introduction

Water is a precious resource and its conservation and management have become increasingly important in recent years. Regional and local water harvesting and water reuse are two methods that can help us conserve water and ensure its sustainable use. Regional water harvesting refers to the collection and storage of rainwater in a particular region or catchment area. This method involves the construction of structures such as dams, reservoirs, and recharge pits to capture and store rainwater. The stored water can then be used for various purposes, such as irrigation, drinking water, and industrial use. Local water harvesting, on the other hand, is the collection and storage of rainwater at the local level. This method involves the installation of rainwater harvesting systems on individual buildings, such as homes, schools, and commercial buildings. The water, which was collected can be utilized for various non-potable purposes, such as washing clothes, flushing toilets, domestic activities, and watering plants.

Water reuse refers to the treatment and reuse of wastewater for diverse purposes, such as domestic usage, irrigational supplement, industrial usage, and even drinking water after proper treatment techniques. Treatment of wastewater can be done using various primary treatment methods such as sedimentation, filtration, and disinfection to remove contaminants and make it safe for reuse. This method can significantly reduce water demand and conserve freshwater resources.

The benefits of regional and local water harvesting and water reuse are numerous. These methods can help reduce water scarcity and ensure the sustainable use of water resources (Chow et al. 1988). By reducing the demand for freshwater, these methods can also reduce the pressure on existing water sources, such as rivers and aquifers. In addition, regional and local water harvesting and water reuse can help mitigate the effects of climate change by reducing the impact of droughts and floods.

The issue of gender has a significant impact on local and regional harvesting and water reuse initiatives. The suggestions on gender differences must be focused on the aspects of sanitation and gender-based approach should be uplifted in design plans and structural planning for implementation (UN Water 2006). Women are disproportionately affected by the lack of access to water resources, and their participation in the decision-making process is often limited (Ahmed 2005a, 2008). Addressing gender issues is essential for the long-term sustainability of water management initiatives.

One of the main challenges faced by women in water management is the lack of access to decision-making processes. Women are often excluded from water management committees and do not have a voice in the planning and implementation of water management initiatives. This exclusion limits the effectiveness of these initiatives and hinders their long-term sustainability. Addressing this issue requires the inclusion of women as a part of decision-making processes and the promotion of gender equality. Policies are need to be developed to address specific needs of both women and men to equip, educate, and empower the women to participate at all levels in water resources development programs, like decision-making and implementation, that are suitable for them (Ahmed 2005b). Gender mainstreaming in management of water resources can be instrumental in helping countries address other national and global challenges of climate change, mitigating poverty, inequalities, and in the implementation of social justice, etc. Less than 50 countries in the world have laws or policies that motivate the participation of women in rural sanitation and water resources administration (UN Water 2021). This integrated approach paves the way for holistic and cross-cutting national policies which contribute to the achievement of the many related Sustainable Development Goals related to water resources (Bhat 2014).

Another challenge facing women in water management is the burden of water collection. In many rural areas, women are the sole responsible for collecting water and lack of access to clean water sources places significant trouble on their daily lives. This burden often prevents women from engaging in other activities, such as education or income-generating activities, further exacerbating gender inequalities. Incorporation of the gender aspects in water resources is of global interest which is to be practiced in the current scenario. Integration of gender with modeling can represent the ideas and representation of behavior and decision-making in modeling (Packett et al. 2020). Addressing this issue requires the provision of clean water sources that are accessible to women, reducing the burden of water collection and promoting gender equality. For example, in some communities, the responsibility for the maintenance and operation of water management systems is assigned to men, which often results in the neglect of these systems (Khandker et al. 2020). Addressing this issue requires the participation of women in the maintenance and operation of these systems, promoting their ownership and long-term sustainability (Suhardiman et al. 2023).

Water accessibility and sanitation are the basic human rights, where female gender lacks the independence by which they are affected with serious health issues, curtailing their economic, social, and educational opportunities (UN Women 2018). Women are the collector, user, and manager of water in households as well as farms for irrigation. Women act as labor for most agricultural activities in low and middle-income countries, their agricultural productivity ranges between 20 and 30% on an average lower than that of male farmers (FAO 2017). Women have a higher knowledge than men in the aspects including the quality of water, sanitation limits, and abundant storage methods. Development of the water resources should have a masculine as well as feminine representative to have a stable and steady water planning. The representation of the female gender in water user associations (WUAs) in the South Asian countries is nominal despite the extensive participation of women in farming and decision-making.

This chapter covers broadly two topics, one is different regional water harvesting structures in Kerala state and another one is the gender issues in these initiatives.

19.1.1 Status of Regional Water Harvesting Structures in Kerala

Kerala is a state in India that has a rich history of water harvesting techniques. Kerala has high population density compared with other states of India, thus becoming a challenge for the government to provide safe drinking water to rural community. The safe drinking water supply still remains impractical for more than 65% of the households in Kerala. Merely 29.3% of the houses in the state are supplied with the tap water supply network and just 34% has access to safe water supply, which is 52% less than the national average. The rural–urban gap is also widening, the whole city population getting safe water to the total population has been decreased from 42.85% in 2001 to 39.4% in the year 2011 (State Planning Board, Kerala 2013). However, in recent years, the state has been facing water scarcity and drought-like conditions during the summer months. Regional and local water harvesting techniques have been implemented in Kerala as a solution to mitigate the impact of water scarcity. Some of the traditional water harvesting structures in Kerala are:

Eris

These are the innovative pond-based water harvesting structures, which can be used for the purpose of capturing rainwater from the impervious surface and storing them for future needs. The advantage of the Eris water harvesting technique is, it utilizes the surface as well as the subsurface runoff flow for the maximum storage efficiency and recharge the groundwater effectively for increasing the water yield than any other traditional methods (Samuel 2010). This system is effective since it has an added advantage of filtration of the stored water, making it suitable for the purpose of drinking with minimal treatment purpose and for all other domestic purposes. These are largely found in Kuttanad region, which stores water from the rivers and canals and uses it for irrigation and domestic use. These structures are based on the local community and have a responsibility by the people of catchment, mainly women who are involved in the household activities. Involving and educating women on this aspect seems to be more effective than involving men. But, it is the responsibility of both the genders to conserve and recharge water. Eris is important in maintaining the lake ecology of the locality by maintaining the microclimate.

Ooranis

Ooranis are similar to Eris water harvesting but they are different in the aspect of their usage. Ooranis are commonly used in villages to take water for drinking purposes only. Ooranis act as a catch can of rainwater only for the purpose of drinking and are not commonly used for irrigation. They are shallower in nature and are smaller in size than the Eris. These are larger, shallow ponds found in Thrissur district that collect and store rainwater for drinking and domestic usage. Present conditions, where extensive use of borewells and overexploitation of groundwater is taking place, this simple old technique which is energy saving and can revive the exploited water table can be extensively utilized by villagers to prevent water scarcity in times of

Fig. 19.1 Eris water harvesting system

reduced rainfall. They play a pivotal role in the feminine gender, who are fetching the Oorani water for drinking purpose. This water harvesting method which has a cleaning mechanism has enhanced the lives of women who waste major time collecting water by walking for hours (Ponnusamy 2013). A research conducted in the three states of India including Kerala, West Bengal, and Tamil Nadu has discussed the role of women in rice farming, where the study explains the local variations in roles that women and men play in rice cultivation and production, where women are subjected to activities like transplanting, weeding, and intercultural practices (Saradamoni 1991) (Figs. 19.1 and 19.2).

Kattas

"Kattas" are temporary check dam structures built across rivulets and streams, which can be commonly sighted in the districts of Karnataka and Kerala (Samuel and Mathew 2008). These traditional temporary check dams were constructed every year during the summer season to utilize water for supplementary irrigation during summer. They can still be found in operation in many streams across Malappuram and Kasaragod districts of Kerala (Nair 2016). However, due to the high density of borewells and modern water conservation methods, the extent of Kattas has declined but are commonly used where paddy is cultivated during off-season and are commonly built by the downstream farmers to escape from the acute shortage of water for crops during the off-season cultivation. Built using *kaccha* stones and locally available materials, kattas are constructed in the post-monsoon seasons mainly during November or December. Interestingly, this method is the only useful water conservation method that can be implemented even after the monsoon subsides (Samuel 2010).

Fig. 19.2 Oorani water harvesting structure

A sequence of kattas erected at the correct time ensures the river runs for a supplementary period of four–six weeks. The cost of constructing kattas is quite high as it requires skilled and experienced labor and fresh soil being an ex-situ water conservation method. They are commonly found in the Malabar regions that are under cultivation, where majorly, the masculine gender dominates as the farmers than the feminine gender. Kattas generally play the work of a check dam and store water for off-season. Since they are on the verge of extinction, the significance of this structure must be educated through NGOs to people and awareness must be created for the construction of such structures and women must actively participate to improve their economic viability through farming activities through this method of water conservation.

Surangams

Surangams are known as tunnel wells and are less common traditional groundwater harvesting structures (More 2020). They can be used to serve as groundwater harvesting structures and at many instances, they can also act as conduits for the groundwater. These are horizontal water wells, structurally similar to Qanats, are constructed in the laterite plateaus and hills for 10 m (Shaji et al. 2020). Surangams are commonly found rectangular in cross-section, with width of 50–80 cm, height of 0.90–1.5 m, and length varies between 3 and 300 m.

The rainwater collected will gradually and steadily seep out through the tunnel wells. The water yield of runoff from a tunnel depends entirely on the thickness of the saturated lateritic zone and its length. These are constructed in the hillocks for the function of freshwater harvesting, which seeps through the overlying mounds. The major types of surangams include, one, well dug in the hillock for freshwater

Fig. 19.3 Operational Katta water harvesting structure

harvesting and the second group are those well dug horizontally at the bottom of the ponds.

The peak summer will have the lowest groundwater yield while the rainy season will have the highest groundwater yield. They eventually drain with groundwater flow to any rivulet or stream. They are mostly constructed in the upstream of a watershed or catchment and they also reduce the erosion through the drilled wells (Fig. 19.3).

There is a critical need to create awareness and educate the community about restoring and stimulating the traditional harvesting structures (Joji and Jacob 2023). These retain sustainable options for a long period than borewells and carry the ancient wisdom of respecting and preserving nature. These are underground tunnels found in Palakkad and Kasargod districts that are used to tap groundwater and store it for irrigation and domestic purposes through Surangams to ponds (Fig. 19.4).

Ponds and Wells

These are commonly used for water harvesting in Kerala, with ponds being used to collect rainwater and wells being used to tap groundwater.

A research organized by the Centre for Water Resources Development and Management (CWRDM), emphasizes that 90% of the open wells in Kerala are subjected to bacteriological contamination, which imposes a threat to the human community. Every human must be responsible for the proper management of water resources, since groundwater is a common property of nature. Reforms are essential to assess and evaluate the groundwater resources, mapping and improving the yield aquifers, monitoring water quality, and in governance of groundwater using legal and institutional aspects. It was found from a study that the state has around 70 lakh

Fig. 19.4 Typical view of Surangam with a storage pond

wells (Central Ground Water Board 2015). The density of wells is considerably very high in Kerala, with a spatial range of 200/km^2, 150/km^2, and 70/km^2 for coastal areas, midland, and highland regions.

Filter-Point Wells

These wells are possible at any place where, the saturated thickness layer exceeds 5 m deep. They are shallow tube wells that are formed by allowing a small diameter pipe, which has a sharp ending with perforations beneath the earth's surface by the force of a water jet, air jet, or manual method of drilling. Loose, sandy water-bearing formations and gravel aquifers are suitable locations for the construction of filter point wells. A mesh or screen is attached to the bottom portion of the pipe to avoid mixing of sand particles or other contaminants into the well water. The withdrawal capacity of filter point wells is expected to be around 10 m from the shallow earth surface (Central Ground Water Board 2013). Filter point wells are mostly seen in the coastal belts, particularly the areas where deep sandy layer is predominant. They are more common in all the coastal districts of Kerala, they are supported by women for the extraction of groundwater and play an important role in water availability in Kerala homes.

Infiltration Gallery

These are used as collector wells, which have horizontal porous pipe-shaped structures made up of gravel-type screen material in highly permeable aquifers, which can be recharged instantaneously with the help of a strong base flow component. They are commonly placed 3–7 m below the soil surface, parallel to the river reaches to intercept the groundwater flow by gravitational force. Irrigation galleries are also used to provide quality drinking water in different seasons irrespective of the quality of river water. These are commonly used to avoid the saltwater intrusion to the groundwater and are very recommended in coastal corridors. They can be found in any river course across Kerala, since the state has a rich river network. They are a community based drinking water supply system rather than a gender perspective.

Kerala typically has more than 3200 mm of precipitation annually, but a significant percentage of this water is becoming useless in the form of direct runoff and evaporation, and rest is recharging into the groundwater aquifer. The major surface water collectors of the state are ponds and lakes, which collect the major portion of the surface water and allow this water to recharge purpose. Approximately, 75% of the total water volume from ponds can be considered as net surface water available for development (Rai 1993). A unique characteristic of temples in Kerala State is the presence of a pond or Theerthakkulam (located on the NE corner of the temple) and Kulappura (located outside the temple wall) in every temple that lacks a nearby river.

Madakas

Madakas serve as traditional rainwater harvesting structures used for supplementary irrigation in rice fields. These structures are generally found in south Karnataka and Kasaragod district of Kerala where there is a chance of a laterite belt. Madakas are basically naturally formed depressions bounded by acute terrain on three different sides, which accumulate rainwater from adjacent sloppy areas of laterite, chiefly runoff from rainfall. To prevent runoff from the slopes, bunds are traditionally built on the fourth side of the depression, which is opened to capture rainwater.

Madakas are suited to agricultural areas with laterite soil due to the soil's fast absorption and depletion of water. They are compared to percolation ponds as they took more time for the water percolation through different cracks of the soil media. The major function of a madaka is to capture most of the runoff from higher elevations i.e., hill slopes and storing them in the depressions. In addition to storing water for seven months, these can help raise the water table of the aquifers below, even though they may dry up during the summer months. Building madakas in higher elevations ensures that water is available in lower areas until the end of summer. The excess flow over the bunds can be used to irrigate fields. Their role in water recharge has only recently gained recognition.

It is necessary to avert the first rains from flowing into the madaka to avoid blockage and clogging of the cracks in the bed with leaves and fine silt and sediments, which reduces percolation and the effective water storage capacity of the structure. Afforestation is another vital aspect of maintaining madakas and can remove the excess accumulation of silt material. Madakkas are not only used for the storage of surface water, recharging of groundwater but also to help in the formation of new ecosystems for several birds, animals, and other aquatic creatures.

Since Kerala has high-density wells, it is essential and it is a duty of women to make their surroundings and their homes nourish with groundwater. This system may decrease the burden of the women to decrease the portability of water from the water bodies by the nourishment of the wells in their homes. These madakas can be used for mitigating floods and can withstand droughts too. Due to the impact of climate change in Kerala, there is an immediate need for these kinds of structures to escape from the imbalance in the hydrological system.

19.1.2 Effectiveness of Regional Harvesting Techniques in Kerala

The traditional water harvesting structures have been used for centuries and have played a significant role in water management and conservation in the state.

Several scientific studies have been done to assess the effectiveness of regional and local water harvesting techniques in Kerala. One such study, conducted by the Indian Institute of Technology (IIT) Delhi, analyzed the groundwater recharge under the influence of water harvesting systems in a small village in Kerala (Liji 2020). The study found that the rainwater harvesting system increased groundwater recharge by 30–40%, leading to a significant improvement in water availability during the summer months. Another study conducted by the Centre for Water Resources Development and Management (CWRDM) in Kerala evaluated the impact of rooftop rainwater harvesting systems in urban areas. The study found that rooftop rainwater harvesting structures were efficient in reducing the demand for freshwater, and in improving groundwater recharge. The study recommended the implementation of such systems in all urban areas in Kerala to improve water availability and mitigate the impact of drought (Fig. 19.5).

Fig. 19.5 Madakka in Kasarcod district of Kerala

19.2 Significance of Women on Water Resources Management

- Domestic activities of women on water resources include farming and agricultural activities like sowing, transplanting, weeding, and intercultural activities.
- Collection of water for drinking and domestic activities from surface water bodies now has decreased due to the borewells, and the knowledge about the groundwater recharge has increased.
- Women make multiple works, which require the most usage of water resources, and effort to make sure that these sources not to be polluted.
- Since they are managing the home they use the grey water for irrigating the home garden and for irrigation activities.
- Inclusion of women in planning of water supply with men is essential to maintain the proper sanitation and water use planning in the home is the need of the hour to conserve water.
- They utilize and conserve and manage the water resources, which are needed for their families, so conservation of water resources in rural areas cannot be done without the involvement and training of women.
- Women involvement in water management should be encouraged through the Reduce, Reuse, and Recycle principle to utilize water effectively.

19.3 Discussion

In addition to scientific studies, regional and local water harvesting techniques have also been implemented in various parts of Kerala with positive results. Palakkad, which is considered as the traditional rice bowl of Kerala, heavily relies on Bharathapuzha, the state's second-largest river. However, the river has now become lean in most blocks of the district, causing significant difficulties for farmers who depend on it to irrigate their paddy fields. With temperatures soaring up to 41 °C, cases of sunburn have also increased. Despite almost all 56 villages in Palakkad facing water scarcity, three villages have managed to find sustainable results to overcome the hard climate conditions. Although the solutions implemented by these villages differ, they share a common feature of women being at the front position of creating water structures. A village called Pookkottukavu, which is in the Bharathapuzha basin and around 40 km from Palakkad district headquarters, boasts of having the largest cluster of skilled women well diggers in the entire country. With a workforce of 278 women, whenever they are called upon to dig wells under the Mahatma Gandhi National Rural Employment Guarantee Act (MGNREGA), they also make sure the construction of rainwater-harvesting structures to balance the groundwater loss.

An implementation of a comprehensive water harvesting program that includes the construction of check dams, farm ponds, and recharge pits in Palakkad has led to a significant increase in water availability, and has also helped in improving soil fertility and agricultural productivity. Overall, regional and local water harvesting techniques and water reuse have proven to be effective in improving water availability and mitigating the impact of water scarcity in Kerala. Scientific studies and practical implementations have shown that these techniques can significantly improve groundwater recharge and reduce the demand for freshwater. However, there is a requirement for continuous investment and promotion of the techniques to guarantee their widespread implementation and long-term existence.

Kochi International Airport has implemented a comprehensive wastewater treatment and reuse system. The system treats wastewater to a high standard, and the treated wastewater is used for irrigation and toilet flushing, significantly reducing the demand for freshwater.

Several initiatives have been implemented to address gender issues in local and regional harvesting and water reuse initiatives. For example, the Women's Water Collective in Kerala, India, is a women-led organization that promotes female participation in decision-making processes related to water systems management. The organization also provides training to women in water management and promotes the inclusion of women in the maintenance and operation of water management systems.

Jalanidhi project in Kottayam district of Kerala, engaged a large population of women in the beneficiary committees and gave opportunities to become for them to obtain excellence in water supply activities. Progressive approach of the project and high-impact data, education, and communication activities in a participatory approach, demand creation, skill development, and training have made the women evolve as water leaders (Table 19.1).

The skills of women in this project has given an opportunity for them to be aware and excel in the implementation from the design and planning stage itself. Several institutions execute trainings to impart water knowledge to women community, to increase the water literacy rates and to instigate their leadership and managerial skills (Fig. 19.6).

19.3.1 Precautions to Achieve Gender Equality

Ensuring gender equality and empowering women in water management is important for sustainable development. When women and girls have equal access to water resources and are able to participate in water management decision-making, it can lead to more efficient and effective water management practices and can also improve social and economic outcomes for women and their communities. Some strategies for promoting gender equality in water management include:

1. Ensuring that women and men have equal access to water resources and infrastructure, such as by building gender-sensitive water facilities that are safe and accessible for all.

Table 19.1 Status of drinking water projects in Kerala

Sl.No	District	No. of urban water schemes	No. of rural water supply schemes		Total
			Multi panchayat	Single panchayat	
1	Thiruananthapuram	6	45	95	146
2	Kollam	3	38	65	106
3	Pathanamthitta	3	19	50	72
4	Alappuzha	7	2	77	86
5	Kottayam	5	55	97	157
6	Idukki	1	106	92	199
7	Ernakulam	14	63	91	168
8	Thrissur	12	147	104	263
9	Palakkad	4	133	51	188
10	Malappuram	5	133	43	161
11	Kozhikode	2	118	128	248
12	Wayanad	0	30	32	62
13	Kannur	6	52	120	178
14	Kasargode	3	64	87	154
15	Total	71	985	1132	2188

Source Jalanidhi/Projects/Drinkingwatersupply

Fig. 19.6 A subsurface tank constructed for drinking water supply

2. Providing education and training opportunities to women and girls in water management, so they have the knowledge and skills to effectively manage water resources.
3. Engaging women in water management decision-making processes, so that their needs and perspectives are taken into account.

4. Supporting initiatives that promote gender equality and women's empowerment more broadly, as this can help to create a more enabling environment for women's participation in water management.
5. No discrimination should be made against women to express their views on water management in sanitation and domestic usage.
6. Men should be educated about water usage, water quality, and sanitation-related issues of water for awareness.

Overall, promoting gender equality in water management is an important part of achieving sustainable development and can help to ensure that water resources are managed in a way that meets the needs of all members of a community.

19.3.2 Awareness of Rainwater Harvesting in Palakkad

A research was conducted to study the impact of RWH systems in Kerala with special reference to Palakkad. Since Palakkad district is a very drought-prone area. Very few studies have been conducted to examine the performance of rainwater harvesting. So, the present study makes an attempt to examine the economics of rain water harvesting. It was found from the study that the water harvesters in Palakkad district used the rooftop method to collect the rainwater with a development cost of below 10,000 rupees, which even can be subsidized. And, it was also found that 92% of harvesters constructed the RWH systems to meet the daily water requirements such as washing and for their domestic usage out of which 57% percentage harvesters have constructed the systems based on their awareness on water resources. The conclusion seems to be healthy regarding the gender aspect since, the rooftop water harvesting is completely managed by women, which was about 36%.

19.3.3 Women and Water of Kuttadampadam Region

The study aims to explain the connection between the "3Ws"—"Women, Water and Well-being" through a case study of the "Kuttadampadam" at Palakkad district in Kerala. The study area women have started to follow the 3R principle—Reduce, Reuse, and Recycle and the construction of "Jalasambharanis" to overcome the water scarcity problems. Major findings show us that the middle-aged women particularly 21–50 years age had awareness and were equipped with various measures and technologies for better water resources management. The conclusion of the study seems to be astonishing that the women in Kuttadampadam region, act as true environmentalists as they follow effective resource management (Shinju and Prasad 2019). It can be summarized that the policies and programs must be implemented to educate and empower them economically, technically, psychologically, and politically.

19.4 Conclusion

In conclusion, regional and local water harvesting and water reuse are important methods for water conservation and sustainable use. These methods can help reduce water scarcity, mitigate the effects of climate change, and ensure the availability of water resources for future generations. However, their successful implementation requires a concerted effort from governments, businesses, and individuals to invest in infrastructure and technology, and to promote awareness and education about the benefits of these methods.

Gender issues have a significant impact on local and regional harvesting and water reuse initiatives. Addressing these issues requires the inclusion of women in decision-making processes, the provision of clean water sources that are accessible to women, and the promotion of women's participation in the maintenance and operation of water management systems. The long-term sustainability of water management initiatives depends on the promotion of gender equality and the inclusion of women in all aspects of water management.

Although the gender approach plays a major role in water resources, it should be a holistic improvement of water resources and their corresponding society. They must create an environment that ensures the security of water for the future generations. Gender approach creates a way to educate women about water harvesting and how they are important in the harvesting cycle. Since it paves way for the aquifer recharge, educating both men and women would tend to create added awareness on the harvesting of water. Rooftop rainwater harvesting accompanied with groundwater recharge shafts could provide an enhancement of groundwater table. Every human is responsible for the climate change and has to harvest rainwater on their own through water harvesting methods to have secured water resources in the future.

References

Ahmed S (2005a) Why is gender equity a concern for water management. In: Flowing upstream: empowering women through water management initiatives in India. Foundation Books with Ahmedabad: CEE, New Delhi

Ahmed S (2008) Gender and integrated water resources management in South Asia. In: Water first: issues and challenges for nations and communities in South Asia. Sage, New Delhi, pp 185–201

Ahmed S (ed) (2005b) Flowing upstream: empowering women through water management initiatives in India. Foundation Books

Bhat LD (2014) Participatory water projects bringing in unequal burden: women's experiences from Kerala

Chow VT, Maidment DR, Mays LW (1988) Applied hydrology. Mc Graw-Hill Book Company, New York, p 572

FAO (2017) The future of food and agriculture: trends and challenges. Food and Agriculture Organization of the United Nations, Rome

Central Ground Water Board (2013) Ground water information booklet of Ernakulam District, Kerala State

Central Ground Water Board (2015) Ground water information booklet of Trivandrum District, Kerala State. Technical Reports: Series 'D'
Joji VS, Jacob RS (2023) Traditional rain water harvesting in India. In: Traditional rainwater harvesting structures. Springer Nature Switzerland, Cham, pp 35–78
Kerala State Planning Board (2013) Economic review
Khandker V, Gandhi VP, Johnson N (2020) Gender perspective in water management: the involvement of women in participatory water institutions of Eastern India. Water 12(1):196
Liji KT (2020) The impact of rain water harvesting system in Kerala with special reference to Palakkad district. Think India J 22(44):82–85
More HB (2020) A review of Indian traditional method of rain water harvesting. Int J Innov Res Sci Eng Technol 9(4):2080–2086
Nair GU (2016) Traditional wisdom in harvesting water. J Tradit Folk Pract
Packett E, Grigg NJ, Wu J, Cuddy SM, Wallbrink PJ, Jakeman AJ (2020) Mainstreaming gender into water management modelling processes. Environ Model Softw 127:104683
Ponnusamy K (2013) Impact of public private partnership in agriculture: a review. Indian J Agric Sci 83(8):803–808
Rai VK (1993) Water resources planning and development. Deep and Deep Publications, New Delhi, p 229
Samuel MP (2010) Evaluation of a traditional community water management structure in Yethadka village of Kasaragod district in Kerala. Indian J Soil Conserv 38(1):28–30
Samuel MP, Mathew AC (2008) Rejuvenation of water bodies by adopting rainwater harvesting and groundwater recharging practices in catchment area-a case study. In: Proceedings of Taal2007: the 12th world lake conference, vol 766. Ministry of Environment and Forests, Government of India Jaipur, India, p 776
Saradamoni K (1991) Filling the rice bowl: women in paddy cultivation
Shaji E, Sarath KV, Prakash P, Abraham AP, Deepchand V, Kunhambu V, Pradeepkumar AP (2020) Tunnel wells, the traditional water harvesting structures of Kasaragod, Kerala: re-visited. Curr Sci (00113891) 118(6)
Shinju V, Prasad A (2019) Women, water & well-being: a case study of 'Kuttadampadam' in Palakkad, Kerala. Humanities 7(1):44–49
Suhardiman D, Raut M, Pradhan P, Meinzen-Dick R (2023) Irrigation systems management in Nepal: women's strategies in response to migration-induced challenges. Water Altern 16(1):244–264
United Nations Women (UN Women) (2018)
United Nations: world water development report 2021: valuing water
UN Water (2006) Gender, water and sanitation a policy brief. United Nations Water

Chapter 20
Using Geochemical Assessments for Basin-Wide Monitoring of Groundwater Quality and Quantity: Case of Omaruru–Swakop Basin

Petrus T. Paulus, Nnenesi A. Kgabi, and Henry Mukendwa

Abstract "Groundwater quality and availability is mostly influenced by the rocks and other materials it passes through from the point of recharge to discharge. When water flows through the geological formation, there are several processes that take place which is responsible for exchange of soluble solids between the rocks and water" (Elango and Kannan 2007). This chapter aims to address the lack of sufficient rock–water interaction data needed to understand the monitoring of quality as well as quantifying the sustainable groundwater resources.

Keywords Groundwater recharge · Storativity · Abstraction potential · Sediments · Groundwater · Geochemical assessments

20.1 Background

Groundwater quantity and quality assessment is key to sustainable development in arid environments due to the lack of direct surface water sources. Ministry of Agriculture, Water, and Forestry (2010a, b) reports that there is a lack of sufficient data

P. T. Paulus
Department of Civil and Environmental Engineering, Namibia University of Science and Technology, Windhoek, Namibia

Department of Natural and Applied Sciences, Namibia University of Science and Technology, Windhoek, Namibia

N. A. Kgabi (✉)
Unit for Environmental Science and Management, North-West University, Potchefstroom, South Africa
e-mail: Nnenesi.Kgabi@nwu.ac.za

H. Mukendwa
Namibia Water Corporation Ltd, Windhoek, Namibia

S. Suriyanarayanan et al. (eds.), *Water Management in Developing Countries and Sustainable Development*, Water Resources Development and Management,
https://doi.org/10.1007/978-981-99-8639-2_20

to quantify the sustainable groundwater resources and related groundwater recharge information in the Omaruru–Swakop Basin of Namibia.

Namibia is a large country with an area of about 825,418 km^2 (Ministry of Environment and Tourism 2010). However, 92% of the Namibian land area is classified by the Food and Agriculture Organization of the United Nations (FAO 2019), as hyper-arid, arid or semi-arid, which indicates vulnerability to the effects of climate change. The rainfall patterns are problematic. According to the Ministry of Environment and Tourism (2011), "the mean annual rainfall in Namibia is less than 250 mm with an upper limit of about 600 mm per year." Of the total rainfall received, only 2% becomes runoff and may be harnessed in surface storage facilities. The groundwater is recharged by one percent, and 97% is lost to evaporation and evapotranspiration.

The Omaruru–Swakop basin is one of the economically important basins in Namibia as it hosts Windhoek (the capital city having many industrial activities, and significant towns as such as Swakopmund), which is a tourism destination for many local and international travelers. The basin also hosts renowned mines among others including the Rossing Uranium Mine, Langer Heinrich Uranium Mine and Navachab Mine. Over the years, the mines have solely depended on groundwater water for their supply. The water was abstracted from the Omaruru River Delta aquifer. The water demand for the whole basin has grown due to population growth, urbanization, and industrialization.

Over the past 15 years, the population has grown exceptionally. According to the population census of Namibia for 2011, the population of the Omaruru–Swakop Basin consisted of about 430,854 persons (Namibia Statistics Agency 2011). Due to this significant population growth, as in other basins of Namibia as well as industrialization, this basin was hit by shortages of water. The consecutive years of droughts worsened the situation.

The Omaruru–Swakop Basin experienced a severe impact of the recurring drought that affected most parts of Namibia (including the Erongo region, where the Omaruru–Swakop Basin is located) during rainy seasons from 2013 to 2016, The (Food and Agriculture Organization of the United Nations 2016). The communities in this basin depend on the Omaruru River aquifer which had experienced poor water recharge. Furthermore, the Minister of Agriculture Water and Forestry (Mutorwa 2016) cautioned on the water shortages and possible water restriction to the central area of Namibia due to the lack of significant inflows from rain to the three dams (Omatako, Von Bach, and Swakoppoort) for the past five (5) years (2012–2016).

The aim of the study was to conduct a geochemical analysis of groundwater in the Omaruru–Swakop Basin in Namibia focusing mainly on the physico-chemical properties of groundwater and sediments, recharge, and storativity of the aquifers, and assessment of the abstraction potential of groundwater from the aquifers. The ultimate goal of the study is thus, to provide a broad understanding of the groundwater quality and its accessibility for abstractions.

20.2 Methods

20.2.1 Description of Study Area

The basin was divided into five sampling stations (Fig. 20.1), namely, Omaruru, Okombahe, Nei-Neis, Omdel Dam, and the Omaruru River Delta (Omdel). Twenty-eight (28) sampling points within the five stations were sampled for groundwater and sediments. The points were sampled as follows; 3 from around Omaruru, 1 from Okombahe, 2 from Nei-Neis, 2 at a farm near the Omdel dam and 20 from the Omaruru River Delta.

Figure 20.2 shows the Namibian water basins, major aquifers, and groundwater potential of rock bodies (Ministry of Agriculture, Water, and Forestry 2010a, b).

20.2.2 Sampling and On-Site/Field Analysis

A sum of 28 water samples was collected from 28 production boreholes and 28 sediment samples were collected from the area surrounding the same boreholes. The pH, conductivity, temperature, dissolved oxygen (DO), and total dissolved solids (TDS) were measured during fieldwork (i.e., on-site) using an HQ40d multi-parameter meter. Salinity was measured using a Hach Pocket Pro + multi 2 Tester, and the coordinates and elevation above sea level The collected samples were acidified for preservation and stored in the refrigerator in preparation for laboratory analysis.

Secondary rainfall data were obtained from the Namibia Meteorological while borehole water levels and production volumes were provided by the Namibia Water Corporation Limited (NamWater) for recharge and storativity assessments.

20.2.3 Laboratory Analyzes of Water and Sediment Samples

The groundwater samples were analyzed for Na, K, Ca, Mg, Fe, Mn, Cd, Cu, Pb, and Zn with the Inductively Coupled Plasma–Optical Emission Spectroscopy (ICP-OES) following method 3120 B on ICP of the Standard Methods for the Examination of Water and Wastewater, SO_4, Cl, NO_3, NO_2, and SiO_2 using Automatic flow analysis method and fluoride ions by Selective Electrode at the Namibia Water Corporation Limited.

The sediment samples were analyzed using X-ray Fluorescence (XRF), Rigaku XRF—CG1384 model. The XRF is particularly suitable for investigations that involve bulk chemical analyses of major elements (Si, Ti, Al, Fe, Mn, Mg, Ca, Na, K, P) and trace elements (>1 ppm; Ba, Ce, Co, Cr, Cu, Ga, La, Nb, Ni, Rb, Sc, Sr, Rh, U, V, Y, Zr, Zn) in rock and sediment (Wirth nd.).

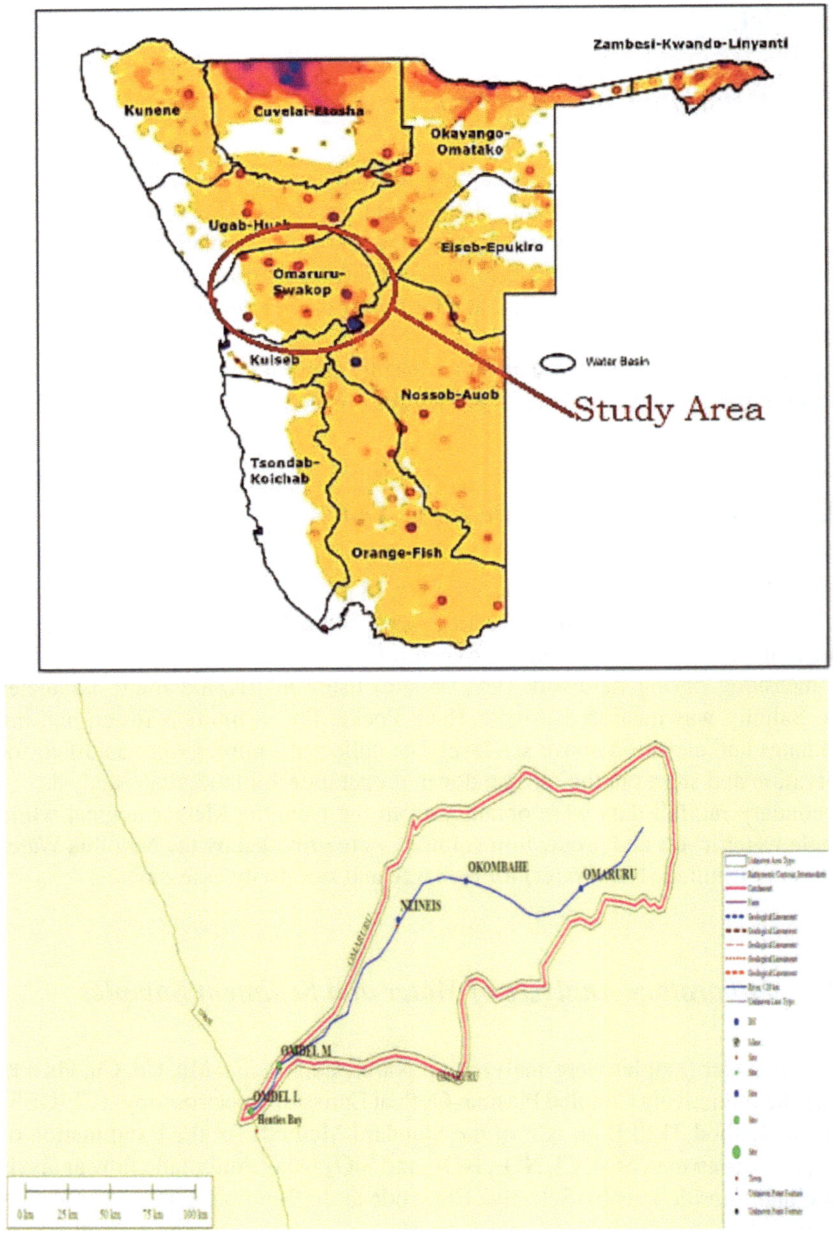

Fig. 20.1 **a** Map of Namibia, showing the Omaruru–Swakop basin (study area); **b** study area in the Omaruru–Swakop basin. *Source* Christelis and Struckmeier (2011)

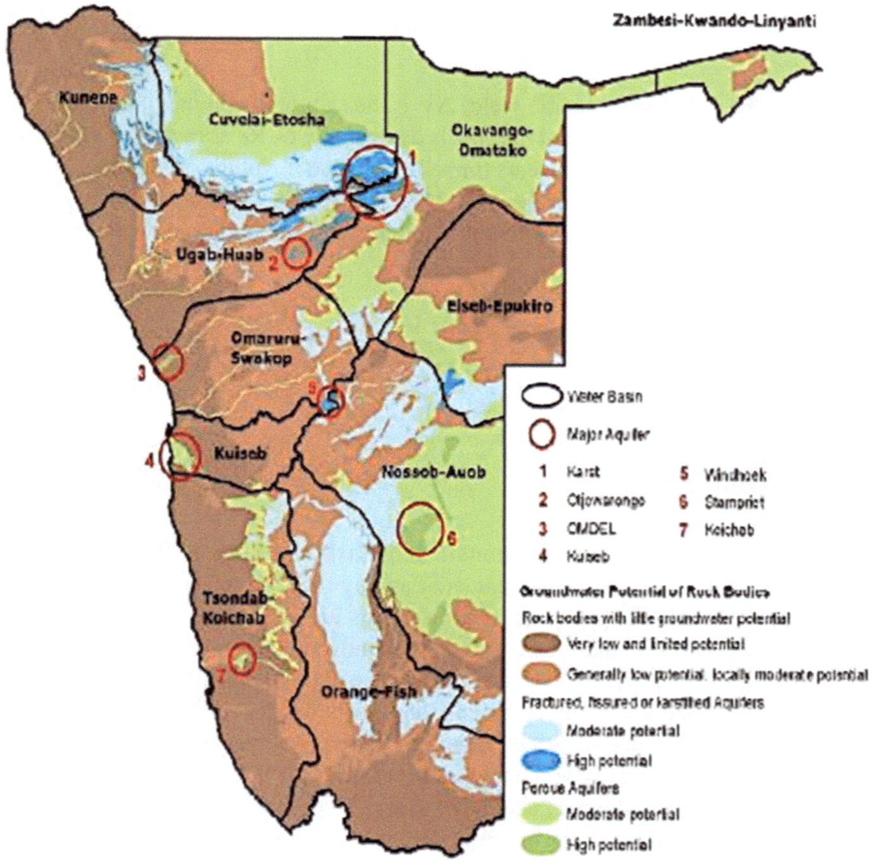

Fig. 20.2 Namibia water basins, major aquifers and groundwater potential of rock bodies. *Source* Christelis and Struckmeier (2011)

20.2.4 Recharge and Storativity

Estimation of percentage recharge was performed using the November 2001 Namibia Version of a programmed Saturated Volume Fluctuations (SVF) method. This Program required an input of the boreholes' production volume, water level, and rainfall data for a given corresponding data. The tables of appendixes *y* and *z* tabulated the data input as used in the SVF program. The program was used to estimate recharge based on the balance between rainfall and abstracted volume from the aquifer by simulating changes in groundwater level over time. The storativity was estimated from aquifer tests and the balance between inflow and outflow in a given section of an aquifer.

Literature shows that the SVF method uses the formulae (Healy and Cook 2002):

$$I - O + \mathrm{Re} - Q = S\frac{\Delta V}{\Delta t} \tag{20.1}$$

where S is the storativity/specific yield, ΔV is the saturated volume of aquifer, and I represents lateral inflow. Lateral outflow is depicted by O, Re as recharge, while Q is the net discharge, and Δt represents time. The change in volume per change in time ($\frac{\Delta V}{\Delta t}$) also represents the water flow rate. Storativity was calculated during periods of no recharge, using the saturated volume fluctuation method as shown in Eq. 20.2 (Healy and Cook 2002):

$$S = \frac{(I - O - Q)}{\frac{\Delta V}{\Delta t}} \tag{20.2}$$

20.2.5 Data Analysis

Descriptive statistics of mean, minimum, and maximum were used to describe the distribution and the relationships that exist among the various parameters from the different boreholes. The anion and cation means are effectively used to order the anions and cations separately based on their abundance (Nur et al. 2012). Microsoft Excel was used to further analyze and present the data graphically.

Correlation analysis of the sediments and groundwater quality was performed using Statistical Package of Social Science (SPSS). A Piper diagram was used to classify the groundwater samples into a "Mixed Ca^{2+}—Na^{+}—HCO_3^- Type." This approach was used by Manoj et al. (2013) to see if there was any change in the water type to assess the ionic stability of the studied water bodies. Finally, the groundwater quality data was critically analyzed for Na^{+}/Ca^{2+} ion exchange (Giménez-Forcada 2010) to see the behavior of these ions across the basin in response to the geology.

It should be noted that, during sampling, hydrochloric acid was used to preserve the samples. As a result, the chlorides and bicarbonates anions could not be measured in the laboratory. However, secondary data was used. The secondary data were sourced from the Namibia Water Corporation Limited's Division of Geohydrology, which samples groundwater every three months for standard chemical analyses. These were the data used for backup because of the lost anions due to acidification.

20.3 Results and Discussion

20.3.1 Physico-Chemical Characteristics of Groundwater

The results obtained from descriptive statistical analysis of the physical parameters of the groundwater are shown in Table 20.1.

Table 20.1 Physical parameters of groundwater in the Omaruru–Swakop basin

Borehole	DO (mg/l)	pH	Temperature °C	TDS (mg/l)	Conductivity (mS/m)	Salinity (ppt)
WW21499	8.25	7.95	23.7	544	110.3	0.71
WW100114	8.01	7.59	24.3	756	152.2	0.93
WW21488	6.95	7.66	25.5	760	152.1	0.94
WW100159	7.88	7.46	27.5	1002	197.9	1.21
WW35336	8.02	7.69	25.2	623	125.4	0.81
WW100111	7.33	7.58	25.6	925	184.3	1.12
WW100153	7.43	7.54	25.5	938	187.7	1.13
WW100157	7.36	7.31	24.9	1046	208.7	1.25
WW100142	8.44	7.3	25.6	935	187.4	1.13
WW22187	7.16	7.32	25.7	895	179.1	1.08
WW100140	6.7	7.2	25.1	920	183.9	1.1
WW100094	7.56	7.66	25.3	715	143	0.91
WW35341	7.41	7.34	25.1	1269	253	1.48
WW35342	6.12	7.43	25.1	1266	253	1.48
WW100135	5.87	7.21	25.9	1040	208.2	1.26
WW22194	6.12	7.31	26.4	1715	342	1.91
WW35343	5.01	7.1	26.3	1123	225	1.32
WW100136	6.06	7.27	25.8	926	14.8	1.12
WW35345	6.79	7.34	25.8	970	194.1	1.17
WW22539	3.12	7.01	26	925	184.9	1.12
Omdel D F BH 1	1.91	6.96	27.7	1326	268	1.59
Omdel D F BH 2	2.92	6.89	27.6	1993	395	2.27
WW25887	5.02	6.73	26.5	245	49.1	0.43
WW26198	6.44	6.82	26.8	791	157.5	0.98
WW33747	6.97	7.01	27	577	115.5	0.71
Omaruru BH 3	7.42	6.96	25.1	131.4	26.2	0.3
Omaruru BH 1	5.98	7.06	24.8	127.3	25.6	0.3
WW27169	6.09	7.01	23.6	150.2	30	0.3
Mean	6.44	7.28	25.69	879.78	169.78	1.07
Min	1.91	6.73	23.60	127.3	14.80	0.30
Max	8.44	7.95	27.70	1993	395	2.27
Rel std dev	25.06	4.10	4.04	48.83	53.42	42.09
Std dev	1.61	0.30	1.04	429.61	90.70	0.45

(continued)

Table 20.1 (continued)

Borehole	DO (mg/l)	pH	Temperature °C	TDS (mg/l)	Conductivity (mS/m)	Salinity (ppt)
Namibian Standards (Water Act, (Act 54 of 1956))		6.0–9.0			150	
WHO Standards		6.5–8.5		1000		

The groundwater of the Omaruru–Swakop basin had values of TDS, conductivity, dissolved oxygen, and salinity that ranged from 127.30 mg/l to 1993 mg/l, 14.80 mS/m to 395 mS/m, 1.91 mg/l to 8.44 mg/l and 0.3 ppt to 2.27 ppt, respectively.

pH

Although the mean pH of the sampled groundwater was 7.28, the minimum was 6.73 recorded for borehole WW 25,887 at Okombahe and the maximum pH was 7.95 for borehole WW21499 of the lower part of the Omaruru River Delta (Omdel). The pH variations with respect to the sampled boreholes are presented in Fig. 20.3.

The aforementioned mean pH of 7.28 shows that there is considerable shift in hydrogen ions in the water to a neutral state. The pH parameter measures the amount of the hydrogen ion concentration in water, and indicates the level of acidity. A pH value lower than 7 is classified as acidic (with the lowest values being more acidic), while a pH greater than 7 (to a maximum of 14) is considered basic (the highest values are more alkaline). The Namibian Drinking Water Quality Standards for the

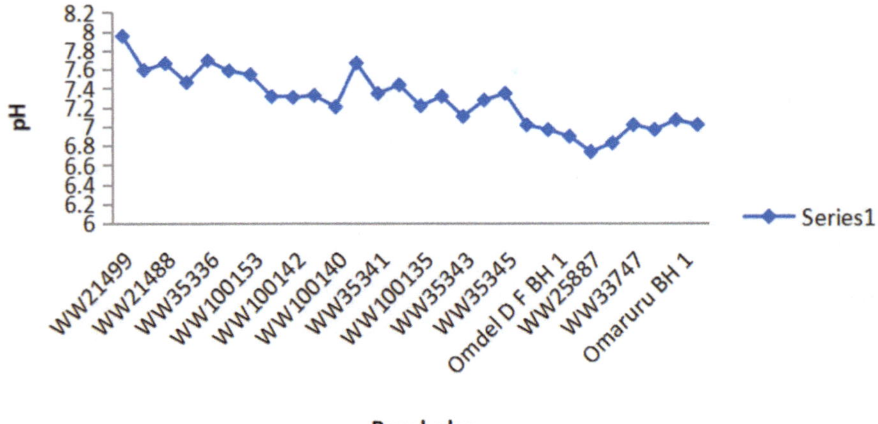

Fig. 20.3 The pH variations in the Omaruru–Swakop basin

Water Act, (Act 54 of 1956) requires the pH of potable (drinking) water to be 6–9 in order for it to be regarded as Group A (excellent water quality).

The mean pH of the in-situ analysis of the groundwater was 7.28, which falls within the parameters of the Namibian Drinking Water Quality Standards. In addition, the minimum and maximum pH of the groundwater which was 6.73 and 7.95, respectively, also fell within the required range. Hence, the pH of the groundwater in the basin is in Group A for excellent water quality.

Temperature

The temperature did not show a major variation as it ranged from a lowest of 23.60 °C to a highest of 27.70 °C.

Conductivity

The conductivity parameter indicates the capability of water to transmit electric current. The high content of ions is an indication that the water can transmit more current. The ions that constitute total dissolved solids are mainly those of dissolved metals and other dissolved materials. The mean conductivity value obtained from the in-situ analysis of groundwater in the basin was 169.78 mS/m while the minimum and maximum were 14.80 mS/m and 395 mS/m respectively (Fig. 20.4). The conductivity therefore ranged from Group A to Group C classifications of the Namibian Drinking Water Quality Standards with Group B set as an indicator for good water quality.

TDS

The level of total dissolved solids (TDS) of the groundwater is shown in Fig. 20.5. There was an observed consistency in the TDS concentrations throughout the basin. The TDS was highest in Omdel Dam Farm BH 2 at 1993 mg/l and lowest at the

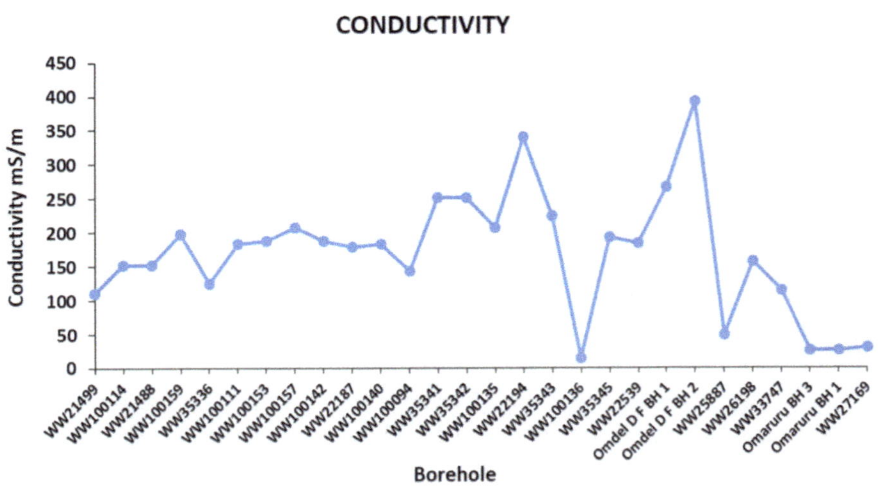

Fig. 20.4 Conductivity variations in the basin

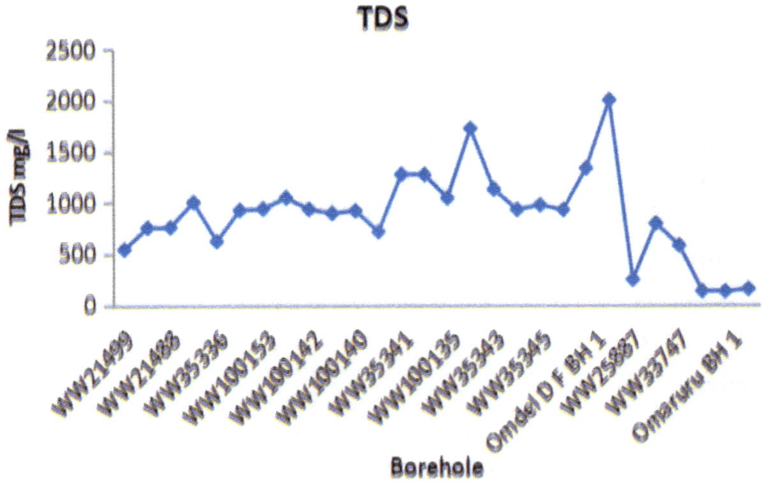

Fig. 20.5 Total dissolved solids (TDS)

Omaruru BH 1 with 127.3 mg/l. It is significant to note that, though the TDS concentration increased expectantly down the basin, it however decreased to 544 mg/l at boreholes WW21499 which are in the lower basin. This deviation from the consistent pattern can be accounted for by the variation in lithology (geology) of the basin.

Salinity

The TDS and conductivity (EC) results described above also give an indication of the salinity of the groundwater (Rusydi 2018).

This study did not include in-situ measurements of salinity however, it can be deduced from total dissolved solids (TDS) which includes all of the disassociated electrolytes that make up salinity concentrations. The TDS and salinity also give an indication of the quality of water. According to Fondriest Environmental, Inc. (2014), if TDS is approximately equal to salinity, the water is considered to be of good quality, i.e., fresh water.

Freshwater sources have low electrical conductivity. The conductivity does not have consistent/set trends per water source because it is affected by geological factors, i.e., the type of rocks and soil.

20.3.2 *Chemical Parameters of the Sediments*

The sediment samples collected around each borehole indicated elevated levels of calcium (36,200 mg/l) and magnesium (10,753 mg/l). The increasing chloride content along the flowline denoted a normal groundwater evolution from 65 mg/l at Omaruru (upland in the basin) to 466 mg/l at the Omaruru River Delta, which is towards the coastal area part of the basin. Selected elements from chemical characterization of sediment samples are presented in Table 20.2. Magnesium was not detected in most samples and the "ND" in the Table was used to show the non-detection.

The sediment analysis shows a minimum and maximum concentration for magnesium, calcium, chloride, and potassium of 4605 mg/l and 10,753 mg/l, 12,500 mg/l

Table 20.2 The elemental composition of the sediments

Sampling point	Mg	Mn	Si	S	Cl	K	Ca	Zn	Fe	Cu
	mg/l	mg/l	mg/l	mg/l	mg/l	mg/l	mg/l	mg/l	mg/l	mg/l
WW100114	ND	245	270,667	227	141	37,633	21,800	22	6897	13
WW22194	ND	230	266,000	717	960	42,100	25,867	28	7007	16
WW35343	ND	259	235,667	1040	142	41,733	38,700	26	7717	14
WW35341	ND	416	244,000	721	296	36,733	36,200	43	15,967	20
Omaruru BH 3	ND	200	293,333	131	50	48,600	12,500	13	4560	12
WW100159	ND	254	250,333	545	171	39,967	36,300	29	10,267	18
WW35342	ND	201	255,000	309	99	39,000	38,867	21	6937	14
WW100135	6377	288	220,000	3587	380	35,067	59,733	27	9473	18
WW100153	ND	290	222,667	1223	222	38,900	49,933	28	7493	15
WW100111	ND	329	233,667	1280	264	38,533	47,867	25	9863	17
WW22539	ND	325	228,667	5543	2430	35,100	55,633	44	9393	33
WW35345	ND	237	286,333	297	493	42,467	18,233	21	7440	16
WW100094	ND	264	253,333	1057	850	35,967	32,533	25	9410	14
WW21488	ND	268	180,000	42,233	807	36,233	73,233	23	7660	15
WW25887	4605	422	208,667	433	161	34,467	34,333	64	21,500	22
WW100142	ND	371	240,667	1967	2113	36,267	51,633	30	10,533	20
WW27169	ND	355	259,000	380	98	38,900	21,467	45	13,367	18
WW100140	ND	265	247,667	11,300	848	38,033	44,100	29	9813	15
WW26198	ND	216	257,667	129	65	38,500	21,633	30	9950	14
WW100136	ND	309	273,667	561	377	39,500	28,033	23	8887	16
WW22187	ND	274	263,667	480	307	37,067	25,767	100	16,867	27
Omdel DF BH1	10,753	648	153,000	256	84	22,800	150,333	50	19,600	29
Omdel DF BH2	7980	491	206,000	339	340	29,900	43,633	74	28,067	27
WW21499	ND	247	273,667	1227	466	42,600	29,567	66	8393	20
Omaruru BH1	ND	189	299,333	297	65	44,900	15,067	15	5027	11

(continued)

Table 20.2 (continued)

Sampling point	Mg	Mn	Si	S	Cl	K	Ca	Zn	Fe	Cu
	mg/l	mg/l	mg/l	mg/l	mg/l	mg/l	mg/l	mg/l	mg/l	mg/l
WW35336	ND	257	260,000	659	226	38,700	35,800	27	8290	14
WW100157	ND	332	279,333	251	118	42,233	20,267	26	9960	19
WW33747	ND	143	296,333	171	63	44,967	14,367	15	4913	11
Mean	7429	297	248,512	2762.9	451	38,460	38,693	35	10,545	18
Minimum	4605	143	153,000	129	50	22,800	12,500	13	4560	11
Maximum	10,753	648	299,333	42,233	2430	48,600	150,333	100	28,067	33

and 150,333 mg/l, 50 mg/l and 2430 mg/l, and 22,800 mg/l and 48,600 mg/l, respectively. The dominant parameters calcium, potassium, and iron showed a pattern of continuity throughout the basin as illustrated on the pie chart in Fig. 20.6a–e.

20.3.3 *Chemical Quality of the Groundwater and Sediments*

The parameters Mg, Si, Cl, and Ca are important as they give an indication of the source groundwater recharge and water quality variations. In this case, local boreholes in the upper basin benefit from local recharge. This is evident from elevated calcium (36,200 mg/l) and magnesium (10,753 mg/l), while increasing chloride content along the flowline denotes normal groundwater evolution (from 65 mg/l at Omaruru to 466 mg/l at Omdel) (Fig. 20.7). Moreover, boreholes in the upper basin exhibit elevated silicate content (299,333 mg/l) compared to those in the lower basin (180,000 mg/l), which signifies that the aquifers in the upper basin exploit proximal sediments while those in the lower basin exploit distal sediments. This implies the water quality of the lower basin is largely inherited from the upper basin and is evident in the silicate content of secondary water quality data which points to a silicate content difference of 0.2 ppm (29 mg/l at Nei-Neis and 29.2 at Omdel).

The piper diagram in Fig. 20.7 shows the distribution of cations and anions of the groundwater and sediments in the basin.

The piper diagram shows that cations and anions in the basin occur abundantly from mid-to-above in the diamond of the piper diagram. The alkaline earth metal content of the basin dominated those of the alkalis earth metal at mean ratio of 74%. This indicates that there is no adsorption of the Mg^{2+} and Ca^{2+} ions by clay. As a result, these ions end up being released into the solution, which therefore accounts for their higher concentration compared to the alkalis earth metals. Furthermore, this implies that the aquifers in the basin are largely unconfined; they take in and store more water. However, they are more susceptible to pollution or contamination due to the lack of confinement. Similarly, Dedzo et al. (2017) found the alkaline earth metal contents to have dominated the alkalis metals in 66.66% of the samples analyzed.

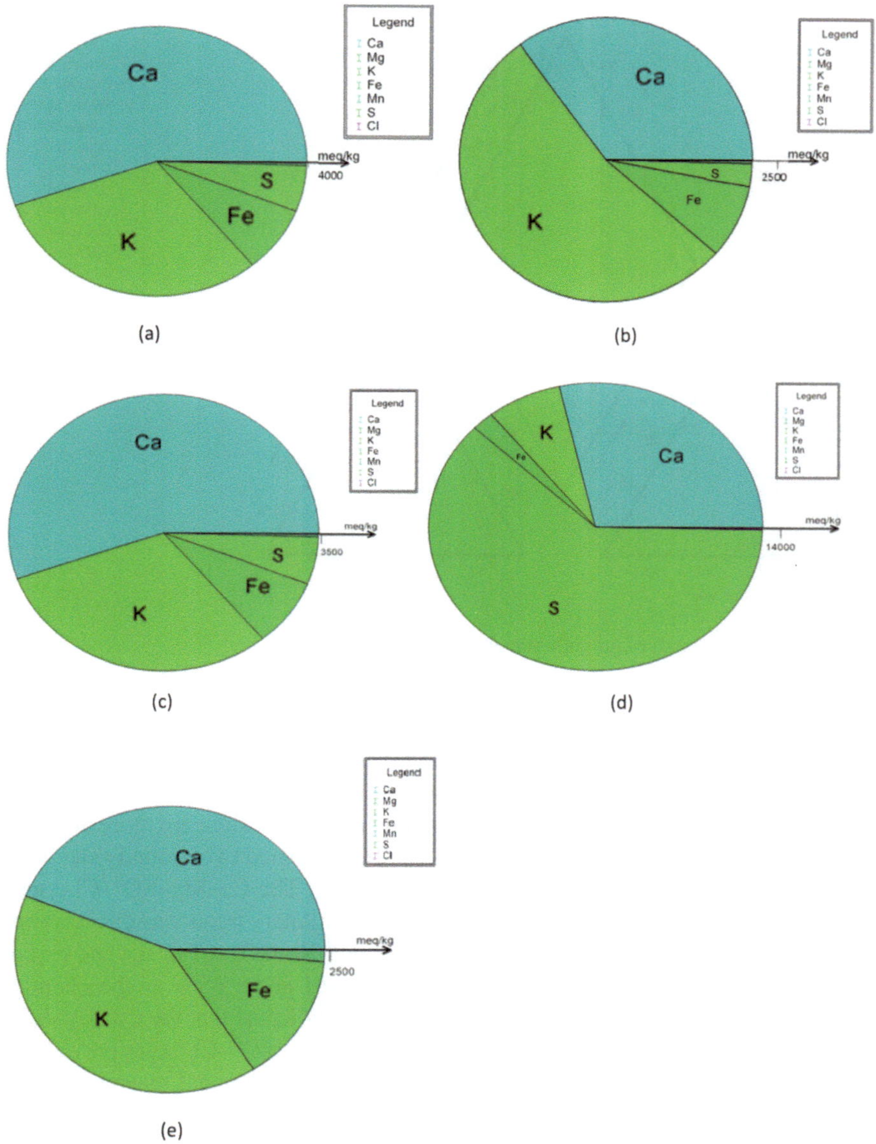

Fig. 20.6 Sediment chemistry for **a** Okombahe, **b** Omaruru, **c** upper Omdel, **d** lower Omdel, and **e** Nei-Neis

Stiff diagrams (Fig. 20.8a–d) were used to confirm trends in concentrations of Mg^{2+} and Ca^{2+} in the basin. The stiff diagrams of various parts of the basin showed cation and anions dominance as follows: Omaruru: Na–Ca–HCO_3 dominance; Okombahe: Ca–Mg–HCO_3 dominance; Nei-Neis: Ca–Mg–Cl dominance;

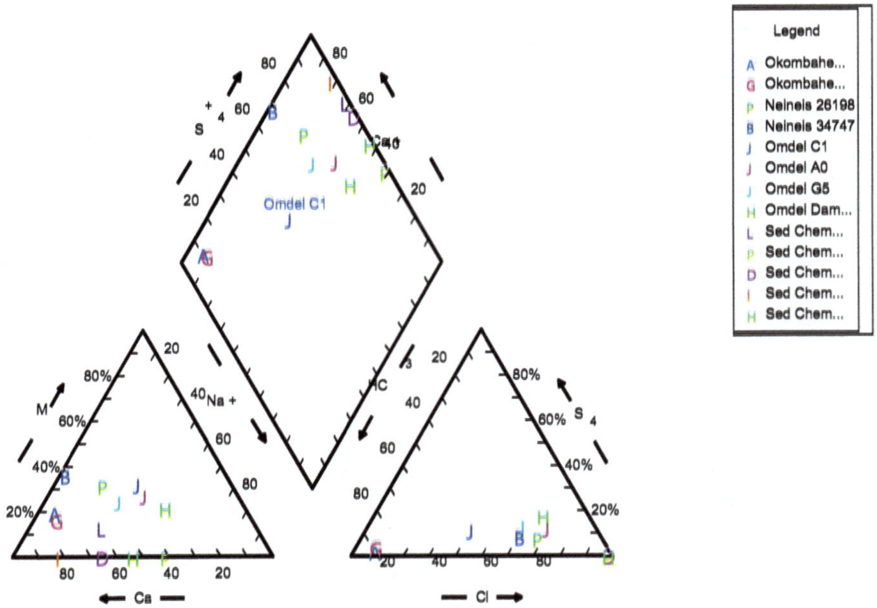

Fig. 20.7 Piper diagram of groundwater and sediment quality

Upper Omdel: Ca–Na–Cl dominance; Middle Omdel: Na–Ca–Cl dominance; and Lower Omdel: Na–Ca–Mg–HCO_3 + Cl dominance.

The dominance of Na in comparison with Ca and Mg in the basin was 4 out of 12 which equates to about 33%. Hence, calcium and magnesium are more dominant with 67% combined. Similarly, Nur et al. (2012), delineated two groundwater zones into: "Ca–Mg–SO^4–Cl and Na–SO^4–Cl water types. The Ca–Mg–SO^4–Cl constituted about 35% of the chemical facies and its evolutionary trend was due to simple hydrochemical mixing between Ca–Mg–HCO^3 and Na–SO^4–Cl facies and reverse cation exchange. The Na–SO^4–Cl facies constituted about 65% of the chemical facies and represented remnant groundwater." This indicated that the Ca–Mg–SO^4–Cl facies were dominant in the recharge areas while Na–SO^4–Cl facies prevailed in discharge areas. In the same vein, the dominance of calcium and magnesium over sodium in the Omaruru–Swakop basin is in the upper part, which is the recharge area while the lower part is mainly for storage and discharge (abstraction, mainly at Omdel).

20.3.4 *Recharge and Storativity of Groundwater in the Basin*

Recharge percentages, storativity, and specific yield for Okombahe and Nei-Neis are shown in Figs. 20.9, 20.10, 20.11, and 20.12.

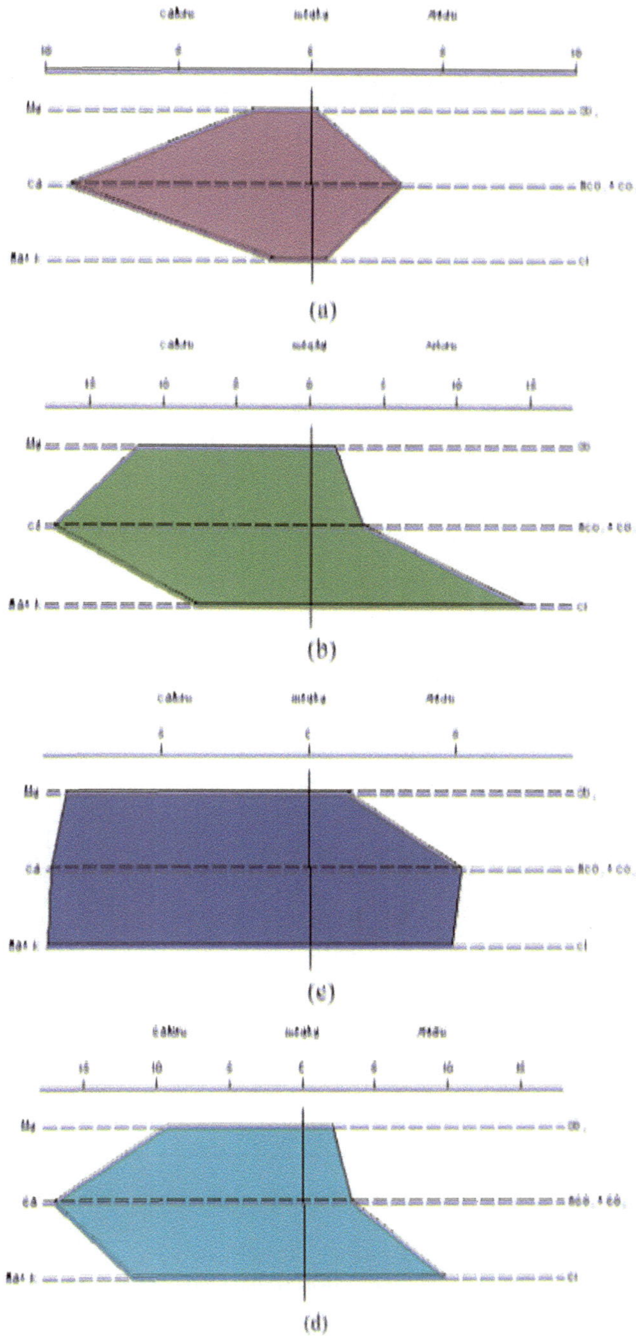

Fig. 20.8 **a** Okombahe primary data, **b** Nei-Neis BH WW 26198, **c** Omdel C1 (lower Omdel), **d** Omdel G5 (upper Omdel)

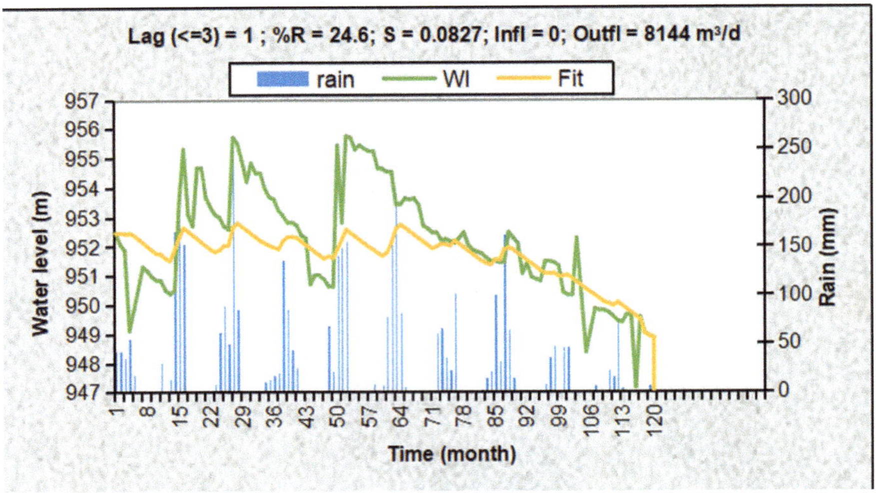

Fig. 20.9 Okombahe recharge percentage and storativity. *Source* Authors

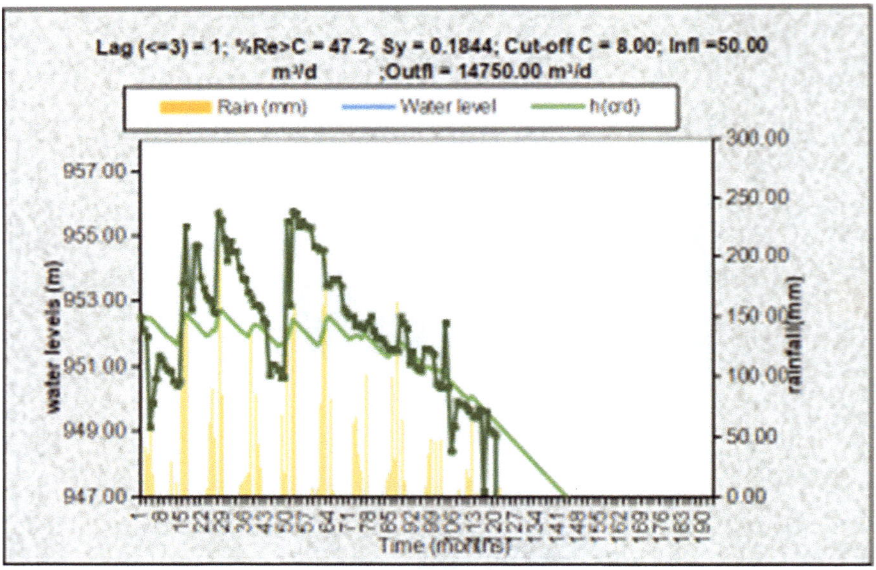

Fig. 20.10 Okombahe recharge percentage and specific yield. *Source* Authors

Although the recharge percentages, storativity, and specific yields could not be determined for the whole basin due to limited data at Omdel; the mean percentage recharges for Okombahe and Nei-Neis were determined as 36% and 60%, respectively (Figs. 20.9, 20.10, 20.11, and 20.12), which implies that about 36–60% of the rainfall recharges the two aquifers.

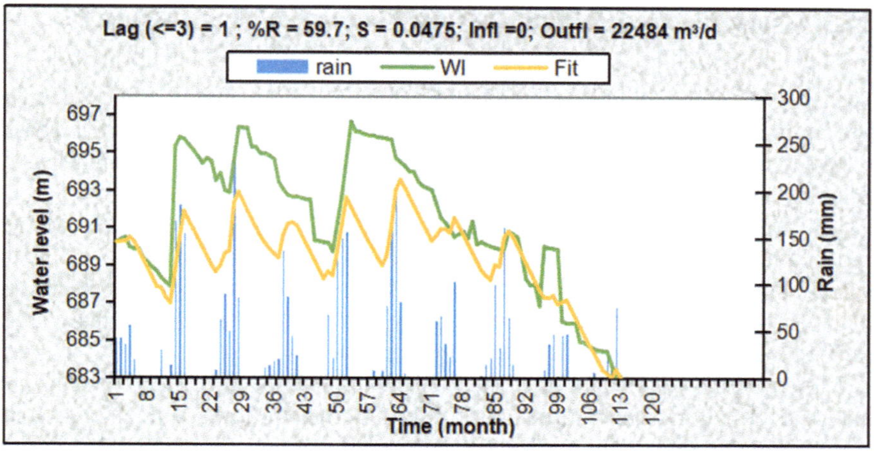

Fig. 20.11 Nei-Neis recharge percentage and storativity. *Source* Authors

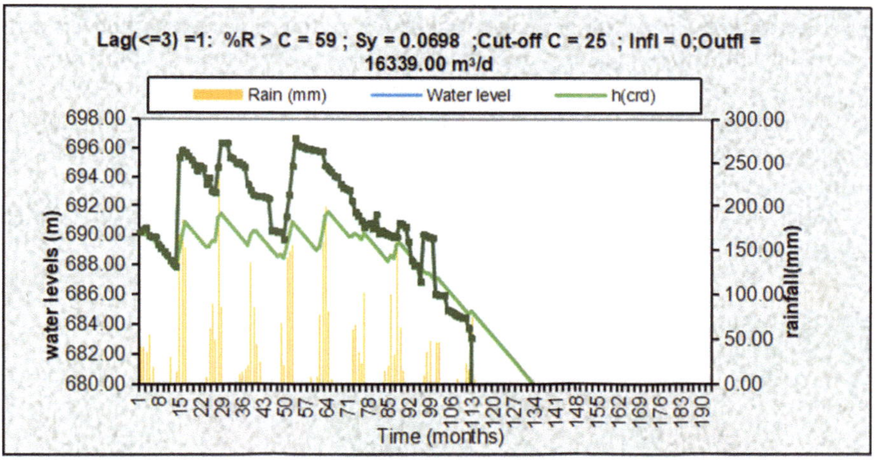

Fig. 20.12 Nei-Neis recharge percentage and specific yield. *Source* Authors

Moreover, the findings of this study are in agreement with the report (Rossouw 2014) that groundwater recharge in the Omdel ranges from 20 to 50%. This was accounted for by the regular flooding of the Omaruru River. Flood waters that mainly flow through the Omaruru River facilitates the infiltration of silt-free flood water that is very instrumental to enhance recharge into the Omdel aquifer just downstream. The construction of Omdel dam as a floodwater capture upstream of the aquifer for subsequent controlled release-enhanced infiltration contributes to successful recharge within the aquifer. "Since construction was completed in 1994, on average 52% of the water stored in the reservoir has been successfully infiltrated each year, to prevent loss by evaporation" (Rossouw 2014).

20.3.5 Abstraction Potential of the Groundwater Aquifers

The lithology of the study area is indicated in Fig. 20.13, which shows the rock types and deposits that constitute the different parts of the basin. The geology of the basin speaks to both the possibility of abstracting groundwater from the basin in terms of quantity, and its quality based on the rock formations.

The upper basin is composed of conglomerate and sandstones, schist and dolomite, and granite formations. This formation indicates that there is a low to moderate potential for groundwater abstraction, and this could be the result of the presence of sandstones and schists that have low porosity. Although the aquifers in these areas can host water, the low porosity affects the permeability.

Moreover, the middle part of the basin is predominantly underlined by schist and dolomite formations. These formations further highlight the low potential for groundwater abstraction of the basin. Lastly, the lower basin has formations comprising schists and dolomites, and limestone. The formation of the lower basin similarly denotes a lower potential for groundwater abstraction. However, infiltration and net recharge vary temporally and spatially by season and it is impacted by factors such as the stormwater intensity, stream stage, vegetation type and cover, slope, and temperature. Even with low permeability, recharge is accelerated by the intensity and duration of surface water that flows through the Omaruru River (Matengu et al. 2019). The unknown areas in the map are granite, limestone, and sandstone types. Overall, the lithology of the Omaruru–Swakop Basin indicates that the potential of groundwater abstraction in the basin ranges from low to moderate.

With regard to hydrochemistry, the chloride–alkalinity ratios performed to relate the geology to the water quality are presented in Table 20.3.

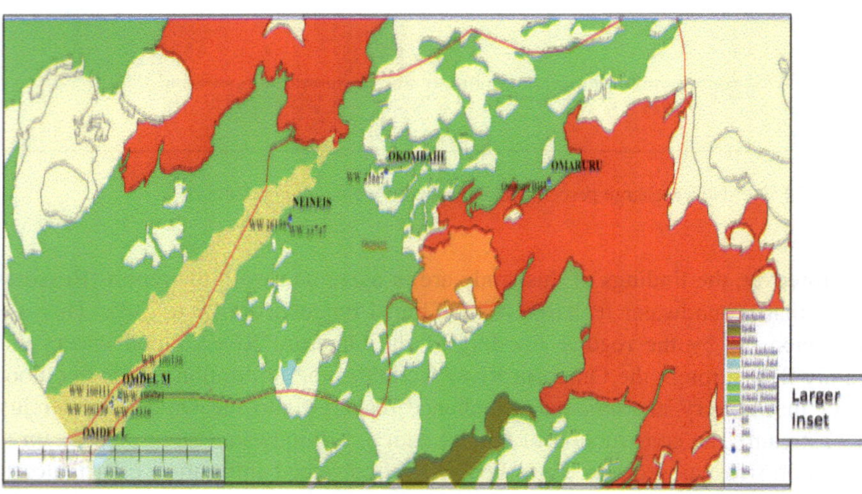

Fig. 20.13 Geology of the study area. *Source* Christelis and Struckmeier (2011)

Table 20.3 Chloride-alkalinity ratios

Place	Chlorine (mg/l)	Alkalinity (mg/l)	Ratio	Description
Omaruru	20	251	0.08	Freshwater
Okombahe	10	300	0.03	Freshwater
Nei-Neis	520	294	1.8	Saline water
Omdel upper	350	250	1.4	Saline water
Omdel middle	425	172	2.5	Saline water
Omdel lower	174	330	0.5	Freshwater

The chloride to alkalinity ratio shows that salinity increases along the flow path down the basin. The upper basin consists more of freshwater having a chloride-alkalinity ratio of 0.03 at Okombahe and 0.08 at Omaruru. The groundwater in the lower part of the basin is saline having a chloride–alkalinity ratio of 1.8 at Nei-Neis and 2.5 at the middle-Omdel. The basin showed carbonate weathering with salinization along the flow path. Some anomalies were observed at upper and lower Omdel. The ratios were 1.4 and 0.5, respectively. The anomaly observed at upper Omdel in the decreased salinity is due to the presence of the Omdel dam in the area. Whereas, the observed anomaly at lower Omdel, the lowest end of the basin (supposed to have been more saline) is due to the limestone in the area.

Furthermore, the effect of limestone was proved by predominant concentrations of calcium, magnesium, and bicarbonate at the lower Omdel, shown in the piper diagram (Fig. 20.7) on the groundwater and sediment quality. This implied that the limestone removed the sodium from groundwater and replaced it with calcium, magnesium, and bicarbonate.

In terms of temporal recharge assessment of the basin, at Nei-Neis, the basin recharges about 5 seasons in 9 years, while at Omdel it is 6 seasons in 26 years as shown in Fig. 20.14 from the Namibia Water Corporation.

The spatial distribution of groundwater in the basin could not be assessed conclusively due to its strong dependence on recharge estimates.

The finding on groundwater potential based on the analysis of the geology of the basin proved to be in line with the Groundwater Namibian Perspective (Van Vuuren 2012). The latter reported that the Omdel Aquifer has a geological formation that is made up of alluvial sand, gravel, and conglomerate, which fills a series of dissected palaeo channels, which include the active channel of the Omaruru River. The aquifer is significantly large as it extends all the way from the coastal areas of Henties Bay town to about 40 km inland in the Omaruru–Swakop Basin. Furthermore, there is a thin silt and clay deposit/layer in the river bed of the Omaruru River.

According to Heine and Völkel (2010), the Omaruru sediments have similar clay mineral compositions as the Swakop sediments (Figs. 20.15 and 20.16). Swakop catchment is dominated by Illite (50–90%) while Smectite only contributes 15–25%, and, kaolinite, 10–20%. Thus, Omaruru and Swakop belong to the same clay mineral province.

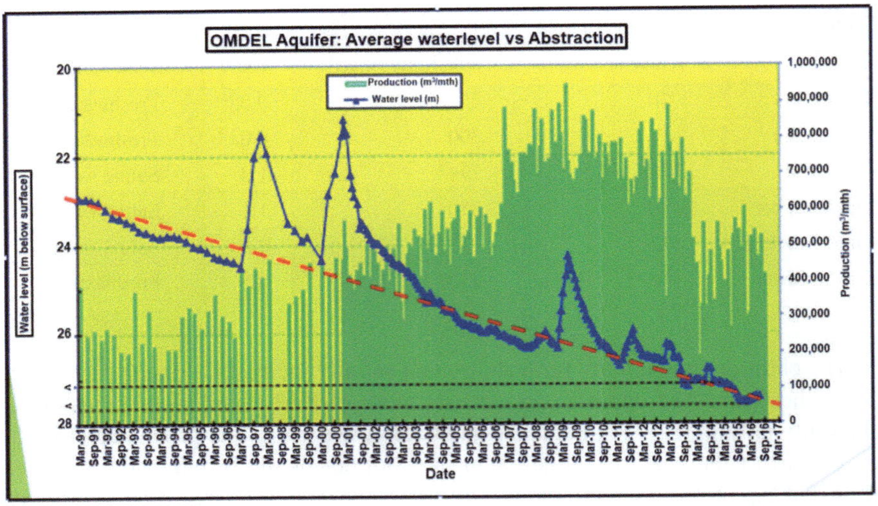

Fig. 20.14 Average water level of the Omdel aquifer. *Source* Authors' analysis and compilation using secondary data from Namibia Water Corporation Ltd

The inset map shows perennial rivers in southern Africa. The clay mineral "provinces" in Fig. 20.15 refer to the areas where certain clay minerals and clay mineral accumulations occur in sediments and soils (colluvia, soils, fluvial and lacustrine sediments, etc., (Heine and Völkel 2010). According to Gingele (1996), "river discharge and aeolian input are the main processes supplying terrigenous material to the southwestern African continental margin." Gingele (1996) confirmed that the two processes suggest that the area is arid and humid.

"The clay mineral compositions are greatly influenced by the geological, geomorphological and (palaeo-) pedological conditions".

The sediment samples were collected in the same area of each borehole. Likewise, the analysis of the rock types and deposits was for the same sampled area of boreholes. The map in Fig. 20.16 shows that the Omaruru–Swakop Basin has a generally low potential and locally moderate potential for abstraction of groundwater based on the rock bodies classification. The lowest part of the basin is classified as very low with limited potential for abstraction. Hence, the basin has a greater part that is composed of rock bodies with little groundwater potential. However, a small part of the basin is comprised of fractured, fissured, and porous formations which have a moderate potential for groundwater abstractions.

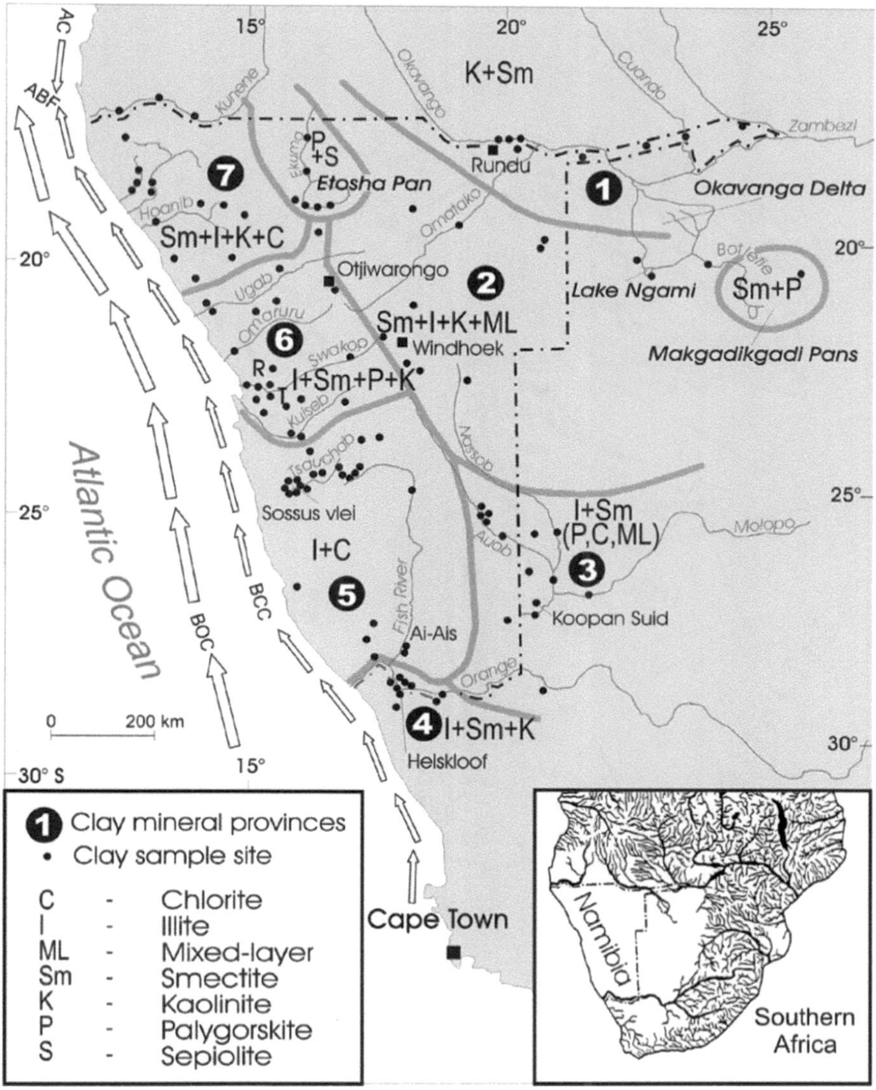

Fig. 20.15 Clay mineral provinces of Namibia and selected sample sites. *Source* Heine and Völkel (2010)

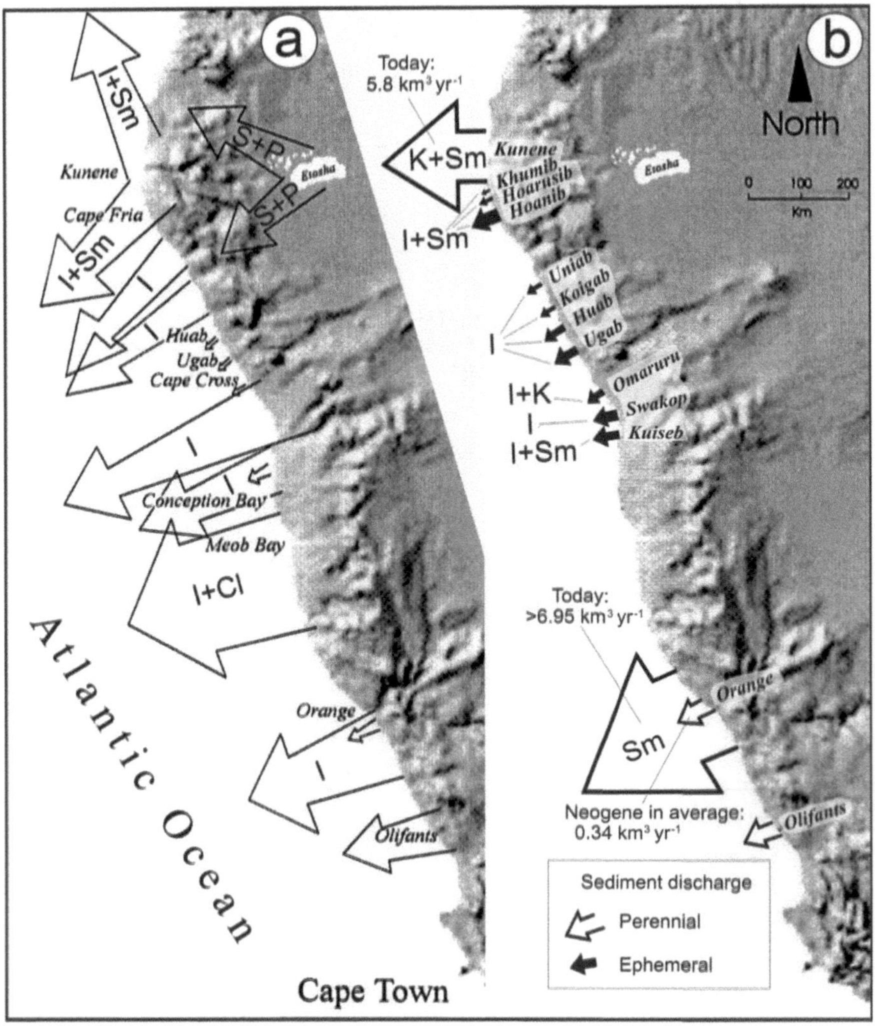

Fig. 20.16 Clay minerals reaching the Atlantic Ocean through wind and water. **a** "Map of dust outbreak corridors for the southwest coast of Africa as identified from space-borne imagery and photography" (Eckardt et al. 2012). See Fig. 20.15 for abbreviations for clay minerals. **b** Conveyance of fluvial sediments (clay minerals) into the southeast Atlantic Ocean. *Source* Heine and Völkel (2010)

20.4 Conclusion

In this study, the quality of the groundwater in the Omaruru–Swakop Omatako Basin and the potential recharge and storativity of the groundwater aquifers were assessed.

The results indicated that the groundwater has a 67% Ca–Mg dominance against Na. The alkaline earth metal content of the basin dominated the chemical composition at a mean ratio of 74% in the basin. This indicated that there is no adsorption of the Mg^{2+} and Ca^{2+} ions and the metals end up being released into the solution. This also implies that the aquifers in the basin are largely unconfined. The study found that the limestone at the lowest end of the basin has a positive effect on the groundwater geochemistry. The possible replacement of sodium by calcium, magnesium, and bicarbonate might also be responsible for a decrease in the total dissolved solids at the lower part of the basin.

The groundwater quality results show that there is a continuity of the dominant parameters from the upper parts of the basin to its lower part through the flow. This is evident in that the upper basin comprising Okombahe, Nei-Neis, and upper Omaruru River Delta (Omdel), which are predominant with calcium bicarbonate while the lower part of the basin is predominant with sodium chloride and sodium bicarbonate. The bicarbonate in the lower basin can be attributed to the effect of the surrounding geology.

Furthermore, the upper catchment showed a dominant effect on the water quality of the basin. The result indicated deterioration in the dominant calcium bicarbonate in the upper basin toward the upper Omdel area, and then to lower basin (mainly middle and lower Omdel) dominant with sodium chloride and sodium bicarbonate.

This area receives more rainfall than the lower and is of a shallow-saturated alluvial of about 15 m. The lower basin is more of groundwater storage with a thick-saturated alluvium of about 40 m. The comparison underscores that both ends of the Omaruru–Swakop Basin are equally important. The upper basin supplies water of good quality while the lower basin stores the water.

The major achievement of this study is the discovery of the limestone at the lowest end of the basin that positively affected the groundwater geochemistry. The limestone removed the sodium and replaced it with more calcium, magnesium, and bicarbonate. This phenomenon also resulted in a decrease in the total dissolved solids at this part of the basin.

It is important to note that, although the recharge percentage was determined using rainfall data, water level, and production volumes of the borehole; stable isotopes are the best determinant of aquifer recharge. The former approach has limitations. Firstly, the balance between inflow and outflow is not known, causing a lot of unknown variables. Secondly, the storativity (S) and specific storage (Sy) are not well established, which would require an analysis of hydraulic pumping which was not covered in this study. Lastly, when using tools other than stable isotopes, it is required that the storativity and specific yield to be known and held constant when determining the percentage recharge.

Acknowledgements The authors acknowledge the support of the 'UNESCO Chair on Sustainable Water Research for Climate Adaptation in Arid Environment" at the Namibia University of Science and Technology (NUST) and the Namibia Water Corporation Ltd.

References

Christelis G, Struckmeier W (2011) Groundwater in Namibia: an explanation to the hydrogeological map. Ministry of Agriculture, Water and Rural Development, Namibia (first edition December 2001; unrevised second edition January 2011)

Dedzo MG, Tsozue D, Mimba ME, Teddy F, Nembungwe RM, Linida S (2017) Importance of rocks and their weathering products on groundwater quality in Central-East Cameroon. Hydrology 1–18

Eckardt FD, Soderbergb K, Coop LG, Muller AA, Vickery KJ, Grandin RD, Jack C, Kapalanga TS, Henschel J (2012) The nature of moisture at Gobabeb, in the central Namib Desert. J Arid Environ. https://doi.org/10.1016/j.jaridenv.2012.01.011

Elango L, Kannan L (2007) Chapter 11 Rock–water interaction and its control on chemical composition of groundwater. Dev Environ Sci 5:229–243

Fondriest Environmental, Inc. (2014) Conductivity, salinity and total dissolved solids. Fundamentals of Environmental Measurements. 3 Mar 2014. https://www.fondriest.com/environmental-measurements/parameters/water-quality/conductivity-salinity-tds/

Food and Agriculture Organization of the United Nations (2016) Assessment of impacts and recovery needs of communities affected by El Niño-induced drought in Kunene, Erongo and Omusati regions of Namibia. Food and Agriculture Organization of the United Nations, Rome

Food and Agriculture Organization of the United Nations (2019) Namibia needs over N$ 500 million to realize its health security plan. https://www.afro.who.int/pt/node/12024

Giménez-Forcada E (2010) Dynamic of seawater interface using hydrochemical facies evolution diagram (HFE-D). Groundwater 48(2):212–216

Gingele FX (1996) Holocene climatic optimum in Southwest Africa—evidence from the marine clay mineral record. Palaeogeogr Palaeoclimatol Palaeoecol 122(1–4):77–87

Healy RW, Cook PG (2002) Using groundwater levels to estimate recharge. Hydrogeol J 10(1):91–109. https://doi.org/10.1007/s10040-001-0178-0

Heine K, Völkel J (2010) Soil clay minerals in Namibia and their significance for the terrestrial and marine past global change research. Afr Study Monogr Suppl.40:31–50

Manoj K, Ghosh S, Padhy PK (2013) Characterization and classification of hydrochemistry using multivariate graphical and hydrostatistical techniques. Res J Chem Sci 3(5):32–42

Matengu B, Xu Y, Tordiffe E (2019) Hydrogeological characteristics of the Omaruru Delta Aquifer System in Namibia. Hydrogeol J 27:857–883. https://doi.org/10.1007/s10040-018-1913-0

Ministry of Agriculture, Water and Forestry (2010a) Development of an integrated water resource management plan for Namibia, theme 2, assessment of resources potential and development needs. IWRM Plan Joint Venture Namibia, Windhoek

Ministry of Agriculture, Water and Forestry (2010b) Integrated water resource management plan for Namibia. IWRM Plan Joint Venture Namibia, Namibia

Ministry of Environment and Tourism (2010) National policy on climate change for Namibia. Ministry of Environment and Tourism, Windhoek

Ministry of Environment and Tourism (2011) Namibia Second National Communication to the United Nations Framework Convention on Climate Change, July 2011

Mutorwa J (2016) Imminent water shortages and possible restrictions in the Central Areas of Namibia (CAN), a policy statement and an update. Windhoek, Khomas Region, Namibia

Namibia Statistics Agency (2011) Namibia 2011 population and housing census main report. Namibia Statistics Agency

Nur A, Ishaku JM, Yusuf SN (2012) Groundwater flow patterns and hydrochemical facies distribution using geographical information system (GIS) in Damaturu, Northeast Nigeria. Int J Geosci 3:1096–1106. https://doi.org/10.4236/ijg.2012.35111

Rossouw T (2014) Conceptual groundwater flow for the Omdel Aquifer. SLR Environmental Consulting (Pty) Ltd.

Rusydi AF (2018) Correlation between conductivity and total dissolved solid in various type of water: a review. IOP Conf Ser: Earth Environ Sci 118:012019. https://doi.org/10.1088/1755-1315/118/1/012019

Van Vuuren O (2012) Groundwater: a Namibian perspective

Wirth K (nd.) X-ray fluorescence (XRF). Geochemical Instrumentation and Analysis. https://serc.carleton.edu/research_education/geochemsheets/techniques/XRF.html

Chapter 21
Water and Sanitation Management in Urban Slum of Bandung City, Indonesia: Assessing the Potential Acceptance and Impact of Composting Toilet Technology in Urban Slum Areas

Neni Sintawardani

Abstract Since 2005, we have developed compost toilets (dry toilets) as alternative toilets used by residents to reduce pollution in rivers and make efficient use of limited clean water. The advantage of this technology is that it is compact, and does not require large space and piping for the flow of clean water or wastewater. Sawdust is used as a natural microbial growth matrix/media to degrade feces and urine into CO_2 and H_2O. After 2–3 months of use, the mixed media with mineral residues can be harvested and used as compost. This study aims to investigate the feasibility and potential impact of introducing composting toilet technology as a sustainable solution for water resource management in urban slum areas of Indonesia. By reducing the reliance on clean water for black water transportation, composting toilets offer a promising alternative that can alleviate the complexities associated with limited sanitation facilities and water availability. The study further explores the acceptance of the user community, with a particular emphasis on the role of women in embracing and promoting composting toilet technology as an effective water-saving toilet system.

Keywords Composting toilet · Water · Sanitation · Acceptance · Women

21.1 Introduction

WHO, in 2013, stated to the media that in 2015, around 2.4 billion people or a third of the world's population will remain without access to improved sanitation (WHO 2013), and it was again reported in 2015 (WHO/UNICEF 2015). Back to year 2001,

N. Sintawardani (✉)
Research Center for Environmental and Clean Technology, National Research and Innovation Agency (BRIN), Central Jakarta, Indonesia
e-mail: neni002@brin.go.id

S. Suriyanarayanan et al. (eds.), *Water Management in Developing Countries and Sustainable Development*, Water Resources Development and Management,
https://doi.org/10.1007/978-981-99-8639-2_21

through the joint global movement of the Millennium Development Goals (MDGs)) used the same projection as a determination to take realistic action steps to improve sanitation to the objected world's population (UN 2001). It turns out that in 15 years, there has been no significant change globally in the number of people without proper sanitation. This prompted the global community at the end of 2015 to launch the Sustainable Development Goals (SDGs) program to ensure concrete action steps that must be taken, which will have an impact on reducing poverty and various other inequalities, as well as reducing climate change by 2030 (UN 2015). The SDGs reinforce the determination "Sanitation for all," which emphasizes human rights to obtain proper sanitation facilities.

Based on the National Socioeconomic Survey (Susenas) conducted annually by BPS-Statistic Indonesia for sanitation facilities, Indonesia's achievements in proper sanitation only increased slightly. In 2004, the proportion of households that have access to proper sanitation facilities, meaning, using septic tanks or pits as final disposal sites, reached two-thirds of all households in Indonesia (67.1%), and in 2021, reached 80.29% households (BPS-Statistics Indonesia 2023). In 2022, the achievement of proper sanitation in Indonesia on average increased only 0.63% compared to 2021. The data does not indicate the quality of the latrines, whether they work properly, meet their purpose, and/or comply with the applicable health and technical standards.

Indonesia as an archipelagic country is very aware of the need for implemented and realistic steps to be able to achieve the SDG goals in all sectors. The establishment of appropriate sanitation has long been presented as a challenge for global development. National Geographic Indonesia (NatGeo Indonesia 2016) reported that 68% of rivers in Indonesia are heavily polluted by domestic waste (from black water). This is a real alarm that calls for solving problems while at the same time, the determination of the nation is improving people's well-being. The results of our study spanning two decades in the densely populated Kiaracondong area of Bandung City confirm the above statement. Population densities, limited space, and economic constraints result in restrictions on public bathing, washing, and toilet facilities (MCK), requiring some residents to use temporary MCK facilities located along a tributary of a river called the Jondol channel (Fig. 21.1). The wastewater from the MCK plant goes directly into the sewer (Sintawardani et al. 2005). Poor sanitation conditions and polluted water bodies affect the health and development of children living in the vicinity and have been shown by the stunting number of children in the area (Otsuka et al. 2019).

Even though Indonesia is now showing rapid growth in economy, some slums in urban areas are still existing in many big cities in Indonesia as one of the negative impacts of this rapid development. Assessment on the urban population in developing countries had been taken by the UN-HABITAT (2010) and showed still 33% of the urban population lived in slum area. Generally, urban slum is defined as the area which has bad living environment due to overpopulation, and insufficient infrastructures. One thing that is very important and related to the primary needs of the community is clean water and sanitation. The lack of infrastructure and limited space worsen the quality of sanitation which can have a direct impact on the health of the people

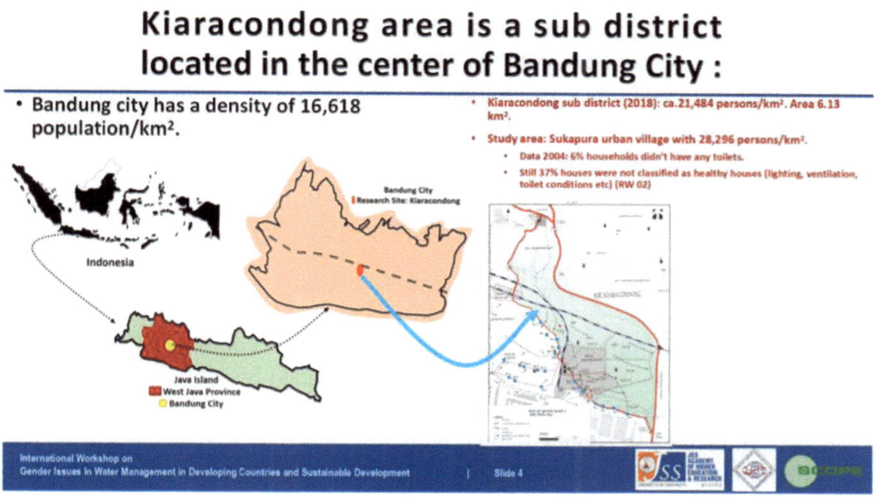

Fig. 21.1 Study area in high populated area in Bandung City, Indonesia

living in the area. Often limited infrastructure causes slum dwellers to pay relatively higher prices for clean water or sanitation facilities than other residents.

We have been investigating this problem since 2003 in an high population area or urban slum area of Kiaracondong in Bandung City, Indonesia (Fig. 21.1). Limited space, intense population movement, and limited infrastructure have made people not really care about the quality of sanitation. Due to limited land and funds, it is difficult for residents in the slum area to be able to build a simple treatment system such as a septic tank. Adequate piping facilities to channel sewerage waste to a centralized treatment site cannot enter the area. Most of the waste is discharged by residents into the nearest waterways or channel, thus having a negative impact on the environment in the downstream areas.

From the results of interviews with 62 households, it could be calculated that the average need for clean water consumption for bathing, defecating, urinating, washing clothes, washing kitchen utensils, cooking, and watering plants was 89 L/person/day. Ironically, 40% of water used was used for defecation purposes (urination, defecation) (Sintawardani et al. 2005).

Starting in 2005, we also started to modify and develop composting toilet technology to be applied in urban slum community. At the beginning of its implementation in the area, we conducted a field study to get an overview of sanitation and clean water problems in the area that are closely related to the acceptance or expectations of the community for the technology to be used. The role of women in understanding and dealing with those problems is something that deserves attention.

This study examines the potential acceptance and impact of composting toilet technology as part of an integrated approach to water resource management in urban slum areas. The introduction of composting toilets aims to address the challenges of limited sanitation facilities and water availability by reducing reliance on clean water for black water transportation. It focused on assessing the acceptance of the user community, particularly the role of women, in embracing this water-saving toilet system. By combining technological innovation with social acceptance, this integrated approach seeks to provide a sustainable solution that promotes efficient water use and improved sanitation practices in urban slum areas.

21.2 Materials and Methods

In order to get a first impression of the technology used, 1 composting toilet system built by us was set up in a narrow courtyard, surrounded by 10 tiny houses, which are lined up in a very small space and are inhabited by a total of 10 households. Each household has 3–5 people. In their daily life, they have to take turns using only one toilet unit attached to one bathroom. There is 1 common well that is shared for cooking, bathing, and doing laundry (Fig. 21.2).

We conducted interviews and distributed questionnaires to the community of potential users and residents. Some of the survey results are presented below.

Some residents usually have to use public toilets in their daily lives. Installing a septic tank is an obligation, and it requires no cheap installation cost. However, for general users, there is no direct benefit by installing a septic tank. As a result of those mismatching, slum dwellers build local infrastructure by themselves. Self-build sewer pipes are a typical example. It is one of the optimal solutions for people in terms of obtaining clean conditions in their houses. However, wastewater discharges directly into the nearby river and causes pollution due to domestic waste and *E.*coli contaminate the wells, which are used for their daily water needs (Soedjatmiko 2006).

Fig. 21.2 The communal toilet, bath, and wash unit for 10 households

21.3 Result and Discussion

Figure 21.3 shows the toilet room for the placement of 1 composting toilet unit used by 10 households. This type of composting toilet does not require piping to distribute clean water or wastewater. By using media from sawdust or other lignocellulosic waste, feces, and urine will be aerobically treated and released into water and CO_2. The rest of the minerals or other nutrients contained in the waste will be left in the media. Periodically, a part of the media can be harvested/taken for further use as compost or soil conditioner. Normally we used sawdust as media, but shredded corn cob could be used as media as well (Sintawardani et al. 2017).

Prior to implementing composting toilets or dry toilets in the community, we have conducted research to analyze the material flow and value flow in the sanitation system in the area. The results show that the cost of garbage and sewage/excreta disposal is evaluated to be 0.7% and 1.1% of income per capita (Ushijima et al. 2013; Sintawardani et al. 2019). These numbers could have significant meaning for people in slum areas whose average income was lower than the average city dweller. The potential use of composting toilets is expected to increase the welfare of some of the environmental acteurs in the area. The size of the population in the urban slum area and the potential for nutrient content in feces and urine which so far has been thrown away, thus polluting rivers or canals in the area have been calculated (Triastuti et al. 2016).

All respondents know that there are several public toilets available free of charge in their environment. It turned out that 43% of the male respondents used public MCK facilities in their daily life, while in the female group, only 8% used public facilities. This indicates that women experience barriers to using public toilets. From

Fig. 21.3 The composting toilet unit was installed just 1 m from the former communal toilet

discussions with several respondents informally, it was known that women's reluctance to use public toilets was more due to distance and convenience, which are closely related to safety. At night, it will be difficult or unsafe to leave the house alone to go to the public toilet. For men, this is not a problem.

For bathing and washing clothes, women do not have a problem doing it in public facilities because they usually wash clothes in the morning and bathe simultaneously. In the afternoon for the purposes of bathing in public facilities is not a problem either. The presence of 1 installed composting toilet unit that can be used by 10 households in the vicinity of the unit gave hope for women to feel more comfortable having toilet facilities near their homes.

For toilet purposes, it was found that men used less water to flush the toilet (squatting type toilet) than women. With sufficient water availability, this difference in water usage seems to be due more to habit. If seen from these results, to change a habit so that it becomes more efficient in water use, apart from technology, women play an important role. Socialization of a technology or policy requires the participation of women. If they can change, then the effect will be on the whole family, so they can actually guarantee the expected change.

Women play an important role in water and sanitation management, making significant contributions at various levels. From the studies conducted, it can be observed that the role of women in the community in slum areas is somewhat different from women in rural areas, because in slum areas with limited facilities and space, water and sanitation rely more on the existing situation. Following are some of the key aspects of their engagement that were observed during the on-site studies:

1. Women managed finances for the daily needs of the family, including buying water and managing toilets. The use of public toilets or shared bathing-washing-toilet (MCK) facilities is also a gathering place for women, so any decision regarding shared facilities will certainly involve women.
2. Women often serve as informal community leaders and mobilizers, promoting hygiene practices and raising awareness about the importance of clean water and sanitation if they get information. Involving women in such initiatives increases their effectiveness, because they are more flexible in their communities in connecting and sharing experiences. Their closeness to children at home and around them also facilitates the transfer of knowledge to young people.
3. Women's participation in water and sanitation management in the observed area also depends on their level of education. The higher the level of education of a mother or woman, the better the way she manages used diapers in her environment (Agestika et al. 2022)

According to custom and religion (Muslims are the majority in Indonesia), people always use water to rinse certain body parts after urinating or defecating. The results of this study, in terms of rinsing the body, men and women have almost the same tendency. There are 60% of respondents need around 3–5 L of clean water to rinse the body once after defecating. 30% of respondents only need less than 3 L. The rest require more than 5 L. It can be observed that those who need more water are those who have relatively higher incomes. So, it reinforces the results of some researches so far that the richer a person tends to be more wasteful of existing resources, in this case, water. Logically, the poor must use their income proportionately far more than the rich. This makes people with low incomes have to save water for their use. From a survey conducted by the Ministry of Public Works of the Republic of Indonesia in 2006 (Directorate of Drinking Water Development, Director-General CiptaKarya), data was obtained on the average water consumption of households in urban areas in Indonesia of 144 L per person per day, of which 45% is used for bathing purposes.

The development of toilet composting technology using lignocellulosic media as a matrix that will assist the black water composting process is carried out to encourage a more sustainable material flow because it produces material that can be sold or reused, namely as a composting media for plants. The application of composting toilets can be a source of organic compost for flower or tea plant farmers who are outside the slum area, but still within the reach of the supply chain area in the Bandung area (Ushijima et al. 2022). On the other hand, research on material flows of sanitation in the area has a challenge in determining which actors can be the driving force in changing the sanitation management system. It is hoped that the application of technology will facilitate the determination of important players, but in fact, social and economic factors must be taken into account (Ushijima et al. 2015, 2022).

In our research, women are easier to get into an idea or technology than men, although sometimes they do not want to understand the technical details. When asked if they had heard of composting toilets, 50% of female respondents said they knew the information from rumors or gossip. But most men said they knew nothing. It could be that this is due to the busyness of men who are more often away from home to earn a living, while women are more often at home so it is easier to exchange information or gossip.

After respondents received an explanation and saw the picture, all respondents indicated that composting toilets should be installed around their homes so that they can become community toilets, which are more accessible than public toilets. All of the women respondents also stated that they were willing to help to clean the toilet. Conversely, 50% of the men stated that they did not want to be involved in cleaning the toilet. The stereotype about who does housework and who is outside for work reasons is still strong there. This image can also be used as a basis to increase the engagement of women to manage composting toilets in the future.

The on-site application generates many new ideas for the development of toilet technology as well as its effective approach to society. Currently, our research focuses on the application of compost toilets in elementary schools, because children can be effective change agents for their families regarding good sanitation. In a long collaboration with several researchers from Japan universities, we tried to find a comprehensive social system design, which contributes to a better life in the aspects of food, water, sanitation, economy, etc. Using a multi-disciplinary approach, the value chain structure should be modeled as a trigger for the implementation of compost toilets in the community.

21.4 Conclusion

The integration of composting toilet technology into water resource management strategies presents a promising solution to the challenges faced by urban slum areas in Indonesia. By reducing water consumption, minimizing pollution, and promoting community engagement, this approach offers a sustainable and efficient alternative that can positively impact water resources and overall environmental sustainability. To achieve successful implementation, understanding the acceptance and involvement of the user community, especially women, is crucial for ensuring long-term effectiveness and scalability.

Acknowledgements This work was supported by the JST (Japan Science and Technology Agency) CREST program 2003–2008. We convey our appreciation and gratitude to Prof. Naoyuki Funamizu for your cooperation and support.

References

Agestika L, Sintawardani N, Hamidah U, Nyambe S, Yamauchi T (2022) Pattern of child faeces management and disposable diaper usage among under-fives in an Urban Slum of Bandung, Indonesia. J Water Sanitation Hygiene Dev 12(1):32. https://doi.org/10.2166/washdev.2021.099

BPS-Statistics Indonesia (2023) Statistical Year Book of Indonesia 2023. ISSN: 0126-2912

National Geographic Indonesia (2016) LimbahDomestik, Musuh Utama Sungai Indonesia, 17 July 2016. https://nationalgeographic.grid.id/read/13305951/limbah-domestik-musuh-utama-sungai-indonesia

Otsuka Y, Agestika L, Widyarani, Sintawardani N, Yamauchi T (2019) Risk factors for undernutrition and Diarrhea prevalence in an urban slum in Indonesia: focus on water, sanitation, and Hygiene. Am J Trop Med Hyg. Mar; 100(3):727–732. https://doi.org/10.4269/ajtmh.18-0063

Sintawardani N, Tri Astuti J, Nilawati D (2017) Evaluation of corn stalk as matrix in Decomposting toilet for treating urine and Feces. Proceedings 1st internationl symposium on green technology for value chains. IOP Conference series: earth and environmental science 60(2017)0 012016. https://doi.org/10.1088/1755-1315/60/1/012016

Sintawardani N, Irie M, Triastuti J, Ishikawa T, Affendi M (2005) Feasibility of introducing the bio-toilet to decrease the water pollution in an open canal: a case study in Kiarocondong Slum Area in Bandung City. Proceedings of the internasional water association (IWA) Conference 2005, Xi'an-China, 18–20 May 2005, pp 101–109

Sintawardani N, Ushijima K, Hamida U, Deguchi Y, Triastuti J, Funamizu N, Irie M, Ishikawa T (2019) Experiment of a resource-oriented agro sanitation. In: Funamizu N (ed) Resource-oriented agro-saitation systems: concept, business model, and technology. Springer, Tokyo, pp 291–314

Soedjatmiko B, Triastuti J, Hamidah U (2006) Pattern of groundwater contaminations in a heavily populated area: a preliminary study at Kiaracondong, East Bandung, Indonesia. Proceedings the 4th international symposium on sustainable sanitation, September 2006, pp 44–53

Tim PenyusunanLaporanTujuan Pembangunan Milenium (MDGs) Indonesia Tahun 2007. Kementerian Negara Perencanaan Pembangunan Nasional/Badan Perencanaan Pembangunan Nasional (2007) Laporan Perkembangan Pencapaian Millennium Development Goals Indonesia 2007, pp 85–94

Triastuti J, Nilawati D, Sintawardani N, Ushijima K (2016) Human excreta discharging system and potential nutrient recovery of ecological sanitation concept is implemented in slum Area of Kiaracondong. Bandung. Teknologi Indonesia 39(2):108–117

Ushijima K, Sintawardani N, Ikemi M (2022) Chapter 13, social relationship and material flow in the co-creation of sanitation systems. In: Yamauchi T et al. (eds), The sanitation triangle, global environmental studies, Springer https://doi.org/10.1007/978-981-16-7711-3_13

Ushijima K, Funamizu N, Nabeshima T, Hijikata N, Ito R, Sou M, Maiga AH, Sintawardani N (2015) The PostmodernSanitation :agro-sanitation business model as a new policy. Water Policy 17:283–298. https://doi.org/10.2166/wp.2014.093

Ushijima K, Irie M, Sintawardani N, Triastuti J, Hamidah U, Ishikawa T, Funamizu N (2013) Sustainable design of sanitation system based on material and value flow analysis for urban slum in Indonesia. Front Environ Sci Eng 7(1):120–126. https://doi.org/10.1007/s11783-012-0460-5

UN-HABITAT (2010) State of the World's Cities 2010/2011: bridging the urban divide. Earthscan, London

UN (2001) Road map towards the implementation of the united nations millennium declaration: report of the secretary-general. United Nations, New York

UN (2015) Transforming our world: the 2030 agenda for sustainable development. A/RES/70/1

WHO (2013) World will miss MDG Target (2.4 billion people will lack improved sanitation in 2015). https://www.who.int/news/item/13-05-2013-2-4-billion-people-will-lack-improved-sanitation-in-2015. Accessed 21 February 2022

WHO/UNICEF (2015) Joint Monitoring Programme (2015) Progress on sanitation and drinking water: 2015 update and MDG assessment. World Health Organization, Geneva

Chapter 22
The Impact of Municipal Waste on Groundwater Quality: Case of Omaruru Municipal Dumpsite

Jesaya N. Andreas and Nnenesi A. Kgabi

Abstract In this chapter, the effect of waste management practices on groundwater quality at the local municipality level is assessed. The case of Omaruru Municipality in Namibia was chosen to show the need for consideration of the interconnectedness of town (settlement) planning, land use, and human activities with groundwater quality, which is often ignored by researchers, municipal planners and water managers in Southern Africa. Based on the findings, i.e. a strong correlation between heavy metals found in soil and heavy metals found in groundwater boreholes, the research area's Contamination Factor (CF) ranging from 0.8 to 2.7, the geo-accumulation Index (Igeo) ranging from 0.003 to 0.7, and the pollution load index (PLI) of 2, which indicated contamination of the study area, the study found that municipal open dumping site contaminates groundwater sources within the river basin.

Keywords Geochemical parameters · Physical parameters · Boreholes · Waste dumpsites · Chemical parameters

22.1 Introduction

Chemical and biological pollutants (UNEP 2002) can compromise the quality of water. When precipitation comes in contact with degraded solid waste, leachate containing significant amounts of organic, inorganic, and heavy metals is produced (Olaoye et al. 2013). Leachate pollutes shallow aquifers as it infiltrates and percolates

J. N. Andreas
Department of Civil and Environmental Engineering, Namibia University of Science and Technology, Windhoek, Namibia

N. A. Kgabi (✉)
Unit for Environmental Science and Management, North-West University, Potchefstroom, South Africa
e-mail: Nnenesi.Kgabi@nwu.ac.za

S. Suriyanarayanan et al. (eds.), *Water Management in Developing Countries and Sustainable Development*, Water Resources Development and Management,
https://doi.org/10.1007/978-981-99-8639-2_22

them. Because of the non-biodegradable nature of heavy metals, leachate migration from municipal dumpsites contaminates soil and water, which worsens the quality of groundwater in the area (Olaoye et al. 2013).

Rapid urbanisation is one of the main causes of changes in land use and associated human activities leading to the accumulation of waste in municipal dumping sites (Adelana et al. 2008). Communities that use groundwater may be impacted by an increase in heavy metal concentration in the water.

Due to over-abstraction and pollution, urbanisation, mining, and industrialisation may seriously jeopardise the quality of groundwater (Bredenhann et al. 2000). Therefore, prevention of groundwater contamination is the best practical method of protecting the quality and quantity of water, because it is difficult and costly to remediate (Bredenhann et al. 2000). To minimise the adverse effects of dumpsites, the flow of leachate to groundwater must be reduced by lining the dumping site with an impermeable material such as clay or plastic (Bredenhann et al. 2000).

Like most municipal waste disposal facilities in Namibia, the Omaruru municipal dumpsite was sited without prior Environmental Impact Assessment (EIA), which is a tool used to identify short- and long-term environmental threats and public health risks, so that mitigation measures can be applied (Department of Water Affairs 2001). The Omaruru river basin supports the riparian habitat, which consists of biota and is vulnerable to pollution from the municipal dumpsite. Groundwater flows North to South-West of the ephemeral Omaruru river sub-catchment. Thus, contamination can occur due to leachate from the open dumpsite, thereby affecting the groundwater quality and quantity, especially during the rainfall season.

This study is aimed to assess the impact of the municipal dumpsite on groundwater quality within Omaruru river basin.

According to the World Health Organisation (WHO 1996), groundwater quality mainly depends on the: (i) composition of infiltrating water (ii) composition of the initial pore water (iii) composition and physical properties of the soil and rock (iv) subsurface inflow that replaces the pore water (v) chemical interaction between the rock, pore water and infiltrating water and (vi) microbiological processes (WHO 1996). During rainfall, water begins to infiltrate and pass through the soil and rock. The host materials dissolve in water and minerals are added to groundwater (Zaporozec et al. 2002).

When a dumpsite is located on the upper gradient surface, leachate that is formed due to precipitation and surface runoff may infiltrate and contaminate the groundwater, causing public health risks during groundwater abstraction (Pellenberg 1997).

22.2 Methods and Materials

22.2.1 *Setting and Physical Features of the Study Area*

The hydrology of Namibia is defined by a semi-arid climate and a relatively limited presence of surface waters due to irregular rainfall patterns (Angula et al. 2011). The perennial rivers in Namibia are the Kunene River, Okavango River, and Zambezi River in the Northern part of the country, and the Orange River and Kwando-Linyati-Chombe River located in the Southern part (Angula et al. 2011). The Omaruru River extends from Mount Etjo in the North, to the Atlantic shore in the South-West part of the country (Angula et al. 2011).

Omaruru municipal open dumpsite is located West of Omaruru town within the Omaruru river sub-catchment at −21.421350 latitude and 15.932237 longitude; at a height of more than 2100 m above sea level, with 200 mm to 450 mm of rainfall per year. The evaporation rate is very high, over 2000 mm per year, exceeding precipitation (Christelis and Struckmeier 2001). The estimated safe yield is 4.5×10^6 and the groundwater abstraction volume is $6.5–8.5 \times 10^6$ (Geyh and Ploethner 1995). The municipal open dumpsite has a size capacity of 156,379 m^2. It is not fenced, poorly managed, it is an un-engineered open dumping site, accommodating municipal solid waste (MSW) from domestic, industrial, commercial, and agricultural activities. According to Geyh and Ploethner (1995), groundwater flowing within the Omaruru river sub-catchment supports the Omaruru Delta (OMDEL) dam.

The river sub-catchment is subjected to a number of land uses, mainly agriculture, industrial and domestic activities. Catchment implications identified are that, the basin management committee was found inactive (in the integrated water management perspective), which serve as a basis for ensuring regulatory management of the basin.

22.2.2 *Climate and Geological Features*

An average rainfall of 350 mm per year is recorded along an east–west gradient of the river sub-catchment (Ministry of Environment and Tourism 2010). The rainy season is very short and lasts from November to April, affecting the amount of water held in the soil significantly. On average, temperatures reach a maximum of 36 degrees in the months of October and November, just before the rainy season (Ministry of Environment and Toursim 2010).

The geology of the river sub-catchment consists of marbles, schists, dolomites and granite rocks. In the schists, the marble beds are the best formation of sources of underground water. The karstic dolomite rocks are permeable, creating groundwater pollution vulnerability that leads to contamination when leachate percolates from the municipal dumpsite. Pollutants can spread onward from the source only as liquid leachates. Rain and river water may dissolve and mobilise contaminating substances infiltrating into the ground. Moreover, the groundwater boreholes in Omaruru sub-catchment are on the banks or broad sandy riverbeds. The beds of the Omaruru river sub-catchment and its tributaries are wide and filled with sand and gravel (Atlas of Namibia 2007).

22.2.3 Sampling

Onsite/Field Sampling

A global positioning system (GPS) was used for geo-referencing of sampling points. A portable HANNA, H19811-5 multi-meter instrument was used to measure the physical parameters onsite, for instance, pH, temperature, turbidity, total dissolved solids (TDS), and electrical conductivity. A digital camera was used for photographic evidence.

The borehole valve or water tap was purged before sampling to obtain water samples that represent the groundwater characteristics. To avoid turbulence and exposing the sample to air during the water sampling process, contact between the sampling line and the sampling container was avoided. In order to avoid turbulence and the disruption of sediments from the borehole yield, the rate of pumping and drawdowns was kept as low as practicable during purging.

Groundwater samples used for measuring heavy metals were collected in 500 g polyethylene containers and were treated with 2 ml concentrated nitric acid (HNO_3). All samples were stored at 4 ^{0}C prior to being sent to the lab for analysis of chemical parameters. The physical and chemical parameters were analysed using standard analytical laboratory methods.

22.2.4 Analyses Methods

Inductively Coupled Plasma Optical Emission Spectroscopy (ICP-OES) technique was used for water analysis and X-ray fluorescence (XRF) technique was used for soil analysis. The following chemical parameters were analysed: potassium, aluminium, barium, cadmium, chloride, copper, magnesium, nickel, manganese, lead and iron. Laboratory analyses were conducted at the Namibian Water Laboratory and the Ministry of Mines and Energy laboratory which are accredited laboratories in Namibia with standard analytical procedures.

22.2.4.1 Analysis of Water Samples Using ICP-OES

Inductively coupled plasma optical emission spectroscopy (ICP-OES) is described in Boss and Fredeen (2004) as a multi-elemental analytical technique that detects trace metals in parts per million (ppm) or parts per billion (ppb). Advantages of ICP-OES include wide dynamic range, multi-element detection capacity, low analytical interference, and low detection limit.

To accommodate various element suites, instrument calibration can be applied in a variety of different ways. Interferences between elements that produce particles with similar mass-to-charge ratios can affect instrument calibration. For instance, interferences between silicon and scandium or between chlorine and arsenic.

22.2.4.2 Groundwater Flow

An understanding of groundwater flow in the area closest to a municipal dumpsite is crucial to addressing the problem experienced with estimating the time taken by pollutants from the dumpsite to reach the groundwater. Though it is not known how long it takes for contaminants from the dumpsite to reach the boreholes, making an assumption that pollutants do not react, and that they move at the same speed as the groundwater helped the authors to determine the time using Eq. 22.1.

$$v = \frac{q}{ne} = \frac{K\frac{dh}{dl}}{ne} \tag{22.1}$$

$$V = \frac{5x10^{-4}x5x10^{-3}}{0.25}$$

$$V = 1 \times 10^{-5} m\,s^{-1}$$

Therefore:

$$t = \frac{200}{1 \times 10^{-5}} = 2 \times 10^{7} s = 230\,days$$

The average linear velocity used in the equation was derived from an illustration (Fig. 22.1) of a municipal waste disposal facility, which is situated in an unconfined aquifer, as presented by (Hiscock and Bense 2014).

Hiscock and Bense (2014) concluded that the waste is a direct contaminant to the water course because it was observed to be directly in contact with the water table.

Groundwater controls the transport of contaminants whilst the hydraulic gradient causes water to move. Therefore, the hydraulic gradient varies as the ratio of the total head and the distance changes.

22.2.4.3 Soil Samples Analysis Using X-ray Fluorescence (XRF)

"X-Ray Fluorescence (XRF) is an analytical method used to ascertain the geochemical composition of the soil, rock, and sediments" (Brouwer 2010). The XRF analyse

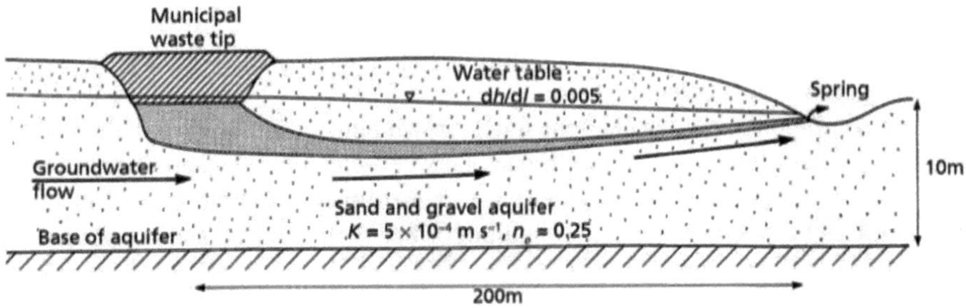

Fig. 22.1 Municipal dumpsite predicting the movement of contaminant plumes. *Source* (Hiscock and Bense 2014)

heavy metals, minerals and petroleum products ray that form part of the magnetic spectrum. The analytical technique measures energy in kilo-electron volts or wavelength in nano-meters (nm). X-RF can analyse different elements, for example, arsenic (As), antimony (Sb), barium (Ba), calcium (Ca), chromium (Cr), cobalt (Co), copper (Cu), iron (Fe), lead (Pb), manganese (Mn), nickel (Ni), potassium (K), titanium (Ti) and zinc (Zn). The concentration levels are measured and range from parts per million to high percentages in solids liquids and powder or natural soil in situ (Brouwer 2010).

22.2.5 Catchment Assessment

Groundwater vulnerability to pollution from the dumpsite can be assessed to identify the impacts of waste disposal on the quality and quantity of the water resources in the river sub-catchment. The geological features such as rock types, landscape and the potential water abstraction, the physical, chemical properties of water within the catchment and ecological functions were assessed. A GIS (ArcView) model was implemented in Geographical Information System (GIS) to identify the geographic area to examine the ecological and hydrological functioning of water within the catchment and its riparian habitat. Every piece of information was entered into the GISsoftware, ILWIS 2.2 and arranged in several thematic georeferenced maps with linked databases. All maps were created in vector format with a coordinate system. The vector maps were georeferenced and rasterised with pixels of 25 by 25 m. The results describe the catchment of the upper and lowland indicating the groundwater flow. The Omaruru river sub-catchment is subjected to different land uses, mainly agriculture, industrial, domestic activities and residential development. Catchment implications identified are that, the basin management committee was found inactive (in the integrated water management perspective), that serve as a basis for assuring the regulatory management of the basin.

22.2.6 Analysis of Data

Statistical Package for Social Sciences (SPSS) software was used to analyse quantitative data and translate the results into useful information. The association between heavy metals in soil and physical and chemical indicators of water quality was ascertained using the Pearson correlation analysis.

It was determined to what extent heavy metals contaminated the soil. The pollution calculation techniques from Eqs. 22.2 to 22.4 were applied to the analytical data. The objective was to generate accurate projections of the level of contamination that would affect the soil near the municipal dumpsite.

22.2.6.1 The Enrichment Factor (EF)

To understand the contribution of metals other than those of lithogenic origin, enrichment factor (EF) is typically determined. For the geochemical normalisation of the heavy metal data in the computation of the enrichment factor, the concentration of either iron (Fe) or aluminium (Al) in the upper continental crust (UCC) can be utilised as a conservative element. For the computations of the Enrichment Factor, iron was utilised as a precautionary element. The enrichment factor is calculated (Likuku et al. 2013) using the following Eq. 22.2.

$$\mathrm{EF} = \frac{(\mathrm{m/Fe})\ \mathrm{sample}}{(\mathrm{m/Fe})\ \mathrm{UCC}} \tag{22.2}$$

where:

(m/Fe) sample is the metal/Fe ratio of the samples and ***(m/Fe) UCC*** is the equivalent element/Fe ratio in the UCC.

The UCC values were sourced from McLennan (2001). When the enrichment factor is <1, the soil is free from pollution, enrichment factor <3 is minor; 3–5 is moderate, 5–10 is moderately severe, 10–25 is severe, 25–50 is very severe and >50 is extremely severe. The enrichment factor estimates and classes calculated are in the range from 0.6 to 20; for Manganese 0.66, Copper 1.41, Lead 2.20, Zinc 0.93, Iron 1.22, Magnesium 2.34, Nickel 1.54 and Potassium 1.59 and Calcium 0.78. This shows that there is minor enrichment, especially for Potassium and Calcium as presented in Table 22.6.

22.2.6.2 The Contamination Factor (CF)

The contamination factor, which was calculated in this study, can be used to gauge the level of general contamination for surface layers at a studied site (Gupta 2014) as shown in Eq. 22.3:

$$\mathrm{CF} = \frac{\mathrm{Cm\ sample}}{\mathrm{Cm\ Background}} \tag{22.3}$$

where:

Cm *sample* is the concentration of a given metal at contamination of a given metal at a contaminated site. Median **Cm *Background*** is the median concentration of an element in the background soil sample.

The ratio of the metals in the samples to the background value of the corresponding metal is the contamination factor. Contamination factor can be categorised into four groups: 1. low (CF > 1), 2. moderate (1 ≤ CF ≤ 3), 3. considerable (3 ≤ CF ≤ 6) and 4. very high (CF > 6).

22.2.6.3 The Pollution Load Index (PLI)

The Pollution Load Index gives an estimate of the level of metal pollution and the appropriate course of action. PLI values below 1 indicate perfection, values of 1 and higher indicate that the quality of the site is deteriorating, whereas PLI values of 1 represent baseline levels of pollutants. The PLI greater than or equals to 1, indicate that an immediate intervention to reduce pollution is required, whilst 0.5 ≤ PLI < 1 requires site monitoring and further study. But, when PLI < 0.5 there is no need for taking remedial action.

22.2.6.4 The Geo-Accumulation Index (Igeo)

Geo-accumulation index (Igeo) can be used to quantify metal pollution in soil and aquatic sediments (Gupta 2014). Igeo can be used to evaluate the impact of anthropogenic heavy metal influx. Geo-accumulation index (Igeo) for soil samples was calculated (Gupta 2014) as in Eq. 22.4.

$$\mathrm{I_{geo}} = \log_2\left(\frac{\mathrm{Cn\ Sample}}{(1.5 \times \mathrm{Cn\ Background})}\right) \tag{22.4}$$

where: ***Cn*** is the metal *(n)* concentration measured in soil samples in the study site, and ***Bn*** is the background value of the corresponding metal *(n)*.

The factor of 1.5 is the background matrix correction due to lithogenic effects. "Igeo has six classes (0–6) namely; (i) uncontaminated (<0), (ii). uncontaminated to moderately contaminated (1), (iii) moderately contaminated (2), (iv) moderately to highly contaminated (3), (v) highly contaminated (4), vi) highly to very highly contaminated (5) and (vii) very highly contaminated (6)" (Rajmohan et al. 2014).

22.3 Results and Discussion

22.3.1 Assessing the Implication of Locating the Dumpsite Near Groundwater Sources

The Omaruru municipal dumpsite was located without an Environmental Impact Assessment (EIA). An EIA is an assessment tool that identifies significant impacts on the socio-economic and the ecological environment, to ensure compliance to environmental requirements. It is an un-engineered municipal open dumpsite. The volume of waste that is disposed of at the dumpsite is not quantified because of the absence of weighing facilities. The type of waste disposed at the dumpsite is Municipal Solid Waste (MSW). Medical waste from health facilities is disposed at the dumpsite, through dig and bury methods when the health facility's incinerator is not functional. Burning practices occur at the dumpsite and the product of combustion becomes part of the leachate process, especially during the rainy season (Fig. 22.2).

Upper gradient water boreholes (dark blue) and down gradient water boreholes (green) of the Omaruru municipal dumpsite within the river basin. The study area (light blue) has been sampled for physico-chemical parameters and the sewerage ponds (yellow), next to the municipal dumpsite within the river sub-catchment.

The municipal dumpsite is located near sewerage ponds which may have an impact on groundwater through seepage or sewerage ponds overflow during the rainy season, creating a public health risk. The down gradient water boreholes namely, Erongo

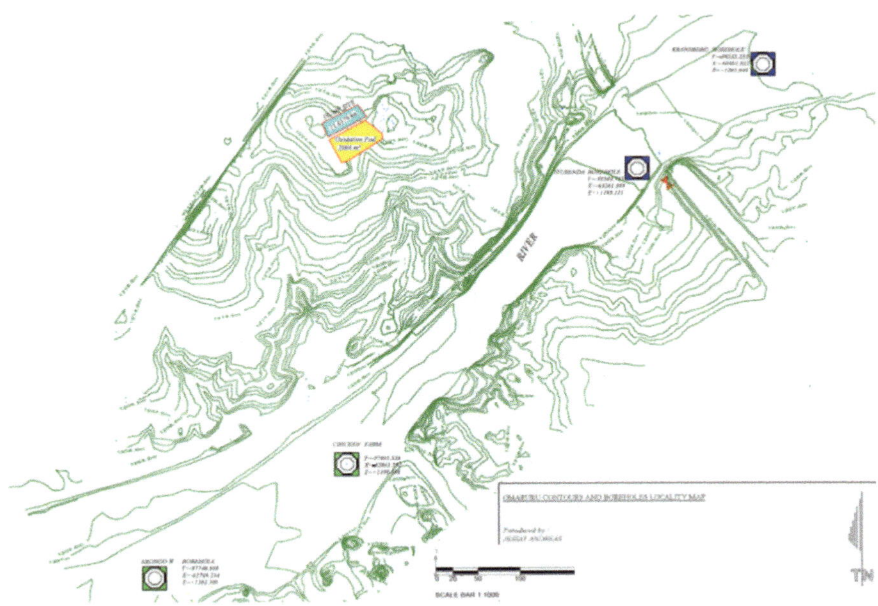

Fig. 22.2 Map indicating the sampled points

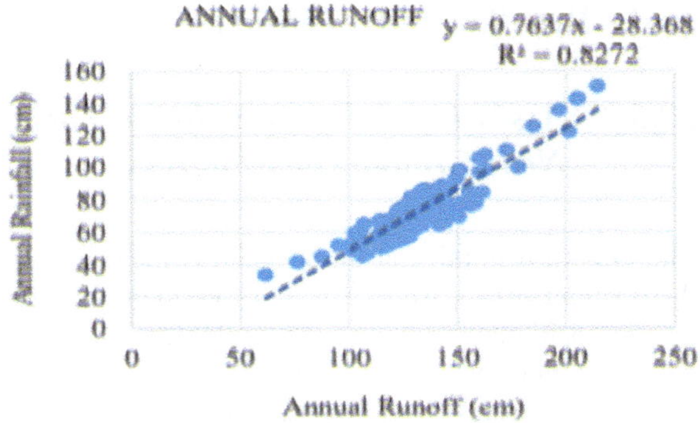

Fig. 22.3 A linear statistical model predicting annual runoff event correlation

mountain winery borehole is situated 3 km from the municipal dumpsite and the Chicken farm borehole is situated 2.5 km from the dumpsite. The two water boreholes located down gradient of the river sub-catchment are privately owned, whilst the other two boreholes located on the up gradient are for the municipality of Omaruru. All four boreholes are in the proximity of the Omaruru municipal dumpsite.

Figure 22.3 illustrates the annual runoff for Omaruru municipality.

The values of rainfall and runoff are largely correlated with ($Y = 0.7637 \times -28.368$ and $R^2 = 0.8272$). The average specific volume was measured as 7.6 (upper gradient) and 0.1 mm/m^2 (down gradient), and the runoff coefficient was found at 2.2 (upper gradient) and 0.0% for down gradient.

22.3.2 The Potential for Groundwater Abstraction of Borehole Water

Omaruru Municipality operates its own bulk water supply scheme. Water is abstracted from Omaruru aquifer by means of production borehole installations. Water is extracted from four boreholes in the immediate vicinity of Omaruru (high yield of boreholes of 0.1 to 80 m^3/h) (Christelis et al. 2001). Two water boreholes (i.e. Kranzberg and Turenda water boreholes scheme east of the town) feed directly into the municipal reservoir. The other two boreholes are privately owned (the chicken farm and Erongo mountain) and, were included in the study, because of their proximity to the dumpsite and the groundwater flow of the aquifer, making a total of four boreholes for this study. The depth, water levels and coordinates of the sampled groundwater boreholes are shown in Table 22.1, giving an indication of the groundwater abstraction potential within the municipality.

Kranzberg and Turenda boreholes have a reserve of 4.4 Mm^3 water but the two boreholes could only supply 55,555 m^3 (according to abstraction figures of the year 2015), which is significantly lower than the historic demand in the area. The present water demand from users, abstraction volume and sustainable yield of the Omaruru

Table 22.1 The depth, water levels and coordinates of sampled groundwater boreholes in Omaruru

	Kranzberg	Turenda	Erongo mountain winery	Chicken farm
Original borehole approx. depth (m)	8	8	6.5	7
Water rest-level 1 (m)	3.13	2.6	3.2	3
Water rest level 2 (m)	2.90	2.40	2.7	2.5
Average water rest-level (m)	2.90	2.40	2.70	2.6
Coordinates	21°25′35.53″S	21°25′30.93″S	21°25′18.24″S	21°26′13.94″S
	15°56′52.98″E	15°54′58.62″E	15°54′28.27″E	15°55′5.25″E

aquifer were reported by (Christoph 2016) as 12 Mm^3/a, 1 Mm^3/a and 3 Mm^3/a, respectively.

The river sub-catchment is subjected to a number of land uses, mainly agriculture, industrial and domestic activities as indicated in Fig. 22.4. Catchment implications identified are that, the basin management committee was found inactive (in the integrated water management perspective), that serve as a basis for assuring the regulatory management of the basin.

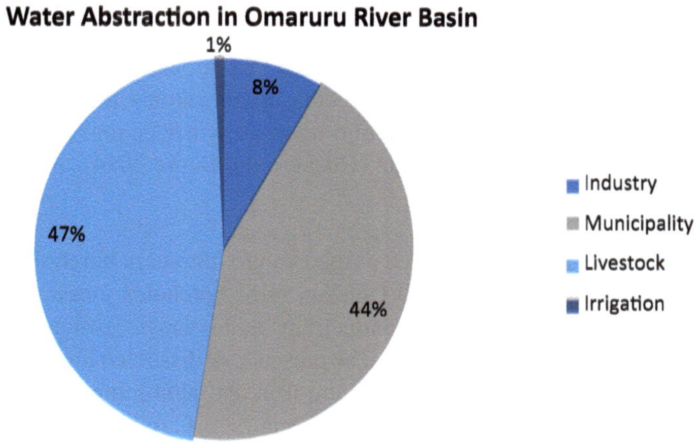

Fig. 22.4 Percentage water abstraction for different activities within the river sub-catchment

22.3.3 *Physical and Chemical Characteristics of Water and Soil from Boreholes*

The physical and chemical parameters were evaluated to identify the correlation in the water and soil. The parameters were benchmarked with the World Health Organisation (WHO) guidelines, Environmental Protection Agency (EPA) standards and the Namibian standards for water quality and the exceedances are highlighted in Table 22.2.

The physical and chemical parameters exceeding the water quality standards were calcium, conductivity, lead, magnesium, potassium and total dissolved solids at all sampled locations.

22.3.4 *Physical Parameters in Groundwater*

(a) The pH

In the watershed area, pH is a crucial indicator of water quality. According to the study, the pH range of the groundwater in the boreholes ranges from 7.9 to 11.7, making it alkaline. The solubility and toxicity of chemicals and heavy metals in water are influenced by the pH's level of acidity. pH in water can also be impacted by anthropogenic (such as waste discharge from industry) and natural (such as precipitation, suspended dust, etc.) sources of contamination (Hiscock and Bense 2014). Natural alterations may also take place because of interactions with the nearby rock, especially carbonate forms and other substances. Additionally, low pH values indicate an increase in the solubility of heavy metals. Metal cations including aluminum, lead, copper and cadmium are released into water as hydrogen ions increase instead of being incorporated into the sediment (Hiscock and Bense 2014).

(b) Electrical Conductivity (EC)

The highest value of EC was observed at the Erongo mountain borehole (1210 mS/cm), whilst Turenda, Kranzberg and Chicken farm boreholes measured the same value (1190 mS/cm). High electrical conductivity values can be attributed to leachate processes from the municipal dumpsite. The presence of dissolved ions in water, such as cations like sodium, magnesium, calcium, iron and aluminium, and anions like chloride, nitrate, sulphate, and phosphate, affects the EC of water. An increase in EC corresponds to an increase in the water temperature. The down gradient water boreholes had an average conductivity of 1200 mS/cm and up gradient water boreholes had an average conductivity of 1190 mS/cm, both exceeding the water quality standards as shown in Table 22.2.

(c) Total dissolved solids (TDS)

Dissolved solids contain calcium, chlorides, nitrate, phosphorus, iron, sulphur and other ions and microparticles. Increased concentrations of total solids make drinking

Table 22.2 The physical and chemical parameters measured at water boreholes near Omaruru municipal dumpsite, Jan 2018 to June 2018

Parameter	Unit	WHO Guide level	EPA Standards	Namibian Water quality Standards	Kranzberg	Turenda	Chicken farm	Erongo
Temperature	^{0} C	*	*	Ambient temp	29 ^{0} C	29 ^{0} C	29 ^{0} C	29 ^{0} C
pH	pH units	*	6.5 - 8.5	6 to 9	11.6	8	7.8	8
Turbidity	NTU	*	*	< 2	0.60	1.39	2.29	1.39
Conductivity	µS cm^{-1} at 20°C	400	*	<300	1190	1190	1190	1210
TTDS	mg L^{-1}	*	500	<2000	3626	610	590	817
Aluminium	mg L^{-1} Al	*	0.05 to 0.2	0.15	0.01	0.01	0.01	0.01
Barium	mg L^{-1} Ba	1.3	0.003	0.5	0.02	0.02	0.04	0.02
Cadmium	mg L^{-1} Cd	0.005	0.003	0.01	0.01	0.01	0.1	0.01
Chloride	mg L^{-1} Cl	0.05	250	0.25	85	88	23	87
Copper	mg L^{-1} Cu	2	1	0.5	0.01	0.01	0.01	0.01
Chromium	mg L^{-1} Cr	0.05	0.1	0.1	0.01	0.01	0.01	0.01
Nickel	mg L^{-1} Ni	0.07	*	0.25	0.01	0.01	0.01	0.01
Calcium	mg L^{-1} Ca	100-300	100	0.15	105	195	193	155
Magnesium	mg L^{-1} Mg	*	*	0.07	63	58	38	63
Potassium	mg L^{-1} K	*	*	0.2	5	4	4	5
Iron	mg L^{-1} Fe	*	0.3	0.1	0.01	0.01	0.01	0.01
Manganese	mg L^{-1} Mn	0.1	0.05	0.05	0	0	0	0
Lead	mg L^{-1} Pb	0.01	0.015	0.05	0.02	0.02	0.02	0.02

* No standards set for the parameters.

Based on WHO, EPA and Namwater Standards

Red – Exceeds all three standards

Brown – Exceeds two standards

Yellow – Exceeds one standard

water unpalatable and might have an adverse effect on the quality of drinking water. The upper gradient boreholes have higher concentrations of Total Dissolved Solids (TDS) than the down gradient boreholes. Total dissolved solids might also indicate the effects of runoff from the municipal dumpsite.

(d) Turbidity.

Turbidity affects the colour of water. Because suspended particles absorb more heat, water temperature rises when there is high turbidity. Analysis shows that the samples from the down gradient water boreholes were more turbid than the samples upper gradient water boreholes of the municipal dumping site.

22.3.5 *Chemical Parameters in Groundwater*

The study found high concentrations of heavy metals at down gradient boreholes. An increase in concentration levels of heavy metals occurred during the rainy season. This also shows that leachate migration occurred underground at the municipal dumping site, thus contaminating groundwater quality. The concentration levels of heavy metals exceeded the local (NamWater) and Environmental Protection Agency (EPA) standards (Figs. 22.5, 22.6 and 22.7).

The down gradient water boreholes had above concentration of chemicals than the upper gradient as indicated on a piper plot (Fig. 22.8). The down gradient water boreholes measured an increased concentration level of Calcium (Ca), Magnesium (Mg), Potassium (K) and Sodium (Na), exceeding the concentration levels of upper gradient boreholes. This can be attributed to leachate processes arising

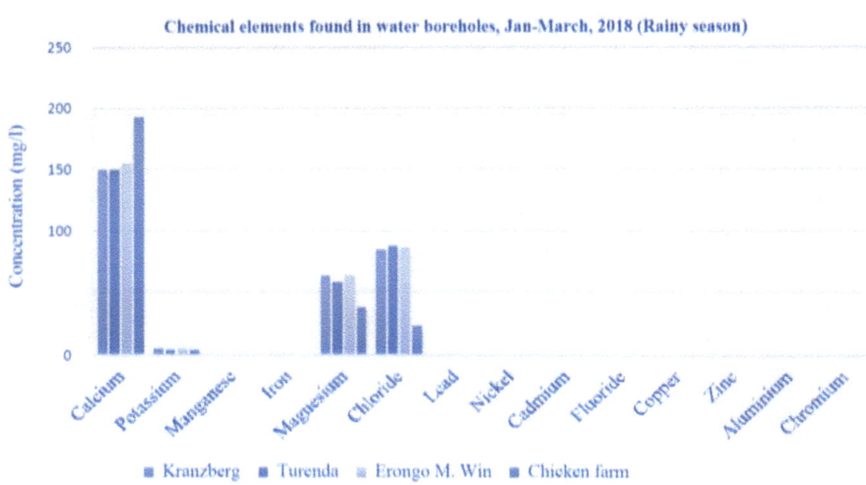

Fig. 22.5 Elemental composition of groundwater from the four boreholes during the rainy season (January to March 2018)

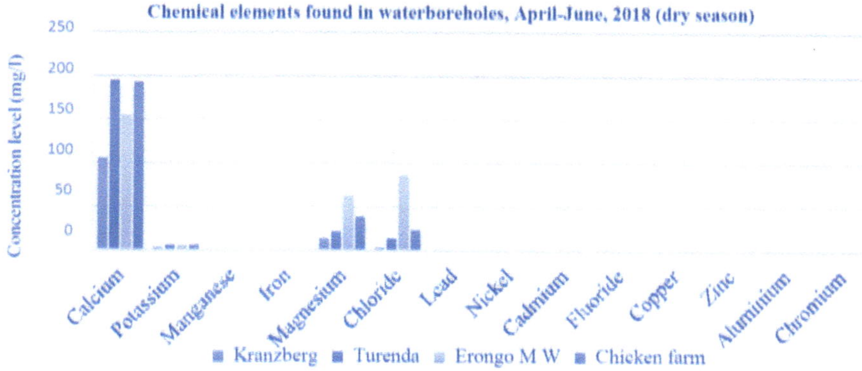

Fig. 22.6 Elemental composition of groundwater in the four boreholes during the dry season, April to June 2018

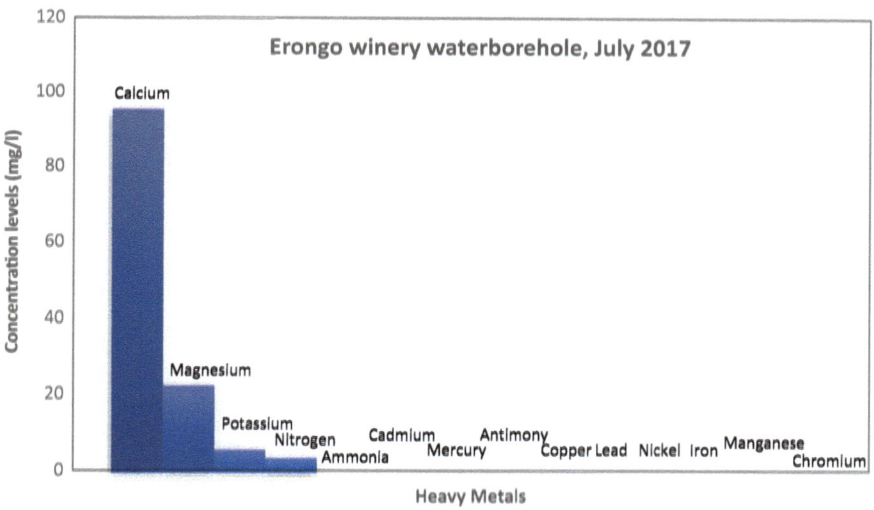

Fig. 22.7 Elemental composition of groundwater from the downstream borehole (Erongo winery borehole), July 2017

from the municipal dumpsite, especially during the rainfall season. An indication of possible contamination of groundwater contamination plumes from the municipal open dumpsite.

The cations and anions are an indication of groundwater chemistry and identify whether any pollution is present. The results are as indicated in Fig. 22.8.

The cations including Sodium–Potassium (Na + K), Calcium (Ca) and Magnesium (Mg) have the concentration level exceeding 5.44 Meq/L in all groundwater boreholes. The downgradient water boreholes had anions such as Chlorine (Cl), Sulfate (SO_4) and Carbonates ($HCO_2 + CO_2$) and can be attributed to the leachate

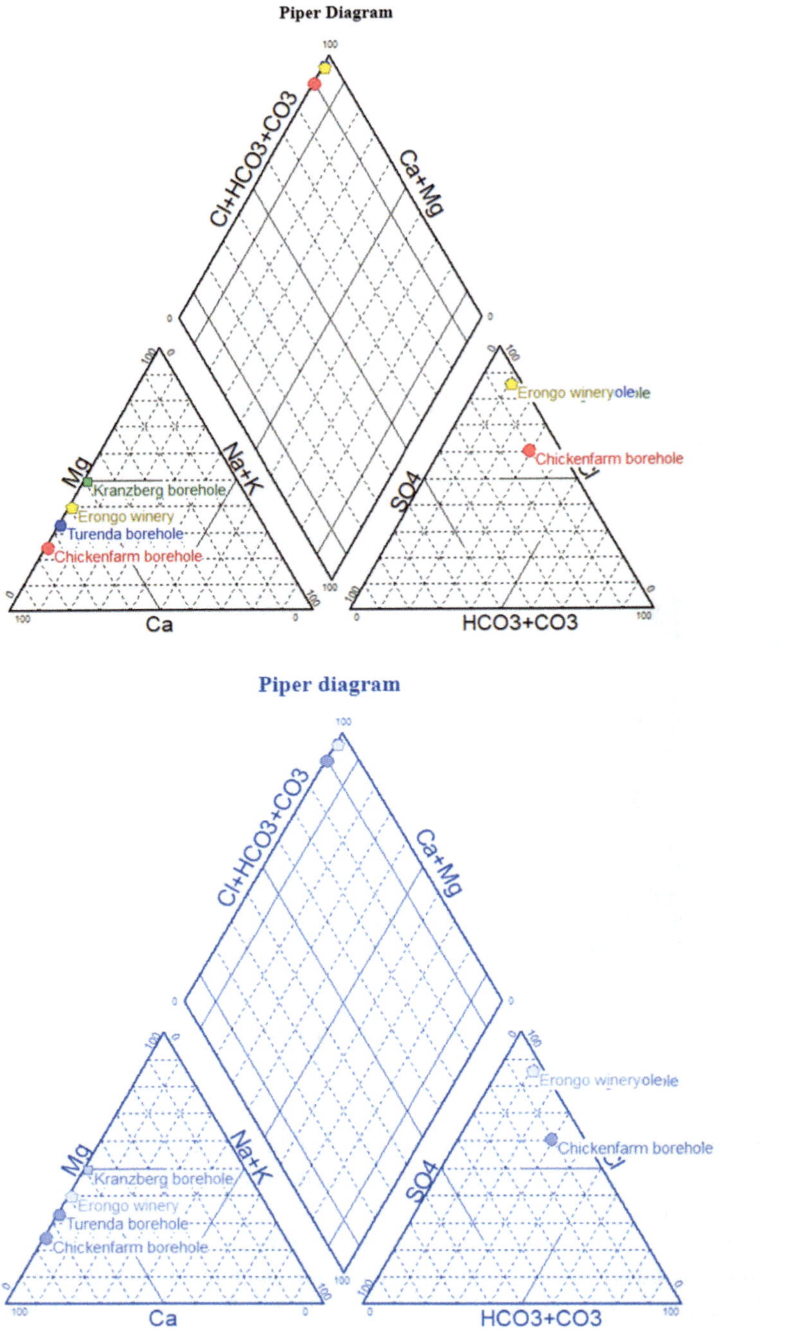

Fig. 22.8 The trilinear graph indicating the concentration of cations and anions of sampled waterboreholes

processes from the municipal open dumpsite, especially during rainfall season. Other activities observed that may have contributed to the chemistry were pesticides applied at the chicken farm as an agricultural activity.

Wastewater from the ponds near the municipal dumpsite were analysed for determination of heavy metal composition in wastewater effluent discharge and further provide evidence on groundwater leachate concentration from the dumpsite. Sodium, Kjeldahl Nitrogen, Ammonia and Sulphates were within permissible limits of effluent discharge into the environment, except chlorine levels with concentrations exceeding the limitations for wastewater as observed in Table 22.3.

22.3.6 Geochemical Parameters of Soil Samples

The results from laboratory analysis indicated the concentration of heavy metals as follows; Aluminium (6.82%), Potassium (5.72%), Calcium (7.87%), Iron (2.46%), Magnesium (2.63%), Phosphorus (0.56%), Manganese (3.57%), Titanium (0.31%), Selenium (0.14%), Chloride (0.09%), Barium (0.04%), Zirconium (0.0118%), Strontium (0.0258%), Chromium (0.0099%), Zinc (0.0073%), Copper (0.0083%), Lead (0.0022%) and Nickel (0.0035%). Silicon was predominant on the upper gradient water boreholes (30.83%) and at the municipal dumpsite (19.92%). Leachate may contain heavy metals from municipal dumpsite, originating from the disposal of waste materials from residential, agricultural, and industrial activities within the river sub-catchment. The upper gradient water boreholes had higher concentration levels of heavy metals than down gradient water boreholes.

Aluminium, Potassium, Calcium, Iron, Phosphorus, Magnesium and Titanium were observed in all sampled areas (Fig. 22.9), the high concentrations were found in the soil from down gradient boreholes.

The heavy metals that exceeded the standard concentration levels were Iron (2.64%), Calcium (4.22%), Magnesium (0.85%), Titanium (0.34%), Phosphorus (0.12%), Zirconium (0.0417%), Manganese (0.0451%), Strontium (0.024%), Barium (0.018%), Zinc (0.0042%), Copper (0.0026%) and Nickel (0. 0027%).

22.3.7 The Correlation Matrix of Heavy Metals in Soil from the Dumpsite, Sewerage Ponds and Soil After the Dumpsite

The correlation coefficient (Pearson coefficient) measures the degree and direction of the linear link between two continuous variables. The strength of the association between the variables increases with the coefficient's absolute value, which is normally between -1 and $+ 1$. Table 22.4 indicates a strong correlation of elements

Table 22.3 Wastewater pond near the municipal dumpsite, with discharge standards into the environment

Phsico-chemical parameter	Measurement unit	Concentration level	Namibian water standards	Environmental protection agency standards
Temperature	°C	29.5 °C	35 °C	<30 °C
pH		6.9	5.5 –9.5	6–9
Turbidity	NTU	61	1	75
Total dissolved solids	mg/l	536	<1000	1500
Total suspended solids	mg/l	90	25	50
Electrical conductivity	mS/cm	100	150	750
Dissolved Oxygen	mg/l	0.1	A saturation of at least 75%	1
Redox potential	mV	−122	*	*
Chemical Oxygen Demand (COD)	mg/l	294	75	250–800
Biological Oxygen Demand (BOD)	mg/l	102	No value given	100–300
Absorbed Oxygen as O_2	mg/l	25	10	*
Sulphate (SO_4)	mg/l	45	200	*
Ortho-Phosphate (PO_4)	mg/l	13		*
Chloride (Cl)	mg/l	55	1.5	*
Nitrate (NO^{-3})	mg/l	0.5	<6	*
Nitrite (NO^{-})	mg/l	0.03	<0.1	*
Ammonia Nitrogen (NH_3-N)	mg/l	49	10	10–30
Sodium (Na)	mg/l	72	<90	<100
Kjeldahl Nitrogen (TKN)	mg/l	50	33	20–85
Fats, oil and grease	mg/l	20		*

*No value standard set

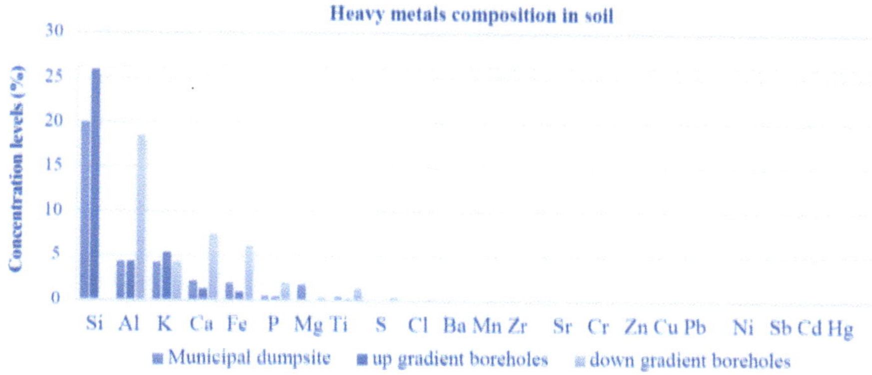

Fig. 22.9 Concentration of heavy metals in soil for sampled points

Table 22.4 Correlation matrix of heavy metals in soil from the dumpsite and sewerage ponds

	Leachate from the dumpsite	Elements from sewerage ponds	Elements in the soil from after the dumpsite
Leachate from the dumpsite	1		
Elements from sewerage ponds	0.950558067	1	
Elements in the soil from after the dumpsite	0.990554275	0.950798	1

from the soil taken after the dumpsite and the leachate from the municipal dumpsite (0.99) and elements from sewer ponds and leachate from the dumpsite (0.95).

Most of the samples plot near the proportionality line and the concentration increases on both the y and x-axis. However, there is a variation of proportionality between leachate from the dumpsite and soil elements from sewage ponds as shown in (Fig. 22.10). There is a strong correlation between leachate from dumpsite and elements in the soil from after the dumpsite (Figs. 22.11 and 22.12).

22.3.8 The Pearson Correlation Coefficient Analysis Between Leachate from the Municipal Dumpsite and Borehole Water

The correlation matrix is used in this study to describe the interrelatedness between the physico-chemical characteristics of groundwater from the boreholes and soil samples from the Omaruru municipal dumpsite.

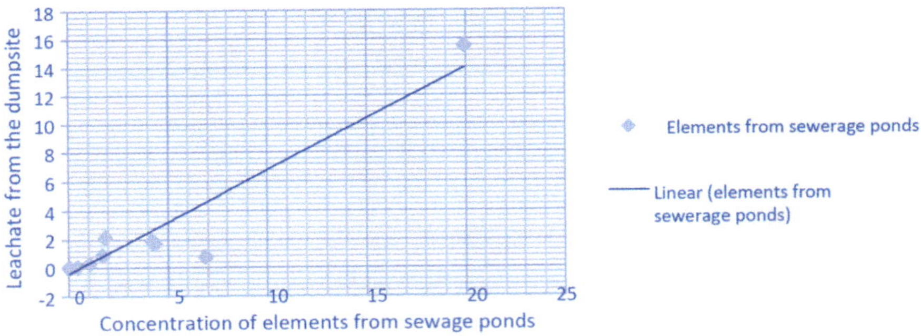

Fig. 22.10 The correlation relationship between leachate from the dumpsite and elements from sewage ponds

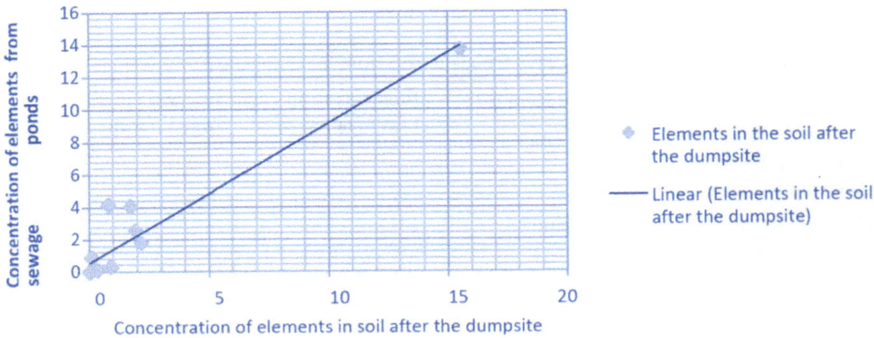

Fig. 22.11 The correlation relationship between elements from soil after the dumpsite and elements from sewage ponds

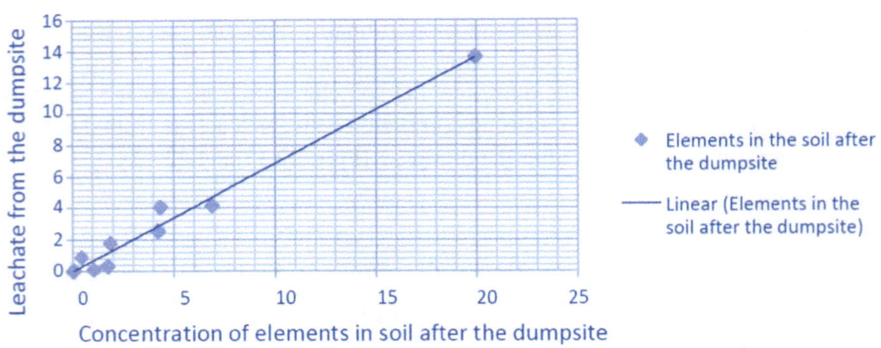

Fig. 22.12 The correlation relationship between leachate from the dumpsite and elements in the soil after the dumpsite

A correlation is a mutual relationship between two variables (Shroff et al. 2015). A direct correlation exists when there is an increase of soil heavy metals in soil which corresponds to an increase of the same heavy metals in groundwater (Table 22.5).

There is a strong correlation between Manganese and Chloride (0.88), Lead (0.87), Calcium (0.18), Fluoride (0.950), Copper (0.98), Zinc (0.96), Aluminium (0.96). A moderate correlation exists between Calcium and Manganese (0.6), Iron (0.61), Chloride (0.56), and Chromium (0.7). Potassium, Nickel, Cadmium and Fluoride showed weak correlation with other parameters. Heavy metals including Arsenic, Cadmium, Chromium, Copper, Lead, Mercury, Nickel and Zinc can be correlated with leachate composition. Heavy metal concentrations in soil can increase because of electronic device disposal. Naturally occurring heavy metals in the environment can contaminate groundwater in excessive concentrations. The down gradient water boreholes are contaminated because concentration levels of most physical and chemical parameters exceed the World Health Organisation (WHO) standards for drinking water. It is evident that the open waste dumping affects local groundwater (Table 22.6).

Igeo values from the study are 0.003–0.7. 0.24 for Manganese, 0.003 for Copper, 0.0025 for Lead, 0.04 for Zinc, 1.35 for Iron, 0.75 for Magnesium and 0.006 for Nickel. At sampling sites, all heavy metals were in the range of uncontaminated to moderately contaminated (1). Heavy metals including Lead, Nickel, Iron and Magnesium have Igeo less than one throughout the depth which indicates moderate concentration of contaminants. Heavy metals have a positive Igeo index, indicating that they were enriched from sources other than their lithogenic origin (Table 22.7).

22.4 Discussion

22.4.1 Implications of the Results Obtained in This Study

The study found that the municipal dumpsite is responsible for contamination of groundwater, through leachate processes. Groundwater contamination is common during the rainfall season than the dry season because of the accelerated biochemical processes occurring within the dumpsite that infiltrates or percolates into groundwater. Open dumping contaminates groundwater through the formation of leachate. It is evident therefore, that groundwater in Omaruru river sub-catchment is contaminated by the municipal dumpsite. This has a direct impact on the quality and sustainability of groundwater supply in the area.

Most contaminants are released during precipitation from waste materials dumped on the surface or below the ground, such as uncollected municipal garbage and unregulated dumpsites. It was found that the down gradient water boreholes, namely chicken farm and Erongo mountain winery were contaminated. Chemical characteristics of groundwater and soil samples from the Omaruru municipal dumpsite have a high association with one another. The chemistry of groundwater can be affected by the concentration and quantity of heavy metals present in the soil and waste.

Table 22.5 Correlation coefficient matrix of leachate from municipal dumpsite and water boreholes within the Omaruru river sub-catchment (Jan 2018 to June 2018)

Parameter	Calcium	Potassium	Manganese	Iron	Magnesium	Chloride	Lead	Nickel	Cadmium	Fluoride	Copper	Zinc
Calcium	0.180403	1										
Potassium	−0.93616	−0.08996	1									
Manganese	−0.70587	−0.01266	0.604295	1								
Iron	0.808471	0.233378	−0.84797	−0.62341	1							
Magnesium	0.628581	0.237595	−0.70774	−0.52415	0.957948	1						
Chloride	−0.81821	0.441408	0.882969	0.698979	−0.70812	−0.56181	1					
Lead	−0.90797	−0.14301	0.863876	0.771342	−0.80136	−0.6736	0.877769	1				
Nickel	−0.26688	−0.1687	0.371735	0.472696	−0.48477	−0.58046	0.225686	0.37964	1			
Cadmium	−0.17625	−0.0.29432	−0.341457	−0.37543	0.232816	0.433499	−0.62661	0.188212	0.228753	1		
Fluoride	−0.94448	−0.13267	0.956446	−0.694067	−0.84448	−0.70728	0.881791	0.947817	0.421913	0.230159	1	
Copper	−0.94449	−0.12779	0.936505	0.779934	−0.84235	−0.70614	0.839512	0.970333	0.429834	0.358438	0.982829	1
Zinc	−0.90418	−0.13092	0.842906	0.850039	−0.79909	−0.6711	−0.856341	−0.978119	0.480922	−0.24503	0.937088	0.962087
Aluminium	−0.899096	−0.10518	0.839255	0.916828	−0.79063	−0.66357	0.80935	0.944357	0.423102	−0.284	0.893687	0.952018
Chromium	0.338553	−0.1448	0.697679	−0.82224	−0.31619	−0.215764	−0.3051	−0.36557	−0.12942	−0.705462	−0.06709	−0.63264

Table 22.6 Igeo values and classes of soil samples collected at different study sites

Geo-accumulation index									
Mean	0.66	1.41	2.20	0.93	1.22	2.34	2.34	1.59	0.78
STD	0.18	0.58	0.99	0.50	0.32	0.94	0.48	0.48	0.45
Contamination factor									
Minimum	0.53	0.45	1.19	0.17	0.79	1.16	0.48	0.37	0.24
Maximum	1.35	4.23	3.99	1.92	2.26	4.29	2.61	1.51	0.52
Mean	0.79	1.65	2.55	1.05	1.46	2.74	1.92	0.85	0.81
STD	0.27	1.65	0.82	0.51	0.44	0.90	0.59	0.31	0.43

Table 22.7 The geo-accumulation index, contamination factor and enrichment factor for soil samples from the study site

	Mn	Cu	Pb	Zn	Fe	Mg	Ni	K	CA
Geo-accumulation index									
Minimum	0.02	0.0017	0.0017	0.0012	0.59	0.78	0.002	3.2	0.91
Maximum	0.03	0.002	0.0357	0.0073	2.46	2.63	0.0035	5.72	7.87
Mean	0.24	0.003	0.0025	0.004	1.35	0.75	0.006	4.5	1.78
STD	5.500	2.748	1.985	2.379	0.66	1.02	1.429	0.825	2.552
Enrichment factor									
Minimum	0.02	0.25	1.18	0.10	0.82	0.78	0.78	0.47	0.26
Maximum	0.06	2.41	5.12	1.81	1.97	4.80	0.94	2.72	1.72

Groundwater pollution can deteriorate water quality, making it unsafe for human consumption.

Given the rising demand for freshwater caused by growing urbanisation and land use, groundwater quality is predicted to continue to deteriorate. The limited and ineffective onsite waste management and disposal techniques also serve as a trigger for groundwater pollution.

Polluted groundwater has a detrimental effect on public health such as carcinogens and cleaning up of groundwater can be an expensive process. Thus, the vulnerability of Omaruru aquifer to groundwater pollution can affect the biota and the socio-economic status of the basin community. It is important to safeguard, conserve and ensure the sustainable utilisation of river basins.

22.4.2 *Landfill/Dumpsite and Leachate Management*

This situation summarised in Sect. 4.1 requires the municipal authorities to put in place waste management and water quality regulations and management plans to protect the quality of groundwater and ensure sustainable water supply in the area.

Namibia and most Southern African countries do not have the resources to convert all conventional waste dumpsites into sanitary landfills with the design to prevent the leakage of leachate and its harmful components. However, even a lined landfill can still experience accidental leachate leakage which, according to Xiang et al. (2019), may be caused by damage to the geomembrane during the construction and operation of the landfill site. According to Chicago Metropolitan Agency for Planning (2013), leachate is a clear environmental liability, and can be managed using suitable siting and sealing methods. However, if leachate percolates into groundwater, it can be the source of, amongst many oxidation-reducing organic compounds leading to an increase in the mobility of toxic metals. Hence, the need for effective methods for treatment of the landfill.

Lin et al. (2021) suggested a combination of membrane bioreactor with nanofiltration as an effective approach for the treatment of landfill leachate, and confirmed its effectiveness in the removal of biodegradable organics and heavy metals to minimise the negative effect of landfill leachate.

In Southern African countries such as Namibia, a waste and water governance and management plan is essential to consider the suitability of the methods for the region, and to ensure reasonable/unbiased and affordable management of the groundwater resource.

Preventative methods such as ensuring reasonable distance between the landfill and the water source (in this case—boreholes) should be included in the management plan. Therefore, monitoring and surveillance may also play an important role in the waste and water governance planning for the country.

22.5 Conclusion

The Omaruru municipal dumpsite poses threats to groundwater within the river basin. Evidence is attained from the observed concentration of physical and chemical parameters in sampled water boreholes. Pollution occurs during the rainy season because of leachate processes. Leachate processes occur during the wet season creating contamination plumes. Soil samples from the dumpsite contained concentration levels of heavy metals that can form part of leachate. A strong correlation exists between heavy metals found in soil and heavy metals found in groundwater boreholes within the river sub-catchment. The research area's Contamination Factor (CF), which indicates the level of contamination, ranged from 0.8 to 2.7. The study's geo-accumulation Index (Igeo) values ranged from 0.003 to 0.7, indicating lithogenic origin, whilst the pollution load index (PLI), which was at 2, indicated contamination of the study region as the PLI was >1. Groundwater is abstracted from Omaruru aquifer by means of borehole installations, with a high yield of 0.1 to 80m^3/per hour. Water demand from users is approximately 12 Mm^3/a. The water boreholes at the lower gradient of the municipal dumpsite indicated higher concentrations of physico-chemical parameters above World Health Organisation (WHO) and Environmental

Protection Agency (EPA) standards for drinking water. The study found that municipal open dumping site contaminates groundwater sources, thereby making the water unsafe for domestic use and the biota that depends on the river sub-catchment.

Acknowledgements The authors acknowledge the support by the "UNESCO Chair on Sustainable Water Research for Climate Adaptation in Arid Environment", and the Omaruru Municipality, Erongo Region in Namibia.

References

Adelana SMA, Tamiru A, Nkhuwa DCW, Tindimugaya C, Oga MS (2008) Urban groundwater management and protection in sub-Saharan Africa. In: Adelana SMA, MacDonald AM (eds.) Applied groundwater studies in Africa. Taylor & Francis, London, pp 231–260

Angula H, Baumle R, Bittner A, Bockmuhl F, Botha P, Dierkes K et al (2011) Groundwater in Namibia: an explanation of the hydrogeological map. Formset (PTY) Ltd., Capetown

Boss CB, Fredeen KJ (2004) Concepts, instrumentation and techniques in inductively coupled plasma optical emission spectrometry, 3rd edn. PerkinElmer Inc., USA

Bredenhann L, Braune E, Wates J (2000) Policy and strategy for groundwater quality management in South Africa. Department of Water Affairs and Forestry, Pretoria

Brower P (2010) Theory of XRF: getting acquainted with the principles, 3rd edn. PANalytical B.V, The Netherlands

Chicago Metropolitan Agency for Planning (2013) Impacts of Municipal Solid Waste, 18 September 2013, Accessed from https://www.cmap.illinois.gov/about/2040/supporting-materials/process-archive/strategy-papers/wast, 16 June 2023

Christelis G, Struckmeier W (2001) Groundwater in Namibia: an explanation to the Hydrogeological Map. Formset (PTY) Ltd., Cape Town

Christelis G, Struckmeier W, Baumle R, Bittner A, Bockmuhl F, Botha P, Abram V et al (2001) Groundwater in Namibia: an explanation of the hydrogeological map. Formset (PTY) Ltd., Cape Town

Christoph EA (2016) Omaruru basin water resources management plan. SLR, Windhoek

Food and Agricture Organization of the United Nations (2016) Assessment of impacts and recovery needs of communities affected by El Niño-induced drought in Kunene, Erongo and Omusati regions of Namibia. Food and Agriculture Organization of the United Nations, Rome

Geyh MA, Ploethner D (1995) Groundwater isotope study in the Omaruru river delta aquifer, central Namib desert, Namibia. In: Application of tracers in arid zone hydrology. IAHS Publisher, Vienna, pp 163–170

Gupta S (2014) Assessment of geo-accumulation index of heavy metal and source of contamination by multivariate factor analysis. Int J Hazard Mater 18–22

Hiscock KM, Bense FV (2014) Hydrogeology: principles and practice, Second Edition. Wiley

Internation Groundwater Resources Assessment Centre (2013) Groundwater Monitoring in the SADC region overview prepared for the stockholm water week 2013. N/A (pp 1–19). International Groundwater Resource Assessment Centre

Lin J, Chen Q, Liu R, Ye W, Luis P, Van der Bruggen B, Zhao S (2021) Sustainable management of landfill leachate concentrate via nanofiltration enhanced by one-step rapid assembly of metal-organic coordination complexes, Water Research, Volume 204, 1 October 2021. https://doi.org/10.1016/j.watres.2021.117633

Likuku AS, Mmolawa KB, Gaboutlo GK (2013) Assessment of heavy metal enrichment and degree of contamination around the Copper-Nickel mine in the Selebi Phikwe Region, Easter Botswana. Environ Ecol Res 32–40

McLennan SM (2001) Relationships between the trace element composition of sedimentary rocks and upper continental crust. Geochem Geophys Geosyst 2. https://earthref.org/ERR/984/

Ministry of Agriculture Water and Forestry (2010) Development of an intregrated water resource management plan for Namibia, Theme 2, Assessment of Reources Potential and Development Needs. Windhoek: IWRM Plan Joint Venture Namibia

Ministry of Agriculture, Water and Forestry (2010) Development of an integrated water resource management plan for Namibia, Theme 2, Assessment of Resources Potential and Development Needs. Windhoek: IWRM Plan Joint Venture Namibia

Ministry of Agriculture, Water and Forestry (2010) Integrated water resource management plan for Namibia. Namibia: IWRM Plan Joint Venture Namibia

Ministry of Agriculture, Water and Rural Development (2000) Water use and conservation: theme report. N/A: Ministry od Agriculture, Water and Rural Development

Ministry of Environment and Tourism (MET) (2003) Atlas of Namibia: a portrait of the land and it's people, viewed n.d., from http://www.uni-oeln.de/sfb389/e/e1/download/atlas_namibia/pics/climate/rainfall-annual.jpg

Ministry of Environment and Tourism (2010) National policy on climate change for Namibia. Windhoek: Ministry of Environment and Tourism

Namibia Statistics Agency (2011) Namibia 2011 population and housing census main report. Namibia Statistics Agency

New Era Publication Corporation (2016) NewEraNewspaper. Retrieved from newera.com.na: https://www.newera.com.na/2016/10/11/severe-water-shortage-hits-omaruru

Olaoye RA, Afolayan DO, Adewale R, Ogundairo TO (2013) Vulnerability assessment of broundwater contamination from an open dumpsite: llabete dumpsite as a case study. IOP conference series: materials science and engineering, 3

Pellenberg N (1997) Groundwater management in the Netherlands: background and legislation. ILRI Workshop, The Netherlands

SERC the Science Education Resources Center and Carleton College (2017) Retrieved from https://serc.carleton.edu/index.html: https://serc.carleton.edu/research_education/geochemsheets/techniques/XRF.html

Shroff P, Vashi RT, Champaneri VA, Patel KK (2015) Correlation study among water quality parameters of groundwater of valsad district of South Gujarat (India). Gujarat J Fundamental Appl Sci

The Namibian (2012) The Namibia. Retrieved from www.namibian.com.na: http://www.namibian.com.na/index.php?id=99072&page=archive-read

UNEP (2002) Children in the new millenium. Library of congress cataloguing, Malta

WHO (1996) Water quality monitoring—a practical guide to the design and implementation of freshwater quality studies and monitoring programmes. UNEP/WHO

Water Reserch Commission (2004) Groundwater recharge assessment of the basement aquifers of central Namaqualand. Water Research Commission, Gezina Petoria

Xiang R, Xu Y, Liu Y-Q, Lei G, Liu J, Huang Q (2019) Isolation distance between municipal solid waste landfills and drinking water wells for bacteria attenuation and safe drinking. Scientific Reports Nat Res 9:17881. https://doi.org/10.1038/s41598-019-54506-2

Zaporozec A, Nonner CJ, Conrad EJ, Romijn E, Hirata R, Weaver MJ et al (2002) Groundwater contamination inventory: a methodological guide. UNESCO, Paris, France

Chapter 23
Security of Food and Water: For a Sustainable Future

S. Nithiya and J. Usha

Abstract Food and water resources are pivotal factors for a sustainable tomorrow as they are centric toward improvising the human health and in the restoration of natural systems needed for support. Concentrating on the water-efficient methodologies in the production of food aids in threatening factors such as reduction of limited water resources, decline in the quality of soil, and unfavorable climatic conditions leading to water scarcity. The food sector has a huge impact directly proportional to the water resources, affecting the social and economic activity more than any other comparatively. Food production is facing a challenge by more than 72% among the community to meet the global population demand. Analysis of Water Foot Prints (WFP) and comparison of common food sources in identifying the water usage and the wastages associated with Blue, Green, and Grey WFP's assists in the understanding of the current methodologies being applied in food processing and preservation and the ways of conserving water resources without compromising on pollution or harmful effects leading to water-borne diseases or water scarcity. Regenerative and organic farming methods in the field of agriculture such as aquaponic, aeroponic, hydroponic, and vertical farming are a few methods that aim to utilize water resources in a wise manner. The consumer needs to require transformational changes in the food and water sector to meet today's needs and ensure a better tomorrow.

Keywords Sustainability · Food resources · WPF · Local · Water scarcity · Organic farming · Vertical farming · Water scarcity

S. Nithiya (✉) · J. Usha
Department of Chemistry, Sir Theagaraya College, Chennai, Tamilnadu, India
e-mail: nithiyas@stcchennai.edu.in

S. Suriyanarayanan et al. (eds.), *Water Management in Developing Countries and Sustainable Development*, Water Resources Development and Management,
https://doi.org/10.1007/978-981-99-8639-2_23

23.1 Introduction

As the whole world is running short of clean and fresh water for feeding and nourishing the rapidly increasing global population, in ensuring a sustainability in development, and in maintaining the health of our planet, there is a need for managing the current scenario where there is no enough water compared to the population of the world that ends in hunger and malnutrition. Therefore, betterment in the management of water is crucial to the availability of food without compromising on nutrition. It is obvious that irrigation is the key to increase the production of food, income from farming, and resilience in spite of the variability in weather and climatic conditions. As the water can also affect food security and nutrition in other ways, in precise, management of irrigation increases not only the volume but also the production of the varieties of food products, including seasonal crops and nutrient-rich food products including fruits and vegetables. Improvements in close proximity, sanity, and cleanliness of the sources of water and the water extraction technologies support the empowerment and the well-being, saving time, and improving health. Effective management of multiple advantages of water and wastewater decreases the exposure towards contamination resulting in the risk of infectious diseases.

In a decisive contribution in ending the poverty and hunger, water management policies and the investments for the implementation must overcome the challenges that daunt. Rise in income and urbanization act as a strong reason in driving the diversified growth in the demand for food and water within the field of agriculture and across the applications, both domestic and industrial. As the global population has a projection in reaching around ten billion in the next thirty years, say by 2050, the problems in food security are expected to be the most severe threat. Rapid increase in incomes and urbanization increases the demand for highly nutritious diets leading to additional requirement of water for the feed of livestock, fruits, and vegetables. The rapid growth in urbanization also boosts the demand for water needed for the domestic household and industrial purposes, leading to competition in the field of irrigation most preferably, in water-scarce agricultural areas, and thus into outright conflict, affecting the local livelihoods and transfer and migration and henceforth paving way for border disputes among places. Solutions for the methods and providing appropriate sources of water in alleviating the competition is a little difficult as the cost required for the development of water for irrigation and other related uses is also increasing, as better accessible sources have already been considered for utilization (IFPRI Blog 2019).

As water, the most needed and important resource for life, act as the central issue on the worldwide agenda for the past few decades, major areas of the world are, in recent years, observed to be affected by the poor quality of water leading to water scarcity (Mancosu et al. 2015). Rapid global population growth rate suggests that a huge demand for food is expected in the near future, with a direct impact on the irrigational and agricultural water usage. In addition, as a result of the increased water scarcity and drought due to climate change (Jiménez Cisneros et al. 2014), extensive irrigational water usage is expected to occur soon, leading to competition between agricultural

Fig. 23.1 Contribution toward scarcity of water

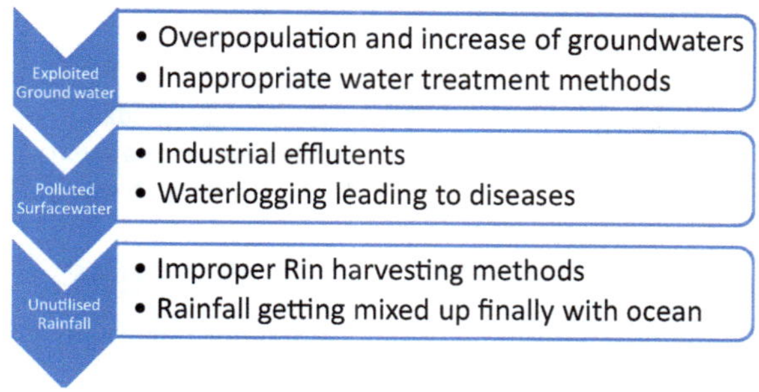

sectors and remaining sectors of the economy. Since an imbalance between the demand for water and the population growth has already started, the need has reached critical levels in many parts of the world, which might also be affecting the demand for food production, as water usage is directly proportional to the above parameters and hence, a sustainable thinking, methodology, and approach towards water resource management in agriculture has become essential. The sustainable water management concept refers to the practices, which improve the yield of crops minimizing the losses of water. In addition, the production of biofuel is increasingly competing with the production of food, that could reduce the resources of water and the availability of food.

In recent days, scarcity is a term that is related to water, leading to more research activities and projects that are focused on the universal assessment of water demand and the availability has been made. The water demand has reached the critical levels in major parts of the world, in particular, in countries with a limited availability of water. Improper usage of the water resources and the lack of infrastructures for the water supply and the climatic changes are a few reasons for water scarcity (Ridoutt 2018) (Fig. 23.1).

23.2 Materials and Methods

Scarcity of water and food are all interconnected with each other. With the increasing demand, the security for food and water are few of the challenges (Noemi et al. 2015) for sustainability in development. Food insecurity is often due to either a lack of natural resources (for example, land or water) in the production of crops, or institutional deficiencies that limit agricultural development (Popp et al. 2014). Water footprint helps us in understanding the availability of our limited freshwater resources and the ways in which they are being polluted during consumption (Correa Porcell et al. 2021). The impact over the issue depends upon where the water is taken from and when. If it comes from a place where there is a water scarcity, the results can be significant and require necessary action (Biba 2015; Khanna et al. 2018) (Fig. 23.2).

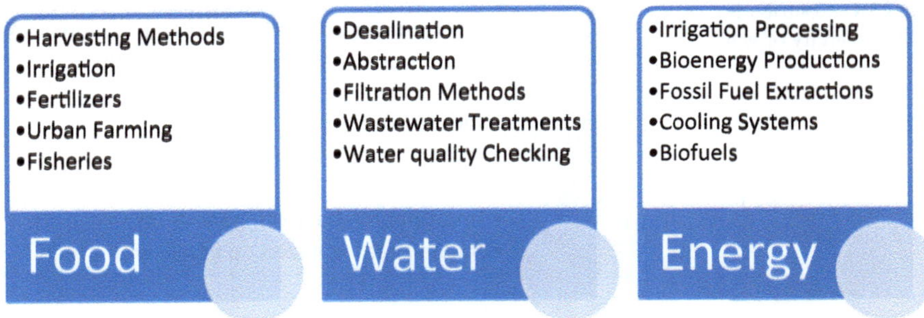

Fig. 23.2 Food, Water and Energy systems

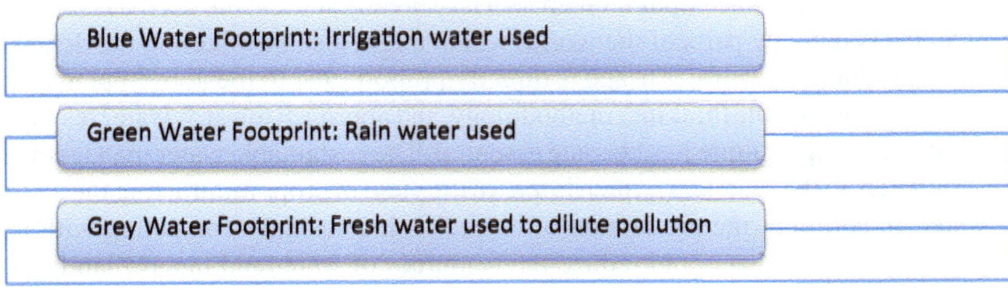

Fig. 23.3 Green, Blue and Grey water footprints

23.2.1 Water Foot Prints

Among the various available footprints, blue water footprint refers to groundwater or surface water usage, green water footprint refers to rainwater, and grey water footprint refers to the amount of water needed to dilute pollutants. (Mekonnen et al. 2011; Bradley et al. 2018) (Fig. 23.3).

23.2.2 Blue Water Footprint

Blue water footprints refer to the amount or the volume of water sourced out from the groundwater resources such as lakes, rivers, wetlands, and aquifers and get evaporated while irrigating crops, or taken from one water source to another, or returned at a different time. Irrigated agriculture, industry, and domestic water usage may all be utilized with a blue water footprint. Blue water footprint is the amount of freshwater utilized which has been evaporated from the entire blue water resources globally in producing the commodities, goods, and services needed by them.

23.2.3 Green Water Footprint

Green water footprints refer to the amount of water that has been precipitated from the root zone of the soil (green water footprint zone) after being stored that is lost by two ways either by evapotranspiration or incorporation by plants, which is mostly relevant for agricultural and horticultural products. Green water footprint is the amount of water, which has been evaporated from all the available green water resources (rainwater stored in the soil) globally.

23.2.4 Grey Water Footprint

Grey water footprints refer to the amount of water required for the dilution of the pollutants such as industrial effluents, seepage from the ponds and other stable water sources, untreated domestic and municipal waters, or nonpoint source pollution such as agricultural runoff or urban runoff) to an extent that the water quality agrees with the water quality according to standards. Grey water footprint is the amount of polluted water, which has been quantified as the volume of water required to dilute the pollutants to an extent such that the quality and purity remains beyond the agreed and the permissible limits of quality standards recommended by the water bodies and associations.

It is calculated as:

$$L / c_{max} - c_{nat}$$

where,

L is the pollutant load expressed as mass flux

c_{max} is the maximum permissible concentration (mass/volume)

c_{nat} is the natural concentration of the pollutant in the receiving water body (mass/volume).

As there exists a strong connection and interlink between the agricultural sector and scarcity of water, agriculture on a global scale being the largest consumer of water, accounts for nearly 70% of water withdrawals in various forms (Mancosu et al. 2015). In addition, the projected increase of more than 15% of the water needs for irrigation initiates and suggests the demand for upgraded technologies for water protection in the near future, which might lead to a sustainable platform (Wu et al. 2022). To further add with, to cope up with the demand for a considerable portion of available water, point, and nonpoint sources for the nutrient runoff from irrigation, cropping, and agriculture act as a major contributor of freshwater sources, further leading to a strain in social, economic, and environmental water bodies (John et al. 2016). Total water footprints of either an individual or community further get classified into three categories: the blue, green, and grey water footprint (Hoekstra 2008).

23.3 Results and Discussions

23.3.1 *Understanding the Importance of Alternate Farming*

The recent practices for protection in farming and agriculture include ways of growing food rather than directly from the soil. By considering and viewing the transcripts received from an FAO report, soil-less cultures are considered as a way of growing agricultural crops without the usage of traditional methods where the utilization of water is more. As a replacement to soil, different growing materials that are inert in nature, also called substrates, are being used, thereby providing enough needed support for crops without compromising on the moisture retention. Benefits and advantages of the alternate farming methods require only limited supply water, say in fractions, with less than 10% of soil reduction, no much need in using fertilizers, insecticides, herbicides, and pesticides, no diseases that are soil-borne, no tilling procures, no much weed growth, simultaneous growing of two or more crops together, an enormous increase in fish stocking density, high yield of crops, no much conventional wastes. Only with the fish fertilizers, food has been found to be secured that can be grown indoors, throughout the year, working even in draught or places with infertile or poor quality soil.

23.3.2 *Greenhouses and Polyhouses*

Climatic-controlled greenhouses have both the cooling and heating systems. These houses are mainly used in cultivating and growing exotic vegetables, off-season vegetables according to the places, horticulture, floriculture, plant material acclimatization, breeding of plants, and even under non-suitable climatic conditions. The degree of utilization of greenhouses includes completely automated systems with roofing using multi-walled polycarbonate sheets, appropriately needed heating and cooling systems, etc. Polyhouses are comparably less cultured and sophisticated than greenhouses that have a naturally ventilated climate-controlled system as that against whole climate-controlled greenhouses (Fig. 23.4).

23.3.3 *Cultivation and Protection*

Cultivation practices are highly protected with various cropping techniques whereas the environment on a micro level is controlled either partially or wholly as per the need of the plant and its growth in maximizing the yield with not much usage of the available resources.

Fig. 23.4 Green House Farming

- **Aeroponics** is the method of growing plants in a misty environment with no requirement of soil as a medium.
- **Hydroponics** is the method of growing plants with using the nutrient solutions from the minerals present in water, again with no need for soil medium.
- **Aquaponics** is the method of growing plants in a symbiotic environment involving both aquaculture and hydroponics (AlShrouf 2017) (Fig. 23.5).

23.3.4 Aeroponics

The alternative cropping aeroponics method is for farmers and agriculturists having limited spaces for plant growth. An aeroponic system includes the usage of entrapped and enclosed air in water and nutrient ecosystem that leads to an enormous plant growth with very little water and direct sunlight that too without soil medium. This system involves high technological type hydroponic system, in which air is utilized as a primary growing medium (Kumari and Kumar 2019). The plant roots hang in the air and get misted with nutrients present in the solutions. When there is an interruption in the misting cycles, the roots would get dried up due to exposure with

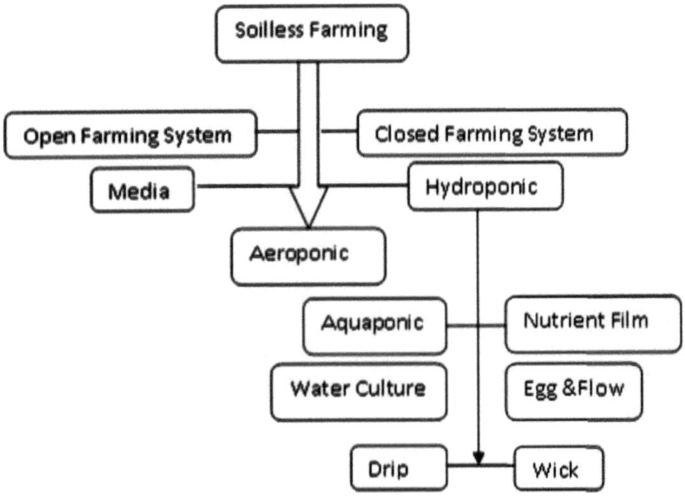

Fig. 23.5 Flow chat for alternate forming

air and hence, a nutrient pump is attached that is controlled by a short cycle timer is attached, where the pump is set to run every second (Tafesse et al. 2021).

Air culture method introduces the nutrient-dissolved mist over the plant roots that are exposed in enabling the growth without soil. The plant roots that suspend are held in a closed chamber where the plant canopy is exposed out. In small plants, foamy cell sheets are compressed, wound, and wrapped around the stem at the bottom and held in to the aeroponic system. In larger plants, trellising method is used, in which wooden or metallic frames are fixed along with the plant body for support. The plants grow fast in an aeroponic system as the environment is maintained sterile and the supply of oxygen is also abundant in the chamber.

23.3.5 Hydroponics

Using water-based nutrients and not soil, the technique of growing plants including a substrate or growing medium, such as vermiculite, perlite, coconut coir, organic-rich products, etc., the hydroponic growing systems are used by farmers, agriculturists, and small-scale enterprises (Richa et al. 2020). The hydroponics method utilizes less water, thereby this soil-less growing method is found to use only 10% water compared to the agricultural, method which uses soil due to its efficient recirculation methods (Velazquez-Gonzalez et al. 2022). The major benefits include that the crops can be grown even in contaminated arable places. These crops can be grown in almost all locations including the greenhouses, apartments, small spaces, terraces, etc., thereby saving a large portion of land. In this method, all kinds of soil-related diseases and also those that arise from pests, insects, and bacteria can be avoided. No nutrient will

Fig. 23.6 Hydroponic Plant with Coir

be left out but can be added to the solution according to the requirement and need, resulting in a 100% controlled environment completely fulfilling the appropriate and specific amounts of nutrients that are needed by the plants (Ali AlShrouf 2017) (Fig. 23.6).

23.3.6 Aquaponics

A symbiotic environment of farming using fishes and/or other aquatic organisms and the hydroponics method in which no soil is needed, for an efficient method of growing both plant and fish crops. Aquatic wastes from aquaculture are broken down by bacteria into dissolved nutrients such as nitrogen and phosphorus compounds, thereby allowing the plants to grow in a hydroponic unit. Aquaponics along with hydroponics provide various advantages over conventional production methods. A few among them are reduction in land area requirements, low water consumption, high yield in plant growth, and round-the-year production in controlled environments (Lennard 2019; Milena Krastanova et al. 2022) (Fig. 23.7).

23.3.7 Vertical Farming

Resulting in a very high production output per unit area, vertical farming saves the soil and water resources more significantly. Producing food in urban areas and environment means being closer to consumers incurs less transportation from the farms to the consumer's hands and therefore reduces the carbon footprint. Additional

Fig. 23.7 Aquaponic Plant

benefits include the protection and crop yield, which not only gets subjected to weather conditions in a controlled production environment, resulting in predictability and consistency. For almost all types of plants that are being produced now, mostly leafy greens, very less pesticides are being used, thereby addressing to an extent, where the consumer concerns on chemical added food, and providing health-based safety (Kalantari et al. 2017).

High degree of transparency arises from the application of satellites, artificial intelligence, IoT, and automation technologies in a monitored and controlled environment. With the safety issues in food production, moving to the forefront and increased consumer awareness, traceable means have now been considered as some important criteria to meet the demands and supplies to sophisticated outlets both wholesale and retail in the countries that consume (van Delden 2021).

Instead of using the major portion of land for growing crops in a traditional manner, vertical farming utilizes even the indoors on walls, shelves, and vertical mounts to grow crops, thus making it highly economical when coming to space. A vertical farm can be set up in any building tall and large enough. Even, unused domestic buildings such as warehouses and factories can be converted into vertical farms for producing food in urban areas.

Vertical farms will also not be affected by weather changes, which means growing of crops is possible in any season. Dubai has built the world's largest vertical farm in the year 2022, codenamed as ECO1, where the facility is nearly 30,000 sq.m. that can produce more than 900 tons of food (weforum.org 2022) (Fig. 23.8).

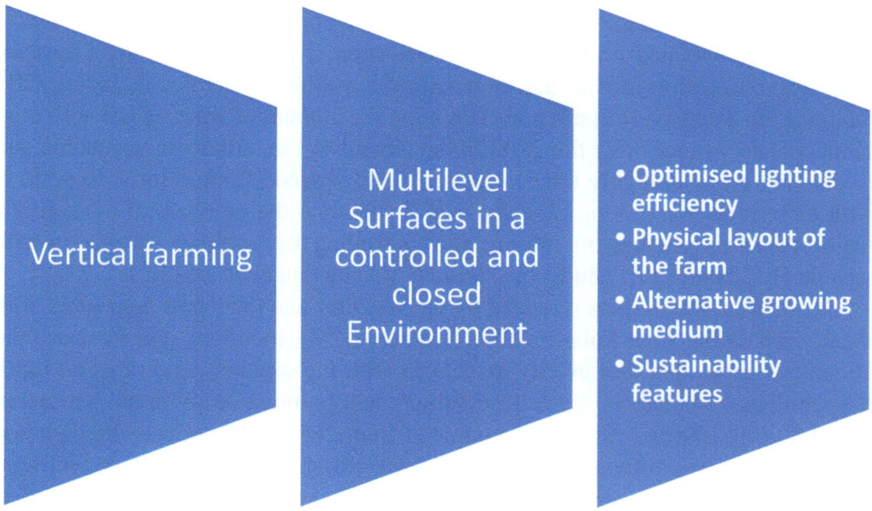

Fig. 23.8 Vertical Farming

23.4 Conclusions

An expected increase in the production of food and the related impacts on the soil gets modified; changes and deviations associated with the biodiversity and climatic changes depend much upon the projections of universal demand for food production and consumption. Researchers from FAO (FAO 2009), have mentioned that the total world food production has to be increased by around 70% in order to feed the total world population in the next thirty years. Even though the percentage was reduced to 60% in a revised study, the original value has been considered to be a reference point by organisations and companies (Feed the World 2015) as well as the articles, journals, and scientific papers (Carvajal-Yepes 2019). Global researchers in their articles expect a drastic increase in the demand for food globally, in particular; say 100–110% in the upcoming three decades.

Those who do not agree with or those who are unable to proceed with the ideas and plans of the agricultural and farming sector, as they are stuck with traditional ad conventional systems, often do not consider or almost dismiss the global projections of the future and the related causes and effects as they believe that the expected and cited demand for food (even when being doubled) understands and frames the global food security as a problem of supply, demand, or even leading to scarcity, closes off the working, planning, discussions, and implementations toward providing solutions that do not primarily rely upon the enormous increase in food production with the benefits of technological advancements and recent innovations.

Applications of the technically modified and upgraded forms of current methodologies are designed, implemented, and planned by the organizations to protect the waters for a sustainable life. Applying the internet of things (Iot), a wide range of bimolecular, thermal, chemical, optical, biochemical, electrical, and biological metrics were obtained on an all-round 360-degree view of the health of agricultural crops. Location monitoring tools with the available satellites including GPS facilities are extensively applied in estimating the amount of water in the soil, crop biomass, and various other things. GPS devices allow the farmers in vegetation and in planting crops in highly efficient patterns and proceed from point A to point B with more precision, saving time and fuel. Evaluating the moisture of the soil is a major issue for evaluating, monitoring and controlling the development of crops with technically advanced AI and IoT technology (Dhanaraju et al. 2022).

The data for the farmers provided by the satellites and the teams associated with them, such as the Joint Polar Satellite System (JPSS), monitors, and forecasts the climatic changes and the severity associated with the weather conditions on Earth (Capabilities-of-Industry). Such observations will be observed with an instrument called Visible Infrared Imaging Radiometer Suite, or VIIRS collects the data and information in two different types of radiations of light such as visible and infrared spectroscopy. In urban regions, in recent days, drones, robots, sensors, satellites, and mobile radio signals from antennas provide the information and the data needed for sustainability, especially among the field of agriculture and farming. If such recent methodologies are being used in rural regions also, and being introduced to the farmers and with the proper monitoring of the activities, the pavement of a sustainable tomorrow will easily be attained by our future generations, for sure.

References

AlShrouf A (2017) Hydroponics, aeroponic and aquaponic as compared with conventional farming. Amer Scientific Res J Eng Technol Sci 27:247–255

Biba S (2015) The goals and reality of the water-food-energy security nexus: the case of China and its southern neighbours. Third World Q 37:51–70

Bradley GR, Michalis H, Martin N, Brett AB (2018) From water-use to water-scarcity footprinting in environmentally extended input-output analysis. Environ Sci Technol 52(12):6761–6770

Carvajal-Yepes MK, Cardwell A, Nelson KA, Garrett B, Giovani DGO, Saunders S, Kamoun JP, Legg V, Verdier J, Lessel RA, Neher R, Day P, Pardey ML, Gullino AR, Records B, Bextine JE, Leach S, Staiger JT (2019) A global surveillance system for crop diseases. Science 364:1237–1239

Correa-Porcel V, Piedra-Muñoz L, Galdeano-Gómez E (2021) Water-energy-food nexus in the agri-food sector: research trends and innovating practices. Int J Environ Res Public Health 18(24):12966. https://doi.org/10.3390/ijerph182412966.PMID:34948576; PMCID:PMC8701236

Dhanaraju M, Chenniappan P, Ramalingam K, Pazhanivelan S, Kaliaperumal R (2022) Smart farming: internet of things (IoT)-based sustainable agriculture. Agriculture 12(10):1745. https://doi.org/10.3390/agriculture12101745

FAO How to Feed the World in 2050 (High-Level Expert Forum, 2009)

Feed the world: a challenge and an opportunity. The John Deere Journal https://johndeerejournal.com/2015/12/smallholder-farmers-big-challenges (2015)

https://www.researchgate.net/figure/Capabilities-of-Industry-40-in-the-medical-field-IoT-Internet-of-things-fig2-332536137 https://www.weforum.org/agenda/2022/05/vertical-farming-future-of-agriculture/

IFPRI Blog: JUNE 12, 2019. Issue Post Food security depends on water security—and we need to act now

Jiménez Cisneros BE, Oki T, Arnell NW, Benito G, Cogley JG, Döll P, Jiang T, Mwakalila SS (2014) Freshwater resources. In: Field CB, Barros VR, Dokken DJ, Mach KJ, Mastrandrea MD, Bilir TE, Chatterjee M, Ebi KL, Estrada YO, Genova RC, et al. (Eds) Climate Change 2014: impacts, adaptation, and vulnerability. Part A: global and sectoral aspects. Contribution of working Group II to the fifth assessment report of the intergovernmental panel on climate change. Cambridge University Press: Cambridge, UK and New York, NY, USA, pp 229–269

John MK, Joseph VD, Dennis M, Scott PS, Scott MS (2016) Sustainable management of Great Lakes watersheds dominated by agricultural land use. J Great Lakes Res 42(6):1252–1259. file:///C:/Users/Comp%20lab%2010/Desktop/Hoekstra-2008-WaterfootprintFood.pdf

Kalantari F, Mohd tahir O, MahmoudiLahijani A, Kalantari S (2017) A review of vertical farming technology: a guide for implementation of building integrated agriculture in cities. Adv Eng Forum 24:76–91.https://doi.org/10.4028/www.scientific.net/AEF.24.76

Khanna M, Swinton SM, Messer KD (2018) Sustaining our natural resources in the face of increasing societal demands on agriculture: Directions for future research. Appl. Econ. Perspect Policy 40:38–59. https://doi.org/10.1093/aepp/ppx055

Krastanova M, Sirakov I, Ivanova-Kirilova S, Yarkov D, Orozova P (2022) Aquaponic systems: biological and technological parameters. Biotechnol Biotechnol Equip 36(1):305–316.https://doi.org/10.1080/13102818.2022.2074892

Kumari R, Kumar R (2019) Aeroponics: a review on modern agriculture technology, pp 286–292

Lennard W, Goddek S (2019) Aquaponics: the basics. https://doi.org/10.1007/978-3-030-15943-6-5

Mancosu N, Snyder RL, Kyriakakis G, Spano D (2015) Water scarcity and future challenges for food production. Water 7:975–992

Mekonnen, Mesfin, Hoekstra, Arjen (2011) The green, blue and grey water footprint of crops and derived crop products. Hydrol Earth Syst Sci Discussions, 15.https://doi.org/10.5194/hess-15-1577-2011

Wu M, Li Y, Xiao J, Guo X, Cao X (2022) Blue, green, and grey water footprints assessment for paddy irrigation-drainage system. J Environ Manag 302(B):114116, ISSN 0301-4797

Noemi M, Richard LS, Gavriil K, Donatella S (2015) Water Scarcity and future challenges for food production 7(3):975–992

Popp J, Lakner Z, Harangi-Rákos M, Fári M (2014) The effect of bioenergy expansion: food, energy, and environment. Renew Sustain Energy Rev 32:559–578, ISSN 1364-0321

Richa A, Touil S, Fizir M, Martinez V (2020) Recent advances and perspectives in the treatment of hydroponic wastewater: a review. Rev Environ Sci Bio/Technology. https://doi.org/10.1007/s11157-020-09555-9

Tafesse EG, Aidoo MK, Lazarovitch N, Rachmilevitch S (2021) Aeroponic systems: a unique tool for estimating plant water relations and NO_3 uptake in response to salinity stress. Plant Direct 5(4):e00312. https://doi.org/10.1002/pld3.312.PMID:33817545;PMCID:PMC8006923

van Delden SH, SharathKumar M, Butturini M (2021) Current status and future challenges in implementing and upscaling vertical farming systems. Nat Food 2:944–956. https://doi.org/10.1038/s43016-021-00402-w

Velazquez-Gonzalez RS, Garcia-Garcia AL, Ventura-Zapata E, Barceinas-Sanchez JDO, Sosa-Savedra JC (2022) A review on hydroponics and the technologies associated for medium- and small-scale operations. Agriculture 12(5):646. https://doi.org/10.3390/agriculture12050646. Ali AlShrouf (2017) Hydroponics, aeroponic and aquaponic as compared with conventional farming. Amer Scientific Res J Eng Technol Sci (ASRJETS) 27(1):247–255

Chapter 24
Future Perception and Opportunities for Sustainable Water Management in Developing Countries

Harikaranahalli Puttaiah Shivaraju, S. Suriyanarayanan, and Jenkins David

24.1 Introduction

Water is a crucial as well as non-renewable resource that is vital for our existence and proper management is necessary for its renewal. Currently, over 1.7 billion people reside in areas where water usage exceeds the natural replenishment rate, and by 2025, two-thirds of the global population will live in countries facing water stress (Cosgrove and Loucks 2015). Effective and fair management of water can pose a significant challenge to sustainable development, but it can also cooperate a crucial part in strengthening the flexibility of social, economic, as well as environmental systems in the aspect of swift and impulsive changes. In Zimbabwe, the pollution of groundwater resources is increasing due to urbanization, industrial expansion, agronomic and mining, inadequate sanitation habits, and overuse to meet the growing necessitate for human and agricultural requirements (Gaye and Tindimugaya 2019). Despite the presence of numerous challenges in managing groundwater, its resources in Africa remain largely undeveloped and can sustainably fulfill various requirements

H. P. Shivaraju (✉)
Department of Environmental Sciences, JSS Academy of Higher Education and Research, Mysuru 570015, India
e-mail: shivarajuenvi@gmail.com

S. Suriyanarayanan
JSS Science and Technology University, JSS TI Campus, Mysuru 570006, Karnataka, India

SCOPE (Scientific Committee On Problems of the Environment), JSS Academy of Higher Education and Research, Salem, India

J. David
Wolfson Nanomaterials and Devices Laboratory, School of Computing, Electronics and Mathematics, Faculty of Science and Engineering, University of Plymouth, Devon 4 8AA, UK

S. Suriyanarayanan et al. (eds.), *Water Management in Developing Countries and Sustainable Development*, Water Resources Development and Management,
https://doi.org/10.1007/978-981-99-8639-2_24

if managed more effectively. Therefore, measures to ensure the sustainable development and management of groundwater must be implemented. These include setting up systems to monitor groundwater levels, understanding the relationship between groundwater and aquatic ecosystems, managing shared aquifers that cross national boundaries, addressing the impacts of climate change on groundwater resources, assessing the sustainability of amplified thrusting from various types of aquifers, and building capacity in management of groundwater. In developing countries, managing water supply and wastewater in urban areas is becoming increasingly challenging for authorities. Implementing strategies for the reuse of water and wastewater can improve the management of urban water resources. Key factors for reducing public health risks have been recognized. The potential for reusing wastewater in agriculture, manufacturing, urban applications, and environmental water development, with recharge of groundwater, is explored through applied examples (*Water and Wastewater Reuse: An Environmentally Sound Approach for Sustainable Urban Water Management | UNEP—UN Environment Programme* n.d.) Effective implementation of wastewater reuse strategies requires capacity developing in policymaking, organized establishment, monetary structures, awareness building, and stakeholder contribution.

24.2 Protecting Our Water: Strategies for Conservation and Pollution Prevention

Despite the potential benefits, the global adoption of water reuse remains limited due to public opposition. In order to achieve economic sustainability in the treatment of water environments through public–private partnership ventures, it is important to shift the focus from initial costs to the costs incurred over the entire life cycle of the project (Li et al. 2019). In Africa territory, groundwater is a vital resource, and its development through techniques such as wells and harvesting of water has remained contributory in decreasing clashes over water access and the sharing of resources across borders. Examples of this can be seen among pastoralist and agro-pastoralist communities that span national boundaries (The World Bank Horn of Africa-Groundwater Initiative 2018).

It is important to continue to focus on sustainable water management and development. This includes implementing strategies to ensure the sustainable development and management of groundwater resources, such as establishing groundwater observation networks, considerate groundwater–aquatic system relations, managing international aquifers, mitigating the effects of climate change on groundwater resources, and building capacity in groundwater management (Bhaduri et al. 2016). In urban areas of developing countries, it is important to continue to improve urban water management through strategies for water and wastewater reuse. This includes minimizing public health risks, exploring possibilities for water reclamation, industry, urban water management, and groundwater replenishment (Russo et al. 2014).

It is also important to continue to focus on institution strengthening, corporate support, monetary systems, outreach and shareholder contribution to execute these approaches for wastewater reclaim. Additionally, addressing emerging threats such as microplastic contamination in urban water systems through risk assessment and mitigation will be crucial. Developing low-cost sustainable nanotechnology approaches for water quality monitoring in developing countries will also be important. Implementing eco-friendly approaches such as phytoremediation, where plants are used to remove contaminants from the environment can be a capable method for sustainable water management (Nedjimi 2021). For example, the Cholistani desert grass Vetiveria Zizanioide could be analyzed for its potential use in water treatment under NaCl salt stress. It will be essential to concentrate on emergent contaminants in water and wastewater by using sustainable treatment techniques (Liu et al. 2016). The potential use of environmental coagulants to decontaminate leachates could also be explored.

24.3 Managing Water: Policy and Governance

Community-based organizations will continue to play a vital role in water governance, particularly in developing countries like Sri Lanka (Jayasiri et al. 2022). The use of renewable energy in water management and irrigation will become increasingly important as the world moves towards more sustainable practices. In India, the support of the community will remain crucial for the successful implementation of continuous water supply systems. The ethical impacts of using nanomaterials for wastewater remediation will need to be carefully considered and monitored. Applied theatre may become an increasingly popular tool for communicating water management challenges and solutions in African communities (Umenyilorah 2014). Desalination technology will likely continue to advance, making it a more viable option for supplementing groundwater supply in arid and semi-arid environments.

Contextual engendering of water and climate adaptation in developing countries will continue to play a crucial role in designing and implementing effective adaptation strategies. By taking into account the specific social, cultural, economic, and environmental contexts of each country, these strategies can be tailored to meet the unique needs and challenges of different communities (Eriksen et al. 2021). Access to safe water for sanitation will remain essential for maintaining health and well-being. The guidelines developed by the World Health Organization (WHO) on sanitation and health will continue to provide valuable advice on maximizing the health impact of sanitation interventions.

In places like Kerala, India, regional water harvesting structures such as rainwater harvesting tanks will continue to provide clean water to communities, reducing medical costs from water-related diseases and eliminating unnecessary physical labour from water collection. The empowerment of women through the management of these tank networks will also continue to be an important aspect of these initiatives. The use of geochemical assessments for basin-wide monitoring of groundwater quality and quantity will also remain an important tool for ensuring sustainable water management. In places like the Omaruru-Swakop Basin, the information obtained from geochemical analyses of groundwater samples will continue to be used to assess the overall quality of the groundwater and to monitor changes over time.

24.4 Conclusion

From a future perspective, it is important to develop innovative solutions to ensure sustainable food and water security for a growing global population. This can include measures such as improving infrastructure, promoting conservation, and increasing public awareness about the importance of water conservation. In developing countries, rapid urbanization and population growth present challenges for sustainable water management. Effective strategies for sustainably managing water resources are crucial to ensure equitable access to water resources and improve living conditions and health. Gender issues also play a role in water management and sustainable development, and it is important to consider the different needs and perspectives of men and women when developing policies and programs. Proper waste management is also crucial to prevent contamination of groundwater resources, which are an important source of drinking water for many communities. Overall, addressing water management issues is essential for a sustainable future.In the context of developing countries; this monograph offers a thorough investigation of the linkages between water management and sustainable development. It focuses on the main issues— Water Resource Management and Challenges, Water Resources, Pollution and Conservation, Advancement in Water Treatment and Infrastructure, Integrated Water Resource Management, Policy and Governance, Climate Change, Water and Sanitation, Case Studies and Future Directions—as well as the opportunity they present. In order to address these issues and advance sustainable water management, it also emphasises viable solutions, best practises, and novel strategies that can be used.

In this monograph, we explore the intricate problems associated with water management in developing countries and the importance of water management in attaining sustainable development in developing nations is highlighted as this book comes to a close. To solve the issues and maximise the advantages of sustainable water management, it emphasises the need for integrated strategies, effective governance, capacity building, and financial support. These chapters aim to provide a thorough overview of the problems, chances, and potential answers associated with water management in the context of sustainable development in developing nations. Developing nations can blaze the trail to a more secure and sustainable future with regard to their access to water by implementing the suggested techniques and learning from successful case studies which is highlighted in this monograph.

References

Bhaduri A, Bogardi J, Siddiqi A, Voigt H, Vörösmarty C, Pahl-Wostl C, Bunn SE, Shrivastava P, Lawford R, Foster S, Kremer H, Renaud FG, Bruns A, Osuna VR (2016). Achieving sustainable development goals from a water perspective. Front Environ Sci 4:182047. https://doi.org/10.3389/FENVS.2016.00064/BIBTEX

Cosgrove WJ, Loucks DP (2015) Water management: current and future challenges and research directions. Water Resour Res 51(6):4823–4839. https://doi.org/10.1002/2014WR016869

Eriksen S, Schipper ELF, Scoville-Simonds M, Vincent K, Adam HN, Brooks N, Harding B, Khatri D, Lenaerts L, Liverman D, Mills-Novoa M, Mosberg M, Movik S, Muok B, Nightingale A, Ojha H, Sygna L, Taylor M, Vogel C, West JJ (2021) Adaptation interventions and their effect on vulnerability in developing countries: help, hindrance or irrelevance? World Dev 141:105383. https://doi.org/10.1016/J.WORLDDEV.2020.105383

Gaye CB, Tindimugaya C (2019) Review: challenges and opportunities for sustainable groundwater management in Africa. Hydrogeol J 27(3):1099–1110. https://doi.org/10.1007/S10040-018-1892-1/METRICS

Jayasiri MMJGCN, Yadav S, Dayawansa NDK, Ingold K (2022) Managing agricultural water resources: addressing the complexity of innovation, social perspectives, and water governance in Sri Lanka. Irrig Drain 71(S1):71–85. https://doi.org/10.1002/IRD.2693

Li H, Xia Q, Wen S, Wang L, Lv L (2019) Identifying factors affecting the sustainability of water environment treatment public-private partnership projects. Adv Civil Eng.https://doi.org/10.1155/2019/7907234

Liu WG, Liu JX, Yao ML, Ma QF (2016) Salt tolerance of a wild ecotype of vetiver grass (Vetiveriazizanioides L.) in southern China. Botanical Studies 57(1). https://doi.org/10.1186/S40529-016-0142-X

Nedjimi B (2021). Phytoremediation: a sustainable environmental technology for heavy metals decontamination. SN Appl Sci 3(3):1–19. https://doi.org/10.1007/S42452-021-04301-4

Russo T, Alfredo K, Fisher J (2014) Sustainable water management in urban, agricultural, and natural systems. Water 6:3934–3956. https://doi.org/10.3390/W6123934

The World Bank Horn of Africa-Groundwater Initiative (2018)

Umenyilorah CU (2014) Theatre as tool for development in Nigeria. IOSR J Humanit Soc Sci 19(6):34–40. https://doi.org/10.9790/0837-19623440

Water and wastewater reuse: an environmentally sound approach for sustainable urban water management | UNEP—UN Environment Programme (n.d.) Retrieved August 7, 2023, from https://www.unep.org/resources/report/water-and-wastewater-reuse-environmentally-sound-approach-sustainable-urban-water

References

Index

S. Suriyanarayanan et al. (eds.), *Water Management in Developing Countries and Sustainable Development*, Water Resources Development and Management,
https://doi.org/10.1007/978-981-99-8639-2

MIX
Papier aus verantwortungsvollen Quellen
Paper from responsible sources
FSC® C105338

If you have any concerns about our products,
you can contact us on
ProductSafety@springernature.com

In case Publisher is established outside the EU,
the EU authorized representative is:
Springer Nature Customer Service Center GmbH
Europaplatz 3, 69115 Heidelberg, Germany

Printed by Libri Plureos GmbH
in Hamburg, Germany